APPROACHES TO GEO-MATHEMATICAL MODELLING

Wiley Series in Computational and Quantitative Social Science

Computational Social Science is an interdisciplinary field undergoing rapid growth due to the availability of ever increasing computational power leading to new areas of research.

Embracing a spectrum from theoretical foundations to real world applications, the Wiley Series in Computational and Quantitative Social Science is a series of titles ranging from high level student texts, explanation and dissemination of technology and good practice, through to interesting and important research that is immediately relevant to social / scientific development or practice. Books within the series will be of interest to senior undergraduate and graduate students, researchers and practitioners within statistics and social science.

Behavioral Computational Social Science

Riccardo Boero

Tipping Points: Modelling Social Problems and Health

John Bissell (Editor), Camila Caiado (Editor), Sarah Curtis (Editor), Michael Goldstein (Editor), Brian Straughan (Editor)

Understanding Large Temporal Networks and Spatial Networks: Exploration, Pattern Searching, Visualization and Network Evolution

Vladimir Batagelj, Patrick Doreian, Anuska Ferligoj, Natasa Kejzar

Analytical Sociology: Actions and Networks

Gianluca Manzo (Editor)

Computational Approaches to Studying the Co-evolution of Networks and Behavior in Social Dilemmas

Rense Corten

The Visualisation of Spatial Social Structure

Danny Dorling

APPROACHES TO GEO-MATHEMATICAL MODELLING

NEW TOOLS FOR COMPLEXITY SCIENCE

Edited by

Alan G. Wilson

Centre for Advanced Spatial Analysis, University College London, London, UK

This edition first published 2016

Registered office
John Wiley & Sons Ltd, The Atrium, Southern Gate, Chichester, West Sussex, PO19 8SQ, United Kingdom

For details of our global editorial offices, for customer services and for information about how to apply for permission to reuse the copyright material in this book please see our website at www.wiley.com.

Library of Congress Cataloging-in-Publication Data applied for

ISBN: 9781118922279

A catalogue record for this book is available from the British Library.

Set in 10/12pt TimesLTStd by SPi Global, Chennai, India.
Printed and bound in Malaysia by Vivar Printing Sdn Bhd

1 2016

Contents

Part III DYNAMICS IN ACCOUNT-BASED MODELS

Part VII AGENT-BASED MODELS

Part VIII DIFFUSION MODELS

Part IX GAME THEORY

Notes on Contributors

Peter Baudains is a Research Associate at the Department of Security and Crime Science at University College London. He obtained his PhD in Mathematics from UCL in 2015 and worked for five years on the EPSRC-funded ENFOLDing project, contributing to a wide range of research projects. His research interests are in the development and application of novel analytical techniques for studying complex social systems, with a particular attention on crime, rioting and terrorism. He has authored research articles appearing in journals such as *Criminology*, *Applied Geography*, *Policing* and the *European Journal of Applied Mathematics*.

Janina Beiser obtained her PhD in the Department of Political Science at University College London. During her PhD, she was part of the security workstream of the ENFOLDing project at UCL's Centre for Advanced Spatial Analysis for three years. Her research is concerned with the contagion of armed civil conflict as well as with government repression. She is now a Research Fellow in the Department of Government at the University of Essex.

Steven R. Bishop is a Professor of Mathematics at University College London, where he has been since arriving in 1984 as a postdoctoral researcher. He published over 150 academic papers, edited books and has had appearances on television and radio. Historically, his research investigated topics such as chaos theory, reducing vibrations of engineering structures and how sand dunes are formed, but he has more recently worked on 'big data' and the modelling of social systems. Steven held a prestigious 'Dream' Fellowship funded by the UK Research Council (EPSRC) until December 2013, allowing him to consider creative ways to arrive at scientific narratives. He was influential in the formation of a European network of physical and social scientists in order to investigate how decision support systems can be developed to assist policy makers and, to drive this, has organised conferences in the UK and European Parliaments. He has been involved in several European Commission–funded projects and has helped to forge a research agenda which looks at the behaviour of systems that cross policy domains and country borders.

Alex Braithwaite is an Associate Professor in the School of Government and Public Policy at the University of Arizona, as well as a Senior Research Associate in the School of Public Policy at University College London. He obtained a PhD in Political Science from the Pennsylvania

State University in 2006 and has since held academic positions at Colorado State University, UCL, and the University of Arizona. He was a co-investigator on the EPSRC-funded ENFOLDing project between 2010 and 2013, contributing to a wide range of projects under the "security" umbrella. His research interests lie in the causes and geography of violent and nonviolent forms of political conflict and has been published in journals such as *Journal of Politics*, *International Studies Quarterly*, *British Journal of Political Science*, *Journal of Peace Research*, *Criminology* and *Journal of Quantitative Criminology*.

Simone Caschili has a PhD in Land Engineering and Urban Planning, and after being a Research Associate at the Centre for Advanced Spatial Analysis (at University College London) and Senior Fellow of the UCL QASER Lab, is currently an Associate at LaSalle Investment Management, London. His research interests cover the modelling of urban and regional systems, property markets, spatial-temporal and economic networks and policy evaluation for planning in both transport and environmental governance.

Minette D'Lima is a researcher in the QASER (Quantitative and Applied Spatial Economics Research) Laboratory at University College London. She was trained as a pure mathematician with bachelors' degrees in Mathematics and Computer Technology, followed by a PhD in Algebraic Geometry. She works in a multidisciplinary group of mathematicians, physicists and economists providing innovative solutions to financial and economic problems. Her research covers a broad range of projects from complexity analysis and stochastic modelling to structuring portfolios in urban investments. She has been a researcher on an EPSRC Programme Grant, "SCALE: Small Changes Lead to Large Effects," and developed a discrete spatial interaction model to study the effect of transport investments on urban space. She has developed a stochastic model for quantifying resilience on a FuturICT-sponsored project, "ANTS: Adaptive Networks for Complex Transport Systems." She has also worked on developing a mathematical model using portfolio theory and agent-based modelling to simulate the agricultural supply chain in Uganda for a World Bank project, "Rethinking Logistics in Lagging Regions." She is currently working in the EPSRC Programme Grant "Liveable Cities," structuring and optimising portfolios for urban investments, and taking into account the socio-environmental impacts of such investments and their interactions.

Toby P. Davies is a Research Associate working on the Crime, Policing and Citizenship project at University College London, having previously been a member of the UCL SECReT Doctoral Training Centre. His background is in mathematics, and his work concerns the application of mathematical techniques in the analysis and modelling of crime and other security issues. His main area of interest is the spatio-temporal distribution of crime, in particular its relationship with urban form and its analysis using network-based methods.

Valerio de Martinis is Scientific Assistant at the Institute of Transport Planning and System (ETH Zurich). He is part of SCCER Mobility (Swiss Competence Center on Energy Research), and his research activities focus on energy efficiency and railway systems. He received his PhD in Transportation Systems in 2008 at the University of Naples Federico II.

Adam Dennett is a Lecturer in Urban Analytics in the Centre for Advanced Spatial Analysis at University College London. He is a geographer and fellow of the Royal Geographical Society and has worked for a number of years in the broad area of population geography, applying quantitative techniques to the understanding of human populations; much of this involves the use of spatial interaction models to understand the migration flows of people around the UK, Europe and the world. A former secondary school teacher, Adam arrived at UCL in 2010 after completing a PhD at the University of Leeds.

Robert J. Downes is a MacArthur Fellow in Nuclear Security working at the Centre for Science and Security Studies at the Department of War Studies, King's College London. Trained as a mathematician, Rob received his PhD in mathematics from University College London in 2014; he studied the interplay between geometry and spectral theory with applications to physical systems and gravitation. He also holds an MSci in Mathematics with Theoretical Physics awarded by UCL. As a Postdoctoral Research Associate on the ENFOLDing project at The Bartlett Centre for Advanced Spatial Analysis, Rob studied the structure and dynamics of global socio-economic systems using ideas from complexity science, with particular emphasis on national economic structure and development aid.

Hannah M. Fry is a Lecturer in the Mathematics of Cities at the Centre for Advanced Spatial Analysis (CASA). She was trained as a mathematician with a first degree in mathematics and theoretical physics, followed by a PhD in fluid dynamics. This technical background is now applied in the mathematical modelling of complex social and economic systems, her main research interest. These systems can take a variety of forms, from retail to riots and terrorism, and exist at various scales, from the individual to the urban, regional and global, but – more generally – they deal with the patterns that emerge in space and time.

Sean Hanna is Reader in Space and Adaptive Architectures at University College London, Director of the Bartlett Faculty of the Built Environment's MSc/MRes programmes in Adaptive Architecture and Computation, and Academic Director of UCL's Doctoral Training Centre in Virtual Environments, Imaging and Visualisation. He is a member of the UCL Space Group. His research is primarily in developing computational methods for dealing with complexity in design and the built environment, including the comparative modelling of space, and the use of machine learning and optimisation techniques for the design and fabrication of structures. He maintains close design industry collaboration with world-leading architects and engineers.

Shane D. Johnson is a Professor in the Department of Security and Crime Science at University College London. He has worked within the fields of criminology and forensic psychology for over 15 years, and has particular interests in complex systems, patterns of crime and insurgent activity, event forecasting and design against crime. He has published over 100 articles and book chapters.

Anthony Korte is a Research Associate in the Centre for Advanced Spatial Analysis at University College London, where he works on spatial interaction and input–output models relevant to the mathematical modelling of global trade dynamics.

Robert G. Levy is a researcher at the Centre for Advanced Spatial Analysis at University College London. He has a background in quantitative economics, database administration, coding and visualisation. His first love was Visual Basic but now writes Python and Javascript, with some R when there's no way to avoid it.

Elio Marchione is a Consultant for Ab Intio Software Corporation. Elio was Research Associate at the Centre for Advanced Spatial Analysis at University College London. He obtained his PhD at the University of Surrey at the Centre for Research in Social Simulation, MSc in Applied Mathematics at the University of Essex and MEng at the University of Naples. His current role consists, among others, in designing and building scalable architectures addressing parallelism, data integration, data repositories and analytics, while developing heavily parallel CPU-bound applications in a dynamic, high-volume environment. Elio's academic interests are in designing and/or modelling artificial societies or distributed intelligent systems enabled to produce novelty or emergent behaviour.

Francesca R. Medda is a Professor in Applied Economics and Finance at University College London. She is the Director of the UCL QASER (Quantitative and Applied Spatial Economics Research) Laboratory. Her research focusses on project finance, financial engineering and risk evaluation in different infrastructure sectors such as the maritime industry, energy innovation and new technologies, urban investments (smart cities), supply chain provision and optimisation and airport efficiency.

Thomas P. Oléron Evans is a Research Associate in the Centre for Advanced Spatial Analysis at University College London, where he has been working on the ENFOLDing project since 2011. In 2015, he completed a PhD in Mathematics, on the subject of individual-based modelling and game theory. He attained a Master's degree in Mathematics from Imperial College London in 2007, including one year studying at the École Normale Supérieure in Lyon, France. He is also an ambassador for the educational charity Teach First, having spent two years teaching mathematics at Bow School in East London, gaining a Postgraduate Certificate in Education from Canterbury Christ Church University in 2010.

Francesca Pagliara is Assistant Professor of Transport at the Polytechnic School and of the Basic Sciences of the University of Naples Federico II. During her PhD course, in 2000 she worked at David Simmonds Consultancy in Cambridge. In 2002 she worked at the Transport Studies Unit of the University of Oxford, and in 2006 she worked at the Institute for Transport Planning and Systems of ETH in Zurich. She was visiting professor at Transportation Research Group of the University of Southampton (2007 and 2009) and at TRANSyt of the University of Madrid (2007 and 2010). She had further research experience in 2013 in France, where she worked at the LVMT of the University of Paris-Est. She is author of academic books, both in Italian and in English, and of almost 100 papers. She has participated at several research projects.

Joan Serras is a Senior Research Associate in the Centre for Advanced Spatial Analysis (CASA) at University College London. He received his PhD in Engineering Design for Complex Transportation Systems from the Open University in 2007. While at CASA, he has been involved in three research grants: "SCALE: Small Changes Lead to Large Effects" (EPSRC), ENFOLDing (EPSRC) and EUNOIA (EU FP7). His research focusses on the development of tools to support decision making in urban planning, mainly in the transport sector.

Frank T. Smith FRS does research on social-interaction, industrial, biochemical and biomedical modelling, as Goldsmid Chair of applied mathematics at University College London. Author of over 300 refereed papers, he collaborates internationally, nationally and within London, and has taken part in many research programmes. Frank has contacts with government organisations, industry, commerce and NHS hospitals, with real-world applications ranging very widely and including consumer choice, social issues and city growth. Recent support has come from international and national bodies and companies. His applications-driven work deals with social applications to help understanding of crime, opinion dynamics, security strategies and hub development, as well as biomedical, biochemical and industrial applications. Frank tends to use modelling combined with analysis, computations and experimental or observational links throughout. He has been on many peer review panels, has contributed to several books is a long-standing Fellow of the Royal Society and is Director of the London Taught Course Centre for doctoral studies in the Mathematical Sciences.

Tasos Varoudis is a Senior Research Associate in the Bartlett School of Architecture at UCL. He is a registered architect, computer scientist, designer and technologist. He studied Architectural Engineering at the National Technical University of Athens and took his doctorate in Computing Engineering at Imperial College, London. His research ranges from architectural computation and the analysis of hybrid architectural spaces to architecture and human–computer interaction.

Sir Alan Wilson FBA FAcSS FRS is Professor of Urban and Regional Systems in the Centre for Advanced Spatial Analysis at University College London. He is Chair of the Home Office Science Advisory Council and of the Lead Expert Group for the GO-Science Foresight Project on the Future of Cities. He was responsible for the introduction of a number of model-building techniques which are now in common use – including 'entropy' in building spatial interaction models. His current research, supported by ESRC and EPSRC grants, is on the evolution of cities and global dynamics. He was one of two founding directors of GMAP Ltd in the 1990s – a successful university spin-out company. He was Vice-Chancellor of the University of Leeds from 1991 to 2004, when he became Director-General for Higher Education in the then DfES. From 2007 to 2013, he was Chair of the Arts and Humanities Research Council. He is a Fellow of the British Academy, the Academy of Social Sciences and the Royal Society. He was knighted in 2001 for services to higher education. His recent books include *Knowledge power* (2010), *The Science of Cities and Regions*, his five-volume (edited) *Urban Modelling* (both 2013) and (with Joel Dearden) *Explorations in Urban and Regional Dynamics* (2015).

Acknowledgements

I am grateful to the following publishers for permission to use material.

Springer: Quantifying the effects of economic and labour market inequalities on inter-regional migration in Europe – a policy perspective, Applied Spatial Analysis and Policy, Volume 7, Issue 1, pp. 97–117, used in Chapter 3; and Space-time modelling of insurgency and counterinsurgency in Iraq, Journal of Quantitative Criminology, 28(1), 31–48, used in Chapter 8.

Sage: Spatial, temporal and spatio-temporal patterns of maritime piracy, Journal of Research in Crime and Delinquency, Volume 50, Issue 4, November 2013, pp. 504–524, used in Chapter 8.

Elsevier: Geographic patterns of diffusion in the 2011 London riots, Applied Geography, Volume 45, December 2013, pp. 211–219, used in Chapter 9; A spatial model with pulsed releases to compare strategies for the sterile insect technique applied to the mosquito *Aedes aegypti*, Mathematical Biosciences, 2014 Jun 11, pii:S0025-5564(14)00107-2, doi:10.1016/j.mbs.2014.06.001, used in Chapter 17; and Static search games played over graphs and general metric spaces, European Journal of Operational Research, Volume 231, Issue 3, pp. 667–689, used in Chapters 20 and 21.

Nature Publishing Group: A mathematical model of the London riots and their policing, Nature Scientific Reports, Volume 3, Article 1303, doi:10.1038/srep01303, used in Chapter 10.

SimSoc Consortium: Modelling maritime piracy: a spatial approach, Journal of Artificial Societies and Social Simulation, Volume 17, Issue 2, p. 9, used in Chapter 13.

Sejong University: Measuring the structure of global transportation networks, Proceedings of the Ninth International Space Syntax Symposium, Edited by Y.O. Kim, H.T. Park, and K.W. Seo, Seoul: Sejong University, 2013, used in Chapter 24.

Pion Ltd.: The evolution and planning of hierarchical transport networks, Environment and Planning B: Planning and Design, Volume 41, Issue 2, pp. 192–210, used in Chapter 25.

I am very grateful to Helen Griffiths and Clare Latham for the enormous amount of work they have put into this project. Helen began the process of assembling material which Clare took over. She has been not only an effective administrator but an excellent proof reader and sub-editor!

I also acknowledge funding from the EPSRC grant: EP/H02185X/1.

We acknowledge the financial support of the Engineering and Physical Sciences Research Council (EPSRC) under the grant ENFOLDing – Explaining, Modelling, and Forecasting Global Dynamics (reference EP/H02185X/1) and the Security Science Doctoral Training Centre (reference EP/G037264/1). We are grateful for the assistance of the Metropolitan Police in the provision of offence data, and thank S. Johnson and P. Baudains for critical discussions. We also thank the anonymous reviewer for his or her particularly helpful comments.

About the Companion Website

This book is accompanied by a companion website:

www.wiley.com/go/wilson/ApproachestoGeo-mathematicalModelling

The website includes:

- Using support vector analysis to predict extinction events in multi-agent models (Thomas Oléron Evans, Steven R. Bishop and Frank T. Smith)
- A spatial diffusion model with pulsed releases to compare strategies for the sterile insect technique applied to the mosquito Aedes aegypti (Thomas Oléron Evans, Steven R. Bishop and Frank T. Smith)
- Results on the optimal mixed strategies of spatial games (Thomas Oléron Evans, Steven R. Bishop and Frank T. Smith)
- Optimal random patrol over spaces of non-uniform value (Thomas Oléron Evans, Steven R. Bishop and Frank T. Smith)

Part One
Approaches

1

The Toolkit

Alan G. Wilson

Geographical systems are characterised by locations, activities at locations, interactions between them and the infrastructures that carry these activities and flows. They can be described at a great variety of scales, from individuals, organisations and buildings, through neighbourhoods, to towns and cities, regions and countries. There is an understanding, often partial, of these entities, and in many case this understanding is represented in theories which in turn are represented in mathematical models. We can characterise these models, with geography as a core, as geo-mathematical models.

In this book, our main examples are models that represent elements of the global system covering such topics as trade, migration, security and development aid. We also work with examples at finer scales. We review this set of models, along with some outstanding research questions, in order to demonstrate how they now form, between them, an effective toolkit that can be applied not only to particular global systems but more widely in the modelling of complex systems.

These examples have been developed in the context of an EPSRC-funded complexity science programme with twin foci: developing new tools and applying these to real-world problems. In presenting the 'tools' here, it is useful to be aware of Weaver's distinction between systems of disorganised complexity and systems of organised complexity. Both kinds of systems have large numbers of elements, but in the first, there are only weak interactions between them; in the second, some strong interactions. This distinction relates to that between *fast dynamics* and *slow dynamics* – essentially, between systems that can return rapidly to equilibrium following a change and those that are slower. It also relates to those that, from a mathematical point of view, can be modelled by using averaging procedures of various kinds and those more challenging systems that demand a variety of methods, many still the subject of ongoing research. Roughly speaking, systems involving large numbers of people – those travelling to work in a city, for example – fall into the first category, while those involving complex organisations within an economy or physical structures, such as buildings, fall into the second.

Approaches to Geo-mathematical Modelling: New Tools for Complexity Science, First Edition. Edited by Alan G. Wilson.

Companion Website: www.wiley.com/go/wilson/ApproachestoGeo-mathematicalModelling

All complex systems involve nonlinearities. In the case of systems of organised complexity, as we will see, path dependence and the possibility of phase changes make the mathematical aspects of this kind of research particularly interesting. It is through these mechanisms that new structures can be seen to 'emerge', and hence the current notion of *emergent behaviour*.

We proceed by reviewing the main elements of the toolkit in this introductory chapter, and then we proceed to illustrate their use through a series of applications. The headings that follow illustrate the richness of the toolkit.

- Estimating missing data: bi-proportional fitting and principal components analysis (Part 2)
- Dynamics in account-based models (Part 3)
- Space–time statistical analysis (Part 4)
- Real-time response models (Part 5)
- The mathematics of war (Part 6)
- Agent-based models (Part 7)
- Diffusion models (Part 8)
- Game theory (Part 9)
- Networks (Part 10)
- Integrated models (Part 11).

There are three kinds of research questions that lead us to new tools for handling issues in complexity science: firstly, the development of particular tools; secondly, new applications of these tools; and, thirdly, the development of new combinations of tools.

The first category includes the addition of spatial dimensions to Lotka–Volterra models – with applications in trade modelling and security, the latter offering a new dimension in Richardson models. Other examples include adding depth to our understanding of the dynamics of the evolution of centres, or the new interpretation of spatial interaction as a 'threat' in building models of security.

Probably the potentially most fruitful area – illustrating Brian Arthur's argument on the nature of technological development – is the development of new combinations. This also illustrates 'new applications'. One of the London riots' models, for example, combines epidemiological, spatial interaction and probability sub-models. In developing the Colonel Blotto model, we have added space and the idea of 'threat' in combination with game theory. By adding dynamics to migration, trade and input–output models, and by incorporating development aid, we have created possible new approaches to economic development.

In Part 2, we show how to expand – by estimating 'missing data' – some sets of accounts. Historically, examples of account-based models are Rogers' demographic model, Leontief's input–output model and the doubly constrained journey to work model developed on a bi-proportional basis by Fratar but later set in an entropy-maximising framework. In this part, we present three examples of account-based models in which bi-proportional fitting is used either to make data from different sources consistent or to estimate missing data. These are examples of well-known techniques being used creatively in new situations. We present three examples. Firstly, we take European migration. There are good data at the intercountry level, and in- and out-totals are available at sub-regional scales. We use bi-proportional methods to estimate the missing data. Secondly, we again apply methods to find missing data on international trade. We have data on total intercountry trade, and we have sector data for exports and imports. We use the model to estimate flow data by sector. Thirdly, we use data on input–output accounts which are rich for a subset of countries, and we use a principal components method to estimate missing data. The results of the second and third of these examples are used in building the integrated model described in Chapter 17.

In Part 3, we describe an account-based trade model integrated with dynamic adjustment mechanisms on both prices and economic capacity. These mechanisms are rooted in Lotka–Volterra types of equations. In this case, therefore, we are demonstrating the power of integrating different 'tools'. Given the complexities of this task, what is presented is a demonstration model. The spatial Lotka–Volterra type of dynamics, as represented in the retail model, can be seen as a more general archetype of centre dynamics. In this part, we explore the dynamics of such models in more depth.

In Part 4, we discuss different statistical approaches to hypothesis testing for spatial and space–time patterns of crime and other events. Methods for examining point processes are presented in the case of insurgency and piracy, and for riots at the area level. We discuss methods for identifying regularities in observed patterns (e.g. spatial statistics, the K-S test, Monte Carlo methods and simulated annealing), methods for testing theories of those patterns (logistic and conditional logit models) and statistical models that may be used to describe and potentially predict them (self-exciting point process models).

Part 5 provides further examples of combinations. Epidemiology provides the model of propensity to riot; spatial interaction modelling answers the 'Where?' question; and we have a third model of probability of arrest. We present two alternative models of the riots: one illustrates the deployment of discrete-choice spatial models, and the other uses an agent-based approach. The differences between such computational approaches and the mathematical models are explored

Models of war, in a broad sense, have a long history. Richardson's 'arms race' model is an excellent example. This model can be seen as a special case of Lotka–Volterra dynamics. Space is not explicit in the original, and this is clearly a critical feature if such models are to be used strategically. In Part 6, therefore, we extend this model to incorporate space.

Agent-based simulations are widely used across a vast range of disciplines, yet the fundamental characteristics of their behaviour are not analysed in a systematic and mathematically profound manner. In Part 7, we present a toolkit of mathematical techniques, using both the rules that govern multi-agent simulations and time series data gathered from them, to deduce equations that describe the dynamics of their behaviour and to predict rare events in such models. In certain cases, the methods employed also suggest the minimal interventions required to prevent or induce particular behaviours.

In Part 8, we introduce diffusion models. These have a long history in disciplines such as physics but less so in the social sciences. We first, following Medda, show how Turing's model of morphogenesis – with two interacting processes generating spatial structure – can be adapted to urban dynamic structural modelling. We also add, in Appendix B, some mathematical explorations of a different kind of diffusion: the control of insect populations, which are also rooted in Lotka–Volterra mathematics.

In Part 9, we invoke concepts from game theory to present a framework for the analysis of situations in which limited resources must be efficiently deployed to protect some region from attacks that may arise at any point. We discuss how the mathematical techniques described may be applied to real-world scenarios such as the deployment of police to protect retail centres from rioters and the positioning of patrol ships to defend shipping lanes against pirate attacks.

We promote Colonel Blotto to Field Marshal in the game of that name by adding space more effectively.

Graph-based analyses, focusing on the topological structure of networks, provide crucial insight into the kinds of activities that occur within them. Studies of small-scale spatial networks have demonstrated conclusive predictive capacity with respect to social and economic factors. The relationship between multiple networks at a global scale, and the effect of one on the structure of another, is discussed in Part 10. It presents models of two different but mutually interdependent kinds of networks at an international scale. The first is a global analysis of the structure of international transportation, including roads, but also shipping, train and related networks, which we intend to be the first such study at this scale. We assess the applicability of centrality measures to graphs of this scale by discussing the comparison between measures that include geometrical properties of the network in space, with strictly topological measures often used in communications networks. The second is the economic structure of national industry and trade, as expressed in recorded input–output structure for individual countries. Flows expressed in these represent a non-spatial or trans-spatial network that can be interpreted by similar measures, to understand both comparative differences and similarities between nations, and also a larger picture of economic activity. The two networks will be analysed as a coupled system of both physical goods through space and non-spatial economic transactions. Due to the relative stability of these networks over time, their use as a background for modelling activity makes them useful as a predictive tool. Visualisations of this network will indicate points most susceptible to shock from economic or physical events, or areas with the potential for greatest impact from investment or aid.

In the concluding chapter which makes up Part 11, we review the progress made in geo-mathematical modelling and discuss ongoing research priorities.

Part Two

Estimating Missing Data: Bi-proportional Fitting and Principal Components Analysis

2

The Effects of Economic and Labour Market Inequalities on Interregional Migration in Europe

Adam Dennett

2.1 Introduction

This chapter employs migration flow data to explore the effects of economic and labour market inequalities on interregional migration in Europe. The data are the estimates obtained from the bi-proportional fitting model described in Chapter 6 of this volume (and also in Dennett and Wilson, 2013). The migration of people is always of interest to governments and policy makers. In countries like the United Kingdom where the balance of net migration is towards immigration, concern can swing between, on the one hand, the benefits brought by migrants such as their skills and contribution to growth and the economy, and, on the other hand, the demand that they might place on finite resources such as housing and services. Where the balance of net migration is towards emigration, different but no less important issues may be of concern, such as the loss of human capital or the benefits accrued by remittances.

The economic crisis, which began in 2008 and which has affected much of the Western world, has forced the issue of migration higher up the political agenda of many countries. In Europe, right-wing anti-immigration parties have been gaining traction in many places (Golden Dawn in Greece, UKIP in the UK and Le Front National in France) and are an unwelcome marker that increased political unease over migration when national resources are being squeezed. Concerns over immigration are not just the preserve of the far right, however, with mainstream politicians from across the spectrum frequently discussing (im)migration and the many economic and social benefits as well as possible drawbacks that it brings. In the UK, the coalition government of 2010–2015 led by the Conservative Party have been more concerned by the negative impacts of migration and pledged to cut positive net-migration rates

Approaches to Geo-mathematical Modelling: New Tools for Complexity Science, First Edition. Edited by Alan G. Wilson.

Companion Website: www.wiley.com/go/wilson/ApproachestoGeo-mathematicalModelling

during the course of the Parliament (May 2012). While the government is able to control much international migration through legislation, they have very little control over migration from other EU member countries, and recent concerns regarding immigration appear to have been amplified by a combination of uncertainty and a lack of agency in this context.

Membership in the European Union means that all countries are bound by conventions which permit the free movement of European citizens. After the accession of the eight former Eastern Bloc countries to the EU in 2004, movement restrictions which had previously applied to these counties were lifted and migrants were able to move as they wished. The UK government had little idea what would happen, and this uncertainty was brought into stark relief when expert predictions of the volume of migration into the UK (Dustmann *et al.*, 2003) were shown, after the event, to be serious underestimates (Fihel and Kaczmarczyk, 2009). In the UK, a temporary veto on migrant movements from the A2 countries (Bulgaria and Romania) has to come to an end (Vargas-Silva, 2013), ushering in feelings of uncertainty and impotence within the government. This particular storm has been played out very publicly in the media with reports (bordering, sometimes, on the surreal) of the UK government considering the launch of an anti-immigration advertising campaign designed to deter would-be migrants from Romania and Bulgaria from even considering a move (http://www.guardian.co.uk/uk/2013/jan/28/campaign-deter-romanian-bulgarian-immigrants-farcical).

The posturing of the government and other political parties and the media interest have been taking place in parallel with serious research into migration into the UK from Eastern Europe (Benton and Petrovic, 2013; MAC, 2011; Rolfe *et al.*, 2013). At least part of this work has tried to address uncertainties around the factors driving migration flows within the EU, but uncertainty does not fit well with policy – policy positions and decisions are made far more straightforward with reliable (often quantitative) evidence. Since the difficultly of producing definitive migration estimates was demonstrated so clearly by the inaccuracy (and poor interpretation of uncertainty by the government) of the Dustmann *et al.* (2003) estimates, this recent work has been far more cautious. Rolfe *et al.* (2013: p43) state that "it is not possible to predict the scale of migration from Bulgaria and Romania to the UK with any degree of certainty", and Benton and Petrovic (2013: p21) comment that "[p]rojections (of intra-European migration flows) are notoriously unreliable", while the report of the Migration Advisory Committee (MAC, 2011) stops short entirely of making predictions other than to say that migration from Romania and Bulgaria is likely to increase.

The difficulty of making reliable migration projections does not mean that policy makers in the UK or other EU member-states should be limited to making decisions without the use of good quantitative evidence – this evidence exists, although accessing it may not be straightforward. If we are able to understand the drivers of migration and the precise effect they have, then it follows that influencing these drivers with policy decisions should have knock-on effects on migration flows. But what are drivers that could be influenced in Europe? Abel (2010) explores a number of migration covariates and highlights the effects that existing stocks of migrants (proxies for existing migrant networks), distance, language and trade links all have on flows of migrants in Europe – none of which could be usefully influenced by policy, except perhaps trade. Amongst migrants in Europe, however, it is economic factors that are the most important influence: 'searching for work' is cited as the primary reason for moving given by most EU migrants (Benton and Petrovic, 2013). This chimes with neoclassical economic theory which suggests that migration is the inevitable outcome of an individual's desire to maximise their wellbeing through the search for higher wages (Borjas, 1989). It could be argued that of all of the drivers of migration within the EU, it is relative economic conditions that have the potential to be most easily influenced by policy, as mechanisms for this already exist.

Within the EU, disparities between countries and regions have led the European Commission, since 1988, to make efforts to increase the economic and social cohesion between the member-states. Since the inception of the EU Cohesion Policy and other EU structural funds, large amounts of money have been made available for investment into regions identified as falling behind on measures of economic and social well-being. During the funding period running from 2007 to 2013, €347 billion were made available in the EU Cohesion Fund for investment in transport, education and skills, enterprise and the environment (INFOREGIO, 2012). Research by Becker *et al.* (2010) examined the effects of inward investment by EU structural funds on GDP per capita growth in target regions and found that, on average, growth was raised by 1.6% in these areas. This is an important finding as it suggests that EU policy mechanisms have a positive effect on the *growth* of regional economies, and are not just in place to arrest decline. Of course, EU structural-cohesion funds are one of the main mechanisms for reducing inequalities; others may also be put in place. Atkinson (2013) highlights two proposals contained in a report by the European Commission (2012). The first is an EU-wide unemployment benefit; the second is an entitlement of every EU citizen to a 'basic income'. Whether ideas such as this would ever make it into EU policy are unclear – Atkinson's (2013: p8) assertion that they are "radical, but not outlandish" suggests they are not close to making it onto the statute books. But when considered in conjunction with the rising popularity of the inequality agenda (Dorling, 2010; Therborn, 2006; Wilkinson and Pickett, 2009), it might be that increased wealth redistribution in Europe leading to reduced inequalities, even in the current economic climate, may not be entirely out of the question.

Given that economic and labour market inequalities are the main drivers of migration flows within the EU, then it might be that economic policy levers relating to the redistribution of wealth actually have the potential to influence migration flows in ways that might not have been expected. This, of course, would lead to an interesting paradox for many of the anti-immigration right-wing political parties within Europe (who are frequently Eurosceptic), suggesting that perhaps one of the most effective ways to reduce levels of intra-European migration would be to increase financial contributions made available to central European funds designed for wealth redistribution. The question that the UK government might be interested in answering is whether increasing contributions to EU central funds for redistribution to poorer member-states can be enough to stop the politically feared migration from the A2 countries.

Until recently, it would have been hard to answer this question effectively. Subnational economic data such as average GDP per person and unemployment rates are readily available from Eurostat and are made available through their online database (http://epp.eurostat.ec.europa.eu/portal/page/portal/statistics/search_database), but the best data on intra-European migration have been at the country level. This is not ideal for policy making as, for example, much of the Polish migration to the UK was concentrated in particular regions like East Anglia. New work, however, by Dennett and Wilson (2013) – and Chapter 6 – has produced a time series of interregional migration matrices for Europe. These new data along with regional economic data already available from Eurostat will enable the study of regional economic inequality and migration interactions for the first time. Questions might be raised about using estimated data in this context, but the highly constrained nature of the estimates (regional flows sum to known country-level flows) means that they are reliable. Consequently, a thorough analysis of these migration and economic data should reveal whether changes to the relative prosperity of regions is likely to have an effect on intra-European migration and provide a better evidence base for European policy makers and governments concerned by migration flows over which they have little direct control.

2.2 The Approach

This chapter will seek to answer the following questions: how much is interregional migration within Europe influenced by the relative economic performance of regions? What impact could the reduction of regional economic inequalities have on migration flows within the EU, and what could this mean for the policies of EU governments concerned by migration flows? Using multinomial logistic regression models to analyse the data, it is shown that migration flows are indeed influenced by these factors, but that the effects are most noticeable at the extremes of migrant behaviour – either high volumes or low volumes – and at the extremes of economic and labour market inequality. A detailed account of these findings and a discussion in relation to the potential for a reduction in regional inequalities to influence migration behaviours are provided in this chapter.

2.3 Data

The data used in this analysis are for Nomenclature of Territorial Units for Statistics, level 2 (NUTS2) areas. The Dennett and Wilson (2013) migration estimates are for 287 NUTS2 regions within 31 European countries (EU27 + Norway, Switzerland, Lichtenstein and Iceland) and are for the 6 years between 2002 and 2007. The period spans the time before and after the accession of the A8 countries to the EU, and so offers the potential to explore economic influences before and after the removal of the political barriers to migration in 2004. The flow estimates are counts of migrant transitions (Rees, 1977) which occurred during the year and are consistent with the intercountry flows published by the MIMOSA project (de Beer *et al.*, 2009).

Economic and labour market data are collected for European NUTS2 regions by Eurostat and made available through their online database already mentioned. Table *nama_r_e2gdp* contains Gross Domestic Product (GDP) data (in Euros and Purchasing Power Standard (PPS)) for NUTS2 regions between 1995 and 2009. In this analysis, PPS will be used rather than strict GDP. PPS is a measure which standardises GDP in Euros by the amount of goods and services it is able to purchase across EU25 countries (for details, see http://stats.oecd.org/glossary/detail.asp?ID=7184). Table *lfst_r_lfu3rt* contains unemployment rates by age and sex for NUTS2 regions between 1999 and 2011. Data from both of these tables are used, although PPS and unemployment data are not available for all of the regions for which migration estimates were produced across all years. Where economic data are not present for origins or destinations, these zones are excluded from the analysis.

Migration flows M are estimated between origin and destination pairs j with 82,369 (287×287) M_{ij} pairs within the system. Internal migration flows (those between NUTS2 regions but within countries) were omitted from the analysis, leaving up to 76,724 pairs to be analysed for any given year. These pairs can be thought of as individual data points, and so for ease of notation we will label these simply M_i. While economic and labour market data correspond to single zones, origin–destination unemployment and PPS data were constructed through computing ratios for each pair of zones in the system. As with the migration data, these ratios are directional so that, for example, if zone A has a PPS per person value of €10,000 and zone B has a value of €8000, the PPS A/B ratio will be 1.25, whereas the B/A ratio will be 0.8. In the subsequent analysis, we use the following ratios: unemployment rates and GDP/PPS (in aggregate and per person), and for ease of presentation and for later analysis, we group them into deciles (from lower to higher). If each of these variables are labelled by $x_1, x_2 \ldots x_m$ and origins and destinations (O/D) by i and j, respectively, then a typical ratio R for variable 1 is:

$$R_{ij}^{x_1} = x_{1i}/x_{1j}$$

Once we have a full set of ratios $R_{ij}^{x_m}$, the values can be grouped into deciles so that we have a new set of variables for each O/D case xd_i, with deciles $k = 1 \ldots 10$: $xd_{i,1,m}, xd_{i,2,m} \ldots xd_{i,k,m}$.

The lowest decile $xd_{i,1,m}$ is very much less than $xd_{i,10,m}$, and vice versa. Examples of the resulting plots are shown in Figure 2.1.

2.4 Preliminary Analysis

Figure 2.1 shows average migrant numbers plotted against variable decile groups for two PPS and three unemployment variables. Decile 1 in each case represents a very low value in the O/D ratio for that variable – for example, for the '15 and above unemployment rate' in 2002, O/D pairs which fell into decile 1 have a ratio of less than 0.3 (e.g. the unemployment rate in the origin was 0.3 times or less than that of the destination). Decile 10, on the other hand, represents a very high value in the ratio of that variable for the O/D pair (e.g. for the '15 and above unemployment rate' variable, the ratio was 10 times or more than that of the destination).

Clear gradients emerge when moving from decile 1 to decile 10 for each of the unemployment variables (first three graphs), with average numbers of migrants increasing across all deciles between 2002 and 2007. We can observe that the lowest migration averages occur in the first decile (where the rates of unemployment are far lower in the origin than they are in the destination). Across almost all years, steadily more migrants can be counted moving towards decile 10, and while there is some small variation in this general pattern of increase from one decile to the next, the pattern is clear – more migration between regions can be observed when the ratio is such that unemployment rates in the destination region are very much lower than they are in the origin. In the years before the A8 countries joined the EU (pre-2004), it appears that – certainly for unemployment in the 15–24 and 15 and above groups (groups where most migrants will be found) – sharper increases can be observed at decile 10 when the ratio between origin and destination rates is largest. This suggests that in these years, employment conditions at the origin need to be significantly worse than at destination for a migration move to be instigated. Post accession, the gradient across the deciles is smoother, demonstrating that with higher rates of migration in general, the potential for moves to occur between origins and destinations with closer unemployment ratios is increased – in other words, labour market drivers for migration might be slightly less important.

Examining the graphs for the PPS variables (the final two in the figure), similar patterns can be observed – the larger the inequality gap, the more likely that migrants will flow from

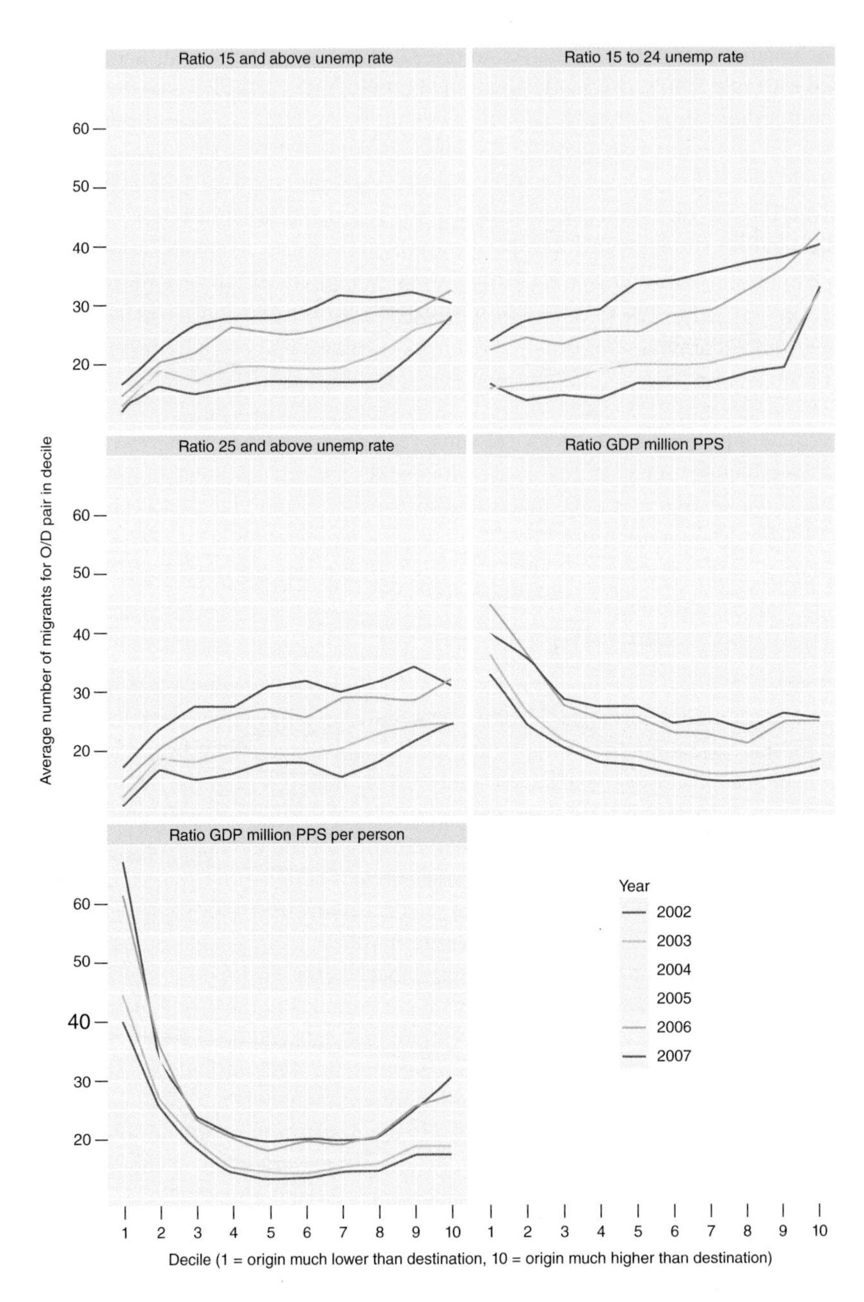

Figure 2.1 Average migration rate per 1000 population by ratio decile

areas where conditions are poor to areas where conditions are good. With PPS, however, the direction of the relationship is apparently reversed, with larger migration volumes experienced in decile 1. This apparent reversal is, of course, due to the direction of the ratio relationship, with the PPS ratio in decile 1 indicating that the PPS at origin is very much lower than it is at the destination. In decile 10, the opposite is true, with the PPS at origin very much higher than it is at destination. The graphs show that when PPS is normalised by population (average PPS per person rather than the total for a region), patterns are somewhat different to when just total PPS is considered. If we study the graph for PPS per person, across all years, migration flow volumes are around twice as large in decile 1 as they are in decile 2 (smaller in 2002, but larger in 2007). In 2007, pairs of regions that fall into decile 1 have a PPS value in the origin less than 0.5 that of the destination. To fall into decile 2, the ratio is between 0.5 and 0.65. This indicates that a potential doubling of personal income might be an important threshold for influencing migration flows. Where the potential is only slightly less than a doubling, then this can have quite a dramatic impact on the number of flows that occur, reducing the volumes by almost 50% in some cases.

Interestingly, if the PPS per person ratios fall between decile 4 and 8, this has very little effect on flow volumes. In 2007, this means that the O/D ratios range between 0.75 and 1.5 of each other. At deciles 9 and 10, however, there is a noticeable increase in flows from areas with origins with much higher PPS values than destinations. Given the patterns shown in decile 1, this is apparently counterintuitive, although with this effect being more pronounced in 2006 and 2007 than in previous years, part of the reason could be the return migration of temporary economic migrants. It may also reflect counter-urbanising moves from city regions (Champion and Vandermotten, 1997), where PPS per person is generally far higher, to rural regions where PPS per person is comparatively lower.

2.5 Multinomial Logit Regression Analysis

Where exploratory analysis has suggested a relationship between variable deciles and rates of migration, it is possible to explore the relationships further using logistic regression models. To make this feasible, with a large number of O/D pairs (cases) to analyse for each economic/labour market variable ($N = 1 \ldots 460,344$ cases for each pair of regions migration can occur over the 6-year period), we group the dependent-variable (M_{ij}) cases into deciles $k = 1 \ldots 10$, which represent 10% of the migration flows. As an indication, in 2002 decile 1 contains pairs of regions where 0 flows occurred, decile 5 contains pairs where 4 or 5 migrants moved, and decile 10 contains pairs of regions where between 39 and 1744 migrants moved. According to the M_{ij} value, each case i is assigned a migration decile value: $Md_{i,1}, Md_{i,2} \ldots Md_{i,k}$. In this analysis, therefore, our goal is to try to explain the relationship between the migration flow deciles $Md_{i,k}$ and the explanatory variable deciles $xd_{i,k,m}$. As we have k possible decile values in both our outcome-dependent and independent-predictor variables, we employ a multinomial logistic regression.

The multinomial logistic regression is similar to other forms of regression model as it uses a linear predictor function, in our case, $f(Md_{i,k})$, to estimate the probability that the O/D case i has the decile outcome k. The full model here would take the following form:

$$f(Md_{i,k}) = \beta_{0,k} + \beta_{1,k} xd_{i,k,1} + \beta_{2,k} xd_{i,k,2} + \beta_{m,k} xd_{i,k,m} \tag{2.1}$$

Or:

$$\ln \frac{Pr(Md_i = k)}{Pr(Md_i = k(ref))} = \beta_{0,k} + \beta_{1,k} xd_{i,k,1} + \beta_{2,k} xd_{i,k,2} + \beta_{m,k} xd_{i,k,m} \tag{2.2}$$

In the formulation of the model here, the outputs are described by odds ratios rather than by the multiplicative coefficients common in ordinary least squares regression models. Odds are simply the probability that an event (high or low migration) occurs to the probability that it doesn't, with the odds ratio always relative to some reference category (or decile k, in our case).

As our migration deciles are on an ordinal scale, an ordinal logistic regression (or ordered logit) model could have been employed instead of the multinomial model (which treats each decile k as a separate category). However, early experimentation with this dataset revealed that the proportional odds assumption underpinning the ordinal model is violated, and so a multinomial model was more appropriate – see UCLA (2012) for a detailed explanation of why this is.

An additional dimension in the model not yet accounted for is *time*. The precise decile thresholds for the migration, GDP and unemployment deciles change in each year of the 2002–2007 dataset, thus the final model which employs two explanatory predictor variables (a ratio decile of over 25 unemployment, xd_1 and a ratio decile of GPD/PPS, xd_2) and disaggregates each model by time, t, is as follows:

$$\ln \frac{Pr(Md_{i,t} = k)}{Pr(Md_{i,t} = k(ref))} = \beta_{0,k} + \beta_{1,k} xd_{i,t,k,1} + \beta_{2,k} xd_{i,t,k,2} \tag{2.3}$$

The model in Equation (2.1) can be run in a variety of common statistical packages; SPSS is chosen in this case as it provides outputs which allow us to assess how well the model fits and the contribution of each independent variable to the model. Taking data for 2002 as a test set, 'Likelihood ratio tests' suggest a significant relationship between the dependent and independent variables, but that there is also a significant difference between the model and the underlying data, indicating a poor model fit. This is partially a function of the volume of data used in the model and also that PPS and unemployment are far from the only factors influencing migration flows within Europe.

Figure 2.2 is a visualisation of the odds ratio outputs from the model described in Equation (2.3) where $t = 2002$. The ratios are all relative to the reference migration decile, which in this case is decile 10 for highest flow volumes (recalling that migration decile 1 has the lowest flow volumes). These migration deciles are represented by each of the nine subgraphs in the figure labelled 1–9 (decile 10, for high flows, is the reference category and so not included).

To recap, the ratios of O/D PPS and over 25 unemployment are also put into decile bins, with deciles 1–5 including all ratios less than 1, and deciles 6–10 including all ratios greater than 1. So, for example, decile 1 for PPS would mean that the origin has very much lower PPS than the destination, whereas decile 10 for unemployment would mean that the origin has very much higher unemployment than the destination.

For each migration decile, the odds ratios of being in a particular unemployment or PPS decile are represented by the points on the graph, with the vertical lines indicating the 95% confidence interval for each point. The confidence interval for each odds ratio is calculated using the following formula:

$$Conf\,Int = OddsRatioEstimate \mp ConfidenceCoefficient * StandardError$$

where the confidence coefficient is taken from the normal distribution and estimated at 1.96 for a 95% confidence interval. If the confidence interval crosses 1 at any point, then this indicates

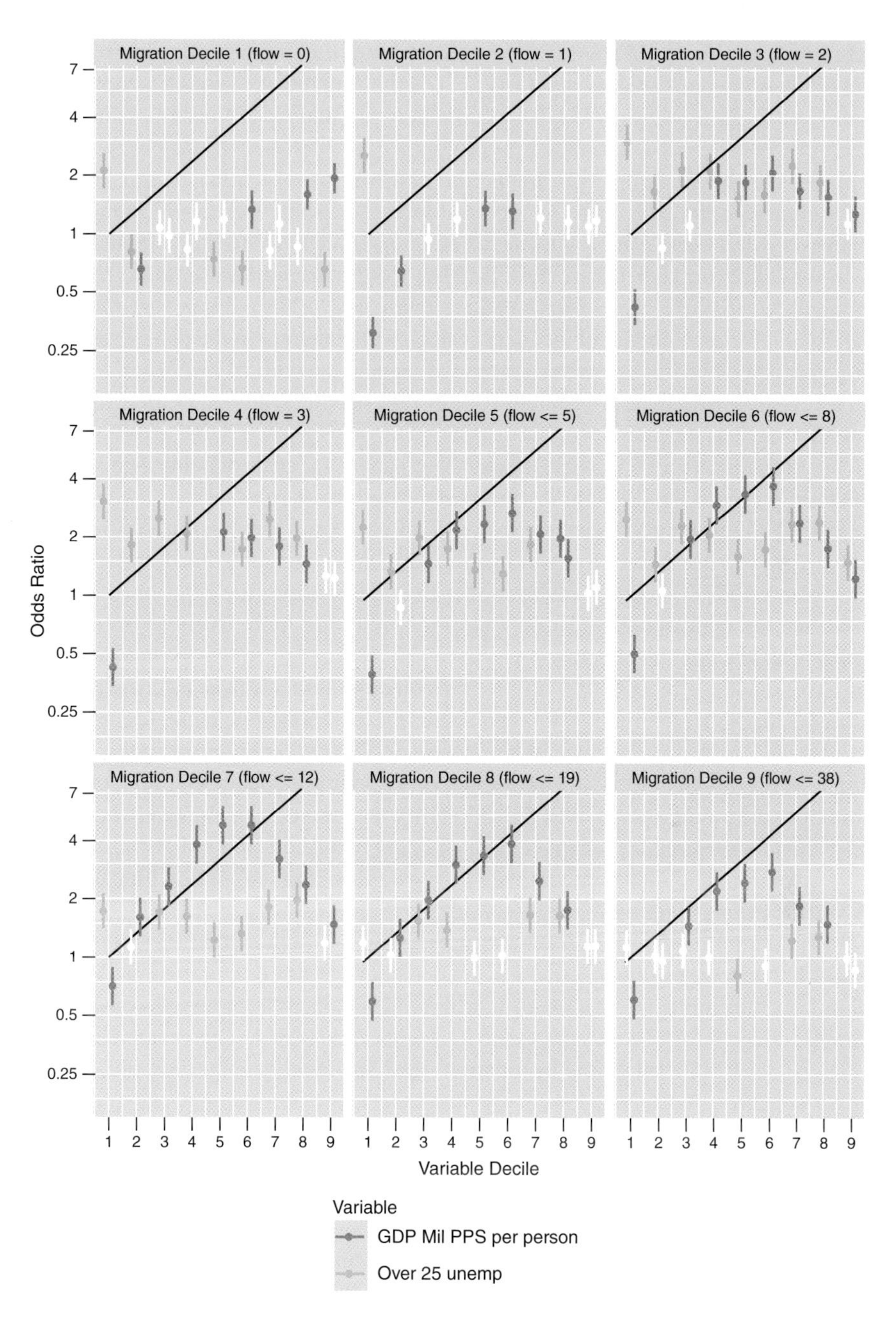

Figure 2.2 Odds ratios for the effect of levels of PPS and over 25 unemployment on O/D migration flow volumes in Europe, 2002 (highest migration level reference)

that there is potentially no difference between the odds of the migration decile in question and the odds of the reference category and the effect of the explanatory variable is not statistically significant (at the 95% or $p > 0.05$ level of confidence). Statistically insignificant odds ratios are shown in white.

Examining the smallest migration flows (the first four panels, from the top left to bottom right, in Figure 2.2), the odds of being a pair of locations where O/D over 25 unemployment ratio is low (Decile 1) versus high (Decile 10, reference) are twice to thrice higher where O/D migration flows are low rather than high. In other words, *low migration flows (those ≤3) are twice to thrice as likely to occur where the unemployment of over 25s is very much lower at the origin than the destination* (subject to the uncertainly inherent in the confidence intervals). Scanning across the blue bars in these panels, we see that the odds of low origin to destination migration reduce as the over 25 unemployment rate becomes higher in the origin than the destination.

Studying odds ratios for PPS across the first four panels, the odds of being a pair of locations where O/D PPS ratio is low (Decile 1) versus high (Decile 10, reference) is a quarter to a half as high where O/D migration flows are low. Put another way, low migration flows are much less likely to occur where the PPS at origin is very much lower than at the destination. Moving to deciles where PPS at origin is much higher than destination (the right-hand side of each panel), we see the opposite effect – low migration flows are more likely to occur where the origin PPS is higher than the destination.

When exploring the effects of PPS differentials on migration, it can be informative to change the contrast variable in the model. In Figure 2.2, the contrast is with the highest migration decile. In Figure 2.3, the contrast is with the lowest decile (or no migration, in this case). Starting at the lower right panel in the figure (which is for migration decile 10, the highest level of migration), it is clear that where the PPS differentials are greatest (i.e. the origin has very much lower PPS than the destination), then the odds of being in this low-origin PPS/high-destination PPS decile are around 4.5 times greater than they would be if the migration flow was zero. Put another way, *high migration flows are far more likely to occur where the O/D PPS inequality is greatest*. This amplified propensity decreases rapidly where the O/D PPS differential is not as severe – in this case, by decile 2 of PPS, the odds have reduced to only around 1.5 times as great. Similarly, this influence of large PPS differentials reduces as the levels of flow decrease. Interestingly, decile 9 for PPS in each of the 9 panels is less than 1. This means that where the origin PPS is much higher than the destination, then any level of migration greater than 0 is less likely to occur.

The patterns of migration associated with PPS do not completely mirror those of unemployment – it is not simply a case of migration flows increasing as the PPS differentials between origin and destination increase. As migration flows increase to between 3 and 38 migrants (panels 4–9 in Figure 2.2), the odds of these levels of flow occurring are increased where the ratios of O/D PPS are closer to parity (deciles 4, 5 and 6 for PPS ratios). Odds drop off again where the ratios are larger. If we make the assumption that places with similar levels of PPS are similar in the lifestyle and other opportunities they offer, this suggests that these moderate moves are motivated by other factors which are more common in these economically similar places – moves up the career ladder, family formation and so on.

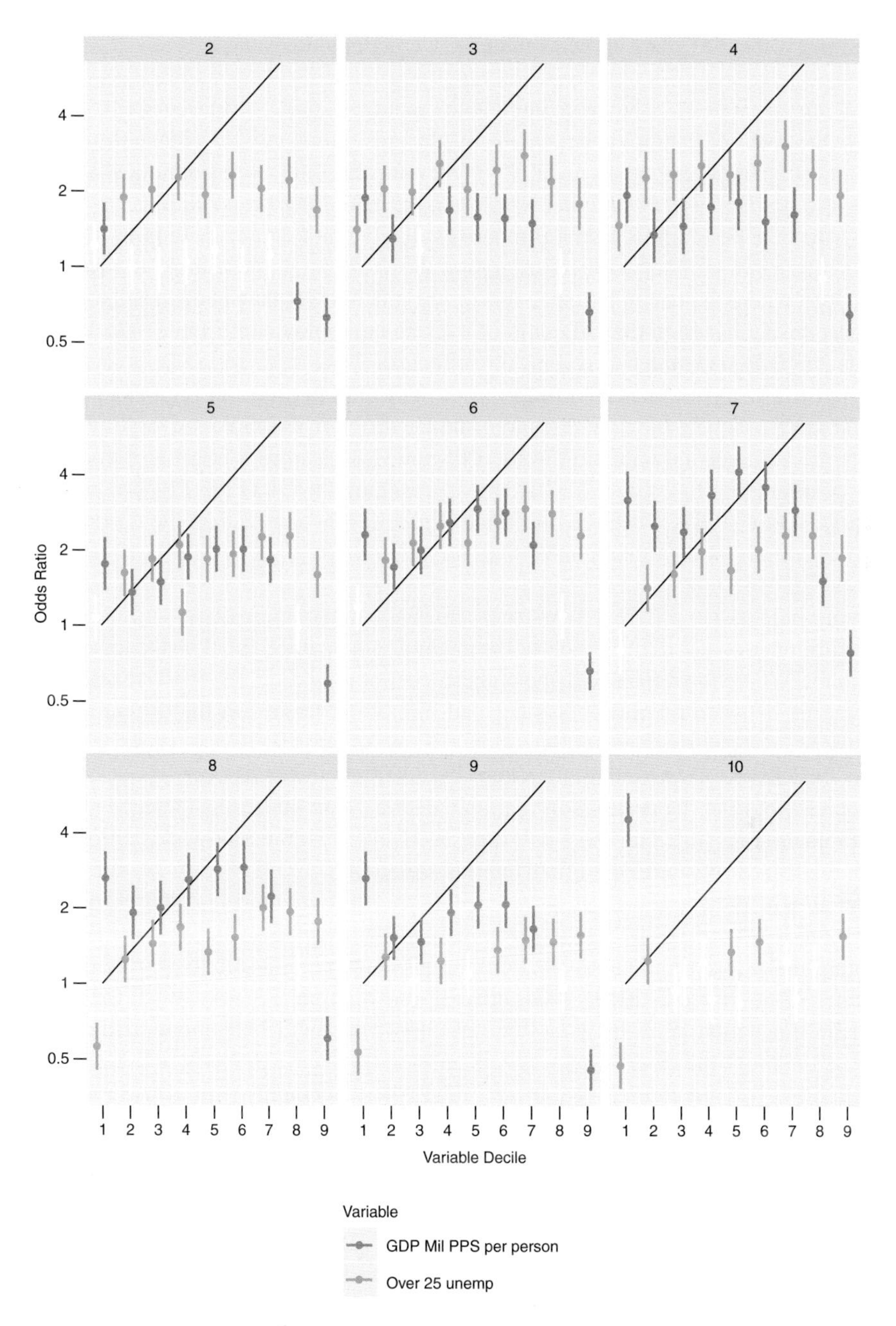

Figure 2.3 Odds ratios for the effect of levels of PPS and over 25 unemployment on O/D migration flow volumes in Europe, 2002 (lowest migration-level reference)

In summary, the 2002 data allow us to draw two main conclusions:

1. Small migration flows, when compared with the largest, are 2–3 times more likely to occur where the levels of unemployment at the origin are significantly lower than the destination, and this effect reduces both as unemployment ratios become more similar and as migration flow volumes increase. Put another way, when the labour market conditions are better at the origin than the destination, then it is much less likely that migrants will move – small flows are 2–3 times more likely. Where migration flows are high, these flows are half as likely to happen where the unemployment at origin is much lower than at the destination – in the first decile. A change in the O/D ratio of over 25 unemployment from decile 1 (≤0.30) to decile 2 (≤0.45) is enough to reverse this effect.
2. Migration flows are influenced by differences in the PPS between the origin and destination, but far more so where the flows are large and the differences in PPS are large. Where flows are large, a one decile reduction in the PPS ratio can lead to a fourfold reduction in the odds of migrating. In 2002, this would mean changing the O/D PPS (PPP per person) ratio from ≤0.43 to ≤0.63. When the origin has higher levels of PPS than the destination, then all levels of migration are less likely than no migration.

Studying outputs for different years (not shown), comparable patterns are observed. One notable difference concerns the effects of the unemployment differential on the lowest volumes of migration. Where the reference is the highest level of migration, for every year in the study, the difference between migration decile 1 and migration decile 2 is the difference between zero migrants and one migrant. In 2002, flows of one migrant were only slightly more likely to occur than flows of zero migrants where the unemployment of over 25s at the origin is very much lower than the destination (variable decile 1). In 2003, this changed slightly, and flows of one migrant were one and a half times as likely as no migrants (odds ratio of 2.09 for zero migrants and 3.62 for one migrant) in this situation. In 2004 (Figure 2.4 for a graphical representation of the data), this jumps noticeably to well over twice as likely (up to an odds ratio of 6.46).

To interpret this, we should really be thinking about the relationship to the reference migration decile. What we see is that in 2004, flows of one migrant (i.e. very low flows – decile 2) are over six times more probable than flows of many migrants where unemployment is lower at the origin than at the destination – this is an increase from 2.5 times in 2002. When low migration flows become more probable over time where unemployment is lower at the origin than the destination, this is a reflection of more moves taking place between origins with high unemployment and destinations with low unemployment. The jump in 2004 coincides with the accession to the EU of many former Eastern Bloc countries with higher over 25 unemployment rates and the known flows from these countries into Western Europe subsequently. The jump being far more noticeable at decile 2 than decile 1 is a reflection of 0 being the most likely value in a system matrix of many origins and destinations. Political change opened the door to these migrants, but it is the inequality gradient which is driving the flows within the system. This effect is maintained in 2005 and 2006, but drops off again in 2007.

This effect is perhaps a little easier to comprehend where the reference category is no migration (migration decile 1). Figure 2.5 shows the odds ratios for the highest levels of migration (vs. no migration) in 2004. Here, we can see that high migration flows are between one-half and one-third as likely where the rates of over 25 unemployment are very much lower at the

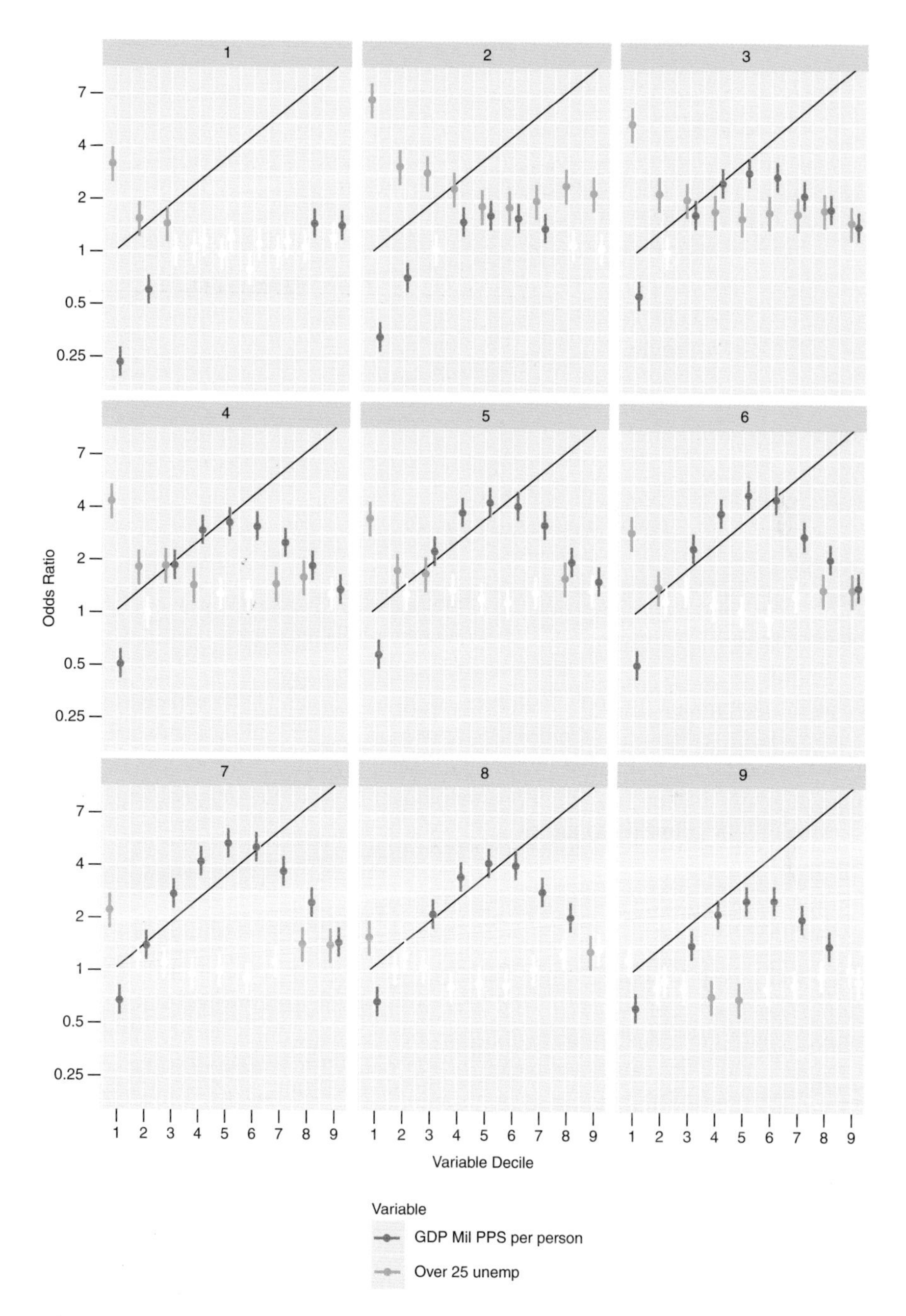

Figure 2.4 Odds ratios for the effect of levels of PPS and over 25 unemployment on O/D migration flow volumes in Europe, 2004 (highest migration-level reference)

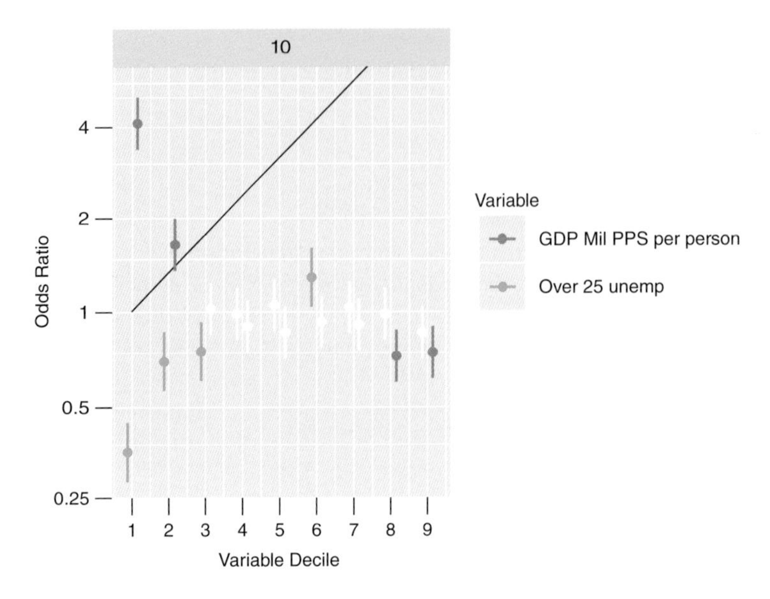

Figure 2.5 Odds ratios for the effect of levels of PPS and over 25 unemployment on highest migration flow volumes (decile 10) in Europe, 2004 (lowest migration-level reference)

origin than the destination. In 2002, high flows were only around half as likely, so by 2004 we can observe a large reduction in the propensity of high migration flows between origins with low unemployment and destinations with high unemployment as the system becomes filled with more migrants flowing from high-unemployment areas to low-unemployment ones after the accession of new countries to the EU.

2.6 Discussion

Before the analysis in Section 2.5, two questions were posed: (i) how much is interregional migration within Europe influenced by the relative economic performance of regions? And (ii) what impact could the reduction of regional economic inequalities have on migration flows within the EU, and what could this mean for EU policy? It is now possible to answer the first of these questions. The fact that migrants are influenced by relative economic and labour market conditions is not new, but the analysis in this chapter has shown that relative economic and labour market conditions are very important for high volumes of migration flows over the 6-year period of analysis. Moving from the highest PPS or unemployment inequality ratio decile to the second highest is enough to effect a significant drop-off in the volume of migration flows. It has also been possible to observe the effects of the accession of the eight former Eastern bloc countries to the EU in 2004, with the removal of political barriers to movement resulting in regional inequalities in PPS and unemployment encouraging even more flows between the areas of greatest inequality than there were pre-accession. This ties in neatly with neoclassical theory and reflects the reality that when the potential gains in earning potential are sufficiently large, then this is a big influence on flows with the EU – when these gains reduce just fractionally, the motivation for moving is seriously curtailed.

But what of the second question? If regional inequalities are reduced, then what impact could this have on migration flows? Returning to the introduction to this chapter – is it feasible, for example, to reduce inequalities between regions in origin countries such as Romania and Bulgaria and destination countries such as the UK to such an extent that the mass migration feared by politicians can be arrested before it begins? Would an absolute increase in GDP or PPS in the region of 1.6% (the average increase which could be attributed to inward investment by EU structural funds, identified by Becker *et al.*, 2010) be anywhere near enough to have an effect, or are attempts to reduce regional economic inequalities unlikely to have a perceptible impact upon migration flows – are the inequality gaps persistently far too large and the potential gains always too great for any real difference to be made? To answer these questions, we need to return to the data used in the first part of the analysis.

From the evidence of migration in Europe between 2002 and 2007, it appears that a big drop-off in the volume of migration would be noticed with a one decile change in O/D inequality. As previously noted, this would mean, in 2002, changing the O/D PPS per person ratio from ≤0.43 to ≤0.63 (or from ≤0.48 to ≤0.65 in 2007) and the ratio of over 25 unemployment from ≤0.30 to ≤0.45 (from ≤0.38 to ≤0.53 in 2007). To understand what these ratios mean in the real world, it is useful to put them into the context of the UK. In 2002, the average O/D ratio for PPS between regions in Poland and regions in the UK was 0.41 – close to the threshold for the decile of highest migration flows. In regional terms, this is close to the PPS ratio of the Zachodniopomorskie (PL42) region in northwest Poland (€9200 PPS pp), and the East Anglia (UKH1) region in the south-eastern UK (€22,600 PPS pp), between which Dennett and Wilson (2013) estimate around 66 migrants moved in 2002. If EU cohesion funding were to increase the average PPS per person in Zachodniopomorskie by 1.6%, this would mean a rise to around €9350 – enough to only improve the PPS ratio to 0.42.

Examining data for Bulgaria and Romania, the inequalities in PPS between the UK are even starker. The average PPS ratio between regions in the UK and regions in both of these countries is around 0.29. Using the East Anglia comparison once again, this is a comparable ratio to the Vest (RO42) region in Romania (€6600 PPS pp). From this comparison, it is clear that modest growth in the economic productivity and average PPS per person, which appears feasible under the recent regimes of cohesion funding, is not going to have enough of an impact upon the growth of regions in countries such as Romania and Bulgaria to arrest the desire to migrate. Indeed, even if some of the inequality-reducing policies mentioned by Atkinson (2013) are implemented, it is difficult to see enough of a chance being affected.

But average ratios and regions indicative of these ratios do not tell the whole story. If we examine the data for the regions that are home to the capitals of Romania and Bulgaria, we see that for the region that is home to Bucharest (RO32, or Bucureşti-Ilfov) the average PPS ratio across all UK regions in 2002 is 0.56 (€12,100 PPS pp), and for the region that is home to Sofia (BG41, or Yugozapaden) the average PPS ratio across all UK regions is 0.43 (€9300 PPS pp) – both high enough ratios to suggest that high levels of economic migration from these regions to regions in the UK would not be likely. Looking next at the improving fortunes of these regions over the 2002–2007 period, significant improvements can be observed: the average PPS per person in Bucureşti-Ilfov almost doubles to €23,000 PPS pp – an improvement to the average ratio with UK regions to 0.92 (decile 5 for PPS ratio in 2007). Similarly, the PPS pp for Yugozapaden almost doubles as well to €16,600 PPS pp – an improvement to the average ratio with UK regions to 0.66 (decile 3 for PPS ratio). At these levels, low migration flows can be almost twice as likely as high migration flows.

Is it likely that other regions might catch up with these capital city regions? That is debatable. As is often the case with capital city regions, especially those of Primate Cities (Jefferson, 1939) such as Sofia and Bucharest, their rates of growth and economic power are rarely matched by other regions in the country. By 2007, the region containing Sofia in Bulgaria had an average PPS per person at least double that of other regions in Bulgaria; similarly, the region containing Bucharest had at least double the average PPS per person of all other regions in Romania. Prospects, therefore, of a turnaround in economic fortunes for the whole country – fuelled by EU funding or otherwise – might be remote.

So where does this leave policy makers looking to influence intra-European migration flows? Given the evidence in this chapter, it appears that the narrowing of the gap between the regions at the top and those at the bottom would need to be quite dramatic to nullify the perceived gains from moving. The current EU policy mechanisms for redistributing wealth from the richest to the poorest regions does have an impact on the growth of those regions, but this growth is unlikely to be large or swift enough to affect real change. That said, there is enough evidence here to show that a gradual narrowing of inequalities over a long period of time will almost certainly have an effect on flows of migrants. Of course, if this were ever to happen, government policy may have reversed by that time as ageing populations and creaking social care systems need propping up by tax revenues from young, economically active migrants.

2.7 Conclusions

Migration within the EU is currently very high on the political agenda of many countries – particularly those who feel they have more to lose than gain from immigration. To a large extent, policy makers within the governments of EU member-states are bound by European law, and as such they have only limited powers to arrest flows of migrants with restrictive legislation. That migrants are heavily influenced by economic and labour market factors is not new, but this chapter has for the first time been able to assess exactly *how much* of an influence regional inequalities have had on flows of migrants over a 6-year period, and it has assessed the extent to which wealth redistribution policies such as the European structural funding might be able to arrest flows of migrants. Logistic regression models have shown that at the extremes of migration behaviour (very high or very low flows – decile 1 or 10), large differences between average PPS per person or over-25 unemployment rates between the origin and the destination can make high or low flows several times as likely, but just a one decile change in the inequality rank can have a dramatic impact on the volume of flows.

Where national governments in this time of economic stagnation are concerned with migration (particularly immigration), especially of migrants over which they have no legislative control (Romanian and Bulgarian migrants moving to the UK from the latter part of 2013), this chapter explored whether existing European mechanisms designed to reduce regional inequalities would offer possibilities for influencing the flows of migrants who are primarily driven by these inequalities. It is shown that for regions in countries with very contrasting economic fortunes, the gaps are probably too large. For the UK government, fearful of migrants from Romania and Bulgaria, it is unlikely that much can be done to reduce the potential gains on offer. That said, these fears are likely to be overblown anyway, given the populations of these countries and their historical migrant links with Southern European countries like Spain and Italy.

Despite this, this chapter has shown that if reductions in migration continue to be one of the main concerns of governments, if greater economic equality could be achieved within Europe, and if the regions at the very bottom in terms of unemployment or wealth generation could be helped to improve and thus move closer to those in their own countries that are faring better – usually the capital city – then volumes of interregional migration within Europe are very likely to fall off swiftly.

References

Abel GJ. (2010). Estimation of international migration flow tables in Europe. *Journal of the Royal Statistical Society: Series A (Statistics in Society)* 173: 797–825.

Atkinson AB. (2013). Reducing income inequality in Europe. In *IZA/VEF Workshop: A European Labor Market with Full Employment, More Income Security and Less Income Inequality in 2020*. Bonn: IZA.

Becker SO, Egger PH and von Ehrlich M. (2010). Going NUTS: the effect of EU Structural Funds on regional performance. *Journal of Public Economics* 94: 578–590.

Benton M and Petrovic M. (2013). *How Free Is Free Movement? Dynamics and Drivers of Mobility within the European Union*. Brussels: Migration Policy Institute Europe.

Champion AG and Vandermotten C. (1997). Migration, counterurbanization and regional restructuring in Europe. In: Blotevogel HH and Fielding AJ (eds), *People, Jobs and Mobility in the New Europe*. London: John Wiley & Sons, 69–90.

de Beer J, van der Erf R and Raymer J. (2009). *Modelling of Statistical Data on Migration and Migrant Populations – MIMOSA: Estimates of OD Matrix by Broad Group of Citizenship, Sex and Age, 2002–2007*. The Hague: NIDI.

Dennett A and Wilson A. (2013). A multi-level spatial interaction modelling framework for estimating inter-regional migration in Europe. *Environment and Planning A* 45: 1491–1507.

Dorling D. (2010). *Injustice: Why Social Inequality Persists*. Bristol: Policy Press.

Dustmann C, Casanova M, Fertig M, *et al.* (2003). *The Impact of EU Enlargement on Migration Flows*. London: Home Office.

European Commission. (2012). *Employment and Social Developments in Europe 2012*. Brussels: European Commission, Directorate-General for Employment, Social Affairs and Inclusion.

Fihel A and Kaczmarczyk P. (2009). Migration: a threat or a chance? Recent migration of Poles and its impact on the Polish labour market. In: Burrell K (ed), *Polish Migration to the UK in the 'New' European Union: After 2004*. Padstow: Ashgate.

INFOREGIO. (2012). *Regional Policy*. Available at http://ec.europa.eu/regional_policy/index_en.cfm

Jefferson M. (1939). The law of the Primate City. *Geographical Review* 29: 226–232.

MAC. (2011). *Review of the Transitional Restrictions on Access of Bulgarian and Romanian Nationals to the UK Labour Market*. Croydon: Migration Advisory Committee.

May T. (2012). *Home Secretary speech on 'An immigration system that works in the national interest'*. London: Home Office.

Rees P. (1977). The measurement of migration from census and other sources. *Environment and Planning A* 9: 257–280.

Rolfe H, Fic T, Lalani M, *et al.* (2013). *Potential Impacts on the UK of Future Migration from Bulgaria and Romania*. London: National Institute of Economic and Social Research.

Therborn G. (2006). *Inequalities of the World*. London: Verso.

UCLA. (2012). *Ordinal Logistic Regression*. Available at http://www.ats.ucla.edu/stat/spss/dae/ologit.htm

Vargas-Silva C. (2013). *Migration Flows of A8 and Other EU Migrants to and from the UK. Briefing*. Oxford: The Migration Observatory, University of Oxford.

Wilkinson RG and Pickett K. (2009). *The Spirit Level: Why More Equal Societies Almost Always Do Better*: London: Allen Lane.

3

Test of Bi-Proportional Fitting Procedure Applied to International Trade

Simone Caschili and Alan G. Wilson

3.1 Introduction

In this chapter, we propose a bi-proportional fitting procedure to estimate bilateral trade between countries for seven macro sectors. We use the 40 countries of the World Input-Output Dataset (WIOD) in order to calibrate and validate our model; our data sources are WIOD (for total imports and exports per sector per country) and the COW Trade dataset (for total bilateral trade). We show that the proposed method is useful for the case of developing countries where there is a lack of data provided by governments, and it might reduce the level of trade information needed to evaluate international trade.

We estimate the international bilateral trade per macro sector, utilising the WIOD in order to calibrate the bi-proportional model for 40 countries. Our methodology can be useful for other cases when there is a lack of information on bilateral trade per sector (but not at the national level), such as evaluation of intra- and intersectoral dynamics at a macro level and how these affect the world economy.

Total bilateral trade at country level has been studied since the early 1960s using gravity models; see Tinbergen (1962), Pullianen (1963), and Linneman (1966), and more recently, Bergstrand (1989), Anderson and van Wincoop (2003) and Baldwin and Taglioni (2006).

Several authors have also paid attention to simultaneous exports and imports within industries between countries (intra-industry trade (IIT)). The seminal papers by Krugman (1979)

Approaches to Geo-mathematical Modelling: New Tools for Complexity Science, First Edition. Edited by Alan G. Wilson.

Companion Website: www.wiley.com/go/wilson/ApproachestoGeo-mathematicalModelling

and Lancaster (1980) have promoted a theoretical framework associating IIT with economies of scale and trade in varieties of (horizontally) differentiated products. Other relevant studies on IIT are Grubel and Lloyd (1975), Greenaway and Milner (1986), Helpman and Krugman (1985) and Lloyd and Lee (2002).

One of the main findings of IIT studies is that "trade integration would not lead to potentially important adjustment costs associated with the displacement of resources from comparatively disadvantaged industries towards a limited number of export-oriented industries (inter-industry trade)" (Fontagne and Freudenberg, 1997).

3.2 Model

Suppose that we have trade T_{ij} between countries i and j. We want to know the trade per sectors between countries (we have decomposed the economies of those countries in m sectors). We can estimate the trade per sector through a bi-proportional fitting model. The input terms of the model are as follows: trade is T_{ij}, $\hat{T}_{ij}$ is the adjusted trade, total exports of sector m is E_i^m, and total imports from sector m to n of each country is I_j^{mn}.

Model constraints:

$$\sum_m \sum_n T_{ij}^{mn} = \hat{T}_{ij} \tag{3.1}$$

$$\sum_j \sum_n T_{ij}^{mn} = E_i^m \tag{3.2}$$

$$\sum_i T_{ij}^{mn} = I_j^{mn} \tag{3.3}$$

We calculate the trade between countries i and j and between sectors m and n as follows:

$$T_{ij}^{mn} = X_{ij}\, A_i^m B_j^{mn} E_i^m I_j^{mn} \hat{T}_{ij} \tag{3.4}$$

where X_{ij}, A_i^m and B_j^{mn} are the following balancing factors:

$$X_{ij} = \frac{1}{\sum_m \sum_n A_i^m B_j^{mn} E_i^m I_j^{mn}} \tag{3.5}$$

$$A_i^m = \frac{1}{\sum_j \sum_n X_{ij} B_j^{mn} I_j^{mn} \hat{T}_{ij}} \tag{3.6}$$

$$B_j^{mn} = \frac{1}{\sum_i X_{ij} A_i^m E_i^m \hat{T}_{ij}} \tag{3.7}$$

In order to satisfy the model constraint in Equation (3.1), we apply a bi-proportional fitting procedure as follows, where T_{ij} and $\hat{T}_{ij}$ are respectively the measured and adjusted values of bilateral trade between countries i and j. The adjusted value $\hat{T}_{ij}$in Equation (3.9) is used in Equations (3.1) to (3.7).

$$\sum_m \sum_i E_i^m = \sum_m \sum_n \sum_j I_j^{mn} \tag{3.8}$$

$$\hat{T}_{ij} = a_i b_j T_{ij} \tag{3.9}$$

$$\sum_j \hat{T}_{ij} = \sum_m E_i^m \tag{3.10}$$

$$\sum_i \hat{T}_{ij} = \sum_m \sum_n I_j^{mn} \tag{3.11}$$

$$a_i = \frac{\sum_m E_i^m}{\sum_j b_j T_{ij}} \tag{3.12}$$

$$b_j = \frac{\sum_m \sum_n I_j^{mn}}{\sum_i a_i T_{ij}} \tag{3.13}$$

3.3 Notes of Implementation

Notation used in this section follows the notation of Section 3.2. The pseudo-code for implementation is:

$$IMPORT\ E_i^m\ \forall\ i,\ m$$

$$IMPORT\ \ I_j^{mn}\ \forall\ j,\ m,\ n$$

$$IMPORT\ \ \hat{T}_{ij}\ \forall\ i,\ j$$

$$X_{ij}\ = 1\forall\ i,\ j$$

$$A_i^m = (Average\ Exports)^{-1}\forall\ i,\ m$$

$$B_j^{mn} = (Average\ Imports)^{-1}\forall\ j,\ m,\ n$$

$$\varepsilon = const$$

$$UNTIL\ convergence\ is\ reached:$$

$$X_{ij}(t+1) = X_{ij}(t) + \varepsilon\left(\frac{1}{\sum_m\sum_n A_i^m(t)B_j^{mn}(t)E_i^m I_j^{mn}} - X_{ij}(t)\right)$$

$$A_i^m(t+1) = A_i^m(t) + \varepsilon\left(\frac{1}{\sum_j\sum_n X_{ij}(t)B_j^{mn}(t)I_j^{mn}\hat{T}_{ij}} - A_i^m(t)\right)$$

$$B_j^{mn}(t+1) = B_j^{mn}(t) + \varepsilon\left(\frac{1}{\sum_i X_{ij}(t)A_i^m(t)E_i^m\hat{T}_{ij}} - B_j^{mn}(t)\right)$$

$$FOR\ \forall\, i, j:$$

$$IF\frac{Abs[(X_{ij})^{t+1} - (X_{ij})^t]}{Abs[(X_{ij})^t]} < 10^{-6}$$

$$THEN\ convergence\ is\ reached$$

$$FOR\ \forall\, i, m:$$

$$IF\frac{Abs[(A_i^m)^{t+1} - (A_i^m)^t]}{Abs[(A_i^m)^t]} < 10^{-6}$$

$$THEN\ convergence\ is\ reached$$

$$FOR\ \forall\, j, m, n:$$

$$IF\frac{Abs[(B_j^{mn})^{t+1} - (B_j^{mn})^t]}{Abs[(B_j^{mn})^t]} < 10^{-6}$$

$$THEN\ convergence\ is\ reached$$

In the pseudo code above, epsilon represents the step size used by the optimisation algorithm to reach a solution. We assume a constant step size but more efficient approaches propose adaptive step size methods which adaptively choose a small step-size or a large step-size at each iteration (Zhou *et al.*, 2006).

3.4 Results

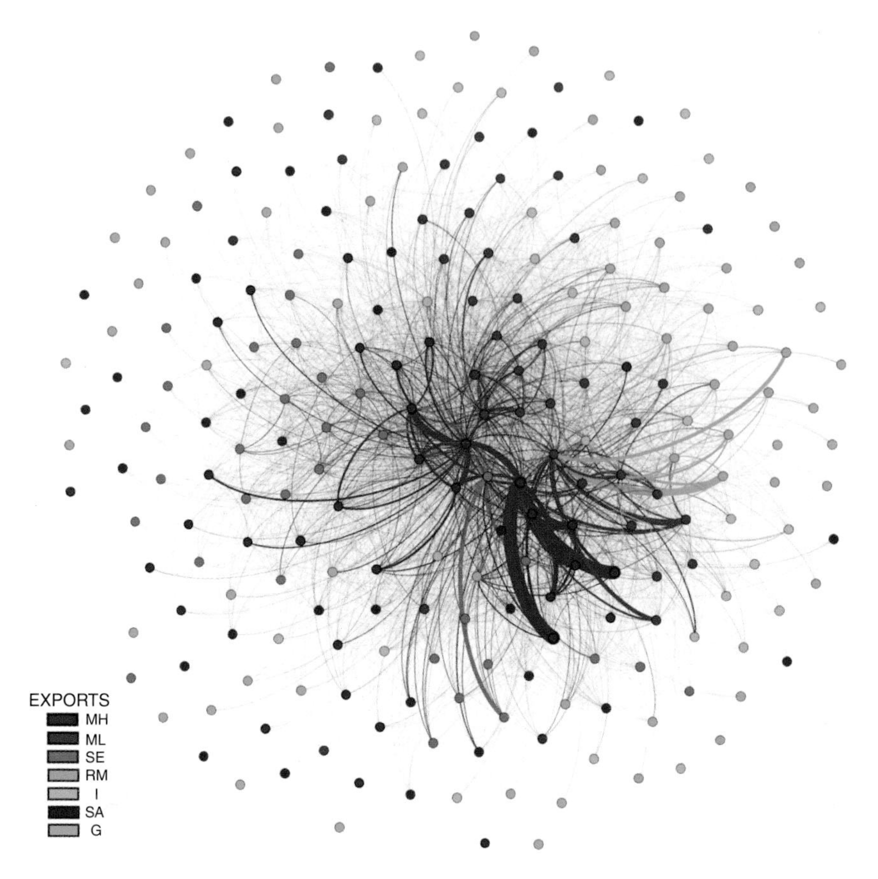

Figure 3.1 Bilateral trade between countries and sectors. Each node represents a sector of a country. Each link has the colour of the exporter country, while link width is proportional to the value of bilateral trade between the two sectors. For ease of visualisation, we have pruned the graph from the links with bilateral trade lower than USD $100 million. Sectors: High Tech Manufacturing (MH), Low Tech Manufacturing (ML), Services (SE), Raw Materials (RM), Industrial (I), Retail Sale (SA), Government (G)

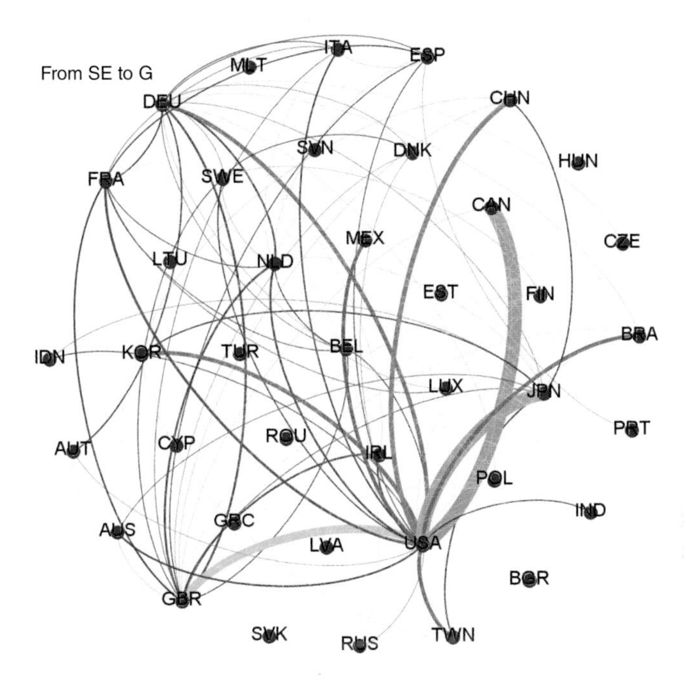

Figure 3.2 Inter-bilateral trade between Service and Government sectors. Width and colour of links are proportional to trade flows. For ease of visualisation, we have pruned the links with bilateral trade lower than USD $3000 million

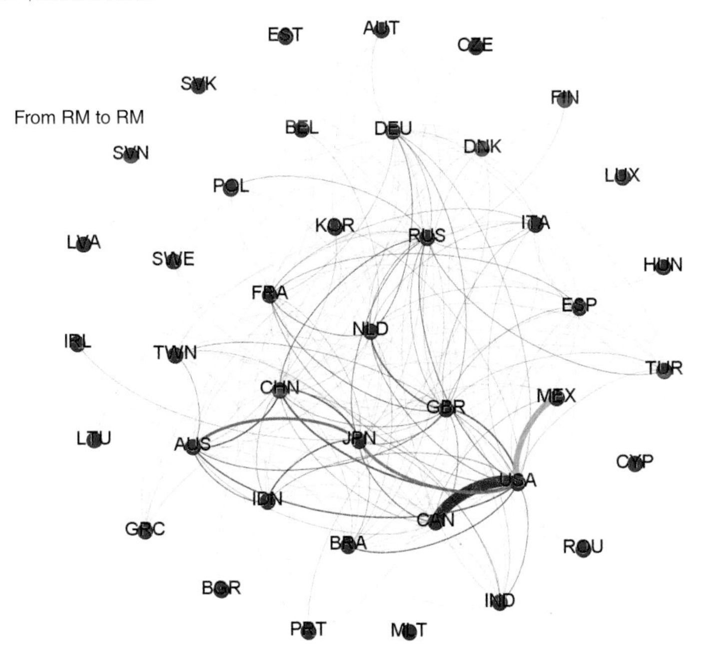

Figure 3.3 Intra-bilateral trade in the Raw Material sector. Width and colour of links are proportional to trade flows. For ease of visualisation, we have pruned the links with bilateral trade lower than USD $2400 million

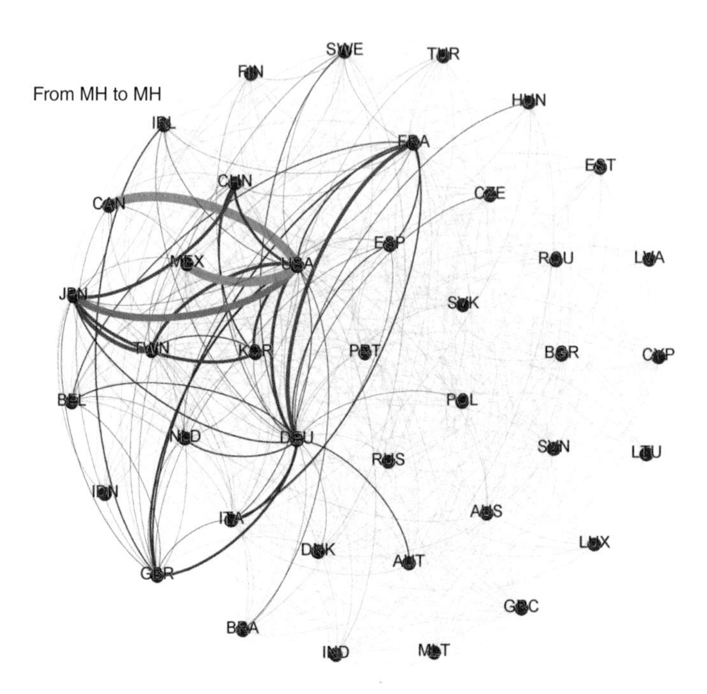

Figure 3.4 Intra-bilateral trade in the Manufacture High Tech sector. Width and colour of links are proportional to trade flows. For ease of visualisation, we have pruned the links with bilateral trade lower than USD $3000 million

References

Anderson JE and van Wincoop JP. (2003). *Measuring the Restrictiveness of International Trade Policy*. Cambridge, MA: MIT Press.

Baldwin R and Taglioni D. (2006). *Gravity for Dummies and Dummies for Gravity Equation*. NBER working paper no. 12516. Washington, DC: National Bureau of Economic Research.

Bergstrand JH. (1989). The generalised gravity equation, monopolistic competition and empirical evidence. *Review of Economics and Statistics* 71: 143–153.

Fontagné L. and Freudenberg M. (1997). Intra-industry trade: methodological issues reconsidered (Vol. 97, No. 1). Paris: CEPII.

Greenaway and Milner (1986). *The Economics of Intra-Industry Trade*. Oxford: Blackwell.

Grubel H and Lloyd PJ. (1975). *Intra Industry Trade: The Theory and Measurement of International Trade with Differentiated Products*. London: Macmillan.

Helpman E and Krugman P. (1985). *Market Structure and Foreign Trade*. Cambridge, MA: MIT Press.

Krugman PR. (1979). Increasing returns, monopolistic competition, and international trade. *Journal of International Economics* 9(4): 469–479.

Lancaster K. (1980). Intra-industry trade under perfect monopolistic competition. *Journal of International Economics* 10: 151–175.

Linneman H. (1966). *An Economic Study of International Trade Flows*. Amsterdam: North-Holland.

Lloyd PJ and Lee H. (2002). *Frontiers of Research on Intra-industry Trade*. London: Palgrave.

Pullianen P. (1963). World trade study: an economic model of the pattern of the commodity flows in international trade 1948–1960. *Ekonomiska Samfundets* 2: 78–91.

Tinbergen J. (1962). *Shaping the World Economy: Suggestions for an International Economic Policy*. New York: Twentieth Century Fund.

Zhou B., Gao L., and Dai Y. H. (2006). Gradient methods with adaptive step-sizes. Computational Optimization and Applications 35(1): 69–86.

4

Estimating Services Flows

Robert G. Levy

4.1 Introduction

In Chapter 4 of the companion volume to this book, *Global Dynamics*, the model presented asks a lot of the data it is provided with. The magnitude of each country-country-sector flow, y_{ijs}, is used to calculate an import propensity, p_{ijs}, which is then fixed and determines, to a large extent, the response of the model to an exogenous change in final demand.

In the case of commodity (i.e. physically tradeable goods) flows, this is a natural way to proceed: due to the need to impose customs taxes/tariffs on physical goods entering or leaving a country, the trade in commodities is extremely well recorded.

The same cannot be said of trade in services, which, since it is not done via border controls, is far less accurately recorded and often not recorded at all (Dietzenbacher *et al.*, 2013, p. 86). Indeed, one could argue that it is not clear what trade in services even *is*, but answering that tricky economics question is outside the scope of this chapter.

Instead, we will content ourselves to follow the World Input-Output Database (WIOD) in assuming that services can be both imported and exported identically to commodities. We will also assume that the UN ServiceTrade database is the best available reflection of the extent to which this trade occurs globally. We will be 'taking the data seriously', despite the fact that we might have cause to suspect that services trade is incompletely reported.

In many cases in the services trade data, country-sector trade totals are reported rather than country-country-sector flows. Or, where point-to-point flows *are* reported, the reported trade partner is often a region, rather than a country. In this chapter, we will therefore assume that only trade totals (i.e. total exports and/or total imports) are known. We will be concerned with methods for estimating country-country-sector flows from these trade totals, which is a fairly well-understood problem. Despite this, as far as we know, none of these approaches has ever been applied to the UN ServiceTrade database.

Approaches to Geo-mathematical Modelling: New Tools for Complexity Science, First Edition. Edited by Alan G. Wilson.

Companion Website: www.wiley.com/go/wilson/ApproachestoGeo-mathematicalModelling

4.2 Estimation Via Iterative Proportional Fitting

As we have seen, this chapter concerns itself with the estimation of point-to-point flows when only flow totals, either in- or out-flows, are known. Iterative proportional fitting can be a true least-assumptions approach to this problem in that it treats the point-to-point flows purely probabilistically if the initial matrix has no information (more on this further in this chapter). It is identical in this situation to a solution via entropy maximisation and the 'null model' in community detection, as we will see. One caveat is that both margins must sum to the same total if the algorithm is to converge.

4.2.1 The Method

Iterative proportional fitting starts with an initial matrix and a set of known row and column totals, which must sum to the same value. The initial matrix may be a set of measured flows which are considered to contain relevant information on the relative sizes despite the fact that they may not sum to the correct margin totals. Alternatively, in the complete absence of point-to-point information, the process starts with a matrix of ones.

In either case, the method is the same and consists of an algorithm which iterates between row operations and column operations. In each row operation, all rows are made to sum to the required margins by simply dividing by the current row sum and multiplying by the margin in question. Thus,

$$f_{ij}^{(1)} = \frac{f_{ij}^{(0)} O_i}{\sum_k f_{ik}} \tag{4.1}$$

where $f_{ij}^{(n)}$ is the matrix entry at ij after the nth iteration, and O_i is the ith row sum. This is repeated for the columns. This row–column interation is then repeated until the row and column sums are both simultaneously within an arbitrary distance from the desired values according to some measure. For example, we might measure the sum of squared deviations.

There are only two further points to make about the method. The first is that since only multiplications and divisions are involved, any zeros in the initial matrix will remain zero throughout the procedure. This will be useful in cases of estimating trade flows if countries/regions and the like are disallowed from trading with themselves, in which case the diagonal elements of the initial matrix can be set to zero.

Secondly, in the case where the initial matrix contains some information (i.e. where the non-zero initial values are not all the same), it is generally the case that, all else being equal, larger initial values tend to result in larger final values. Therefore, IPF can, in some very weak sense, be seen as rank-preserving. To demonstrate this empirically, we take an initial matrix of random values between zero and one. We then run IPF against randomly chosen row and column totals. The only constraint on the row and column totals is that they were both forced to sum to the total sum of the randomly selected initial matrix values.

Figure 4.1 shows the relationship between initial value and final value in this experiment with 10,000 runs and a random initial matrix of 5×5. Each matrix value had its initial value tracked, and paired with its value after iteration had completed. These initial and final pairs can then be scatterplotted with the initial value on the x-axis and the final value on the y-axis. For visual clarity, Figure 4.1 is not a scatterplot, but rather a density plot, where the xy space is binned into hexagonal regions. The darker hexagons represent more data points falling into that area of the space.

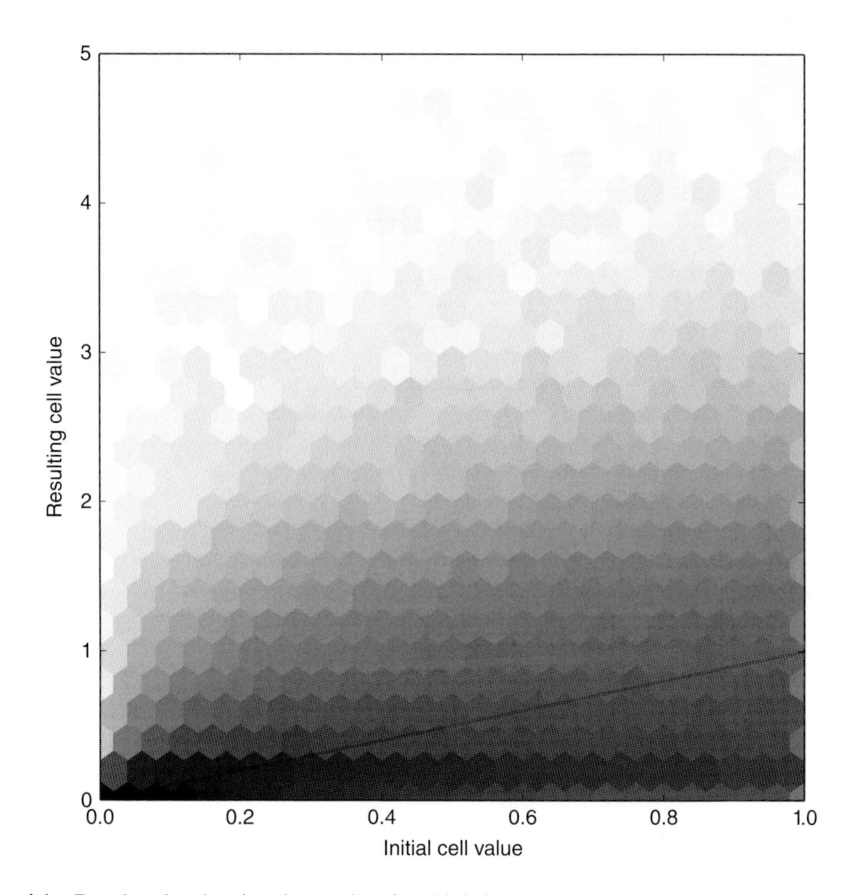

Figure 4.1 Density plot showing the results of multiple iterative proportional fitting runs. During each run, the initial value of every matrix element was tracked and paired with its value after the iteration completed. This graph shows these pairs of initial (x-axis) and final values (y-axis), having first binned the initial/final space into hexagonal-shaped regions for visual clarity. Darker hexagons represent more matrix elements occupying that space. The initial matrix was uniformly distributed between zero and one with shape 5×5. 10,000 runs were performed. The straight line shows $y = x$

To show that larger values generally end up larger after iteration, y should be correlated with x and we should expect to see an upward-sloping distribution with more values at the bottom left than at the top left, and more values at the top right than at the bottom right. This particular run has a Pearson correlation coefficient of 0.3 (with a very small p value, since n is of the order 10^5) between the initial and final values, demonstrating that larger initial values tend to lead to larger final values.

4.2.2 With All Initial Values Equal

If there is no initial information on which to base an estimate of point-to-point flows, the initial matrix is usually set to a matrix of ones. To see how this affects the result, we solve equation

(4.1) when all $f_{ij} = 1$. The equation for the first iteration is then:

$$f_{ij}^{(1)} = \frac{O_i}{n}$$

where n is the dimension of the square matrix. We then run the column step of the iteration:

$$f_{ij}^{(2)} = \frac{f_{ij}^{(1)} D_j}{\sum_k f_{kj}^{(1)}} = \frac{O_i D_j}{\sum_k O_k} \tag{4.2}$$

Since this method has a closed-form solution[1], the algorithm always completes in a single-row/column operation. This result has some interesting equivalences in seemingly unrelated fields.

4.2.3 Equivalence to Entropy Maximisation

Since the row and column margins are defined as summing to the same value, we can represent both $\sum O_i$ and $\sum D_j$ as a single value a. We can then rewrite equation (4.2) as

$$f_{ij}^{(2)} = \sum_k D_j \frac{O_i D_j}{\sum_k O_k \sum_k D_j} = a \frac{O_i D_j}{\sum_k O_k \sum_k D_j} = aA_i B_j O_i D_j$$

where we define $A_i = \frac{1}{\sum O_i}$ and $B_j = \frac{1}{\sum B_j}$.

From this last representation, it is clear that when the initial matrix contains no information, IPF is identical to entropy maximisation (up to the multiplicative factor a) as presented in, for example, Wilson (1969). Furthermore, if we define the proportional limits as being:

$$p(M_i) = \frac{M_i}{a}$$

where M_i is a margin total, either O or D, then equation (4.2) can also be written as:

$$f_{ij}^{(2)} = ap(O_i)p(D_j)$$

which is the same as the second choice of null model proposed by Reichardt and Bornholdt (2006, p. 3) for assessing the fitness of a particular community grouping in network analysis. In their context, O_i is a node's *in-degree*, the number of directed edges pointing into the node, and D_j is the *out-degree*. Crucially, it is this choice of null model which makes the Reichardt and Bornholdt (2006) method identical to that of Newman and Girvan (2004), who first defined the critical concept of *modularity* which is at the heart of much subsequent work in that literature. Rosvall and Bergstrom (2008, p. 1122) also refer to this choice of null model and describe it as "the most general form" of the modularity maximisation approach.

[1] Provided, as mentioned here, that the margins sum to the same number. If not, either the rows or the columns will sum to the correct margin depending on which is done last. The other margin will sum to an arbitrary number.

4.2.4 *Estimation with Some Known Flows*

When some flows are known, but not a complete matrix, the simplest approach recommended by Lahr and de Mesnard (2004, p. 125) is simply to subtract the known flows from the relevant margin totals and set the entries in the initial matrix to zero.

By doing this, the known flows remain zero, and the remaining flows are proportionally fit to the margins *without* the known flows such that the known flows can simply be added back in at the end.

Notice that this process makes explicit the fact that a known flow has no influence over any other point-to-point flow other than through its effect on the relevant margins. A more sophisticated approach to dealing with known flows would require a model for how one point-to-point flow affects other such flows.

4.2.5 *Drawbacks to Estimating Services Flows with IPF*

As discussed here, IPF is very much a 'no-information' solution to the problem of estimating point-to-point flows from flow totals. Services flows estimated in this way will contain none of the trade patterns we might expect to see in a global commodities trade network. For example, if the Netherlands is a large exporter of a particular service, then this will be reflected equally in the import patterns of both the UK and China. But this disallows for the possibility that countries within the EU have greater-than-proportional trade when compared with that which occurs across EU borders.

The problem with using IPF to estimate services flows is that we might be discarding important information about patterns of global trade contained in other data sources. Section 4.3 seeks to address this problem by making some assumptions about similarities between the services trade network and that of trade in commodities.

4.3 Estimating Services Flows Using Commodities Flows

In the area of global trade, we are far from being in a no-information situation. As mentioned, trade in physical commodities is extremely well documented. In this section, we will use the information contained in the commodity flows to make better estimates than we were able to using IPF. We will begin with the simplest possible exposition, and add further complexity until we are satisfied that we have used as much of the information as we have available as feasible.

4.3.1 *The Gravity Model*

In Section 4.2.5, we discussed the confounding effect which political trade relationships will tend to have when estimating flows purely proportionally: countries within the trade relationship will trade more with one another than they 'should' if trade were determined proportionally.

Similarly, the most commonly cited factor which distorts trade from pure proportionality is geographical distance. For example, Disdier and Head (2008) perform an excellent meta-analysis of the literature on how distance affects trade. When combined with a bi-proportionality related to the GDP of the importer and exporter countries, as does McCallum (1995), this model is called the gravity model and remains a standard tool in trade analysis. For example, Novy (2013) uses it in its simplest form to estimate trade costs for the World Bank. The model is usually presented as:

$$y_{ij} = \frac{Y_i^{\alpha} Y_j^{\beta}}{d_{ij}^{\gamma}} \tag{4.3}$$

where Y_i is the GDP of country i, and d_{ij} is the geographical distance between the two countries. For the present purposes, we will use minimum distances (such that neighbouring countries have $d_{ij} = 0$) from the CShapes database (Weidmann *et al.*, 2010). Zeros in the data present a problem for multiplicative formulations such as this one, since division by zero is not defined, and multiplication by zero always results in zero overall. We will therefore replace d_{ij} with $e^{d_{ij}}$ throughout the remainder of this treatment. When using centroid distance, this replacement will not, in general, be necessary.

The parameters α, β and γ are to be estimated from some other known dataset, for example data from a previous year. Due to the multiplicative nature of this equation, the estimation is easy to carry out by ordinary least-squares regression by taking logs.

Usually, the formulation in equation (4.3) is extended to include other variables which affect levels of trade. Written more generally, if there are K variables thought to affect trade, we have:

$$y_{ijs} = \prod_{k=1}^{K} x_{kijs}^{\beta_k} \tag{4.4}$$

where x_{kijs} is the kth variable relating trade from i to j in sector s. There are then K parameters to be estimated. Denominator parameters such as distance in equation (4.3) are represented by having a negative parameter. Finally, variables which do not relate to i or j are fixed to be the same for all values. For example, in the case of exporter GDP:

$$x_{kij} = x_{ki} = Y_i \quad \forall \ k, j$$

Table 4.1 shows the results of using equation (4.4) with a variety of specifications[2]. The results of the formulation in equation (4.3) are shown in column 1.

The minimum distance term has the expected negative sign (the regression was run multiplicatively, i.e. with d_{ij} in the numerator), and is significant at 1%. The point estimate of −0.35 is somewhat smaller than those obtained by Baldwin and Taglioni (2006), McCallum (1995) and Anderson and Wincoop (2003), who get around −0.8 using a different data set and definition of distance. This discrepancy is not explained by using exponentiated distance: the point estimate is even closer to zero when using log distance (results not shown).

Both the total production figures have the expected positive sign and are highly significant. They are both much closer to the McCallum (1995) estimates of around 1.1 for x_i and 1.0 for x_j.

[2] Note a few country–sector combinations have, presumably erroneously, all zero technical coefficients. This would imply that they are able to produce output with no intermediate input. Since this clearly cannot be the case, these country–sectors have been excluded from the analysis. They are CYP Fuel, LUX Fuel and Leather, LVA Fuel and SWE Leather.

Table 4.1 Results of using equation 4.4

	Dependent variable					
	log(y_ijs)					
	(1)	(2)	(3)	(4)	(5)	(6)
Dummies	none	none	none	none	c	c and s
mindist	-0.345^{***}	-0.346^{***}	-0.344^{***}	-0.342^{***}	-0.348^{***}	-0.353^{***}
	(0.005)	(0.005)	(0.005)	(0.005)	(0.005)	(0.005)
log(x_i)	1.049^{***}	0.564^{***}	0.576^{***}	1.031^{***}		
	(0.011)	(0.015)	(0.015)	(0.011)		
log(x_j)	0.890^{***}	0.893^{***}	0.892^{***}	0.889^{***}	0.896^{***}	0.904^{***}
	(0.010)	(0.010)	(0.010)	(0.010)	(0.010)	(0.009)
log(f_is)		0.485^{***}	0.463^{***}		0.511^{***}	0.368^{***}
		(0.011)	(0.011)		(0.012)	(0.022)
log(v_is)			-0.486^{***}	-0.870^{***}	-0.160^{***}	-0.267^{***}
			(0.049)	(0.049)	(0.051)	(0.066)
Constant	-22.671^{***}	-19.894^{***}	-20.401^{***}	-23.353^{***}	-12.113^{***}	-11.792^{***}
	(0.206)	(0.208)	(0.214)	(0.208)	(0.190)	(0.269)
Observations	23,913	23,913	23,913	23,913	23,913	23,913
R^2	0.310	0.329	0.359	0.347	0.382	0.650

*p<0.1; **p<0.05; ***p<0.01.

In columns (2) and (3) of Table 4.1, additional independent variables are added which are not available in ordinary trade datasets, since they come from the input–output tables at the heart of our model of trade. The first is the final (domestic) demand for sector s in the exporter country. Krugman (1980, p. 956) provides theoretical reasons why a country with a large domestic market for a product might be a larger exporter of that product. Thus, we would expect a positive coefficient on f_{is}. The coefficient is indeed positive and highly significant. Interestingly, the inclusion of domestic demand greatly reduces the parameter estimate on total production in the exporter country, suggesting that some of the production effect in models such as equation (4.3) might really be due to the fact that domestic demand for that product is high (which, of course, will have a direct effect on total production).

Another determinant of the attractiveness of an exporter might be its ability to produce good s cheaply. A proxy for this, added in column (3), is the value added per unit output, defined from the technical coefficients for the relevant sector as:

$$v_{is} = 1 - \sum_{r} a_{rs} \tag{4.5}$$

This makes intuitive economic sense since, under perfect competition assumptions, price is equal to the marginal cost of production, defined as the price to create a single unit of a good at current production levels. The technical coefficients represent the intermediate costs of producing a unit of good, but they do not include labour costs. It is in this sense that they are a proxy for production costs rather than being the production costs themselves.

All things being equal, we would expect countries with a lower production cost to export more, since they are more efficient producers. The negative sign on v_{is} in Table 4.1 is therefore surprising. Various different specifications have shown that this negative coefficient on v_{is} is very robust, showing that this is not merely a misspecification problem. We will investigate this further in Section 4.3.2.

Columns (4) and (5) of Table 4.1 show the effects of adding country (c) and sector (s) fixed effects. Of course, with country fixed effects included, x_i must be excluded since it is also fixed per country. The general result, though, is that the results discussed here are all robust to the addition of sector and country fixed effects: no point estimate is changed hugely, and none of the variables loses its significance level.

4.3.2 Splitting Up Value Added

We can investigate the negative sign on v_{is} by splitting v_{is} into its constituent r parts $1 - a_{rs}$ in equation (4.5). Table 4.2 shows the result of including each of the 'from sector' technical coefficients (restricted to the commodity sectors). The regression run was identical to that in column (6) of Table 4.1, except without sector dummies (for obvious reasons). Each row of the table is then a measure of how much of the given sector was used in production per unit in the exporting country.

We might expect that lower technical coefficients lead to more trade, since this implies a more efficient production technology[3]. The picture from the per-sector analysis is rather mixed. Of the sectors with significant ($p < 0.01$) parameters, Leather, Fuel, Metals, Minerals, Paper, and Textiles all have the expected negative sign: a production technology which is efficient in these sectors encourages export. Particularly important here are Metals, Minerals and Paper, which all have relatively large point estimates. It is perhaps encouraging that the coefficient on Fuel is significant and negative. This suggests that fuel efficiency is indeed a factor when importers are choosing who to trade with.

But there are also sectors with a positive and significant ($p < 0.01$) parameter. These are Plastics, Machinery, and Chemicals, and the point estimates on these sectors are also comparatively large. Perhaps we might conclude that the difference comes from the extent to which these sectors require inputs themselves. Metals, Minerals and Paper are all close to being raw materials, requiring little input in their manufacture. We will refer to these as primary products. But Plastics, Machinery and Chemicals are secondary products, requiring more effort to produce. It seems that using comparatively more primary product in a production process is a *discouragement* to export, but that using comparatively more of a secondary product is an *encouragement* to export. Perhaps what we are observing here is a preference for refined goods over unrefined, certainly in the dollar value terms which all these flows are measured in.

4.4 A Comparison of The Methods

The basic approach will be to compare estimates using the three methods above (bi-proportional fitting, the gravity model and the extended gravity model), using only

[3] It does in intermediate terms at least. Labour is not included in this measure and is implicitly assumed to be equal across all sectors and countries, a highly questionable assumption.

Table 4.2 Regression similar in structure to Table 4.1 but with v_{is} divided into its constituent sectors

	Dependent variable	
	log(y_ijs)	
	(1)	(2)
log(food)	0.088***	
	(0.021)	
log(leather)	−0.046***	
	(0.015)	
log(fuel)	−0.081***	
	(0.026)	
log(plastics)	0.104***	
	(0.024)	
log(metals)	−0.181***	
	(0.027)	
log(machinery)	0.127***	
	(0.025)	
log(vehicles)	−0.043*	
	(0.024)	
log(agriculture)	−0.031***	
	(0.011)	
log(wood)	−0.029	
	(0.018)	
log(minerals)	−0.107***	
	(0.020)	
log(manufacturing)	0.027	
	(0.021)	
log(paper)	−0.236***	
	(0.028)	
log(electricals)	0.084***	
	(0.027)	
log(textiles)	−0.059***	
	(0.021)	
log(mining)	0.015	
	(0.015)	
log(chemicals)	0.124***	
	(0.030)	
log(total)		−0.012
		(0.061)
log(x_j)	0.893***	0.893***
	(0.009)	(0.009)
log(f_is)	0.328***	0.350***
	(0.023)	(0.022)
Observations	23,019	23,019
R^2	0.617	0.612

*p<0.1; **p<0.05; ***p<0.01.

the information contained in the import and export totals (row/column sums). We can calculate a mean squared error from the point-to-point flows which *are* contained in the UN ServiceTrade database and use these to compare the methods.

Additionally, we will restrict ourselves to concentrating on the 40 countries of the WIOD to make the analysis manageable in terms of number of countries.

4.4.1 Unbalanced Row and Column Margins

Since any balancing procedure only makes sense if the row and column margins sum to the same number, and this would not be expected, in general, to be the case in a set of trade data from disparate reporters, throughout this analysis we will adopt the following procedure to ensure that the row and column margins balance.

- Each set of margins will be given an additional element, labelled RoW (for 'rest of world').
- If the row margins sum to more than the column margins, the RoW element of the row margins will be set to zero, and that of the column margins set to (the absolute size of) the difference between the two sums.
- If the column margins sum to more than the row margins, the opposite procedure will be applied.

4.4.2 Iterative Proportional Fitting

Since we do not have a complete set of prior information about the point-to-point flows in the services sectors, we will start the procedure with a matrix of ones. In this case, and given the adjustment described in Section 4.4.1 to ensure the margins sum to the same number, there is a closed-form solution as per equation (4.2).

The column labelled ipf of Table 4.3 shows the root mean squared error (RMSE) for the iterative proportional fitting method applied to each of the services sectors. The columns of the table are ordered by total RMSE left-to-right from lowest to highest. Iterative proportional fitting is in the middle column.

The sectors vary a lot in terms of their RMSE, but this is likely due to the fact that the total dollar values of trade of each sector are very different, a point we return to in Section 4.5.1.

4.4.3 Gravity Model

We will explore each of the specifications given in Table 4.2 to see which produces the set of flows which most closely matches those point-to-point flows which *are* in the dataset. This is an *in-sample* comparison. Specification (6) is not included here since the sector dummies in Table 4.2 are for commodities sectors and we are now dealing with services sectors.

The results are shown in the columns labelled 'gravity1' to 'gravity5' in Table 4.3. We will start with specification (1), the first and simplest, including only distance, and the total production of the importer and exporter. This is the gravity model in what might be termed its 'purest' form. It is the second most effective specification behind only (4), which we will look at in more detail in this chapter. Along with all the other specifications, this specification varies greatly in how well it predicts the flows of each sector. The most effectively predicted is

Table 4.3 The root mean squared error (RMSE) for each services sector when point-to-point flows are estimated using the method associated with each column. RMSE is calculated as $\sqrt{\sum_i (\bar{x}_i - x_i)^2 / n}$. The specification number refers to the column of Table 4.1 used as the gravity model coefficients

	gipf4	gipf3	gipf1	gipf2	gipf5	ipf	gravity4	gravity1	gravity3	gravity5	gravity2
Air Transport	372	373	374	374	375	330	555	535	538	531	532
Business Services	787	787	787	787	787	1206	2201	2127	2091	2090	2021
Communications	130	130	130	130	130	156	269	348	423	558	579
Education	40	40	40	40	40	70	185	308	554	1012	880
Financial Services	294	295	295	295	296	323	840	838	900	1011	1071
Health	9	9	9	9	9	13	145	265	824	1233	1222
Hospitality	1113	1111	1111	1110	1109	1634	2625	2570	2406	2393	2326
Inland Transport	218	218	218	218	218	237	436	468	492	507	573
Other Services	988	989	989	990	991	1253	2148	2084	1934	2011	1862
Public Services	94	95	95	95	95	101	549	586	1345	1704	1962
Real Estate	29	29	29	29	29	42	96	207	601	920	991
Retail Trade	644	644	644	644	644	732	1356	1314	1302	1201	1404
Transport Services	250	249	249	249	249	333	516	530	505	524	505
Utilities	70	71	71	71	71	75	217	282	393	460	516
Vehicle Trade	67	68	68	68	68	67	140	250	271	304	398
Water Transport	578	579	579	580	582	413	770	771	810	818	808
Wholesale Trade	64	64	64	64	64	67	142	282	551	830	832
Total RMSE	5748	5751	5751	5753	5755	7051	13,191	13,765	15,941	18,108	18,482

Real Estate (RMSE = 207) which, as will be seen in Section 4.5.1, is also the sector with the smallest fraction of total flow value in point-to-point flows, at 14%. This relationship does not hold in general however, since the highest fraction, at 23%, is in the Education sector (308), which is fifth-best estimated, behind only Health (265) and Utilities and Wholesale Trade (both 282).

The most poorly estimated sector is Hospitality (2570), which has only 15% of its dollar value in point-to-point flows. This, along with Business Services (2127) and Other Services (2084), has the highest RMSE by a very wide margin. This ordering from best to worst estimated is broadly consistent across all specifications, perhaps suggesting that some sectors (namely Real Estate, Health, Utilities and Wholesale Trade) behave more like commodities sectors than do others (namely, Hospitality, Business Services, Other Services and Retail Trade).

Specification (2) adds exporter final demand to (1), which makes the estimation very much worse for almost all sectors. The situation is somewhat improved by the addition of per-unit value added in (3), but not completely salvaged until specification (4), which keeps value added but removes final demand. This latter is the best specification overall and, for some sectors, by a wide margin. The addition of country dummies in (5) does nothing to improve specification (3) to which it is identical bar the dummies, and therefore, for considerations of space, we do not test specification (4) in the presence of country dummies.

Recall that the parameters on these variables are taken from a regression run on commodity flows. The fact that the additional variables in specifications (2), (3) and (5) make the predictions worse than what we might term the *baseline variable set* indicates either that services flows have a fundamentally different relationship to the regressors or that a gravity model is, in general, not doing a good job of estimating flow magnitudes.

In this context, it is interesting to see that the gravity model performs worse in every specification other than IPF. Certainly, the unconstrained nature of the gravity model makes it naturally a poorer predictor of flow *magnitudes* than IPF, although it may do a better job of *ranking* the flows by magnitude. The current experiment does not test this. Specification (4) has an RMSE of 11,851, making it the best of the gravity model specifications, but it is still a long way short of a simple IPF of 7051. We therefore turn to a method of combining the rankings of the gravity model with magnitudes of IPF.

4.4.4 Gravity Model Followed by IPF

The results of the gravity model do not use the row and column totals which are considered to be known. This is a considerable waste of the available information. We might therefore usefully use the outcome of the (unconstrained) gravity model as the input to an iterative proportional fitting routine. Recall from Section 4.2.1 that flows which are larger in the initial matrix tend, in general, to be larger in the final fitted result. Thus, the output of the gravity model provides relative differences in point-to-point flow values, and the subsequent IPF ensures that the row and column margins balance while retaining "some" of the information from the first stage. If this turns out to be more efficient than IPF alone, it would confirm our suspicions in Section 4.4.3 that the gravity model is better than IPF at estimating the relative sizes of the flows, but poor at estimating the magnitude. This is simply because IPF takes care of the magnitude but will broadly retain the relative flow sizes from the gravity model.

We can think of this two-stage procedure as being an "operationalised" version of the doubly constrained spatial interaction model of Wilson (1967) where, rather than explicitly calculating the balancing factors, A_i and B_j, we are simply operating on the flows predicted from the main section of that model's equation, until the constraints are met.

The results are shown in Table 4.3 with the columns labelled 'gipf' (for 'gravity IPF'). Here we see that the five specifications of the gravity model are almost identical, suggesting that the IPF has "smoothed out" the anomalies in the flow magnitudes of the pure gravity approach. But the important result is that these 'gipf' specifications are all far more effective than either gravity alone or IPF alone, in most sectors by a very wide margin. Business Services sticks out as a particularly strong example of this.

This is an interesting result and not one, as far as we know, that has been documented anywhere else.

4.5 Results

We will now turn our attention to an analysis of particular services flows, comparing estimated values with values from data.

4.5.1 Selecting a Representative Sector

Since all the methods outlined here take each sector separately, it makes sense to pick a sector for the analysis which has either the largest proportion of point-to-point flows to check against or, alternatively, the largest proportional dollar value of point-to-point flows. Section 4.5.1 shows the situation for each of the services sectors. The first group of three columns shows the number of records against each sector. The first column shows the number of records which are point-to-point ("P2P"), which means that they are recorded as being both from and to a *country*, in the sense of an entity with a three-letter ISO code[4]. The inverse of this is "Regional" flows, shown in the second column. These are flows recorded as being from, to or (in some cases) from *and* to non-country regions. These include geographical areas (such as "Eastern Europe n.e.s." [5]), trade areas (such as "Southern African Customs Union") and catch-all unknown categories (such as the very common "World" and the generic "Areas n.e.s."). The third column shows the fraction of all records for that sector which are point-to-point ("Frac. P2P").

Interestingly, the range of P2P fractions is very small, from 79% to 88%. It is also interesting that the fractions of P2P flows are all so high by this measure. Because of this, we turn to the total dollar value of those flows. We want to choose the sector with the largest proportion of dollar value contained within P2P flows.

Columns 4, 5 and 6 show the same measures as the first group but for the dollar value, measured in $US billions. Here again, the range is fairly small, just nine percentage points, but, this time, the fraction contained in P2P flows is much smaller, between 14% and 23%. Since the sector with the highest proportion of P2P flows by this measure (Education) is comparatively small in magnitude, we might compromise between a large proportion of P2P and a large

[4] GBR for Great Britain, DEU for Germany, etc.

[5] "Not otherwise specified".

Table 4.4 The services sectors categorised according to two measures. The first is the number of records in each category, and the second is the total dollar value of all the flows in each category. In both cases, the two categories are point-to-point and regional, where *regional* is defined as any flow which originates or ends at a region, such as 'Europe' or 'World', rather than a country

	Number of Records			$ Value (billions)		
Sector	P2P	Regional	Frac. P2P	P2P	Regional	Frac. P2P
Education	1382	184	0.88	31	106	0.23
Other Services	3092	404	0.88	744	2965	0.20
Construction	1674	302	0.85	98	398	0.20
Financial Services	2623	401	0.87	300	1258	0.19
Business Services	3421	405	0.89	997	4498	0.18
Water Transport	2976	424	0.88	449	2159	0.17
Vehicle Trade	2790	422	0.87	88	432	0.17
Utilities	2791	423	0.87	94	458	0.17
Retail Trade	2794	280	0.91	546	2737	0.17
Air Transport	3085	429	0.88	348	1697	0.17
Inland Transport	3104	428	0.88	248	1217	0.17
Communications	2586	372	0.87	107	507	0.17
Wholesale Trade	2790	422	0.87	88	432	0.17
Transport Services	2290	224	0.91	241	1229	0.16
Health	896	163	0.85	5	27	0.16
Hospitality	2914	430	0.87	1211	6619	0.15
Public Services	1700	391	0.81	64	380	0.14
Real Estate	2148	222	0.91	27	165	0.14

volume of trade overall. By this combined measure, we select Financial Services as a good compromise. It is a comparatively large trade, in both number terms and value terms, and is also comparatively easy to interpret as a traded sector (unlike, say, Communications, whose interpretation as a traded sector would require significant additional interpretation as to what is and is not included). Therefore, the rest of this analysis will be performed on the Financial Services sector.

4.5.2 *Estimated in-Sample Flows*

The flow data for the Financial Services sector contains 2623 P2P flows (those where the origin *and* destination are known). We can use these to test the effectiveness of each of the estimation methods outlined here by checking their values against the values estimated by each method-where, of course, the methods have not had access to these values, only the row and column totals.

Table 4.5a shows the 10 best estimated Financial Services flows using the "gipf4" specification. The error is simply the difference in dollar terms between the flow in the data and the estimated flow. Flows smaller than $US 10 million are excluded. Perhaps predictably, since the errors are reported in absolute, not relative, terms, the best estimated flows tend to be smaller ones. This can be seen by comparing with the worst estimated flows in Table 4.5b.

Table 4.5 The 10 best estimated and most poorly estimated Financial Services flows using gravity model specification (4) followed by IPF. Error shown is simply the difference between the estimated flow and the flow in data. Only flows greater than $US 10 million are included

(a) The 10 Best Estimated

		Trade Val. ($US M)	Error ($US M)
ESP	CZE	10.12	0.13
BGR	AUT	10.32	0.15
ESP	SWE	10.65	0.23
DNK	BEL	16.55	−0.29
EST	DEU	13.21	0.37
AUT	FIN	10.71	0.43
AUT	TUR	13.65	−0.67
DNK	AUT	10.00	−0.80
ITA	IDN	12.73	0.99
BEL	CZE	23.40	−1.09

(b) The 10 Worst Estimated

		Trade Val. ($US M)	Error ($US M)
USA	GBR	15574	−5778
GBR	USA	15668	−3659
GBR	LUX	1590	3027
USA	JPN	4261	2752
ITA	IRL	3265	−2435
DEU	USA	6907	−2252
GBR	JPN	2677	−2218
FRA	GBR	1508	2189
FRA	LUX	3538	−2173
LUX	ITA	2931	−1956

4.5.3 Estimated Export Totals

To have some faith in the estimation procedure, it would be reassuring to see that no country is consistently over- or underestimated. If this was the case, we might have reason to suspect that some important variables are being omitted in the specification used at the gravity stage of the estimation in Section 4.4.4. In Section 4.5.2, we saw that the estimation error seemed to be broadly related to the magnitude of the flow itself.

We might similarly expect that an exporting country's total estimation error (that is, the sum of the estimation errors across all flows emanating from that country) is correlated with the total exports of that country. If there are marked deviations from any such relationship between total error and total export, this might be evidence of a systematic bias in the estimation process.

Section 4.5.3 shows a scatterplot of logged total exports against logged total *over* estimation, the latter defined as the sum of all estimation errors greater than zero. The relationship is clearly

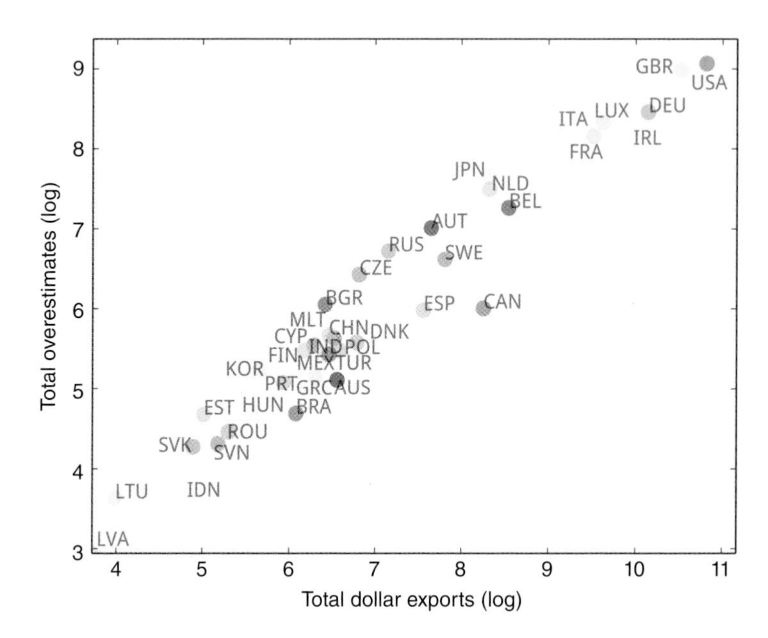

Figure 4.2 The total *gipf4* overestimate (the sum of all errors greater than zero) for each exporter, plotted against that country's total exports

linear and is remarkably unscattered. This gives us additional evidence that the *gipf4* estimation process, while of course not being precise, is at least treating each country equally in terms of overestimation. The only country which might be considered an outlier is Canada, which is slightly less overestimated than its total export size would suggest, given the trend with all other countries. Visually speaking, though, this is not a hugely noticeable outlier.

If countries are generally overestimated to a similar extent, once we take into account their export size, we might be interested in whether the extent to which a country's underestimation is closely related to its overestimation. Again, if some countries are more likely to be under- than overestimated, this might suggest a bias in our estimation procedure. Section 4.5.3 shows a summary of the relationship between overestimation and underestimation. The countries are presented in order of absolute total error, defined as the absolute value of total overestimate minus total underestimate. The total under- and overestimates are also shown, to the nearest $US million.

It is striking how closely the under- and overestimates are balanced. This result is not seen in using IPF alone, where the total errors are far from zero in most cases. It seems that a combination of gravity model and IPF results in an extremely even balance between under- and over-estimation.

The most consistently underestimated country in terms of Financial Services flows is the USA, with a total underestimation of just $US 1000.

We have therefore shown that the *gipf4* specification is neither biased towards a particular country, nor towards under- or overestimation.

Table 4.6 The five most and five least balanced exporting countries in terms of overestimation versus underestimation. All figures are in $US millions and relate to the Financial Services sector. The errors are from the *gipf4* specification

Country	Total export	Overestimate	Underestimate	Absolute total error
USA	49,745	8671	−8671	1.00×10^{-3}
GBR	37,146	8002	−8002	3.96×10^{-4}
IRL	21,213	3820	−3820	1.10×10^{-4}
DEU	25,441	4724	−4724	9.50×10^{-5}
CAN	3838	405	−405	7.30×10^{-5}
⋮	⋮	⋮	⋮	⋮
SVK	133	72	−72	5.29×10^{-7}
IND	608	244	−244	4.83×10^{-7}
IDN	124	38	−38	3.12×10^{-7}
LVA	41	19	−19	2.58×10^{-7}
LTU	54	38	−38	2.39×10^{-7}

4.6 Conclusion

In this chapter, we have shown how to deal with the fact that, in many cases, services flows are given only as import or export totals, not as point-to-point flows.

We introduced the iterative proportional fitting (IPF) method as a "no information" simplest case. We showed that IPF is zero retaining and broadly rank preserving.

We then went on to show how commodities flows, which are far better recorded and far more often point-to-point, can be used to better estimate services flows. To do this we introduced a generalised gravity model specification, which we then estimated in several specifications. In doing this, we uncovered the surprising fact that value added per unit has a negative exponent in the multiplicative gravity model which seeks to explain trade. We investigated this puzzle further by splitting value added into its constituent components and found that the use of primary products such as metals and minerals is an active discouragement to trade, but that use of secondary products such as plastics, machinery and chemicals encourages trade.

Turning back to services flows, we performed an in-sample comparison of the estimates obtained by each specification of the gravity model and chose an optimum specification. We additionally found that IPF performed better than any of the gravity specifications and hypothesised that a combination of these approaches might be best.

We showed that this hypothesis was indeed true and introduced the *gipf* (for gravity IPF) approach as an operationalised version of the doubly constrained spatial interaction model. This was shown to be the best in terms of in-sample comparison and was also shown to be unbiased in its estimation of each exporter. The error was very closely correlated with the total export of a given exporter. Finally, the *gipf* specification was shown to produce almost exactly as much underestimation as overestimation, a result which further encourages its use in estimating point-to-point services flows from trade totals.

References

Anderson JE and Wincoop EV. (2003). Gravity with gravitas: a solution to the border puzzle. *The American Economic Review* 93(1): 170–192. http://www.jstor.org/stable/3132167

Baldwin R and Taglioni D. (2006). *Gravity for Dummies and Dummies for Gravity Equations*. Working Paper 12516. Washington, DC: National Bureau of Economic Research. http://www.nber.org/papers/w12516

Dietzenbacher E, Los B, Stehrer R, Timmer M and de Vries G. (2013). The construction of world input-output tables in the WIOD project. *Economic Systems Research* 25(1): 71–98. doi:10.1080/09535314.2012.761180, http://www.tandfonline.com/doi/abs/10.1080/09535314.2012.761180

Disdier AC and Head K. (2008). The puzzling persistence of the distance effect on bilateral trade. *The Review of Economics and Statistics* 90(1): 37–48. https://ideas.repec.org/a/tpr/restat/v90y2008i1p37-48.html

Krugman P. (1980). Scale economies, product differentiation, and the pattern of trade. *The American Economic Review*: 950–959. http://www.jstor.org/stable/1805774

Lahr M and de Mesnard L. (2004). Biproportional techniques in input-output analysis: table updating and structural analysis. *Economic Systems Research* 16(2): 115–134. doi:10.1080/0953531042000219259, http://www.tandfonline.com/doi/abs/10.1080/0953531042000219259

McCallum J. (1995). National borders matter: Canada-U.S. regional trade patterns. *The American Economic Review* 85(3): 615–623. http://www.jstor.org/stable/2118191

Newman MEJ and Girvan M. (2004). Finding and evaluating community structure in networks. *Physical Review E* 69(2): 026–113. doi:10.1103/PhysRevE.69.026113, http://link.aps.org/doi/10.1103/PhysRevE.69.026113

Novy D. (2013). Gravity redux: measuring international trade costs with panel data. *Economic Inquiry* 51(1): 101–121. doi:10.1111/j.1465-7295.2011.00439.x, http://onlinelibrary.wiley.com/doi/10.1111/j.1465-7295.2011.00439.x/abstract

Reichardt J and Bornholdt S. (2006). Statistical mechanics of community detection. *Physical Review E* 74(1): 016 110–. doi:10.1103/PhysRevE.74.016110, http://link.aps.org/doi/10.1103/PhysRevE.74.016110

Rosvall M and Bergstrom CT. (2008). Maps of random walks on complex networks reveal community structure. *Proceedings of the National Academy of Sciences* 105(4): 1118–1123. doi:10.1073/pnas.0706851105, http://www.pnas.org/content/105/4/1118

Weidmann NB, Kuse D and Gleditsch KS. (2010). The geography of the international system: the CShapes dataset. *International Interactions* 36(1): 86–106. doi:10.1080/03050620903554614, http://dx.doi.org/10.1080/03050620903554614

Wilson AG. (1967). A statistical theory of spatial distribution models. *Transportation Research* 1(3): 253–269. http://www.sciencedirect.com/science/article/pii/0041164767900354

Wilson AG. (1969). The use of entropy maximising models, in the theory of trip distribution, mode split and route split. *Journal of Transport Economics and Policy* 3(1): 108–126. http://www.jstor.org/stable/20052128

5

A Method for Estimating Unknown National Input–Output Tables Using Limited Data

Thomas P. Oléron Evans and Robert G. Levy

5.1 Motivation and Aims

In Chapter 4 of the companion volume to this book, *Global Dynamics*, we define an economic model (the ENFOLDing model) combining data on international trade from the United Nations COMTRADE database with data representing the transfer of goods, services and money between economic sectors within national economies, drawn from the World Input-Output Database (WIOD), to produce a demand-driven model of global trade.

When introducing the ENFOLDing model, we noted that WIOD contains data on only 40 countries. However, the ENFOLDing model was designed to be a comprehensive model of world trade, making full use of the COMTRADE data to comprehensively track trade flows between all the countries of the world.

We originally addressed this through the introduction of an entity labelled ‘Rest of World’, which acts as a source and sink for all trade that occurs between those countries that are explicitly included in the model and those that are not. Although WIOD countries represent some 70% of global economic activity, this nevertheless means that a huge amount of activity cannot be represented. However, it should be noted that a Rest of World entity would always be necessary, even if input–output data were available for every country.

There are two main reasons for this. Firstly, many of the regions included in the COMTRADE database as separate trading entities are not independent nations, meaning that COMTRADE data and country-level data from other sources cannot generally be combined without discrepancies and complications. Secondly, given that COMTRADE is an extremely large and

Approaches to Geo-mathematical Modelling: New Tools for Complexity Science, First Edition. Edited by Alan G. Wilson.

Companion Website: www.wiley.com/go/wilson/ApproachestoGeo-mathematicalModelling

complex database, involving data collected and reported by different agencies across the world, internal inconsistencies are inevitable. Since the trade patterns predicted by the ENFOLDing model are determined with reference to the trade patterns that are observed in the COMTRADE data, these inconsistencies lead to consequent disparities in import and export figures. The Rest of World entity is therefore necessary to absorb such imbalances.

For the first iteration of the ENFOLDing model, as described in Chapter 4 of *Global Dynamics*, only the 40 WIOD countries were modelled explicitly. However, if the ENFOLDing model is truly to function as a comprehensive representation of world trade, the proportion of economic activity accounted for by the Rest of World entity should be as small as possible. One way of achieving this would be to estimate input–output tables for countries that are not included in WIOD.

In this chapter, we start by discussing the main obstacles to a successful estimation of this nature. We then present a method of estimation which – in common with the ENFOLDing model itself – uses only the available COMTRADE data and WIOD data (alongside national GDP figures drawn from the World Bank) to inform its estimates. Finally, we attempt to determine the quality of the estimated input–output tables that are obtained by following this procedure.

Throughout this chapter, we make reference to the 35 economic sectors of WIOD. For the purposes of the mathematics, these sectors will be referred to by the numerical labels 1 to 35, though their specific ordering is arbitrary.

5.2 Obstacles to The Estimation of National Input–Output Tables

Our intention is to use the information contained in WIOD to learn about the possible structural features of a national input–output table and, given a country whose input–output table is unknown for a particular year, to use this information to deduce a set of values for the table that is consistent, both with these observed structural features and with relevant import and export data drawn from the COMTRADE database, and with the GDP figures collected from the World Bank.

By *structural features*, we mean the shape and location of the region of $\mathbb{R}^{2660}$ in which vectors corresponding to potential national input–output tables, those that describe possible national economies, could lie. Each national input–output table in WIOD corresponds to a vector $\mathbf{x}$ (see Section 5.3), whose location in $\mathbb{R}^{2660}$ provides information about the shape of this theoretical region.

At the time of writing, the authors were able to access WIOD input–output tables for 40 countries for each year from 1995 to 2011. COMTRADE data and GDP figures for each of these years were also available. This means that a total of 600 national input–output tables for real economies were available for analysis, the aim being to deduce the degree of variation that is possible across different input–output tables and hence the range of possible forms that an estimated table should take.

Critically however, these 600 tables do not represent 600 independent depictions of national economies. We would clearly expect there to be a high degree of correlation between the input–output tables for a particular country over several different years. Therefore, it could be argued that each set of 15 input–output tables relating to a single country only provides useful information on one possible form for a national economy. For the purposes of estimating the possible form of unknown input–output tables, the extra value contributed by each additional year covered by the data is therefore limited.

Effectively, then, the WIOD data provide just 40 'independent' snapshots of possible forms for a national economy, presumably representing a very small sample of the possible economies that could exist, particularly given the extremely high dimensionality of the space $\mathbb{R}^{2660}$ in which the vector input–output tables (represented by $\mathbf{x}$) lie.

The combination of high-dimensional data and a small sample size means that the risk of overfitting is particularly acute when using the available data to predict the location of new points within the space. Specifically, given any k points in $\mathbb{R}^n$ for $k \leq n$, it is always possible to find at least one $(k - 1)$-dimensional linear manifold passing through all the points. To suppose that this linear manifold provided a thorough and accurate representation of the spread of the data would be to imply that there was a perfect linear relationship between the coordinates of the points, a relationship that would, in many circumstances, be entirely spurious. Any reasonable estimation procedure must be carefully designed to take this issue into account.

The problems associated with the small sample size are exacerbated by the fact that the countries included in WIOD are very far from being a representative sample of all the countries of the world. The countries of WIOD are generally more wealthy, with a disproportionate representation of European countries. Unfortunately, we would expect that many of the countries whose input–output tables we would particularly like to estimate to have economies that are structured very differently from those of the countries that appear in WIOD. For example, the fact that WIOD contains no African countries means that estimated input–output tables for African nations that are derived from WIOD data may diverge very significantly from economic reality. This is a particularly problematic issue, since even determining the scale of this divergence is essentially impossible, given the available data.

Given these serious issues, the purpose of the estimation procedure described in this chapter must be stated very carefully. The estimated input–output tables that are produced in this chapter should not be seen as attempts to infer the true structures of the economies of those countries that are not included in WIOD. As discussed here, such a goal would be impossible to achieve given the paucity of available data. Instead, the estimated input–ouput tables presented here should be seen as mathematical objects whose structure is broadly in line with the structure of those national input–output tables for which data are available, and which reproduce certain key features of the non-WIOD countries for which data do exist, namely import and export figures for each of the 35 sectors in WIOD (from the COMTRADE data) and GDP (from the World Bank data). The aim of the procedure is primarily to allow additional countries to be included explicitly in the ENFOLDing model, rather than implicitly through the Rest of World entity, thus allowing for a more detailed and comprehensive analysis of the model and its value as a meaningful representation of global trade.

5.3 Vector Representation of Input–Output Tables

It is necessary to outline the mathematical form in which the WIOD country–year input–output tables will be presented for the purposes of the estimation procedure that will be described in Section 5.4. Throughout the remainder of this chapter, in order to demonstrate and evaluate the method, we will restrict our consideration to one particular (arbitrarily chosen) year: $Y = 2005$.

Each country-year input–output table is represented as a 2660-dimensional column vector $\mathbf{x} = (x_1, \dots, x_{2660})^\top \in \mathbb{R}^{2660}$. The interpretation of the terms of this vector is as follows:

- $x_{35(i-1)+j}$ is the value of domestic goods from sector i consumed by sector j (1225 terms).
- $x_{1225+35(i-1)+j}$ is the value of imported goods from sector i consumed by sector j (1225 terms).
- x_{2450+i} is the final demand for domestic goods from sector i (35 terms).
- x_{2485+i} is the final demand for imported goods from sector i (35 terms).
- x_{2520+i} is the net value of domestic goods that are either stored for future use or drawn from existing stores, for sector i (35 terms).
- x_{2555+i} is the net value of imported goods that are either stored for future use (positive) or drawn from existing stores (negative), for sector i (35 terms).
- x_{2590+i} is the value of domestic goods from sector i that are exported (35 terms).
- x_{2625+i} is the value of imported goods from sector i that are exported (35 terms).

More detailed explanations of the meanings of these terms may be found in Chapter 4 of *Global Dynamics* and in Timmer *et al.* (2015). All values are measured in current-value US dollars.

Recall that, by definition, the GDP of a country is equal to the sum of the 'value added'[1] across all sectors, as discussed in Chapter 4 of *Global Dynamics*. In the context of our input–output table vectors, it is straightforward to demonstrate that this quantity is equal to the sum of domestically supplied final demand, domestically supplied exports and the net value of domestic goods that are either stored for future use or drawn from existing stores, minus the value of all intermediate demand (i.e. supplied to the 35 sectors, rather than to final demand, export or storage) on imported goods:

$$\mathrm{GDP} = \sum_{i=1}^{35} \left(x_{2450+i} + x_{2520+i} + x_{2590+i} - \sum_{j=1}^{35} x_{1225+35(i-1)+j} \right)$$

For the purposes of the estimation procedure, certain constraints will be imposed on the terms of the vector input–output tables to be estimated. These constraints will be set out in Section 5.4.

5.4 Method

5.4.1 Concept

The basis of the estimation procedure is to use linear principle component analysis (see e.g. Ringnér, 2008) to identify an h-dimensional linear manifold, L, embedded in $\mathbb{R}^{2660}$, representing the main directions of variation in the known input–output table vectors, $\mathbf{x}_1, \dots, \mathbf{x}_{40}$, for the 40 countries in WIOD, for a particular year. Although considering tables from a single year significantly limits the amount of data available to inform the estimation procedure, it also ensures that all these data points represent independent snapshots of the structure of a possible national economy.

[1] The value added of a sector is equal to the total production of that sector, minus its total consumption of domestic and imported goods, drawn from all sectors.

Given L, an estimated input–output table for a new country is produced by choosing a point in $\mathbb{R}^{2660}$ that is consistent with the import and export figures for that country, drawn from the COMTRADE data, and the appropriate GDP figure, drawn from the World Bank data, and which lies 'as close as possible' to the mean point, $\bar{\mathbf{x}}$, of the known vector input–output tables. However, the distance from $\bar{\mathbf{x}}$ to the new estimate is not the standard Euclidean distance, but is weighted in such a way as to allow more variance in directions parallel to the major principal components of the data than is allowed in directions perpendicular to these components. More specifically, the distance from $\bar{\mathbf{x}}$ measured parallel to L is weighted component-wise by the inverses of the observed standard deviations in each of the relevant principal component directions, while the distance from $\bar{\mathbf{x}}$ measured in any component direction orthogonal to L is weighted by the inverse of the minimum standard deviation across all the principal component directions used to define L.

This dual approach allows for the artificially small standard deviations associated with principal components relating to directions in $\mathbb{R}^{2660}$ for which the data provide little to no information – a consequence of the small sample size and high dimensionality of the data (see Section 5.3) – to be prevented from having a disproportionate influence on final estimates. The efficacy of this strategy relies on a careful consideration of the number of components h that should be used to define L itself, with those that can be confidently judged to provide useful information about the range of variation of possible national input–output tables being included, while all those that cannot be so judged are omitted. This value is determined by a cross-validation procedure which is described in Section 5.4.3.

5.4.2 Estimation Procedure

We now present a mathematical description of the estimation procedure. Our goal is to estimate the input–output table vector $\mathbf{z}$ for a particular country A, for a given year Y, given the input–output table vectors for C other countries, also for year Y, along with some key economic data relating to country A.

We begin by introducing notation to represent the data that will be used to inform the estimation.

- Known input–output table vectors:

$$\mathbf{x}_1, \dots, \mathbf{x}_C \in \mathbb{R}^{2660}$$

- Import totals (in US dollars) for country A, in year Y, in each of 35 sectors:

$$m_1, \dots, m_{35} \in \mathbb{R}$$

- Export totals (in US dollars) for country A, in year Y, in each of 35 sectors:

$$e_1, \dots, e_{35} \in \mathbb{R}$$

- GDP (in US dollars) for country A, in year Y:

$$G \in \mathbb{R}$$

We also suppose that we have chosen a value for the following parameter:

- The number of principle components to consider:

$$h \in \mathbb{N}, h < C$$

Note the specification that h should be less than the number of data points C. This condition is imposed because a maximum of $C - 1$ principle component directions may be defined from C vectors.

We now make some additional definitions, which will lead to greater clarity in later expressions:

- The mean data vector:

$$\bar{\mathbf{x}} = [\mathbf{x}_1 + \ldots + \mathbf{x}_C]/C$$

- The data matrix:

$$\mathrm{X} = \begin{pmatrix} \uparrow & & \uparrow \\ \mathbf{x}_1 & \ldots & \mathbf{x}_C \\ \downarrow & & \downarrow \end{pmatrix}$$

- The eigenvalues of $\mathrm{XX}^\top$:

$$\lambda_1 \geq \ldots \geq \lambda_{2660}$$

These are the variances of the data measured in each of the principle component directions, numbered in order of decreasing magnitude.
- The corresponding unit eigenvectors of $\mathrm{XX}^\top$ (the principal components):

$$\mathbf{v}_1, \ldots, \mathbf{v}_{2660}$$

- The matrix of the first h principal components:

$$\mathrm{V} = \begin{pmatrix} \uparrow & & \uparrow \\ \mathbf{v}_1 & \ldots & \mathbf{v}_h \\ \downarrow & & \downarrow \end{pmatrix}$$

- The vector of the first h eigenvalues:

$$\boldsymbol{\lambda}^{(h)} = (\lambda_1, \ldots, \lambda_h)^\top$$

The estimation procedure consists simply of solving the following quadratic optimisation programme (see e.g. Hoppe, 2006, Chap. 3) for the required vector input–output table $\mathbf{z}$.

5.4.2.1 Vector Input–Output Table Estimation Procedure

- Minimise:

$$\mathbf{z}^\top[2\mathrm{M}]\mathbf{z} - [2\bar{\mathbf{x}}^\top\mathrm{M}]\mathbf{z}$$

- Over:

$$\mathbf{z} \in \mathbb{R}^{2660}$$

- Where:

$$M = V[\mathrm{diag}(\boldsymbol{\lambda}^{(h)})^{-1} - \lambda_h^{-1} I_h]V^{\top} + \lambda_h^{-1} I_n$$

- Subject to:
  - Import constraints:

$$z_{2485+i} + z_{2555+i} + z_{2625+i} + \sum_{j=1}^{35} z_{1225+35(i-1)+j} = m_i \, , i \in \{1, \ldots, 35\}$$

  - Export constraints:

$$z_{2590+i} + z_{2625+i} = e_i \, , i \in \{1, \ldots, 35\}$$

  - Constraints that imported goods and services should not be re-exported[2]:

$$z_{2625+i} = 0 \, , i \in \{1, \ldots, 35\}$$

  - A GDP constraint:

$$\sum_{i=1}^{35} \left(z_{2450+i} + z_{2520+i} + z_{2590+i} - \sum_{j=1}^{35} z_{1225+35(i-1)+j} \right) = G$$

  - Non-negativity constraints[3]:

$$z_l \geq 0 \, , \forall l \in \{1, \ldots, 2660\}$$

  - Value-added constraints[4]:

$$z_{2450+i} + z_{2520+i} + z_{2590+i} + \sum_{j=1}^{35} [z_{35(i-1)+j} - (z_{1225+35(j-1)+i} + z_{35(j-1)+i})] \geq 0$$

$$\forall i \in \{1, \ldots, 35\}.$$

5.4.3 Cross-Validation

This part of the procedure is performed before the estimation of any unknown input–output tables for countries that do not appear in WIOD and is used exclusively to determine the number of principle components h that should be used to define the linear manifold L that will represent the observed variation in the WIOD data.

[2] This constraint is consistent with the data from WIOD.

[3] Across the 15 years of WIOD data, from 1995 to 2009, the only negative values corresponding to terms in our input–output table vectors relate to domestically supplied exports and to the net values of domestically produced goods stored for future use or drawn from existing stores. The first of these two cases is irrelevant here, since exports are explicitly constrained to equal known positive values by the export and re-export constraints that have been stated. Non-negativity constraints may therefore be imposed without affecting the outcome of the estimation procedure. For the sake of simplicity, we have decided to impose non-negativity constraints in the second case also, since negative values of this nature in the WIOD data are both rare and small in magnitude. Only around 0.1% of such values are both negative and greater than 1.5% of the relevant country's GDP (in absolute terms).

[4] Observe that the left-hand side of these inequalities are equal to the 'value added' of sector i. Therefore, the constraints ensure that the total value of intermediate goods consumed by a particular sector in a given year does not exceed the total value of the goods produced by that sector. This condition is required for the ENFOLDing model to function and is satisfied by all sectors in all countries in the WIOD data.

To perform the cross-validation, the estimation procedure described in Section 5.4.2 is performed 40 times for each value of $h \in \{1, \ldots, 38\}$. In each of these 40 procedures, 39 of the 40 known vector input–output tables, $\mathbf{x}_1, \ldots, \mathbf{x}_{40}$, are used to estimate the remaining vector that has not been included. For each value of h, the (standard Euclidean) distance of each estimate from its corresponding known vector is calculated and the mean of these distances is calculated as a rough measure of the quality of the estimates produced using that particular number of principle components. The value of h for which this mean is minimised is then used for the purposes of estimating unknown input–output tables.

Performing such a cross-validation on the 40 WIOD vector input–output tables for $Y = 2005$ yielded an optimal value of $h = 16$, indicating that 16 principal components should be used in the estimation procedure.

5.5 In-Sample Assessment of The Estimates

Using Python (Python Software Foundation, 2012) with the packages NumPy (Numpy Developers, 2012) and CVXOPT (Andersen *et al.*, 2014), estimated vector input–output tables were produced for 193 countries (including the 40 countries of the WIOD). Data on input–output tables were taken from WIOD (Timmer *et al.*, 2015). Data on global trade in commodities and services were taken from COMTRADE (UN, 2013). The figures for purchasing-power-parity adjusted GDP per capita and population were taken from the World Bank's World Development Indicators (World Bank, 2015).

Each estimate represents either an intermediate flow from one sector to another within a particular country[5] or a final demand flow within that country. The estimation process "estimated" the 40 countries of the WIOD as well as the other 153 non-WIOD countries. This allows us to make a simple analysis of the estimates of the 40 countries with their "true" values as reported in WIOD.

5.5.1 Summary Statistics

A summary of the results of this analysis is shown in Figure 5.1. The WIOD value for each sector-to-sector flow or sector-to-final-demand flow (hereafter, we will simply call these *flows*) is plotted against the equivalent estimated flow, with both axes being logged.

If the estimates were a perfect reconstruction of the data, we would expect every point to lie along $y = x$. Any deviation from this represents a deviation of the estimated flow from the flow in data. Since this is a log–log plot, deviations at the right-hand side (larger range) are orders of magnitude more severe than deviations at the left-hand side (smaller range). It is therefore encouraging to see that as we reach the higher magnitudes, the points do seem to lie fairly close to the $y = x$ line and, in general the plot shows a broad balance between under- and over-estimation with a slight bias towards over-estimation at the smaller scales of magnitude.

[5] Within a particular year, but since all estimates are from 2005, this will be implicit in everything which follows. An interesting direction for future research could be to repeat the analysis of this chapter across the 16 years for which WIOD has data.

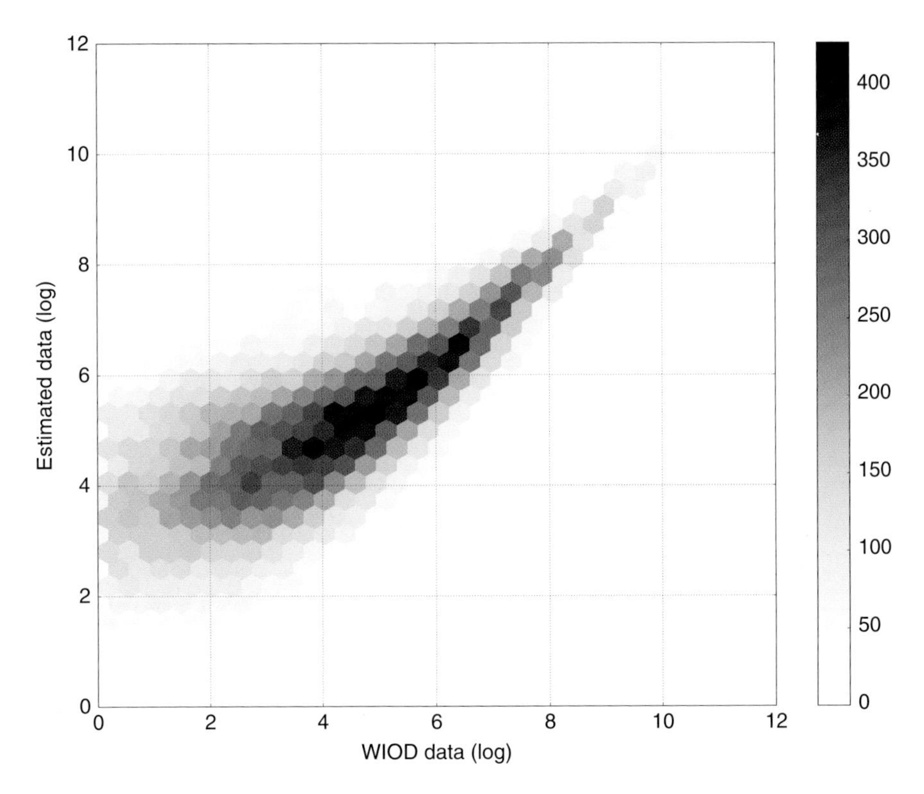

Figure 5.1 A density plot showing the relationship between the estimates of the 40 WIOD countries and the WIOD data itself. Darker shades indicate the presence of more data points. All data points are flows, measured in $US millions. A little over 100,000 flows are plotted.

We can look in more detail at the estimates by breaking down the estimation error (estimated value minus the original data value) by country and by "from-sector" (i.e. the sector from which the flow emanates.) Figure 5.2 shows summary statistics broken down by both variables. Estimation error is measured in the same units as the flows themselves, $US millions. In each case, the elements of the graphs are as follows: the box represents the inter-quartile range (IQR) of the estimation error, from the 25th percentile to the 75th percentile. The "whiskers" reach from the 10th to the 90th percentiles. The mean is represented by a small filled-in box, and the median is an unbroken horizontal line.

Figure 5.2a shows summary statistics for the 10 biggest countries by total flow magnitude in the data. There is a general tendency towards over-estimation, but this tendency seems to be exacerbated by some considerably over-estimated 'outliers'. To see this, observe that the mean across all errors lies within the IQR of only three countries: Korea, Spain and the USA. In all countries, the median error lies relatively close to zero. The broadest 10th–90th percentile ranges of errors are those of the USA ($4.7 billion), Germany ($2.5 billion) and Japan ($1.9 billion). The smallest are those of Korea ($680 million), Spain ($720 million) and China ($960 million), this last example showing that estimation error is not simply a function of total GDP.

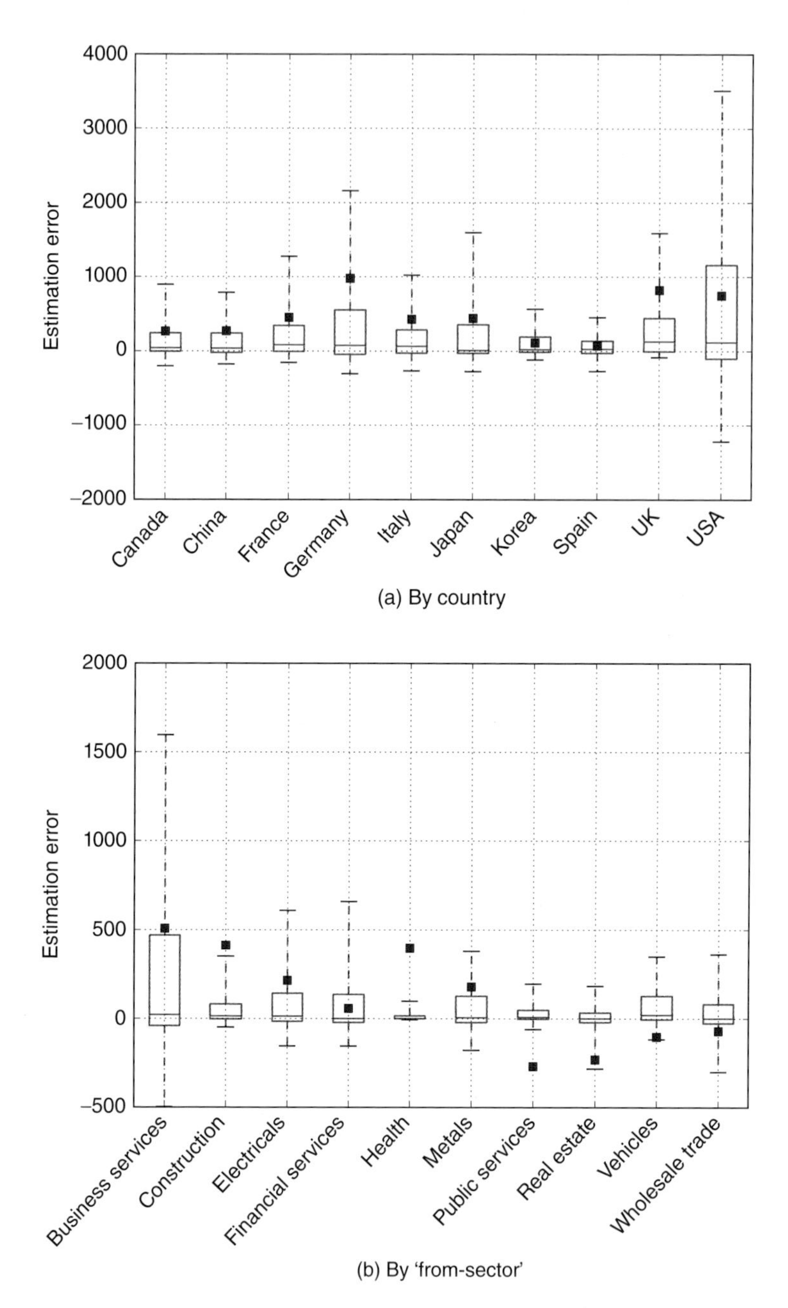

Figure 5.2 Boxplots showing the estimation error of flows within the 10 largest countries and the 10 largest sectors by total flow magnitude. The dotted vertical lines show the 10th to the 90th percentiles. The box shows the interquartile range. The mean is represented as a small filled square, and the median by a horizontal line. Estimation errors are measured in millions of $US.

5.5.2 *Visual Comparison*

In order to look in more detail at some of the best and worst estimated countries from Section 5.5.1, we can show the input–output table for a particular country as a heatmap. This involves a grid of sector–sector flows where the colour of the grid element is related to the size of the flow. This allows us a visual overview of the "shape" of a particular input–output table, and allows us to visually compare, in a necessarily broad-brush fashion, the WIOD input–output tables with the equivalent estimated tables.

We will restrict ourselves to comparing two well-estimated countries from Section 5.5.1, Spain and China, and two more poorly estimated countries, Germany and the USA. For visual clarity, we will also roll the 35 WIOD sectors into a more manageable seven 'super-sectors'. This is done in a fairly arbitrary fashion by simple inspection of the sectors. The decision of which sectors to put into which super-sector will, to some extent, affect the conclusions in all which follows, but the arbitrariness of the categorisation is not in itself problematic as we will simply be using it for visual inspection. The super-sectors are composed from WIOD sectors, as shown in Table 5.1.

Figure 5.3 shows heatmaps for the four countries mentioned (the USA, Germany, Spain and China) and has the WIOD data input–output tables side-by-side with the equivalent set of estimates. Flows are from the row sector to the column sector. Note that flows to Final Demand are not included in these images. It should also be noted that each heatmap has its own colour scale, which is not shown. Thus, this exercise tells us only about the shape of each input–ouput table, *not* the absolute magnitudes of the flows inside it.

An encouraging initial observation, is that the estimates are visually broadly similar to their data counterparts. We can therefore conclude that the estimation process produces what we might term 'broadly realistic' input–output tables, by which we mean that the tables "look" like real input–output tables. Also encouraging is the fact that it is visually, immediately clear which estimate map goes with which WIOD data map: although the estimates and the data maps are not identical, there is, at least among these four countries, no danger of mistakenly attributing an estimate set to the wrong dataset.

Visually, the most well-estimated country is China, a finding which agrees with the summary statistic finding in Section 5.5.1. In fact, the estimate map is almost indistinguishable from the data map. Flows to services and trade from services and secondary are slightly under-estimated, but the estimation successfully finds that primary, secondary and raw are the main sectors in China along with flows to (but not from) public.

Table 5.1 The seven "super-sectors" into which each of the 35 WIOD sectors has been aggregated in order to facilitate visual analysis of the estimates

Transport	Inland Transport, Water Transport, Air Transport, Transport Services
Primary	Agriculture, Food, Leather, Paper, Textiles
Secondary	Chemicals, Electricals, Plastics, Manufacturing, Machinery, Vehicles
Raw	Wood, Mining, Metals, Fuel, Minerals
Trade	Vehicle Trade, Wholesale Trade, Retail Trade, Real Estate
Services	Business Services, Financial Services, Hospitality, Private Households
Public	Education, Health, Other Services, Communications, Construction, Public Services, Utilities

The other three countries have at least one visually obvious over-estimate. In Spain, the over-estimate is secondary–secondary, and in Germany and the USA it is public–public and, to a lesser extent, services–public. The flows from services are generally slightly over-estimated in both Spain and Germany, and flows from transport are overestimated generally in Germany and the USA.

Generally speaking, though, the shapes of the maps match surprisingly well. In Spain, the right half of the map, flows to services, trade and transport, is reproduced almost exactly, which agrees with our assessment in Section 5.5.1 that Spain was among the most well-estimated countries.

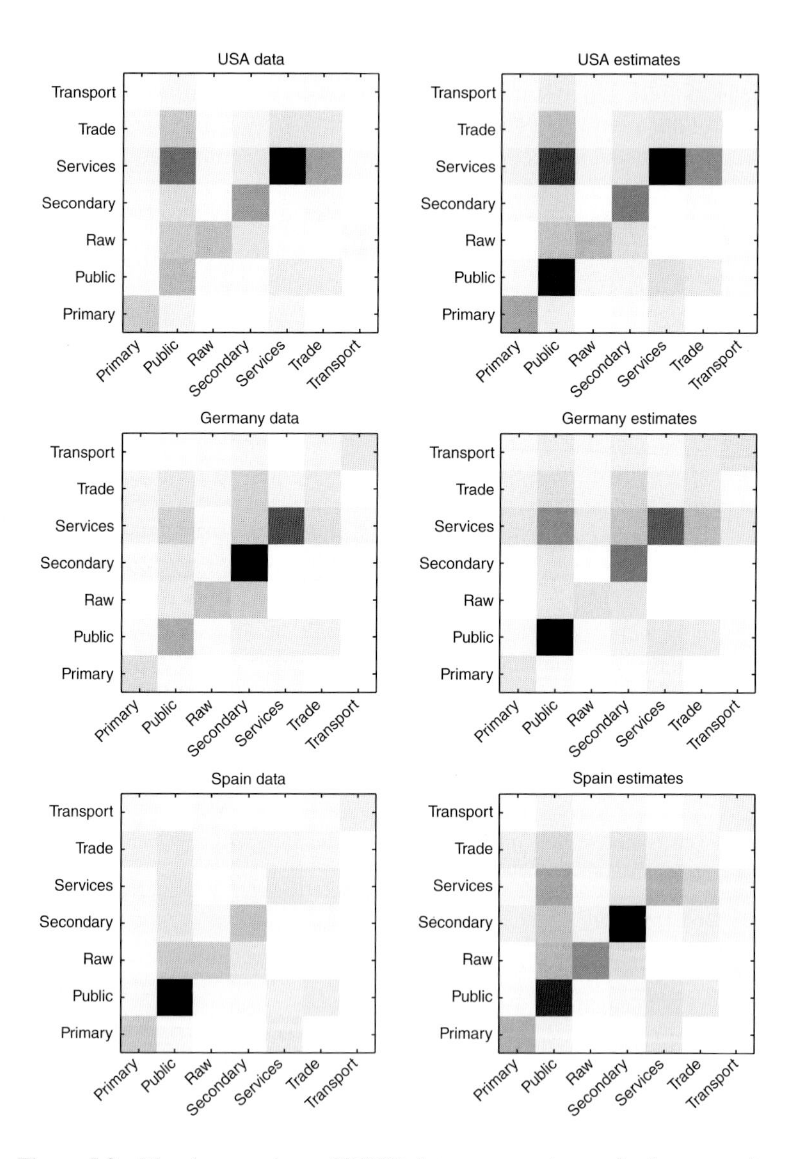

Figure 5.3 Visual comparison of WIOD data versus estimates for four countries

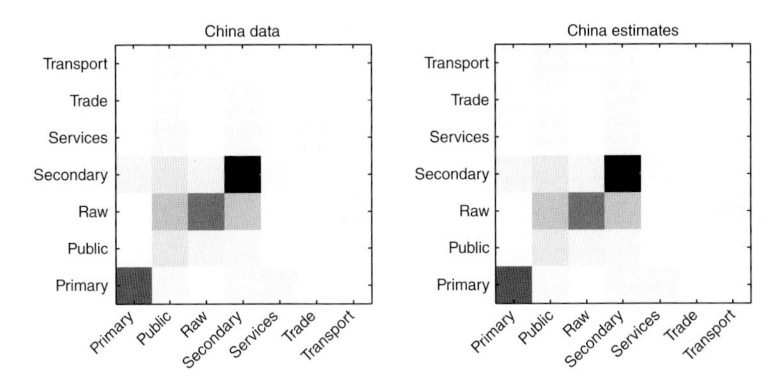

Figure 5.3 (*continued*)

With the exception of the overestimations mentioned here, the USA and Germany are also remarkably well estimated. This shows that, while the magnitudes of these countries are not well estimated, as we found in Section 5.5.1, the shape of the table still may be. This is a particularly important finding in this context, since the flows will largely be normalised by the conversion of these flows to technical coefficients in the usual input–output manner. It is therefore the shape, far more than the magnitudes, which is of importance.

5.6 Out-of-Sample Discussion of The Estimates

The discussion in Section 5.5 has hopefully gone some way towards convincing us that the estimation algorithm estimates reasonably well those countries included in its training set, in relative terms (what we referred to as the *shape* of the resulting input–output table) if not always in absolute terms.

But our purpose in running the algorithm in the first place is of course not to estimate the countries we already know, but to estimate those we do not. Beyond the fact that the estimation procedure produces input–output tables which, in some way, 'look' like real input–output tables, rating the validity of the countries estimated out-of-sample is a more difficult task and will inevitably involve subjectivity in our assessment of the quality of the estimates.

We will start with developing a simple distance measure which will allow us to compare countries to one another. In this way, we will be able to see which WIOD country an estimated country most resembles and see if the results are in any way intuitively appealing.

5.6.1 Final Demand Closeness

We introduce of a method of categorising estimated countries based on the estimated final demand for domestic goods and comparing each to the same set of final demands in WIOD countries. For simplicity of exposition, we will only examine countries with populations of more than 10 million. This arbitrary cut-off includes 52 countries, which is a similar number to the 40 WIOD countries.

We begin by treating the final demand for domestic goods as a vector, defined as follows:

$$\mathbf{f}_i = \begin{bmatrix} f_{i1} \\ f_{i2} \\ \vdots \\ f_{iS} \end{bmatrix} \tag{5.1}$$

where f_{is} is the final demand for domestically produced sector s and S is the total number of sectors. This then allows us to think of the direction in which each of these S-vectors 'points', and of the concept of their being an angle between any two such vectors.

By normalising the vectors, thereby setting their lengths to one, we can find the angle by using a dot product and some trigonometry:

$$\theta_{ij} = \arcsin(\bar{\mathbf{f}}_i \cdot \bar{\mathbf{f}}_j) \tag{5.2}$$

where a bar shows a vector has been normalised.

This leads us naturally to a categorisation of the estimated countries based on which WIOD country has the closest final demand vector by this measure. Table 5.2 shows the results of this experiment for the 52 large-population countries, where each estimated country is placed with the WIOD country for which equation (5.2) is a minimum.

There are eight WIOD countries which are a non-WIOD country's closest neighbour: China, Estonia, India, Indonesia, Lithuania, Mexico, Russia and Turkey. All these countries are at the lower end of the GDP-per-capita spectrum, which might be considered encouraging for this exercise, since WIOD countries are generally richer, European and OECD countries.

Also encouraging is the fact that the two countries in the China group are the Philippines and Thailand, both South-east Asian countries. Bangladesh, Cambodia, Sri Lanka and Vietnam align with India in a similarly encouraging geographic fashion, although Pakistan does not. Malaysia aligns with Indonesia. Egypt, Pakistan and Syria align with Turkey. By this geographic reasoning, we might perhaps hope that the South American countries on the non-WIOD list would align to either Brazil or Mexico, but this is not the case.

Instead, we might justify some of these results in economic terms. There is a set of oil-producing countries which align with Russia: Algeria, Iran, Nigeria, Saudi Arabia and

Table 5.2 The 52 non-WIOD countries with a population of greater than 10 million, grouped by the WIOD country to which the final demand vector is closest in angle, defined by equation (5.2)

WIOD country	Estimated countries
China	Philippines, Thailand
Estonia	Argentina, Colombia, South Africa
India	Bangladesh, Cambodia, Sri Lanka, Morocco, Tunisia, Vietnam
Indonesia	Afghanistan, Angola, Burkina Faso, Ivory Coast, Cameroon, DRC, Ghana, Kazakhstan, Madagascar, Mali, Mozambique, Malaysia, Nepal, Sudan, Senegal, Tanzania, Uganda, Uzbekistan, Yemen
Lithuania	Chile
Mexico	Iraq, Malawi, Niger, Chad, Ukraine, Zambia, Zimbabwe
Russia	Algeria, Iran, Nigeria, Saudi Arabia, Venezuela
Turkey	Cuba, Ecuador, Egypt, Ethiopia, Guatemala, Kenya, Pakistan, Peru, Syria

Venezuela, all of which are members of OPEC. The only other OPEC members on this list are Angola and Iraq, who align with Indonesia and Mexico, respectively. Argentina, South Africa and Chile are among the richest countries on the list and align with Estonia and Lithuania, the two richest of the WIOD category countries. The largest group by far is that aligned with Indonesia, the poorest of the category countries after India. In this group are Afghanistan, Kazakhstan, Nepal and Uzbekistan, along with all the sub-Saharan countries on the list apart from Malawi, Niger, Chad, Zambia and Zimbabwe which align with Mexico; and Ethiopia and Kenya which align with Turkey.

In summary, we may find some sources of encouragement in Table 5.2, and also some cause for puzzlement. The grouping around Mexico in particular is difficult to explain. But it should be borne in mind that this experiment gives a precedence to the final demand vector (and only final demand for domestic goods) over all other aspects of the estimates, which the estimation process itself does not share. Thus, by focussing in this way on the final demand vector, we may be excluding encouraging similarities with WIOD countries or, alternatively, further puzzling alignments.

5.6.2 Technical Coefficient Clustering

The analysis of Section 5.6.1 grouped estimated countries with their "nearest" WIOD neighbour. We might also ask how the estimated countries group to one another, and whether any geographic or macroeconomic justification might be found which will encourage us that the estimation process has done a reasonable job.

It is important to focus on the shape of the estimated input–output tables, as we did in Section 5.5, not the magnitude of any particular flow, and this brings us naturally to studying technical coefficients, which are unit-less and constrained to sum to less than one down a particular column. In similar fashion to the ENFOLDing model, we will aggregate the intermediate flows from domestic and imported goods, thus focussing more on the technical aspects of the production process and less on the importance to that process of imports.

The first check of a sensible input–output table is that none of the columns of technical coefficients sum to more than one. In the case of this set of estimates, this is true in all cases except for four sectors in Bermuda. For this reason, Bermuda will be excluded from all the analysis which follows.

To group the countries by the similarity of their technical coefficients, we will use the well-understood k-means clustering algorithm which is very simple and is implemented in many statistical packages. One drawback is that it involves specifying a number of clusters *a priori* which is difficult to do with any rigour. Additionally, k-means has an element of stochasticity in its output, with multiple runs of the algorithm producing different end results. To deal with both these problems, we will run k-means many times, and aggregate the results into a "co-occurrence" matrix. This is a square symmetric matrix with columns and rows indexed by the countries in the analysis. A cell ij records when a particular run of the clustering algorithm places country i in the same cluster as country j. In this way, the co-occurrence matrix represents a picture of the various runs of the clustering algorithm which we can visualise as a heat map as shown in Figure 5.4, with darker colours representing a more frequent pairing of the row and column countries in a cluster.

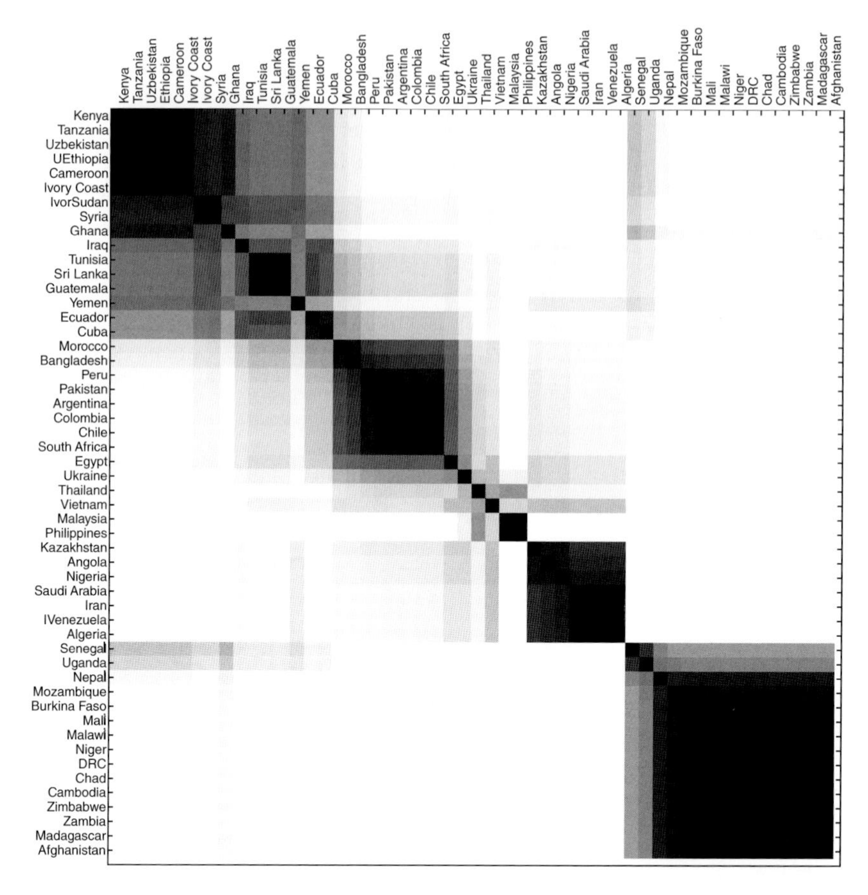

Figure 5.4 A co-occurrence matrix showing the result of 100 runs of the k-means clustering algorithm, set to split the technical coefficients of the countries shown into five clusters. Darker colours indicate a more frequent co-occurrence of the pair of countries in the same cluster

A mixed picture emerges with three well-defined clusters in the bottom-right corner of the matrix, and perhaps three more poorly defined overlapping clusters in the top left. The large cluster at the far bottom right is what we might term the *sub-Saharan Africa cluster*, consisting of Burkina Faso, Mali, Malawi, Niger, DRC, Chad, Zimbabwe, Zambia and Madagascar. Also in this cluster are Nepal, Cambodia and Afghanistan. Senegal and Uganda are weakly in this cluster as well as weakly in the other major African cluster at the top left, but they are most often placed in a cluster of their own.

The OPEC cluster that we identified in Section 5.6.1 is very well defined at second from the bottom left, containing Kazakhstan, Angola, Nigeria, Saudia Arabia, Iran, Venezuela and Algeria. Vietnam is also weakly in this cluster.

A South-East Asia cluster containing Malaysia, the Philippines and Thailand and, more weakly, Vietnam is the third of the well-defined clusters of the bottom right.

The picture in the top left is more complex and difficult to categorise. We have a strong South American cluster of Argentina, Columbia, Chile and Peru, but it also contains Bangladesh,

Morocco and Egypt. Ecuador and Cuba also belong weakly to this cluster, although they more often appear as a cluster of their own.

Another strong cluster of African countries at the top left is made up of Kenya, Tanzania, Ethiopia, Cameroon, Ivory Coast, Sudan and Ghana, although this cluster also strongly features Uzbekistan and Syria. Within the larger weak cluster of the top left is an entirely non-geographic cluster of Tunisia, Sri Lanka and Guatemala.

It is particularly intriguing that the countries of Africa are divided very sharply into two completely non-overlapping clusters, with Senegal and Uganda being the only 'bridge countries'. On one side, followed by numbers indicating their position in the World Bank richest country rankings[6], are Burkina Faso (166), Chad (157), DRC (182), Mali (169), Malawi (183), Madagascar (173), Mozambique (179), Niger (180), Zambia (136) and Zimbabwe (163). On the other side are Cameroon (149), Ethiopia (176), Ghana (134), Ivory Coast (142), Kenya (150), Sudan (140) and Tanzania (154). The two bridge countries are Senegal (156) and Uganda (167). It appears that the African countries are thus broadly split by GDP per capita into a richer cluster at the top left and a poorer cluster at the bottom right. By this categorisation, only Zambia and Ethiopia are in the "wrong" cluster.

5.7 Conclusion

In this chapter, we have described an estimation procedure that allows us to produce estimated input–output tables for countries that do not feature in WIOD, using only data from WIOD and COMTRADE, and with GDP figures from the World Bank, to inform the estimates.

As we have discussed, this is a particularly difficult task, owing to the small number of truly independent data points available in WIOD to represent the range of possible forms for a national input–output table and also to the high dimensionality of these data points. The fact that the countries covered by WIOD do not provide a representative sample of all national economies, being skewed towards wealthier nations and lacking any input–output tables for countries in Africa, is another significant issue.

We have noted that, given these issues, it would be unreasonable to expect that any estimated input–output table could be expected to closely match the true structure of a corresponding national economy that does not feature in WIOD. There is insufficient information contained in the WIOD data and the COMTRADE data to do this with any meaningful degree of accuracy.

For this reason, it should be understood that the estimated national input–output tables produced by the method presented here should not be seen as attempts to accurately predict the true values of an unseen input–output table, but rather as synthetic tables which match key values from the data (such as import and export figures for each sector from the COMTRADE data), and which have a structure that is sufficiently close to those of the known input–output tables that they are suitable for use in the ENFOLDing model of Chapter 4 of *Global Dynamics*. However, this does not preclude the possibility that additional sources of data could be introduced to further inform the estimates, producing input–output tables that would be a closer match to reality.

Despite these caveats, we have seen that the estimates produce feasible-looking input–output tables, and that a clustering exercise reveals that some geographically and economically intuitive patterns are present in the estimated tables. We hope that the creation of these estimated

[6] Richest countries measured in purchasing-power-parity adjusted GDP per capita for the years 2011 to 2014.

input–output tables will make it possible to use and analyse the ENFOLDing model as a more comprehensive model of global trade, allowing for its continued development as an increasingly versatile and powerful research tool.

References

Andersen MS, Dahl J, Vandenberghe L. (2014). CVXOPT: Python software for convex optimisation. Available at http://cvxopt.org

Hoppe HW. (2006). Optimization theory. Course notes, Department of Mathematics, University of Houston. Available at http://www.math.uh.edu/~rohop/fall_06/

Numpy Developers. (2012). NumPy. Available at http://numpy.scipy.org

Python Software Foundation. (2012). Python Programming Language. Available at http://www.python.org

Ringnér M (2008). What is principal component analysis? *Nature Biotechnology* 26(3): 303–304.

Timmer MP, Dietzenbacher E, Los B, Stehrer R and de Vries GJ. (2015). An illustrated user guide to the world input–output database: the case of global automotive production. *Review of International Economics* 23: 575–605.

United Nations Statistical Division. (2013). UN COMTRADE. Available at http://comtrade.un.org

World Bank. (2015). *World Development Indicators 2014*. Washington, DC: World Bank. doi: 10.1596/978-1-4648-0163-1

Part Three

Dynamics in Account-based Models

6

A Dynamic Global Trade Model With Four Sectors: Food, Natural Resources, Manufactured Goods and Labour

Hannah M. Fry, Alan G. Wilson and Frank T. Smith

6.1 Introduction

In economics and policy, it is important to understand the mechanisms of international trade, both in analysing existing links and in testing the impacts of, for example, changing technologies and network resilience. Ideally, this would be done on the basis of input–output models for each of the 200 or so countries, each of which would include import and export flows by sector and country of origin or destination, respectively. This can be done in theory, for example by using the methods of Wilson (1970) as applied in an interregional context by Rho *et al.* (1989). However, the data sources are to say the least imperfect, and it is an enormous task to assemble what is known in these terms.

An objective of research in these areas, therefore, is to create a model of national economies and trade flows that is feasible in scale and capable of replicating the principle phenomena of the full system. In recent years, several such models have been presented with varying success. Gravity models, for example, which concentrate on the influence of spatial structure have been validated empirically but may not take comparative advantage into account. The alternative Heckscher–Ohlin model of Ohlin (1935), which is based on comparative advantage, is capable of handling country-specific capital and labour markets, but cannot include unemployment or wage discrepancies across countries and demonstrates poor predictive power on an international level (see Trefler and Zhu, 2000).

Approaches to Geo-mathematical Modelling: New Tools for Complexity Science, First Edition. Edited by Alan G. Wilson.

Companion Website: www.wiley.com/go/wilson/ApproachestoGeo-mathematicalModelling

Table 6.1 A summary of the inputs and outputs of the model

	Farming	Mining	Manufacturing
Product	Food	Resources	Goods
Consumed by	Population	Manufacturing	Population
Unit of production constraint	Volume	Volume	Money
Paid for by	Income	Goods prices	Income after food purchases
Cost prices depend on	Labour costs	Labour costs	Labour and natural resources costs

In this chapter, we present a novel approach intended to tackle some of these issues, following the principles established in Wilson (2010). A key feature of the model is the distinction between countries that are rich and poor in terms of Gross Domestic Product (GDP) per capita, and those that are rich and poor in terms of natural resources such as oil or precious metals. This produces a 2×2 classification of countries and allows us to incorporate country-specific dynamics of labour, unemployment, income and production capacity.

We assume that the population needs food, but that food consumption is different for each country, reflecting income. Individuals also consume manufactured goods, and for the purposes of simplification, we assume that the balance of expenditure by the population on food is spent on manufactured goods.

The manufacturing industry, on the other hand, is the sole consumer of natural resources, while the three production sectors (food, natural resources and manufacturing) all require labour inputs. Thus, the four sectors are defined, and the implicit input–output model is very simple. A summary of the system may be seen in Table 6.1.

At the core of the trade model is a set of spatial interaction models for the three production sectors. These are built on entropy maximisation – see, for example, Harris and Wilson (1978) as applied to retail models of consumer spending flows; Willekens (1999), Fotheringham *et al.* (2004) and Stillwell (1978) as models for inter/regional migration flows; and Rho *et al.* (1989) and Wilson (1970) as applied to trade flows. A description may also be found in Wilson (2010), and the model presented here is both an extension and simplification of that work. This standard spatial interaction model, however, requires an adaptation to allow for varying production levels, and we introduce a variable selling price as a mechanism to do this.

At each time step, each sector of each country enters the global marketplace with a production capacity: a volume which they may not sell above, and a cost price which they may not sell below. The global competition then takes the form of an iterative algorithm: increasing the selling price whenever demand exceeds capacity and reducing the production whenever a sector cannot sell at capacity.

At the end of the trading process, the national government reinvests any profits into proportionally increasing the capacity of their best performing sectors, thus adding a dynamic element to the model. A schematic diagram of how the various elements of the model interact is given in Figure 6.1.

To put this work into context, the motivation for this model is as a first step in a global demonstration model, incorporating the coupled dynamics of migration, international aid and security. Thus, the structure is built with future disaggregation in mind, and designed to cope with varying populations and national and international investment.

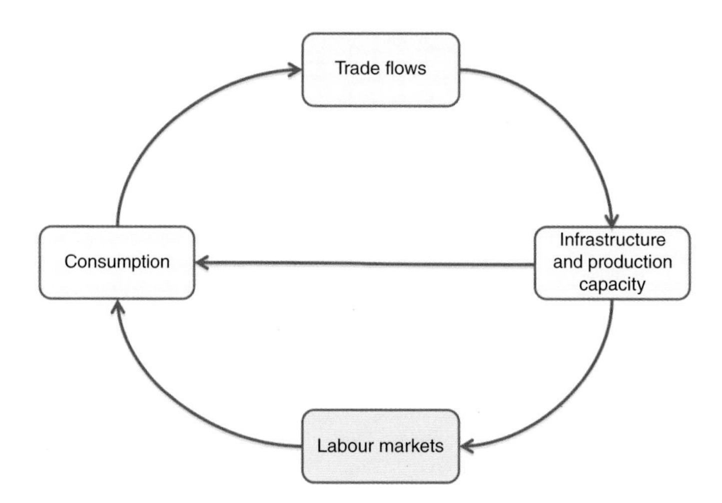

Figure 6.1 A schematic diagram of how various elements of the model interact

In Section 6.2, we define all the variables used to describe the system of interest. The pricing and trade flows algorithm which forms the core of the trade model is presented in Section 6.3. This algorithm requires inputs of production capacity, cost prices, a measure of the quality of goods and transport costs. Theoretically, these measures should be available from data and treated as exogenous. For the purposes of our demonstration model, however, we have created estimates of these quantities which are outlined in Section 6.4. Once these values are known, the pricing and trade flows algorithm, with minor modifications, may be applied to each of the producing sectors. The process follows in detail in Sections 6.5, 6.6 and 6.7. Finally, the dynamics of the model are demonstrated in Section 6.8, and the results of an experimental run are described in Section 6.9.

6.2 Definition of Variables for System Description

The three producing sectors, as we have seen, are food, natural resources and manufacturing. These are labelled by a superscript $n = 1, 2$ and 3, respectively.

The countries are each given a label denoted by subscript i when the country is acting as a seller, and subscript j when acting as a buyer. Thus, at each time step (between t and $t + \delta t$), food, resources and manufactured goods flow from i to j. The variables of interest are in Table 6.2.

Using this notation, the diagonal of the trade flow matrix $Y_{ii}^{(n)}$ will give the volume of n, consumed and produced in i. In this chapter, we refer to this as local consumption.

6.3 The Pricing and Trade Flows Algorithm

Within each country and sector, the dynamics of income, global pricing and production levels are all governed by the trade flows which form the heart of the model. To simplify the explanation of the algorithm, the mechanism of the adapted spatial interaction model which

Table 6.2 Parameters used in base case simulation

$\{P_i\}$	Population of country i
$\{u_i\}$	Number of unemployed in i
$\{\iota_i\}$	Income per capita of country i
$\{L_i^{(n)}\}$	Work force of sector n in i (so that $\sum_n L_i^{(n)} + u_i = P_i$)
$\{\bar{X}_i^{(n)}\}$	Capacity of production of n in i
$\{X_i^{(n)}\}$	Actual production of n in i over δt
$\{\chi_i^{(n)}\}$	Monetary value of $X_i^{(n)}$
$\{Y_{ij}^{(n)}\}$	Trade flows of n between i and j
$\{\psi_{ji}^{(n)}\}$	Money flows from sales of n between j and i/Monetary value of $Y_{ij}^{(n)}$
$\{Z_i^{(n)}\}$	Consumption of n in i
$\{\zeta_i^{(n)}\}$	Monetary value of $Z_i^{(n)}$
$\{q_i^{(n)}\}$	Relative quality of product n in i
$\{v_i\}$	Volume of food required per person in i
$\{m_i\}$	Raw material cost per unit of goods production in i
$\{\widehat{\phi}_i^{(n)}\}$	Cost price per unit of n in i during δt
$\{\phi_i^{(n)}\}$	Sale price per unit of n in i during δt
$\{D_{ij}\}$	Distance between i and j
$\{\theta^{(n)}\}$	Transport costs of n per unit volume per unit distance
$\{C_i^{(n)}\}$	Total cost of maintaining capacity of n in i
$\{\Pi_i^{(n)}\}$	Profit (or loss) made by sector n in i over δt
Δ_i	National debt (if population cannot afford to feed itself)
S_i	Total surplus in δt

derives these trade flows is described in a separate section. The principles outlined here are identical for all sectors, so, for ease of notation, the superscripts (n) are temporarily dropped.

Assume, for the purpose of this section at least, that consumption Z_j, production capacity $\bar{X}_i$, product quality and transport costs θD_{ij} of each country are known.

If the price ϕ_i were also known, the flow of goods Y_{ij} between seller i and buyer j could be calculated directly and without iteration from a singly constrained spatial interaction model:

$$Y_{ij} = \frac{Z_j q_i^{\alpha} \exp[-\beta(\theta D_{ij} + \phi_i)]}{\sum_k q_k^{\alpha} \exp[-\beta(\theta D_{kj} + \phi_k)]}. \tag{6.1}$$

Using equation (6.1), the sum over all buyers would then yield the total demand in i:

$$X_i = \sum_j Y_{ij}. \tag{6.2}$$

In some cases, this demand X_i will exceed the production capacity $\bar{X}_i$ of i. Where this occurs, an alteration to the trade flows is needed. Using the usual theory of supply and demand, a variable price ϕ_i is introduced as a mechanism to deal with this adjustment. The aim is to increase the unit price ϕ_i of a product in any country where demand exceeds supply until the quantity demanded by consumers is balanced by the production capacity: $X_i = \bar{X}_i$. All other countries may not increase their prices and will sell at, or below, capacity. Assume also, for the time being, that a cost price, or price which a country cannot sell below, $\widehat{\phi}_i$, is known. For any country, then, either the price is known (as $\widehat{\phi}_i$) and the production must be found (wherever $X_i < \bar{X}_i$), or the production is known as $X_i = \bar{X}_i$, and the price, $\phi_i > \widehat{\phi}_i$, must be found.

The algorithm to determine the trade flows will proceed as follows:

1. The spatial interaction model is run once using cost prices $\widehat{\phi}_i$ to determine the demand on each country X_i.

$$X_i = \sum_j \frac{Z_j q_i^\alpha \exp[-\beta(\theta D_{ij} + \widehat{\phi}_i)]}{\sum_k q_k^\alpha \exp[-\beta(\theta D_{kj} + \widehat{\phi}_k)]}. \tag{6.3}$$

2. In any country where demand exceeds supply, the optimal prices – such that demand equals capacity – are calculated separately by rearranging equation (6.1) and taking $X_i = \bar{X}_i$. More formally:

$$\forall m \;\; \text{s.t.} X_m > \bar{X}_m, \tag{6.4}$$

$$\phi_m = \frac{1}{\beta} \ln\left[\frac{1}{\bar{X}_m}\right] + \frac{1}{\beta} \ln\left[\frac{Z_j q_i^\alpha \exp\left[-\beta\theta D_{mj}\right]}{\sum_k q_k^\alpha \exp[-\beta(\theta D_{kj} + \phi_k)]}\right] \tag{6.5}$$

which is solved iteratively since ϕ_m appears in the denominator of the last term.

3. The new prices contribute to the latest pricing vector ϕ_i, on which the spatial interaction model is re-run:

$$X_i = \sum_j \frac{Z_j q_i^\alpha \exp[-\beta(\theta D_{ij} + \phi_i)]}{\sum_k q_k^\alpha \exp[-\beta(\theta D_{kj} + \phi_k)]}. \tag{6.6}$$

4. The new prices have the ability to push the demand in other countries over the capacity of production. If this is the case, the process is repeated until a global equilibrium is established and all production levels and prices are known.

This algorithm relies on knowing consumption Z_j, production capacity $\bar{X}_i$, product quality q_i, transport costs θD_{ij} and cost price $\widehat{\phi}_i$. Once complete, it returns production X_i, sale price ϕ_i and the trade flows Y_{ij}.

To obtain a complete description of trade evolution, these inputs and a mechanism for dynamics based on the outputs must be determined for each sector. We present this in Sections 6.5, 6.6 and 6.7 for the farming, natural resources and manufacturing sectors, respectively. However, first we briefly discuss some estimates of the exogenous variables required within the simulation.

6.4 Initial Setup

To apply the trade flows algorithm, we must determine the capacity, cost prices, income, quality and transport costs for each country. Theoretically, these could all be found from data sources, particularly if the the model was disaggregated to apply to individual products, such as oil, wheat and so on. In the present setting, however, our main interest lies in demonstrating the workings of the system, rather than in obtaining realistic results, and so we proceed by outlining some simple functions to approximate the above variables.

Working through the variables in the order listed above, we take one unit of food, raw materials or goods as the amount one worker can produce or extract in a given time period. Using this formulation, the production capacity of a given sector $\bar{X}_i^{(n)}$ will be equivalent to the labour force of that sector $L_i^{(n)}$. Knowing the population and the number of unemployed u_i from data, an initial guess of the workforce of each sector in each country may be made, providing an initial $\bar{X}_i^{(n)}$. Cost prices are sector specific but may be found from the production capacity using a simple expression. The exact formulations are outlined for each sector within the relevant sections.

The demonstration model does incorporate two readily accessible data sources: Gross Domestic Product (GDP) and distances between the countries of consideration calculated between centroids using the haversine formula. From these, we may approximate the remaining variables. Assuming that only the workers draw a salary and that income is uniform across the population, an initial mean per-capita income may be found:

$$\iota_i = \frac{\text{GDP}_i}{P_i - u_i}. \tag{6.7}$$

This value (6.7) will change during the simulation, as countries with a strong economy will invest in improving income. The method applied to facilitate this is within the dynamics section; see equation (6.57). It is worth noting at this stage that the simple expression (6.7) could easily be replaced with an income distribution which would not effect the workings of the algorithm. It is for simplicity in this case that this particular form has been chosen.

A measure of product quality within a country could also be derived from the relative income. We again propose a simple expression which normalises (6.7) by lowest income of the countries included:

$$q_i = \frac{\iota_i}{\max(\iota_k)}. \tag{6.8}$$

With this formulation, k is used as a dummy variable for i and $0 < q_i \leq 1$ for all i. Thus, the richest country included within the model has $q = 1$.

We leave the transport costs, $\theta^{(n)}$, as a parameter to be calibrated within the modelling process. Generally speaking, however, to ensure the argument of the exponential in equation (6.1) is dimensionless and $O(1)$, we would expect $\theta^{(n)}$ to roughly take the form:

$$\theta^{(n)} \sim \frac{1}{\beta \max(D_{ij})} \tag{6.9}$$

As mentioned in this chapter, the distances D_{ij} between each country may be found from data, while β is another parameter in equation (6.1) to be determined. Physically, β quantifies the relative importance of price, rather than quality, to the buyer.

These variable definitions provide enough information to proceed to the individual sector algorithms and flows, although one final adjustments is needed for the numerics. With the current definition, the pricing and trade flows algorithm shown in equations (6.3) through (6.6) has no mechanism to stop global production capacity from dipping below global demand. On occasions where this occurs, demand must be scaled back. To do so, we propose the additional assumption that in cases of global shortage, it is the poorest countries which see the largest reduction in consumption. Reduction in global demand is found by scaling according to income. To put this more formally, whenever:

$$\sum_j Z_j^{(n)} > \sum_i \bar{X}_i^{(n)}, \tag{6.10}$$

the reduction in consumption $Z_j^{(n)}$ in country j will be relative to:

$$r_j = 1 - \frac{\iota_j}{\max(\iota_k)}. \tag{6.11}$$

This sets $r = 0$ for the richest country under consideration, and allows the new consumption vector to be found from the original consumption minus the relative reduction, times the total reduction:

$$Z_j^{(n)} = Z_j^{(n)} - \frac{r_j}{\sum_k r_k}\left(\sum_l Z_l^{(n)} - \sum_i \bar{X}_i\right), \tag{6.12}$$

It now remains to detail the full structure of the system for each individual sector. In the interests of clarity, we do so separately for each of the three sectors, in Sections 6.5, 6.6 and 6.7.

6.5 The Algorithm to Determine Farming Trade Flows

The method to determine the farming trade flows is largely based around the pricing and trade flows algorithm of Section 6.3. First, we present some estimates for the sector-specific variables of consumption $Z_j^{(1)}$ and cost price $\widehat{\phi}_i^{(1)}$. These combined with $\bar{X}_i^{(1)}$ and $\theta^{(1)}$ discussed in Section 6.4, may be passed to the trade flows algorithm shown in equations (6.3) through (6.6) to provide values for the actual production $X_i^{(1)}$ (bounded above by $\bar{X}_i^{(1)}$), sale price $\phi_i^{(1)}$ (bounded below by $\widehat{\phi}_i^{(1)}$) and trade flows matrix Y_{ij}. To proceed with food consumption in country j, we define a parameter v_j to describe the volume consumed per capita, per unit income. This sets the total consumption in j as:

$$Z_j^{(1)} = v_j \iota_j P_j. \tag{6.13}$$

Although v_j is country specific, taking it as a constant across all models, as we do in our experimental run, implies that richer countries will consume more food than their poorer counterparts. The cost price per unit volume of food produced in i is calculated from the total labour costs of farming infrastructure $\iota_i L_i^{(1)}$, divided by the expected production output or sales:

$$\widehat{\phi}_i^{(1)} = \frac{\iota_i L_i^{(1)}}{X_i^{(1)}(t - \delta t)}. \tag{6.14}$$

We base the expected production output on the actual output of the previous time period. This removes the need for an extra iteration, and seems reasonable when the time periods under consideration are small. Indeed, $X_i^{(1)}(t) = X_i^{(1)}(t - \delta t)$ in the limit as $\delta t \to 0$.

As discussed, the current formulation defines one unit of food production as the amount which one worker can produce in one time period. Thus, it follows that the production capacity of the farming sector is equal to the labour force of the sector.

$$\bar{X}_i^{(1)} = L_i^{(1)}. \tag{6.15}$$

It is this sector-specific quantity, production capacity, which determines whether the country is resource rich or poor and allows for our 2×2 classification of countries. Equation (6.15) has the direct result that equation (6.16) may be rewritten

$$\widehat{\phi}_i^{(1)} = \frac{\iota_i \bar{X}_i^{(1)}}{X_i^{(1)}(t - \delta t)}. \tag{6.16}$$

Thus, cost price is income scaled by the ratio of the sales of the previous time period, to the capacity at the present time. Given this form for equation (6.16), it is necessary to define a minimum expected production value. Otherwise, a country with little or no sales in the previous time period will have a very large or infinite cost price, making sector growth extremely difficult.

The now known values of $Z_j^{(1)}$, $\bar{X}_i^{(1)}$, $\widehat{\phi}_i^{(1)}$ and $\theta^{(1)}D_{ij}$ are fed into the pricing algorithm of Section 6.3, providing results for $X_i^{(1)}$, $\phi_i^{(1)}$ and $Y_{ij}^{(1)}$:

$$Z_j^{(1)}, \bar{X}_i^{(1)}, \widehat{\phi}_i^{(1)}, \theta^{(1)}D_{ij} \rightarrow X_i^{(1)}, \phi_i^{(1)} Y_{ij}^{(1)} \quad \text{via equations (6.3)–(6.6).} \tag{6.17}$$

With a set of farming flows $Y_{ij}^{(1)}$ found, the price which the buyer j pays, including transport costs for imports, is given by:

$$D_{ij}\theta^{(1)} + \phi_i^{(1)}, \tag{6.18}$$

and the total spent by the population on food may be determined from the per unit price of each import, times the volume imported and summed over all countries (including one's own):

$$\zeta_j^{(1)} = \sum_i (\theta^{(1)} D_{ij} + \phi_i^{(1)}) Y_{ij}^{(1)}. \tag{6.19}$$

As discussed in Section 6.1, any money left over will be spent on goods. All remaining expenditure is used, regardless of the volumes of goods the money buys. Thus, we may define each country's goods consumption for the manufacturing sector in units of money:

$$\zeta_j^{(3)} = \text{GDP}_j - \zeta_j^{(1)}. \tag{6.20}$$

There will be some $Z_j^{(3)}$ consumption term associated with equation (6.20) in volume units, although, as may be seen in Section 6.7, this does not feature explicitly in the problem.

If the population does not have enough money to feed itself, we want to avoid the computationally expensive additional iterative procedure of reducing consumption and repeating the pricing and trade flow calculations. To do so, we introduce national debt, which acts to keep track of any deficit:

$$\Delta_i = -\frac{\zeta_i^{(3)}(\text{sgn}(\zeta_i^{(3)}) - 1)}{2}. \tag{6.21}$$

This structure is chosen as it is zero while $\zeta_i^{(3)}$ is positive, and equal to $-\zeta_i^{(3)}$ if $\zeta_i^{(3)}$ is negative. National debt is paid for by the profits of the three producing sectors; details of how this mechanism is applied are in Section 6.8.

6.5.1 The Accounts for the Farming Industry

With all flows and sales prices determined, it is now possible to determine all costs, takings and profits for the farming industry as follows.

Costs of production (only labour costs within this sector) are given by

$$C_i^{(1)} = \iota_i \bar{X}_i^{(1)}. \tag{6.22}$$

Money taken/sales (sale price per unit times number of units):

$$\chi_i^{(1)} = \phi_i^{(1)} X_i^{(1)}. \tag{6.23}$$

Profit from farming sector (sales−cost) is

$$\Pi_i^{(1)} = \phi_i^{(1)} X_i^{(1)} - \iota_i \bar{X}_i^{(1)} \tag{6.24}$$

6.5.2 A Final Point on The Farming Flows

It is worth noting that the total sales $\chi_i^{(1)}$ given in equation (6.23) will take one of two less general forms due to a subtlety in the pricing and trade flows algorithm of Section 6.3. The mechanism of that section relies on two types of seller. In the first type, country i will sell below capacity, $X_i^{(1)} < \bar{X}_i^{(1)}$, at cost price $\phi_i^{(1)} = \hat{\phi}_i^{(1)}$. In this case, given equation (6.16), the total sales of i (6.23) become:

$$\chi_i^{(1)} = \iota_i \bar{X}_i^{(1)} \frac{X_i^{(1)}}{X_i^{(1)}(t - \delta t)} \tag{6.25}$$

In the second type, i will sell at capacity $X_i^{(1)} = \bar{X}_i^{(1)}$, at a higher price than cost, $\phi_i^{(1)} > \hat{\phi}_i^{(1)}$. In this case, equation (6.23) becomes:

$$\chi_i^{(1)} = \phi_i^{(1)} \hat{X}_i^{(1)}. \tag{6.26}$$

This implies two corresponding forms for profit, originally defined in equation (6.24). The profits of the first type, like equation (6.25), will take the form

$$\Pi_i^{(1)} = \iota_i \bar{X}_i^{(1)} \left(\frac{X_i^{(1)}}{X_i^{(1)}(t - \delta t)} - 1 \right), \tag{6.27}$$

and are positive if sales exceed those of last year, but negative otherwise. Meanwhile, the second type uses equation (6.26) to make equation (6.24):

$$\Pi_i^{(1)} = \hat{X}_i^{(1)} (\phi_i^{(1)} - \iota_i), \tag{6.28}$$

and profits are built from actual sales prices above the national income (theoretically equivalent in this case to the actual per-unit cost price as in equation (6.22). In both cases, equations (6.27) and (6.28) are linearly related to the size of the current infrastructure. It is this feature which gives the dynamics a logistic, or Lotka–Volterra form, as discussed in Section 6.8.

6.6 The Algorithm to Determine The Natural Resources Trade Flows

The method to determine the natural resources flows largely mirrors the algorithm applied in Section 6.5 to farming flows. For the sake of clarity, we include a brief description of the full process here. To begin, we must derive expressions for the consumption and cost prices. Since the sole customer of the natural resources sector is the manufacturing sector, the raw materials consumption in country j is linked to manufacturing production capacity. We introduce m_j so that the consumption is equal to the volume required per unit production times manufacturing production capacity.

$$Z_j^{(2)} = m_j \bar{X}_j^{(3)} \tag{6.29}$$

The per-unit cost price of raw materials in i, in units of money, is the total labour cost of maintaining the mining infrastructure, divided by the expected production output (based on last year's actual production).

$$\phi_i^{(2)} = \frac{\iota_i \bar{X}_i^{(2)}}{X_i^{(2)}(t - \delta t)} \tag{6.30}$$

Again, as in the farming flows, a minimum expected production value is found to be necessary to avoid a large cost price in small sectors. The trade flows are calculated in an identical procedure as before: $Z_j^{(2)}$, $\bar{X}_i^{(2)}$, $\hat{\phi}_i^{(2)}$, $\theta^{(2)} D_{ij}$, are fed into the spatial interaction model outlined in Section 6.3, providing results for $X_i^{(2)}$, $\phi_i^{(2)}$ and $Y_{ij}^{(2)}$.

$$Z_j^{(2)}, \bar{X}_i^{(2)}, \hat{\phi}_i^{(2)}, \theta^{(2)} D_{ij} \rightarrow X_i^{(2)}, \phi_i^{(2)} Y_{ij}^{(2)} \text{ via equations (6.3)–(6.6).} \tag{6.31}$$

Once the flows and prices have been determined, the total per-unit price to the buyer, including transport costs, is:

$$D_{ij}\theta^{(2)} + \phi_i^{(2)}, \tag{6.32}$$

which gives the money spent on raw materials by the manufacturing industry in j as the price per unit for each import times the volume imported, summed over all countries including one's own:

$$\varsigma_j^{(2)} = \sum_i (D_{ij}\theta^{(2)} + \phi_i^{(2)}) Y_{ij}^{(2)}. \tag{6.33}$$

The total spend on raw materials, given by equation (6.33), will feature in the manufactured goods pricing of Section 6.7.

6.6.1 The Accounts for The Natural Resources Sector

Again, this follows the workings of the farming sector, so that costs of production are determined only by labour costs:

$$C_i^{(2)} = \iota_i \bar{X}_i^{(2)}. \tag{6.34}$$

Money taken/sales is sale price per unit times number of units:

$$\chi_i^{(2)} = \phi_i^{(2)} X_i^{(2)}. \tag{6.35}$$

Profit from mining sector (sales−cost) is

$$\Pi_i^{(2)} = \phi_i^{(2)} X_i^{(2)} - \iota_i \bar{X}_i^{(2)}. \tag{6.36}$$

As in equations (6.27) and (6.28), this profit will take one of two forms, both linearly related to $\bar{X}_i^{(2)}$.

6.7 The Algorithm to Determine Manufacturing Trade Flows

Goods consumption in country j is calculated from the per capita income minus food purchases. We assume that people spend all their disposable income on goods, regardless of what volume of goods this buys them. The result, in units of money, is given in equation (6.20) and repeated here for clarity:

$$\zeta_j^{(3)} = \text{GDP}_j - \zeta_j^{(1)}. \tag{6.37}$$

In the situation where a country cannot afford to feed themselves, there will be no remaining income to spend on manufactured goods. In such a case, the volume of food required is bought regardless, and a national debt term is introduced (see equation (6.21)). To allow for this scenario, we rewrite the amount of money each country has to spend on goods as:

$$\zeta_j^{(3)} = (\text{GDP}_j - \zeta_j^{(1)})(1 - \text{sgn}(\Delta_i)), \tag{6.38}$$

where, by definition, debt Δ_i may be only zero or positive.

Within the manufacturing sector, the consumption (6.38) is expressed in terms of money, so that volumes do not feature explicitly in the problem. Thus, instead of applying the pricing and trade flows algorithm to solve the system (equations (6.3)–(6.6)) as before, an adapted version must be applied to take into account this special case of consumption in units of money. A per-unit cost price is still required, however, where the form differs from that seen in farming (6.16) and mining (6.30) to allow for the natural resources costs to maintain the manufacturing infrastructure. This cost was found in the mining flows in equation (6.33). Thus, the per-unit cost price of goods in i, in money units, is:

$$\widehat{\phi}_i^{(3)} = \frac{\iota_i \bar{X}_i^{(3)} + \zeta_i^{(2)}}{X_i^{(3)}(t - \delta t)}. \tag{6.39}$$

The adapted pricing and trade flows algorithm determines an initial demand on sales, by applying the standard spatial interaction model:

$$\chi_i^{(3)} = \sum_j \frac{\zeta_j^{(3)} q_i^{\alpha} \exp[-\beta(\theta^{(3)} D_{ij} + \widehat{\phi}_i^{(3)})]}{\sum_k q_i^{\alpha} \exp[-\beta(\theta^{(3)} D_{ik} + \widehat{\phi}_k^{(3)})]}. \tag{6.40}$$

Any countries where demand exceeds capacity are entitled to raise their sales prices. This price adjustment takes the form:

$$\forall m \text{ s.t } \chi_m^{(3)} > \widehat{\phi}_m^{(3)} \bar{X}_m^{(3)} \tag{6.41}$$

$$\phi_m^{(3)} = \frac{1}{\beta}\ln\left[\frac{1}{\phi_m^{(3)}\bar{X}_i^{(3)}}\right] + \frac{1}{\beta}\ln\left[\sum_j \frac{\zeta_j^{(3)} q_i^{\alpha}\exp[-\beta\theta^{(3)}D_{ij}]}{\sum_k q_i^{\alpha}\exp[-\beta(\theta^{(3)}D_{ik} + \widehat{\phi}_k^{(3)})]}\right], \tag{6.42}$$

which is solved iteratively as $\phi_m^{(3)}$ appears on the right-hand side of equation (6.42).

These adjusted prices form part of the latest pricing vector $\{\phi_i^{(3)}\}$, on which the spatial interaction model is re-run:

$$\chi_i^{(3)} = \sum_j \frac{\zeta_j^{(3)} q_i^{\alpha}\exp[-\beta(\theta^{(3)}D_{ij} + \phi_i^{(3)})]}{\sum_k q_i^{\alpha}\exp[-\beta(\theta^{(3)}D_{ik} + \phi_k^{(3)})]}. \tag{6.43}$$

The new prices $\{\phi_i^{(3)}\}$ have the ability to push other countries over production capacity thus, the process is repeated until a global equilibrium is established and all prices are known. Finally, the flow in money units from j to i is given from:

$$\psi_{ji}^{(3)} = \frac{\zeta_j^{(3)} q_i^{\alpha}\exp[-\beta(\theta^{(3)}D_{ij} + \phi_i^{(3)})]}{\sum_k q_i^{\alpha}\exp[-\beta(\theta^{(3)}D_{ik} + \phi_k^{(3)})]}, \tag{6.44}$$

and actual production may be found from:

$$X_i^{(3)} = \frac{\chi_i^{(3)}}{\phi_i^{(3)}}. \tag{6.45}$$

Despite being a money flow, the price term and quality term allow the flows to differentiate between large-volume, cheap, low-quality exports and expensive, low-volume, high-quality exports – the buyer is always driven towards the best deal. What is meant by 'best', of course, is dictated by the calibration of the Lagrangian multipliers α and β.

6.7.1 *The Accounts for The Manufacturing Industry*

Costs of production (within this sector, costs include both labour and raw materials):

$$C_i^{(3)} = \iota_i \bar{X}_i^{(3)} + \zeta_i^{(2)}. \tag{6.46}$$

Money taken/sales (here, of course, $\chi_i^{(3)}$ is known from the flows, but the following should balance nonetheless):

$$\chi_i^{(3)} = \phi_i^{(3)} X_i^{(3)}. \tag{6.47}$$

Profit from manufacturing sector (sales−cost):

$$\Pi_i^{(3)} = \phi_i^{(3)} X_i^{(3)} - \iota_i \bar{X}_i^{(3)} - \zeta_i^{(2)}. \tag{6.48}$$

The last term here is not quite linearly related to $\bar{X}_i^{(3)}$ and the profits of the manufacturing sector do not take the same form as those seen previously.

6.8 The Dynamics

The profit of all sectors is now known, and it remains to reinvest any profits to increase the capacity of each nation's best performing sectors. Before we do so, however, any national debt accrued must be accounted for, as other countries have already been paid for the food which the population of i bought. The total surplus of all industry in i is the profit from each sector (which could be negative) plus any national debt:

$$S_i = \Pi_i^{(1)} + \Pi_i^{(2)} + \Pi_i^{(3)} - \Delta_i. \tag{6.49}$$

Although we use the term *surplus*, S_i could be negative. As all surplus is reinvested, it would be possible at this stage to introduce national or international investment, and the main structure of the dynamics would remain unchanged. For the purposes of our demonstration model, we neglect this additional variable and leave surplus as defined in equation (6.49).

Assume that each country, and all sectors within it, is governed by one central body which reinvests in a manner best for everyone by splitting the surplus between sectors according to the performance in the latest time step. This reinvestment takes the form of both additional employment and wage increase. Such assumptions could well be adjusted at a later date, to include shareholders or allow for firms, for example. However, this simplifying assumption works well for the purposes of our demonstration model.

With static populations, a country may not employ more people than the current unemployed and cannot fire more people than the employed. So, the extra number of people who may be employed (or fired) is given by:

$$E_i = \frac{1}{2}(1 + \mathrm{sgn}(S_i))\min\left[\frac{\varepsilon_1 S_i}{l_i}, u_i\right] - \frac{1}{2}(1 - \mathrm{sgn}(S_i))\min\left[\frac{-\varepsilon_2 S_i}{l_i}, P_i - u_i\right], \tag{6.50}$$

where ε_1 and ε_2 are parameters to be calibrated.

The algorithm to assign this change in labour force to each sector is based on the sector's contribution to the overall profits:

$$\delta L_i^{(n)} = \delta t \frac{\Pi_i^{(n)}}{\sum_k \Pi_i^{(k)}} E_i, \tag{6.51}$$

although naturally this must be adjusted if it yields any negative results since a country cannot have a negative workforce. This translates directly to a change in production capacity in the current model, since one unit of volume is equivalent to the amount one worker can produce in a time period. Thus:

$$\delta \bar{X}_i^{(n)} = \delta L_i^{(n)}. \tag{6.52}$$

Note that the extra employment E_i in equation (6.50) takes the same sign as the country's overall profits $\sum_k \Pi_i^{(k)}$. If both are positive (i.e. if the country is in profit overall), any sector with losses $\Pi_i^{(n)}$ will still see a shrinking workforce, and vice versa. Even if a country does badly, the mechanism (6.51) allows growth in a thriving sector.

In addition, given the two forms of profits discussed in Section 6.5, equations (6.27) and (6.28), the dynamics of production capacity (eqution (6.51)) will take one of two forms within the farming sector. These will be:

$$\delta \bar{X}_i^{(1)} = \delta t \iota_i \bar{X}_i^{(1)} \left(\frac{X_i^{(1)}}{X_i^{(1)}(t-\delta t)} - 1 \right) \frac{E_i}{\sum_k \Pi_i^{(k)}} \tag{6.53}$$

$$\delta \bar{X}_i^{(1)} = \delta t \bar{X}_i^{(1)} (\phi_i^{(1)} - \iota_i) \frac{E_i}{\sum_k \Pi_i^{(k)}} \tag{6.54}$$

to correspond with equations (6.27) and (6.28), respectively. Both of these, equations (6.53) and (6.54), are therefore in Lotka–Volterra form. This is also true for the natural resources sector, as equations (6.27) and (6.28) apply when $n = 2$. For manufacturing, however, the change in unit for the constraint on production capacity, and the interaction with the farming sector, leads to a slightly different form. Specifically, combining equations (6.39), (6.48) and (6.51) gives the two possible forms as:

$$\delta \bar{X}_i^{(3)} = \delta t \frac{E_i}{\sum_k \Pi_i^{(k)}} \left\{ \iota_i \bar{X}_i^{(3)} \left(\frac{X_i^{(3)}}{X_i^{(3)}(t-\delta t)} - 1 \right) + \frac{\zeta_i^{(2)}}{X_i^{(3)}(t-\delta t)} \right\} \tag{6.55}$$

$$\delta \bar{X}_i^{(3)} = \delta t \frac{E_i}{\sum_k \Pi_i^{(k)}} \{ \bar{X}_i^{(3)} (\phi_i^{(3)} - \iota_i) - \zeta_i^{(2)} \} \tag{6.56}$$

These expressions may be considered as ‘forced’ Lotka–Volterra.

The national income also has a dynamic element which in turn affects both consumption and quality of products:

$$\delta \iota_i = \delta t \varepsilon_3 S_i \tag{6.57}$$

where ε_3 is a final parameter to be calibrated. Finally, latest GDP may be found as the sum of all incomes

$$GDP_i = \iota_i \sum_n L_i^{(n)} \tag{6.58}$$

and relative quality q_i may then be recalculated. The entire process may be repeated for subsequent time steps.

6.9 Experimental Results

The model generates an interesting balance of factors between countries. A poor country has the benefit of low cost of production due to lower wages, but (with the current assumptions on quality (6.8)) lower quality goods. The relative importance of these two factors to the buyer is determined by the weighting parameters α and β. In line with the derivation of the spatial interaction model of equation (6.1), this benefit to a potential buyer may be quantified as follows:

$$\text{Benefit} = \exp(\alpha \ln q_j - \beta \phi_j - \beta \theta D_{ij}). \tag{6.59}$$

Exploring the relationship between price and quality further, some plots are presented in Figure 6.2 of total benefit for a range of α values, at a fixed β. In the plots, we have taken $\theta = 0$, so that distance is deemed unimportant to the buyer. In reality, these curves would

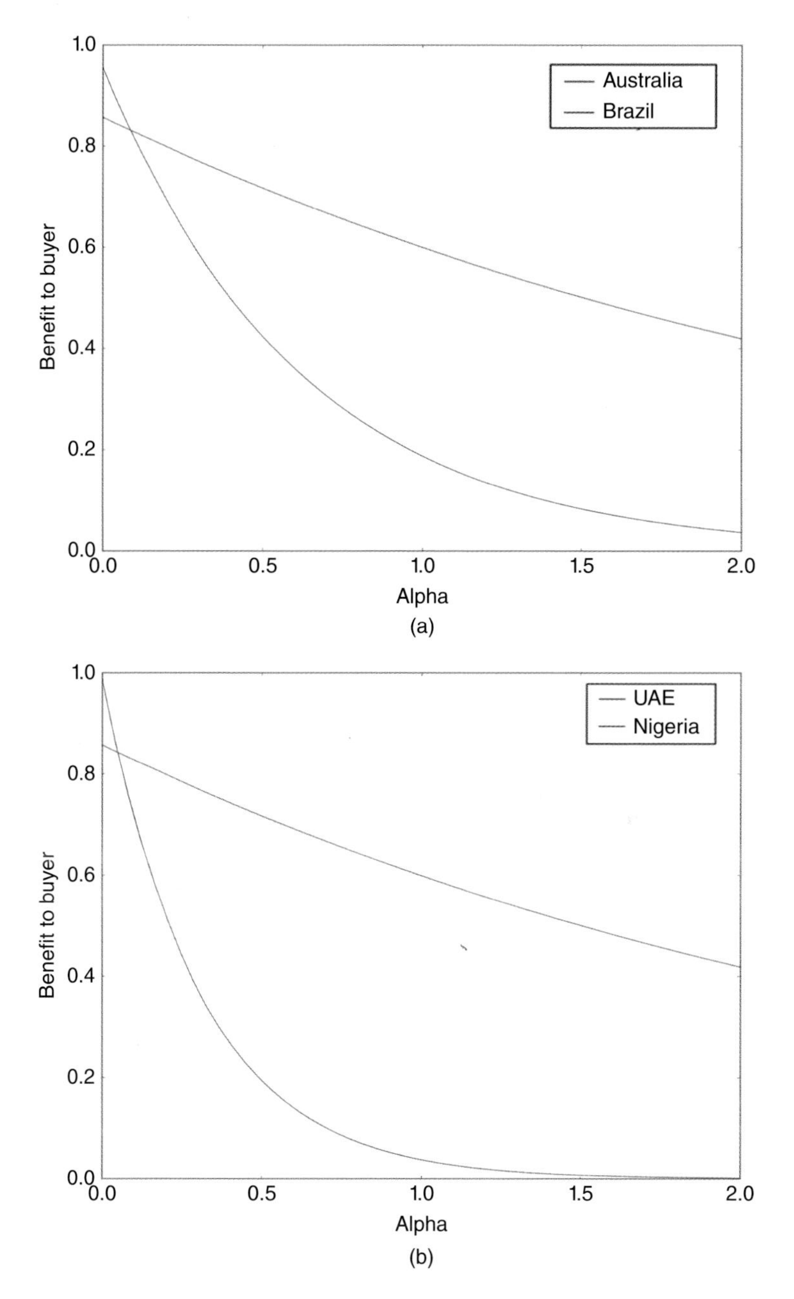

Figure 6.2 A quantitative measure of benefit to a potential buyer against the parameter α. See equation (6.59)

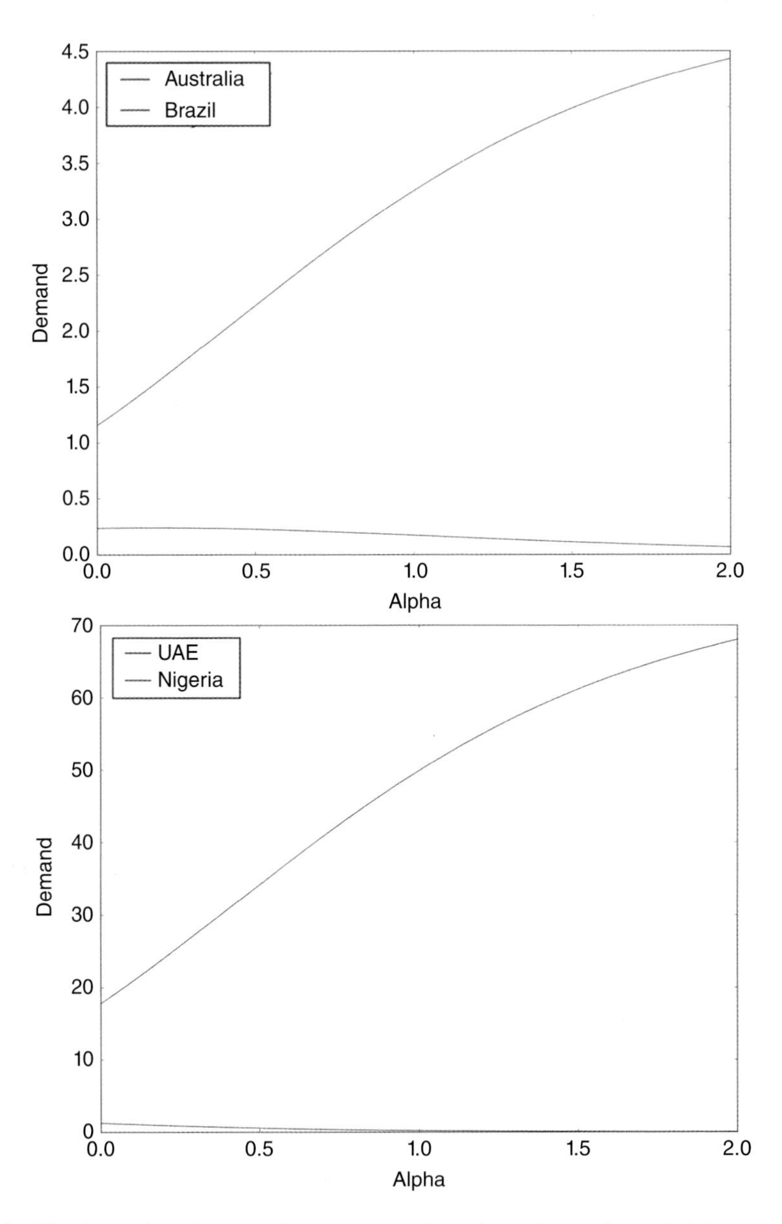

Figure 6.3 The demand on the natural resources products for various values of alpha, normalised by production capacity. In these plots, $\theta = 25$, $\beta = 5 \times 10^{-6}$

be distorted by θ, and different for each potential buyer. The examples chosen illustrate that there is a value of α below which cheaper items are valued, and above which quality becomes more important. The countries selected form two pairs: Brazil and Australia, which are both big exporters of coal, timber and iron ore; and Nigeria and UAE, which both export petroleum products. The mean income of UAE is 18 times that of Nigeria, while Australian citizens earn an average of 3.55 times more than Brazilians. Thus, the four demonstrate examples of GDP in rich (Australia, UAE) and poor countries (Brazil, Nigeria). In contrast, the production capacity (or, equivalently, workforce) of Brazil is five times that of Australia: resource-rich and resource-poor countries, respectively. By our measure of relative quality given in equation (6.8), Australia has $q = 0.7$, Brazil $q = 0.20$, UAE $q = 0.7$ and Nigeria $q = 0.03$. Of course, a large variance in the quality of oil such as these is unrealistic, although in the context of our toy model the results serve the purpose of demonstrating the interesting potential bifurcations which the algorithm generates. In all plots, benefit tends to zero with large α, although this happens much faster for low-quality products, such as those from Nigeria and Brazil.

Benefit is not the only factor in the trade model. Aside from distance between buyer and seller and cost to transport goods, consumption – both local and global – plays a key role in contributing to the demand on a nation's products. In Figure 6.3, we present a plot of the demand (as a ratio of capacity) on the natural resources products of the same four countries listed above, when offering goods at cost price. This is equivalent to the sum over j of the trade flows at the very first step in the pricing and trade flows algorithm – equation (6.1). In this example, $\theta = 25$, and so distance is taken into account in the buying process. The plots again demonstrate that high values of α correspond to quality being favoured, with the products of UAE and Australia increasing in demand. Given that quality and cost price of UAE and Australia are similar, the bigger demand on UAE is probably due to the large distances between Australia and its potential consumers. The relative demand of Nigeria and Brazil is at a maximum when $\alpha = 0$, and price dominates the decision process. The demand in both cases, however, tends to zero with increasing α.

If demand exceeds supply, the countries may raise their sale price until production equals capacity. Our final plot of this type then is given in Figure 6.4 and shows the ratio of sale price to cost price for the same range of α and values of β, θ. As demand on Brazil never exceeds capacity in this simulation, they continue to sell at cost price, regardless of the value of α. This is not the case for Nigeria, where other countries are prepared to pay up to 140 times the cost price for low values of α. As the importance of quality along with α increase, demand in Nigeria drops to below capacity, and they must sell at cost price. For UAE and Australia, however, the relative sale price, like demand, increases with α.

To empirically validate this model, its pricing algorithm, sector accounts and dynamicsm these parameters α, β and θ will have to be calibrated against data. In practice, the dynamics will be determined by a number of adjustments; the time path will be very much a function of the initial conditions, and variables at each stage. To explore the system, along with model refinements and extensions, is our next task, but to demonstrate the goal of the demonstration model we include a map of simulated flows for the farming sector in Figure 6.5. The key is given within the caption of that figure. As one would expect, countries with higher GDP tend to produce a smaller percentage of their own food, and consume more per capita. In addition, trade links tend to form over shorter distances which matches with the results of other models, for example the gravity model. In general, then, a great deal of work needs to be done to explore the feasible solutions; however, this first-stage simulation looks to be very promising.

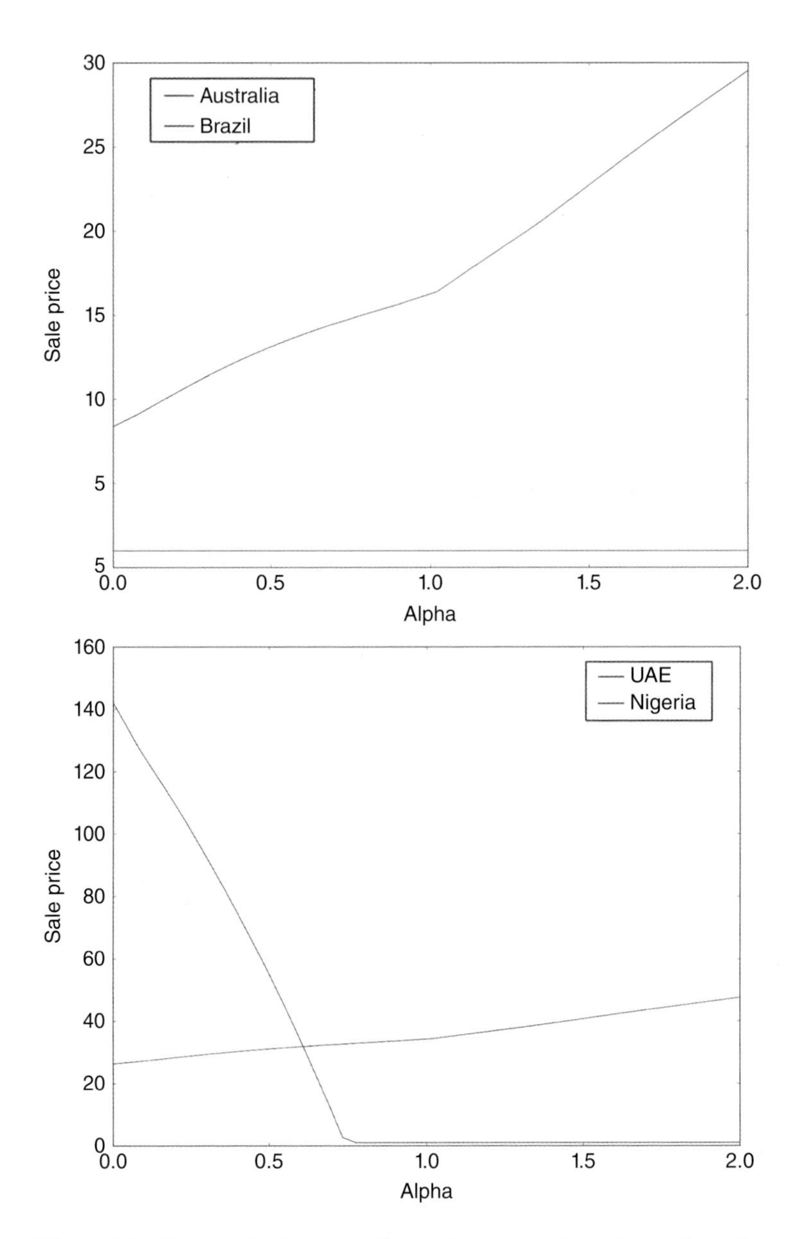

Figure 6.4 The actual sales price of natural resources for various values of α

6.9.1 *Concluding Comments*

We believe that the model of trade flows presented here offers a good alternative to existing models of trade, not least for the various extensions and adaptations which the method can handle. Perhaps most importantly, the formulation allows us to pose the dynamic programming question: if we add, say, World Bank investment by country, and an objective is formulated to

Figure 6.5 Some first results of simulation for the food flows. Here, the node size is the consumption per person, edge width is the size of the inter country flows and node colour is the percentage of consumption met locally (see scale)

increase GDP per capita in a number of poor countries, what is the optimal path towards the objective? Indeed, is there a feasible solution?

The method also allows for variable migration, and connecting to an equally well-developed model of migration should be relatively straightforward. Given the growing impact of piracy on trade flows, the model here could also be adjusted to test various scenarios, including the resilience of global flows to attacks on a given shipping route. Along these lines, the total money spent on shipping within the model is given by the relatively simple expression:

$$\sum_{ij} Y_{ij}^{(1)} D_{ij} \theta^{(1)} + Y_{ij}^{(2)} D_{ij} \theta^{(2)} + \frac{\psi_{ji}^{(3)}}{\phi_i^{(3)}} \theta^n D_{ji} \tag{6.60}$$

so that introducing a global shipping company into the model would be possible. Indeed, any number of firms or stakeholders could be introduced – the only adjustment being in the reinvestment dynamics of Section 6.8.

These extensions demonstrate the flexibility of the model, but the next step remains clear: to empirically validate this model, calibrate the parameters against real data and explore the dependence of initial conditions on the dynamics.

References

Fotheringham AS, Rees P, Champion T, Kalogirou S and Tremayne AR. (2004). The development of a migration model for England and Wales: overview and modelling out-migration. *Environment and Planning A* 36: 1633–1672.

Harris B and Wilson AG. (1978). Equilibrium values and dynamics of attractiveness terms in production-constrained spatial-interaction models. *Environment and Planning A* 10: 371–388.

Ohlin B. (1935). *Interregional and International Trade*. Cambridge, MA: Harvard University Press.

Rho JH, Boyce DE and Kim TJ. (1989). Comparison of solution methods for Wilson's interregional commodity flow model. *Geographical Analysis* 21(3): 259–267.

Stillwell J. (1978). Interzonal migration: some historical tests of spatial-interaction models. *Environment and Planning A* 10: 1187–1200.

Trefler D and Zhu S. (2000). Beyond the algebra of explanation: HOV for the technology age. *American Economic Review* 90(2): 145–149.

Willekens F. (1999). Modeling approaches to the indirect estimation of migration flows: from entropy to em. *Mathematical Population Studies* 7: 239–278.

Wilson AG. (1970). Interregional commodity flows: entropy maximizing approaches. *Geographical Analysis* 2(3): 255–282.

Wilson AG. (2010). Urban and regional dynamics from the global to the local: 'hierarchies', 'DNA', and 'genetic planning'. *Environment and Planning B, Planning and Design* 37(5): 823–837.

7

Global Dynamical Input–Output Modelling

Anthony P. Korte and Alan G. Wilson

7.1 Towards a Fully Dynamic Inter-country Input–Output Model

There is a wealth of literature on input–output (I/O) modelling – indeed, the model has iconic status. However, research challenges remain. The most common models are of a comparative static nature and are for a single country (or region or city). There is an extensive literature on making the models dynamic and, separately, on making them multi-country. The latter initiative involves modelling trade. The challenge in making the model fully dynamic involves modelling investment. Our objective in this chapter is to articulate a model that is multi-country, multi-sector and dynamic. We build on Duchin's single-country dynamic model (Duchin, 2004) and Wilson's multi-country dynamic model (Wilson, 2015), and in each case show how to generate versions of a fully multi-country, multi-sector, dynamic model. We proceed as follows. In section 7.2, we articulate the accounts for a single country and show how to include endogenous final demand. We then extend the single-country model to be inter-country and show how to incorporate investment (section 7.3). In section 7.4, we further extend the argument to show how investment can be treated as capacity building on a planning basis in a global model; and in section 7.5, we offer an alternative model in terms of investment growth. In each of these sections, we explore model dynamics with hypothetical cases.

A static Leontief I/O global trade model has been elaborated in Levy *et al.* (2014) and Wilson's *Global Dynamics* (2015, chap. 4), utilising published global I/O coefficients from the World Input-Output Database (WIOD) (see Timmer *et al.*, 2012). It is static in time and is presented as an iterative scheme. In the appendix to this chapter, it is shown that the scheme can be cast as an inhomogeneous set of linear equations which can be solved directly using matrix inversion or least-squares iteration. A static (in time) iterative I/O scheme can be interpreted dynamically, as it can be recognised as a Leontief expenditure lag-type model

Approaches to Geo-mathematical Modelling: New Tools for Complexity Science, First Edition. Edited by Alan G. Wilson.

Companion Website: www.wiley.com/go/wilson/ApproachestoGeo-mathematicalModelling

(Takayama, 1985), but this is not elaborated in Levy *et al.* (2014). The appendix also shows the equivalence of the model of Levy *et al.* (2014) and the Leontief–Strout formulation and entropy maximisation in Wilson (1970).

Adequate modelling of the investment process remains one of the main methodical problems of I/O analysis within a dynamic framework (see e.g. Edler and Ribakova, 1993).

Capital stock consists of infrastructure, buildings, machinery and equipment that are essential for production and consumption (see e.g. Duchin, 2004). These durable goods require energy, materials and other resources for their production, and after their economic or physical lifetime is exhausted, they are a major source of waste and a secondary source for materials. Capital can be defined as goods and services used beyond the standardised period, which accumulate in the economy as a stock, only a portion of which gets 'consumed' during the current period. The depreciation of capital investment is referred to as a capital consumption allowance. The dynamic I/O model represents the demand for capital goods on the part of each producing sector and provides sectoral detail for the input requirements for resources and products to produce these goods.

The exogenous vector of investment, formerly part of final deliveries, is replaced by an expression where a matrix of stock requirement coefficients is multiplied by the anticipated increase in capacity between the present time period and the subsequent period. Therefore, investment is made endogenous, driven by depreciation and expected future output.

7.2 National Accounts

7.2.1 Definitions

We begin by articulating the accounting side of the model. Productive activities purchase land, labour, capital inputs and intermediate inputs from commodity markets, and use these to produce goods and services. These are supplemented by imports and then sold through commodity markets to households, the government, investors and the rest of the world. In a circular flow diagram, each institution's expenditure becomes another institution's income. For example, household and government purchases of commodities provide the incomes that producers need to continue the production process. Additional inter-institutional transfers, such as taxes and savings, ensure that the circular flow of incomes is closed. Therefore, all income and expenditure flows are accounted for, and there are no leakages from the system.

The conventions in Wilson (1970) are adopted: indices i,j refer to countries, and m,n to sectors, with $i,j = 1, \ldots, P$ and $m,n = 1, \ldots, N$, with the total number of countries and sectors given by P and N, respectively.

The account flow variables are defined for each country in Table 7.1.

These account relationships are shown in figure 7.1, where $A \rightarrow B$ means B supplies A or A pays B. Flows into, out of and in between the red boxes must be equal and contribute to the opening liabilities or assets, which therefore contribute to the net worth of the economy.

These variables are coupled in the balances for the national accounts for each country, where the left-hand sides of the equations are payments and the right-hand sides receipts. These are given further throughout this section, and serve to define savings of households, government spending and the current account balance of a country with respect to the rest of the world. This in turn determines the net worth of a country.

Table 7.1 Variables used to define the national accounts at a country level

Variable	Description
L_i	Household income (= gross national income (GNI))
X_i	Exports
M_i	Imports
Xf_i	Intra-country sales of all commodity goods
Zd_i	Intermediate demand
C_i	Household consumption (final, non-investment, demand)
I_i	Capital goods purchases (gross capital formation)
G_i	Government expenditure
S_i	Household savings (dis-saving, if negative)
T_i	Taxes
R_i	Net remittances received
W_i	Welfare transfers
BP_i	Current account balance (total capital inflows from abroad)
B_i	Fiscal surplus (public savings, or deficit if negative)
A_i	Net aid receipts
W_{1i}	Opening net worth
W_{2i}	Closing net worth
Rv_i	Revaluations

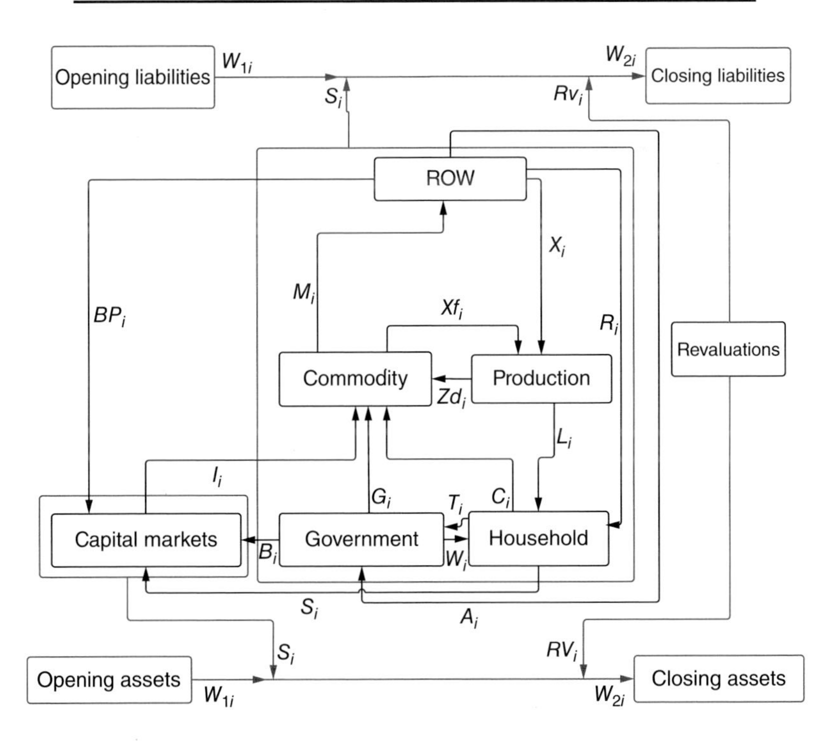

Figure 7.1 National circular account flows and net worth of the economy

7.2.2 The Production Account

Production activities produce goods and services by combining the factors of production with intermediate inputs. The proceeds from exports and intra-country sales of commodities pay for wages and intermediate demand:

$$L_i + \mathrm{Zd}_i = \mathrm{Xf}_i + X_i. \tag{7.1}$$

In this presentation, the production account is equivalent to the supply pool of Wilson (1970).

7.2.3 The Commodity Markets Account

Commodities are either supplied domestically or imported. A number of economic entities purchase commodities to use them as intermediate inputs for production. Final demand for commodities consists of household consumption spending, government consumption, or recurrent expenditure, gross capital formation or investment, the proceeds of which, along with intermediate demand pay for intra-country sales and imports, are:

$$M_i + \mathrm{Xf}_i = C_i + I_i + G_i + \mathrm{Zd}_i. \tag{7.2}$$

In this presentation, the commodity markets account is equivalent to the demand pool of Wilson (1970).

7.2.4 The Household Account

Households can be considered as the ultimate owners of the factors of production, and so they receive the incomes earned by factors during the production process. They also receive transfer payments from the Government (e.g. social security and pensions) and from the rest of the world (e.g. remittances received from family members working abroad). Households then pay taxes directly to the government and purchase commodities. The remaining income is then saved (or dis-saved if expenditures exceed incomes), giving:

$$C_i + S_i + T_i = L_i + R_i + W_i. \tag{7.3}$$

7.2.5 The Capital Markets Account

The capital market holds personal savings on behalf of consumers and lends it to businesses to purchase capital goods as investments. In general, the total savings in the economy is equal to total investment (personal savings, fiscal surplus and net lending from abroad):

$$I_i = S_i + BP_i + B_i. \tag{7.4}$$

7.2.6 The Rest of the World (RoW) Account

The current account balance (which gives a country's balance of payments) is defined so that the circular flow for the RoW account balances – it is the net lending from abroad (Miller and

Blair, 2009; i.e. the current account balance is the net capital inflow from abroad):

$$X_i + BP_i + A_i + R_i = M_i. \tag{7.5}$$

7.2.7 *The Government Account*

The government receives transfer payments from the rest of the world (e.g. as foreign grants and development assistance). This is added to all of the different tax incomes to determine total government revenues. The government uses these revenues to pay for recurrent consumption spending and transfers to households, with anything left over being a fiscal surplus:

$$G_i + B_i + W_i = A_i + T_i. \tag{7.6}$$

7.2.8 *The Net Worth of an Economy and Revaluations*

During any given period, the national accounts do not include capital gains or losses – the creation or destruction of value not coming from real economic output or income generated. Revaluations resulting from price changes provide the principal connection between periods but are excluded from valuation during any given period.

With a national system of accounts, however, transactions or flows in the economy are accounted for, as well as capital accumulation, but the total value of accumulated assets and liabilities or stocks in the economy has been ignored. This can be accomplished by incorporating the balance statement into an accounting balance sheet with a valuation of opening net assets in the economy, that is, the depreciated value of tangible assets held plus the excess of any financial claims held as assets over financial claims issued as liabilities, which is defined as the net worth of the economy.

For any period of time, the transformation of an opening balance sheet into a closing balance sheet can be traced in two equivalent ways: (1) the net assets at the end of the period are equal to the sum of the net assets at the beginning of the period, net domestic and foreign investment during the period, and revaluations needed to adjust the value of assets previously acquired or liabilities previously issued to the prices in place at the closing date; and (2) the net worth at the end of the period is equal to the net worth at the beginning of the period plus new savings accumulated during the period and revaluations resulting from price changes.

The net worth of an economy is therefore described by:

$$W_{2i} = W_{1i} + S_i + Rv_i. \tag{7.7}$$

Equations (7.1)–(7.6) account for flows in the economy, whereas equation (7.7) accounts for the total value of stocks in the economy.

7.2.9 *Overview of the National Accounts*

Combining equations (7.1) and (7.2) one obtains $L_i = C_i + I_i + G_i + X_i - M_i$ (i.e. GNI equals gross national product (GNP; see e.g. Miller, 2009). This identity reflects the balance of flows entering and exiting a box drawn around the production and commodity accounts in figure 7.1.

Combining equations (7.4), (7.5) and (7.6), one obtains $G_i + I_i + R_i + X_i + W_i = S_i + M_i + T_i$, equivalent to the flows entering and exiting a box drawn around the capital markets, RoW and government accounts, or equivalently the flows entering and exiting a box drawn around the production, commodity markets and household accounts. Then, if one uses equations (7.1) and (7.2), equation (7.3) is identically satisfied. This is equivalent to the statement that if a box is drawn around the household account, and a box around the remaining accounts, once flows in and out of those remaining accounts are specified, then flows in and out of the household account are fully determined by conservation of flow.

Here we take the expenditures approach to measuring GDP (see e.g. Econport, 2006), that is, $GDP_i = C_i + I_i + G_i + X_i - M_i$. For the history of GDP and its use as a measure, see Coyle (2014).

7.2.10 Closing the Model: Making Final Demand Endogenous

One can close the model to household final demand by including it as another sector in the economy in the standard way. In most developed countries, consumption (consumer, or household, purchases – the amount spent on consumption goods) is the largest component of final demand. In the United States in 2003, personal consumption expenditure was 71%, gross private domestic investment (producers' durable equipment, plant construction, residential construction and net inventory change) was 16% and government purchases were 19% (Miller and Blair, 2009).

The alternative to closing the model in this way is to make household consumption depend on current or expected income (see Burress *et al.*, 1988). Define $T_i = g_i^T(L_i + W_i)$ and $W_i = g_i^W L_i$, $g_i^W \geq 0, 0 \leq g_i^T \leq 1$. For the taxation of remittances, see for example Mohapatra (2010). For the distribution of welfare payments W_i in the United Kingdom, see for example Bentley (2015). One can allow for the possibility that consumers (both household and government) can over-spend by borrowing by using a modulus function. If its argument is negative, the consumer is borrowing. Household consumption is therefore defined as:

$$C_i = g_i^C |L_i + R_i + W_i - T_i| + \tilde{f}_i^H, \tag{7.8}$$

$g_i^C \geq 0$; and $\tilde{f}_i^H$ is additional exogenous final household demand. Similarly, define where government spending:

$$G_i = g_i^G |A_i + T_i - W_i| + \tilde{f}_i^G + Z_i, \tag{7.9}$$

where $g_i^G \geq 0$; Z_i is military spending; and $\tilde{f}_i^G$ is additional exogenous final government demand. Aid receipts are negative if a country is a net donor. Then, equation (7.5) defines BP_i, equation (7.6) defines B_i, equation (7.4) defines S_i, equation (7.7) defines net worth in the economy and equation (7.3) is identically satisfied.

We disaggregate C_i and G_i across sectors into household and government non-investment final demand, f_i^H and f_i^G respectively:

$$f_i^{H,m} = p_i^{H,m} C_i, \qquad \sum_m p_i^{H,m} = 1, \tag{7.10}$$

$$f_i^{G,m} = p_i^{G,m} G_i, \qquad \sum_m p_i^{G,m} = 1, \tag{7.11}$$

such that total final demand is $f_i^m \equiv f_i^{H,m} + f_i^{G,m}$. The proportions $p_i^{H,m}, p_i^{G,m}$ can be specified exogenously or given as a particular (e.g. uniform) probability distribution. They simply partition total consumption in a country into different available sectors. In the standard parlance, the dynamical models in this chapter are open.

In the remainder of this chapter units are in notional dollars.

7.3 The Dynamical International Model

7.3.1 Supply and Demand

In Table 7.2, we extend Table 7.1 by adding variables for the international system.

According to Table 7.2, for production x_i^m, inter-industry trade flows x_{ij}^{mn} (production flow of sector m in country i to sector n in country j), and final demand Y_i^m, the trade flows can be estimated in a spatial interaction model, derived from an entropy-maximising method, that is, maximising $-\sum_{ijmn} x_{ij}^{mn} \log (x_{ij}^{mn})$, with constraints:

$$x_i^m = \sum_{jn} x_{ij}^{mn} + Y_i^m, \tag{7.12}$$

$$\sum_i x_{ij}^{mn} = a_j^{mn} x_j^n, \tag{7.13}$$

$$\sum_{ijn} x_{ij}^{mn} c_{ij}^m = C^m, \tag{7.14}$$

where a_j^{mn} is the usual I/O technical coefficients (see the appendix to this chapter); c_{ij}^m is the unit cost of the flow of commodity m in the given time period; and C^m is the total expenditure in the system of the flow of commodity m; see Wilson (1970) in the same time period. This gives:

$$x_{ij}^{mn} = a_j^{mn} \exp(-\mu^m c_{ij}^m) x_j^n / \sum_k \exp(-\mu^m c_{kj}^m). \tag{7.15}$$

Table 7.2 Indexed variables for the international system.

Variable	Description
$x_i^m(t)$	Production of sector m in country i in period t
x_{ij}^m	Flow of production of sector m from country i to country j
z_i^{mn}	Intermediate demand: flow of sector m to sector n in country i
t_{ij}^m	Trade coefficients: fraction of the total commodity flow of sector m into country j supplied by country i
$K_i^m(t)$	Production capacity of sector m in country i in period t
$k_i^m(t)$	Increase of production capacity of sector m in country i in period t
a_i^{mn}	Technical coefficients: amount of the mth good used per (dollar) unit of the nth good in country i
$b_i^{mn\theta}(t)$	Capital produced in period t by sector m per unit increase of capacity of sector n in country i in period $t + \theta$

Comparing this with the trade coefficients t_{ij}^m in Campisi *et al.* (1991) and also Campisi and Gastaldi (1996), one sees that:

$$t_{ij}^m = \frac{\exp(-\mu^m c_{ij}^m)}{\sum_k \exp(-\mu^m c_{kj}^m)}, \tag{7.16}$$

where $\sum_i t_{ij}^m = 1$. As in (Wilson *et al.*, 1981; Pooler, 1994; Miller and Blair, 2009; Wilson and Fry, 2012), one can introduce an attractiveness weighting term w_i^m, a measure of country j's demand for commodity m, such that $t_{ij}^m = (w_i^m)^{\alpha_m} \exp(-\mu^m c_{ij}^m)/\sum_k (w_k^m)^{\alpha_m} \exp(-\mu^m c_{kj}^m)$.

In Wilson *et al.* (1981), the weighting term is used as a device which effectively removes (neutralises) the I/O (supply = demand) identity (equation 7.12) which is how equation 7.15 is obtained.

The use of the 4-index trade flow variable was used here for convenience. One could instead work with the 3-index flow variable x_{ij}^m of Wilson (1970). In the entropy maximisation of that paper, Wilson uses the Leontief–Strout equation (see the appendix) as a constraint along with the analogue of equation (7.14) ($\sum_{ij} x_{ij}^m c_{ij}^m = C^m$) to obtain $x_{ij}^m = \delta_i^m \epsilon_j^m \exp(-\mu^m c_{ij}^m)$. By eliminating ϵ_j^m, one obtains:

$$x_{ij}^m = t_{ij}^m \sum_k x_{kj}^m, \tag{7.17}$$

with trade coefficients t_{ij}^m given by:

$$t_{ij}^m = \frac{\delta_i^m \exp(-\mu^m c_{ij}^m)}{\sum_k \delta_k^m \exp(-\mu^m c_{kj}^m)}, \tag{7.18}$$

where δ_i^m is obtained from an implicit equation involving the known parameters appearing in the constraints of the maximisation. From equation (7.17), one sees that the trade coefficient t_{ij}^m is simply the fraction of the total industry trade flow (from all countries) country j receives from country i in the commodity m. In the same way as before, neutralising the Leontief–Strout equation gives $\delta_i^m = 1$, or equation 7.16 . This is not necessary (equation (7.18) is the general result), however, to reach the remarkable conclusion that Wilson's entropy maximisation of the probability distribution associated with trade flows naturally gives rise to trade coefficients (equations (7.17) and (7.18)). (In a similar way, one can also see that trade coefficients also (implicitly) arise in Wilson and Fry (2012), where c_{ij}^m is expressed in terms of distance and price.)

To see how trade coefficients can be written interchangeably in terms of the import ratios d_i^m and import propensities p_m^{ij} of Levy *et al.* (2014), see the appendix. The literature (see e.g. Miller and Blair, 2009; Campisi *et al.*, 1991) generally uses trade coefficients as opposed to import ratios and propensities. For an early reference to the use of import coefficients, see Redwood *et al.* (1987, p. D.4) and the appendix to this chapter. Wilson goes one step further, however, because equation (7.18) expresses the trade coefficients in terms of spatial interaction quantities: price, distance and the relative importance of price, rather than quality, to the buyer.

Following Duchin and Szyld (1985), Edler and Ribakova (1993) and Johansen (1978) for country i, denote as $b_i^{mn\theta}$ the amount of capital produced by sector m to increase the capacity of sector n by one unit and which must be delivered θ periods in advance before the piece of production capacity, of which it is a part, is to be ready for use (i.e. $\theta = 1, \ldots \tau$, where τ is the longest gestation period).

The N-sector, P-country, dynamical I/O equations, expressing the relationship supply = demand, are then:

$$x_i^m(t) = \sum_{jn} t_{ij}^m a_j^{mn} x_j^n(t) + \sum_j t_{ij}^m f_j^m(t) + \sum_{\theta=1}^{\tau} \sum_{jn} t_{ij}^m b_j^{mn\theta} k_j^n(t+\theta) \tag{7.19}$$

The technical coefficients a_i^{mn} and capital coefficients $b_i^{mn\theta}$ are in general functions of t, but unless otherwise stated are taken as constant in simulations. In general, equation (7.19) is an example of a difference equation with advanced argument (a mixed difference equation), arising because of the investment decision regarding anticipated capital stock, as in Kaddar and Alaoui (2008). The first two terms correspond to the I/O model in Levy *et al.* (2014) and say nothing more than that x_i^m supplies a fraction t_{ii}^m of its own intermediate and (noninvestment) final demand and a fraction t_{ij}^m of every other country's ($j \neq i$) intermediate and (noninvestment) final demand.

Block diagonal matrices A, B^θ are assembled by forming P blocks of $N \times N$ matrices (i.e. $B^\theta = \text{diag}(B_1^\theta, \ldots, B_P^\theta)$, with B_i^θ defined in the same way as A in the appendix). As in the appendix, T is formed by $T_{ij} = \hat{t}_{ij}$ where $\hat{t}_{ij} = \text{diag}(t_{ij}^1 \ldots t_{ij}^N)$. Thus, all matrices are of size $NP \times NP$. Therefore, equation (7.19) becomes:

$$x(t) = T(Ax(t) + f(t)) + \sum_{\theta=1}^{\tau} TB^\theta k(t+\theta). \tag{7.20}$$

Note that the left-hand side of this equation is supply, the right-hand side demand. In order to solve this equation in an efficient manner, one must be careful to use sparse matrices, pre-allocation of memory, indexing for the creation of sparse matrices and an iterative method for the inversion.

7.3.2 The National Accounts Revisited

Recalling that quantities have a country i subscript, one has for the production account (equation (7.1)):

$$L_i^m = x_i^m (1 - \sum_n a_i^{nm}), \tag{7.21}$$

$$X_i = \sum_{mn, j \neq i} t_{ij}^m (a_j^{mn} x_j^n + \sum_\theta b_j^{mn\theta} k_j^n(t+\theta)) + \sum_{m, j \neq i} t_{ij}^m f_j^m, \tag{7.22}$$

$$\text{Xf}_i = \sum_{m,n} t_{ii}^m (a_i^{mn} x_i^n + \sum_\theta b_i^{mn\theta} k_i^n(t+\theta)) + \sum_m t_{ii}^m f_i^m. \tag{7.23}$$

Along with the expression given for Zd_i in the appendix, and since by definition $z_i^{mn} = a_i^{mn} x_i^n$, the production account equation (7.1) is therefore identically satisfied. Note that $L_i^m > 0$ because of the Brauer–Solow relation $\sum_m a_i^{mn} < 1$; see Danao (2007). For the commodity markets account, one has that:

$$M_i = \sum_{mn, j \neq i} t_{ji}^m (a_i^{mn} x_i^n + \sum_\theta b_i^{mn\theta} k_i^n(t+\theta)) + \sum_{m, j \neq i} t_{ji}^m f_i^m \tag{7.24}$$

and also:

$$I_i = \sum_{mn\theta} b_i^{mn\theta} k_i^n(t+\theta), \tag{7.25}$$

so that the commodity markets account equation (7.2) is also identically satisfied.

There are accounting relations that must be satisfied by the whole system and are useful as a consistency check on numerics. By equation (7.19), summing over all countries, one has:

$$\sum_i x_i^m = \sum_{in} a_i^{mn} x_i^n + \sum_i f_i^m + \sum_{in\theta} b_i^{mn\theta} k_i^n(t+\theta), \tag{7.26}$$

and from equations (7.22), (7.23) and (7.24) one can also derive $\sum_i X_i = \sum_i M_i$.

7.4 Investment: Modelling Production Capacity: The Capacity Planning Model

7.4.1 The Multi-region, Multi-sector Capacity Planning Model

In order to proceed, one needs to model production capacity, which is defined as the maximum production level if all the resources originally endowed are utilized.

For sector m in country i, denoting $k_i^m(t)$ as the increase in productive capacity in period t, $K_i^m(t)$ as the output capacity in period t, and $K_i^{*m}(t)$ as the projected capacity requirements for future period t, the capacity planning approach uses the following algorithm for the decision function for investment with respect to anticipated capital stock as an input to solving equation (7.19), which generalises the capacity planning model in (Johansen, 1978; Duchin and Szyld, 1985; Edler and Ribakova, 1993) to more than one country:

$$K_i^{*m}(t+\tau) = \min\left(1+\epsilon_i^m, \frac{x_i^m(t-1)+x_i^m(t-2)}{x_i^m(t-2)+x_i^m(t-3)}\right)^{\tau+1} x_i^m(t-1) \tag{7.27}$$

$$k_i^m(t) = \max(0, K_i^{*m}(t) - K_i^m(t-1)) \tag{7.28}$$

$$K_i^m(t) = (1-\delta_i^m)K_i^m(t-1) + k_i^m(t) \tag{7.29}$$

where ϵ_i^m is the maximum admissible annual rate of expansion for sector m in country i (set by policy) and δ_i^m is the rate of depreciation of production capacity K_i^m. Equation (7.27) is a decision function, set by policy, which can therefore be replaced by any other desired function. The geometrical form of the equation is to postulate that anticipated production capacity τ periods into the future depends on a fraction of the production in the previous time period. This fraction is raised to the power of $\tau + 1$ because this is the number of time periods between the previous time period and τ periods into the future.

The advantages of this model are that it does not require full capacity utilisation ($K_i^m = x_i^m$), which involves perfect foresight of future stock requirements, and also that it does not involve reversibility of the capital stock. There is, however, no mechanism to prevent over-utilisation of capacity. This could be remedied by allowing t_{ij}^m to change dynamically to redistribute production demand to countries and sectors that have the capacity.

Using the same method as in the appendix for assembling indexed quantities into vectors and matrices, one can cast these equations in matrix form:

$$K^*(t+\tau) = \min(1+\epsilon, (x(t-1)+x(t-2))\emptyset$$

$$(x(t-2)+x(t-3)))^{\circ(\tau+1)}\circ x(t-1) \tag{7.30}$$

$$k(t) = \max(0, K^*(t) - K(t-1)) \tag{7.31}$$

$$K(t) = (1-\delta)K(t-1) + k(t), \tag{7.32}$$

where $\circ$ and $\emptyset$, respectively, denote Hadamard product and division, that is, for matrices A, B, $(A\circ B)_{ij} = A_{ij}B_{ij}$ and similarly for division. Vectors x, k, K, K^*, f are defined of size $NP \times 1$, for example $x^T = (x_1^1, \dots, x_1^N, \dots, x_P^1, \dots, x_P^N)$. The matrix $\delta = \text{diag}(\delta_1^1, \dots, \delta_1^N, \dots, \delta_P^1, \dots, \delta_P^N)$ and similarly for ϵ. These vector equations show that the algorithm interrogates the sectors from each country altogether, albeit at different time periods.

Initial conditions are to specify $x(t)$ for $t = t_0 - \tau - 2, \dots, t_0 - 1$ and $K(t_0)$. Then equations (7.30)–(7.32) are solved in order, followed by equation (7.20).

One can display the non-linearity of these equations by solving for $k(t)$:

$$k(t) = \max[0, \min[1+\epsilon, (x(t-\tau-1)+x(t-\tau-2))\emptyset(x(t-\tau-2)+x(t-\tau-3))]^{\circ\tau}$$

$$x(t-\tau-1) - ((1-\delta)^{t-t_0-1}K(t_0) + \sum_{i=t_0+1}^{t-1}(1-\delta)^{t-i-1}k(i))],$$

for $t \geq t_0 + 2$, with $x(t)$ given in terms of $k(t)$ by equation (7.20) .

Shown in Figures 7.2 and 7.3 are time sequences for a trade model with $P = 6$ countries, $N = 9$ sectors and maximum gestation lag $\tau = 3$ for the 3rd and 5th countries as illustrated. The black curves are capacity, and blue curves production. In Figure 7.2, depreciation of the capital equipment is zero, with production in the initial conditions set at random values. The technical, capital and trade coefficients are set to random values; the final demands are all set to be 10 units; and $\epsilon = 0.02$. In Figure 7.3, the model reaches a fixed point from random initial conditions in just a few time steps.

In Figure 7.1, depreciation is set to zero, and the long-term solution is used as an initial condition, except that a 50% jump is introduced in the sixth sector of the first country for

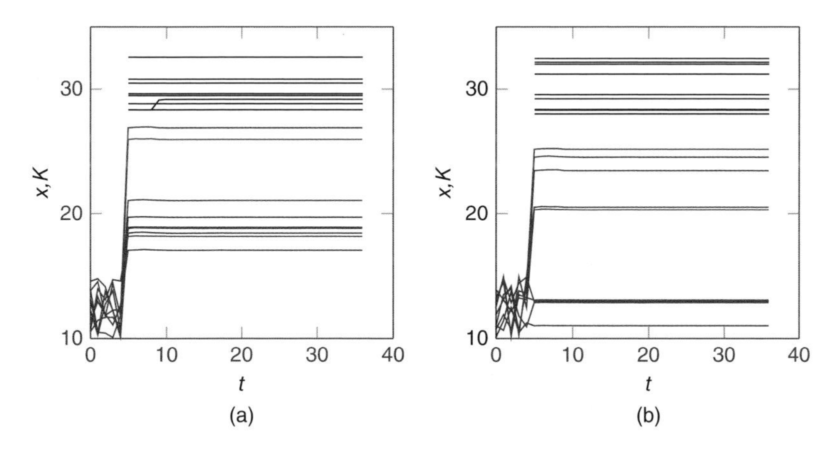

Figure 7.2 Production and its capacity, $P = 6$, $N = 9$, $\tau = 3$. (Left) 3rd country; (right) 5th country

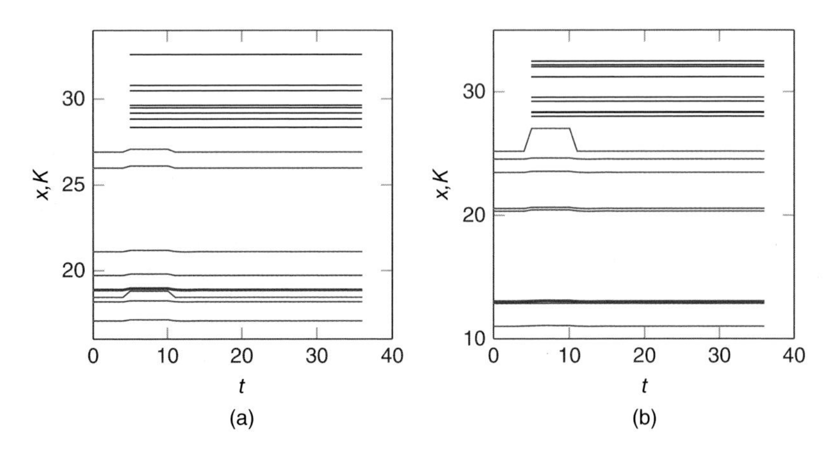

Figure 7.3 Production and its capacity, $P = 6$, $N = 9$, $\tau = 3$. (Left) 3rd country; (right) 5th country

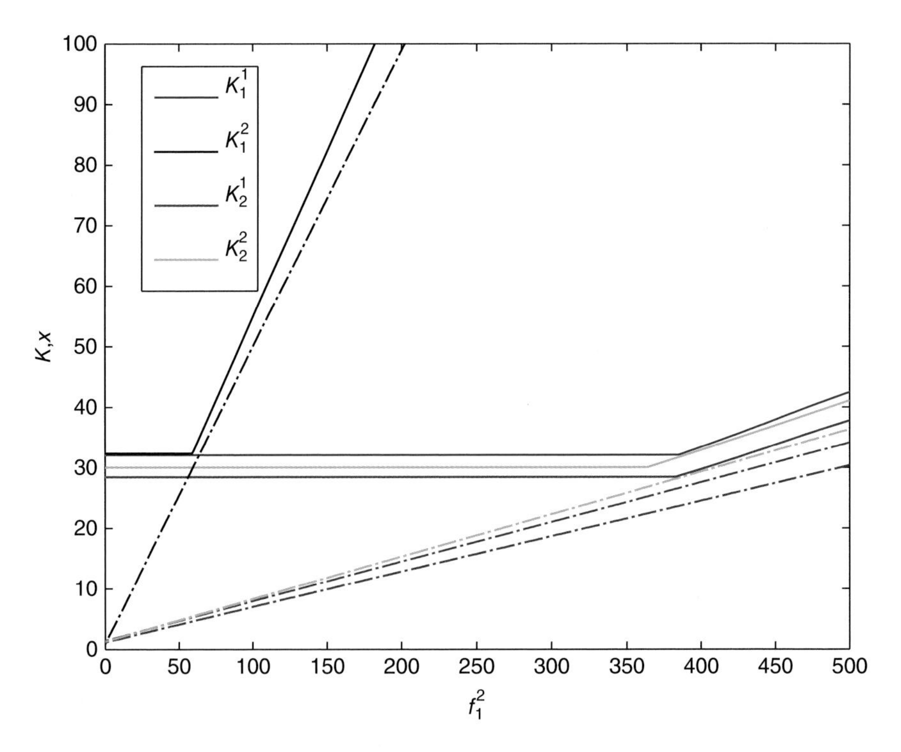

Figure 7.4 $P = 2, N = 2, \tau = 3$ capacity planning model; production capacity (solid lines), production (dash-dot lines); same colours for x as for K in legend

$t = 6$ to $t = 11$. The model reaches a fixed point in just a few time steps. This can be seen by considering the solution to the differential equation $dK/dt = \max(0, \alpha - K)$ for constant α. Approximating the max function by $\max(0, x) = x(1 + \tanh(2gx))/2$, with $g \to \infty$, one can solve this equation exactly in terms of hyperbolic sine and cosine integrals, which when taking the limit $g \to \infty$ gives $K(t) = \alpha - (\alpha - K(0)) \exp(-t)$ (i.e. $K(t)$ approaches the constant value α exponentially quickly – a few time steps, say). The max function essentially truncates the dynamics exponentially quickly, but without necessarily destroying non-linear behaviour, as will be seen in later sections of this chapter.

In this generalisation of the Duchin model, the solutions are clearly non-linear, but no bifurcation behaviour could be found by varying the technical coefficients or by varying the trade coefficients. In the results of Figure 7.4, the capacity K does not change as $K^* < K$, but the response to the jump in final demand can be seen in the changes to production. Moreover, a constant value for the variables is reached after just a few time steps, which could be used for forecasting. This quick response is due to the max function.

Shown in Figure 7.4 is the response of the Duchin model to increasing final demand, taking $f = 1$ and varying as a bifurcation parameter f_1^2. It can be seen that the production response to final demand is linear, although there is a bifurcation in the production capacity. This is not unexpected because when x increases, the capacity planning approach will respond when production approaches maximum capacity from below.

7.5 Modelling Production Capacity: The Investment Growth Approach

7.5.1 *Multi-region, multi-sector Investment Growth Models with Reversibility*

Alternatively, we generalise to N sectors the model in Wilson (2010), where economic activity is taken as the engine of growth for production capacity expansion in a model which has the form of Lotka–Volterra equations. This form of equations arises because the fractional change in production capacity is taken to be proportional to the difference between economic activity and the cost of maintaining production capacity, with a positive difference increasing capacity growth and vice versa. Suppose we also take economic activity to be sector-specific production x_i^m. The constraints $x \geq 0, K \geq 0$ are easily imposed in the numerics. Full-capacity utilisation is not assumed, in that solutions are sought where $x \leq K$, but where this is not possible, the standard accounting device is used whereby inventories are a reservoir that is a slack or surplus variable for production. Thus, if demand exceeds supply, there is an inventory drawdown and vice versa. Similarly, if supply exceeds capacity, there is an inventory drawdown. In the model of equation (7.20), k is positive, but here k can be negative, so one has that supply$(t) = x(t)$ and

$$\text{demand}(t) = \max\left(0, T(Ax(t) + f(t)) + \sum_{\theta=1}^{\tau} TB^{\theta}k(t+\theta)\right), \tag{7.33}$$

which allows for the reversibility of capital stock. Stock is reversible if unused capacity in a particular sector is freely transferable to other uses within the economy. This occurs when $k < 0$. As before, defining vectors of size $NP \times 1$ for K and x, and defining $r^T = (r_1^1, \dots, r_1^N, \dots, r_P^1, \dots, r_P^N)$, and similarly for p, one can express non-linear products between vectors as Hadamard products, as before. A Hadamard product between two vectors can, however, be expressed as ordinary matrix multiplication between two diagonal matrices. The equations in Wilson (2010), assuming $\tau = 1$ but generalising to N sectors, are:

$$k(t+1) \equiv K(t+1) - K(t) = \mathrm{diag}(r \circ K)(x(t) - \mathrm{diag}(p)K), \tag{7.34}$$

where $x(t)$ is given by minimising the absolute difference between supply and demand and p_i^m is the unit cost of maintaining capacity of sector m in country i per unit price.

7.5.2 *One-country, One-sector Investment Growth Model with Reversibility*

To gain insight, consider a one-country, one-sector, reversible model (one-country, multiple-sector models abound in the literature (e.g. Duchin and Szyld, 1985; Miller and Blair, 2009). The equations for this truncated model, assuming economic activity to be equivalent to production, are:

$$x(t) = \max(0, ax(t) + f + bk(t+1)) \tag{7.35}$$

$$k(t+1) \equiv K(t+1) - K(t) = rK(t)(x(t) - pK(t)), \tag{7.36}$$

where p, f are assumed to be constant; and $\theta = 1$. Assuming $x, K, x - pK \geq 0$, then $x(t)$ can be eliminated (with a non-linear dependence on K), and rescaling $K \to fK, x \to fx$, one obtains the non-linear map $K(t+1) = f(K(t))$, where

$$f(K) = K + \frac{\lambda K \left(\frac{1}{1-a} - pK\right)}{1 - \frac{b\lambda K}{1-a}}, \tag{7.37}$$

where $\lambda = rf$.

Shown in Figure 7.5 are bifurcation diagrams for (scaled) production capacity and production for $p = 0.5$ and $p = 0.9$. A smaller cost of maintaining production capacity p gives a higher capacity (shown on the left figure). A policy maker has no control over the structural form of equation (7.37) as it reflects an internal mechanism of capacity investment by actors not under his control (external investment here is set to zero), but he could influence and change the production capacity growth rate r. The non-linear form of the function $f(K)$ is such that it has a single maximum. The maximum value of the parameter λ (so as to not cross the location of the strange attractor) is such that the function evaluated at the maximum of the function must be less than the zero of the function (May, 1976), giving:

$$0 < \lambda < \frac{(1-a)^2 p(4b + 3(1-a)p)}{(2b + p(1-a))^2}. \tag{7.38}$$

This maximum value corresponds to the location of the strange attractor of the map, beyond which the map diverges to minus infinity, which in this model can be interpreted as leading to a collapse to zero of both production and its capacity when the growth rate is too high, just as in the logistic map. (A strange attractor is the fractal set of numerical values towards which the system evolves dynamically at a particular value of the bifurcation parameter.) To prevent this from happening, a policy maker can reduce the growth rate when the final demand

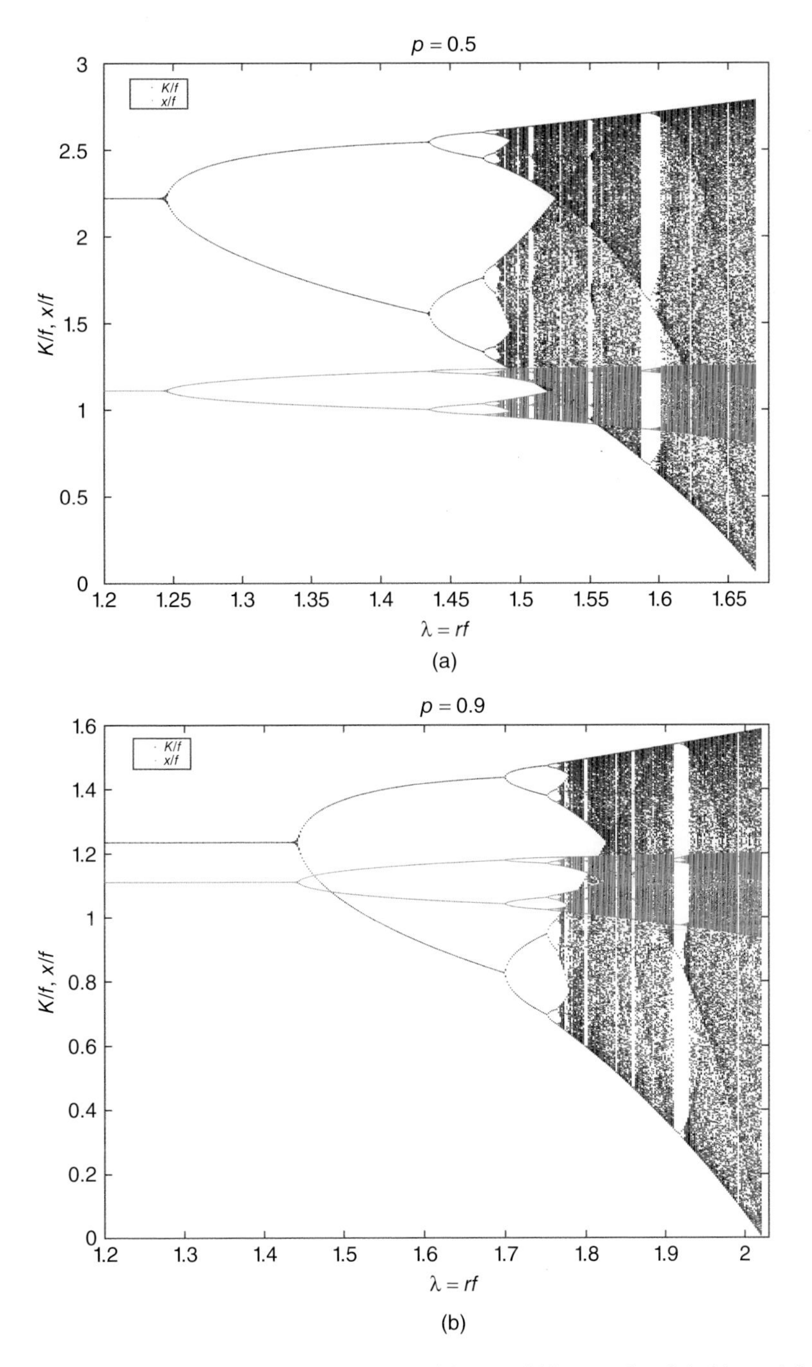

Figure 7.5 Logistic, one-country, one-sector model with reversibility, $a = b = 0.1$. (a) $p = 0.5$; (b) $p = 0.9$. Production capacity (black)/production (red) versus $\lambda \equiv rf$

is increasing to keep to a desired level of λ, as shown in Figure 7.6. Note that the production variables are scaled linearly by f. The onset of period doubling is given by:

$$\lambda \geq \frac{2(1-a)^2 p}{2b + (1-a)p}. \tag{7.39}$$

For $a = b = 0.1$, the requirement that $x \leq K$ gives $K \geq 1/(0.45 + 0.05\sqrt{(}81 - 20\lambda))$. Note that p can be used as a mechanism for implementing this constraint. For $p = 0.5$, the strange attractor condition gives $\lambda_{max} \approx 1.67751$, so that the capacity constraint gives approximately $K \geq 1.2$. This will be reflected in the initial conditions: if K is below this threshold initially, it will take typically one time step before the capacity exceeds production. The requirement $x \geq 0$ gives $K \leq 1/\sqrt{(}bp\lambda)$ or $K \geq (1-a)/b\lambda$. These inequality constraints apply only for the case where supply matches demand. In practice, the difference between supply and demand need only be minimised for the given parameter values, so that these inequalities can be avoided.

For reasons of stability, a policy maker might want to avoid period doublings and chaotic behaviour of production and its capacity. As can be seen from Figure 7.5, where only stable branches are shown, this can be done by reducing the growth rate of capacity so that one keeps to the left-hand side of the graphs. Note that the vertical axes are scaled linearly by the final (non-investment) demand f. Thus, when final demand increases, so will production and its capacity, but a policy maker can reduce the growth rate of capacity to compensate so that the country remains at a desired λ, preventing any period doublings from occurring.

By using AUTO (Doedel and Oldeman, 2012) to detect period doublings and changes of stability, one finds that the limit of period doubling is $\lambda_\infty \approx 1.484$. The AUTO bifurcation diagram for capacity is shown in Figure 7.6, with dashed lines corresponding to unstable branches and solid lines to stable branches.

The corresponding non-reversible model is shown in Figure 7.7, where r is varied as the bifurcation parameter.

7.5.3 Two-country, Two-sector Investment Growth Model with Reversibility

A numerical non-linear solver (trust-region-reflective) is used to solve equation (7.33) for x as well as to implement the constraints $x, K \geq 0, x \leq K$. The solution for x is sought to minimise the absolute difference between supply and demand. Shown in Figure 7.8 are production capacity and production (bifurcation, i.e. fixed point) curves for a two-country, two-sector (logistic) model where the final demand f_1^2 is varied as a constant parameter (i.e. the final demand for country 1, sector 2). Only stable branches are shown. The final demands for the other sectors are held constant at 1 unit, with non-zero elements given by $r = 0.3, p = 0.5$ and $A = B = 0.1$. The trade coefficients (assumed to be constant) are: $t_{ij}^1 = [0.4656, 0.9916; 0.5344, 0.0084]$, $t_{ij}^2 = [0.1940, 0.6483; 0.8060, 0.3517]$. Note that when either of t_{ij}^m, N or P, is changed, the location of the strange attractor inevitably changes. This is ultimately because the generalisation of equation (7.38) to a P-country, N-sector model would involve both these quantities as well as the trade coefficients, which are not present in the simpler model. The general features of the bifurcation diagrams are similar to the one-dimensional model in the "One-country, one-sector investment growth model with reversibility" section.

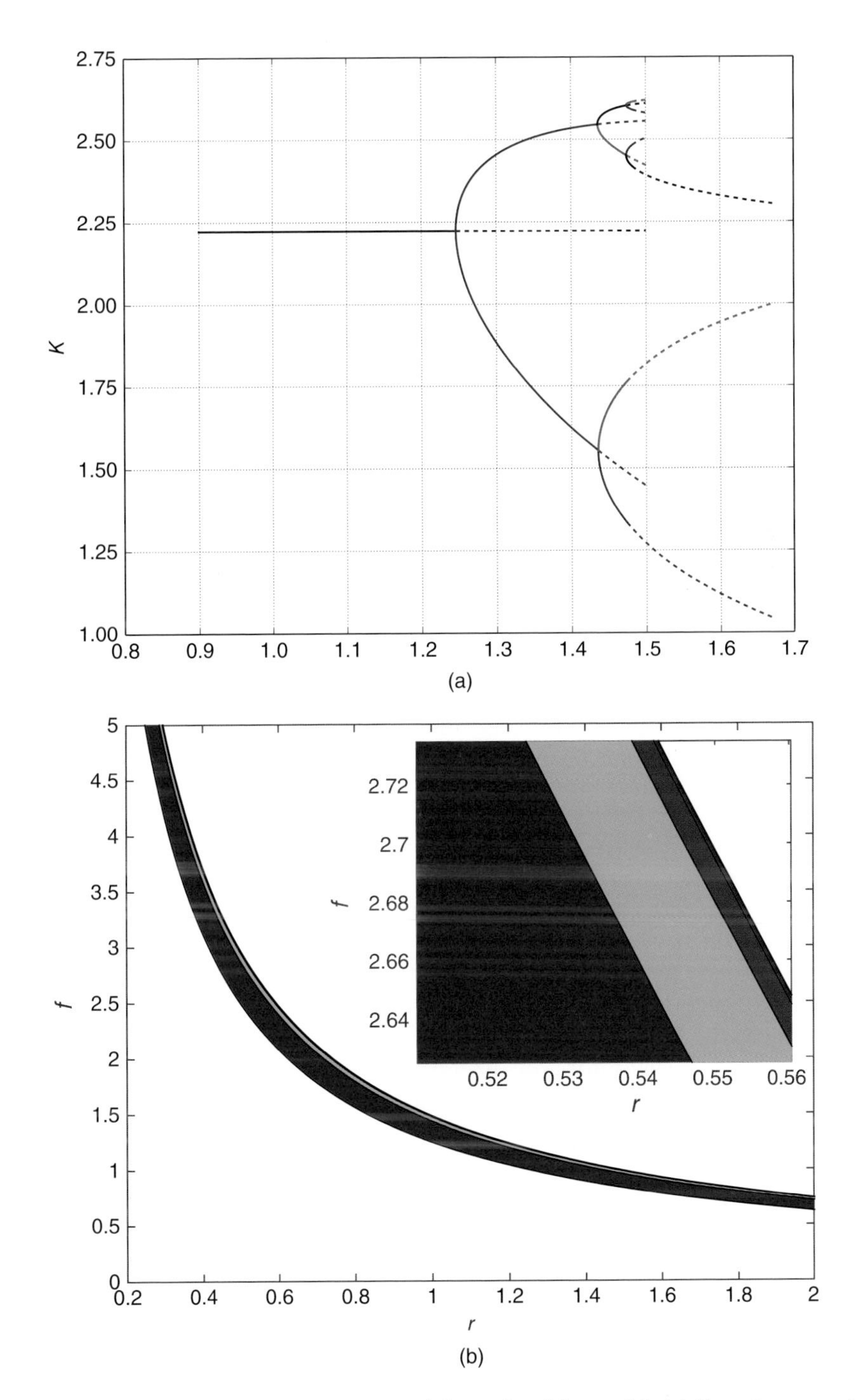

Figure 7.6 Logistic, one-country, one-sector model, $a = b = 0.1$, $p = 0.5$, (a) Capacity versus $\lambda \equiv rf$; (b) final demand and production capacity growth rate. Inset: period doubling contours

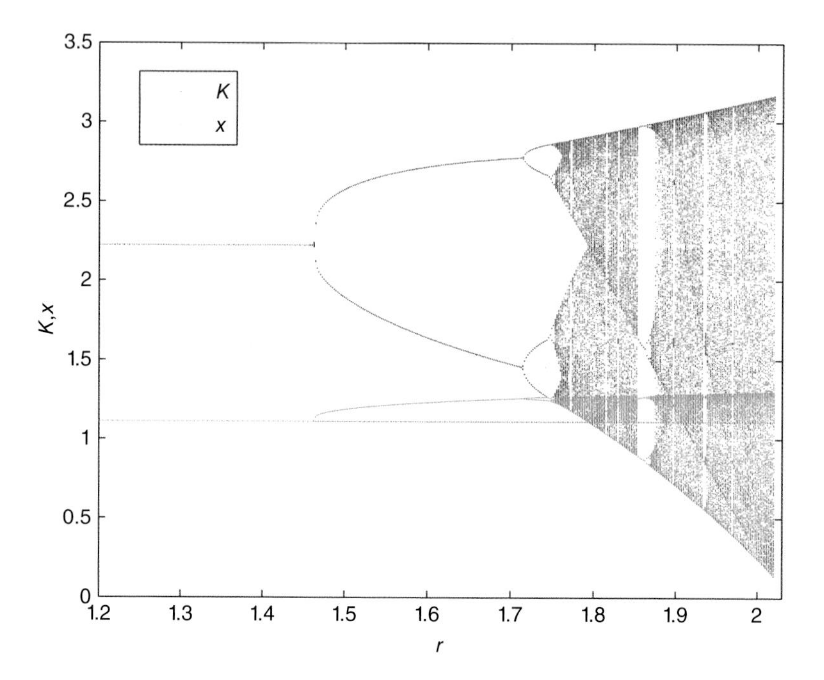

Figure 7.7 Logistic, one-country, one-sector non-reversible model, a=b=0.1, p=0.5, f=1, production capacity (solid lines), production (dash-dot lines) versus r

Before period doubling occurs, production and its capacity vary linearly with f_1^2 corresponding to $x = p \circ K$ in the capacity equation, equation (7.34), but this linear branch becomes unstable once period doubling occurs. In this model, even though the constraints $x, K \geq 0, x \leq K$ are satisfied, supply does not meet demand for $f_1^2 \geq 5.9772075$, that is, only well within the period doubling region, with demand>supply, for these parameters. This could be ameliorated by allowing T to vary dynamically (see the "A multi-region, multi-sector, investment growth model" section), but even in this case there is no guarantee that supply can meet demand due to fluctuations in the logistic equation. This model predicts non-linear logistic behaviour in production and its capacity for varying final demand of one sector five times more than any other, with otherwise linear behaviour (where supply = demand).

7.5.4 A Multi-region, Multi-sector, Investment Growth Model without Reversibility

Reversibility is removed if, instead of equation (7.20), one has

$$x(t) = T(Ax(t) + f(t)) + \sum_{\theta=1}^{\tau} TB^{\theta}\max(0, k(t+\theta)), \tag{7.40}$$

where x represents production, $\tau = 1$ is taken, and the capacity equation is by equation (7.34). Shown in Figure 7.8 are production capacity and production curves for the same data as in

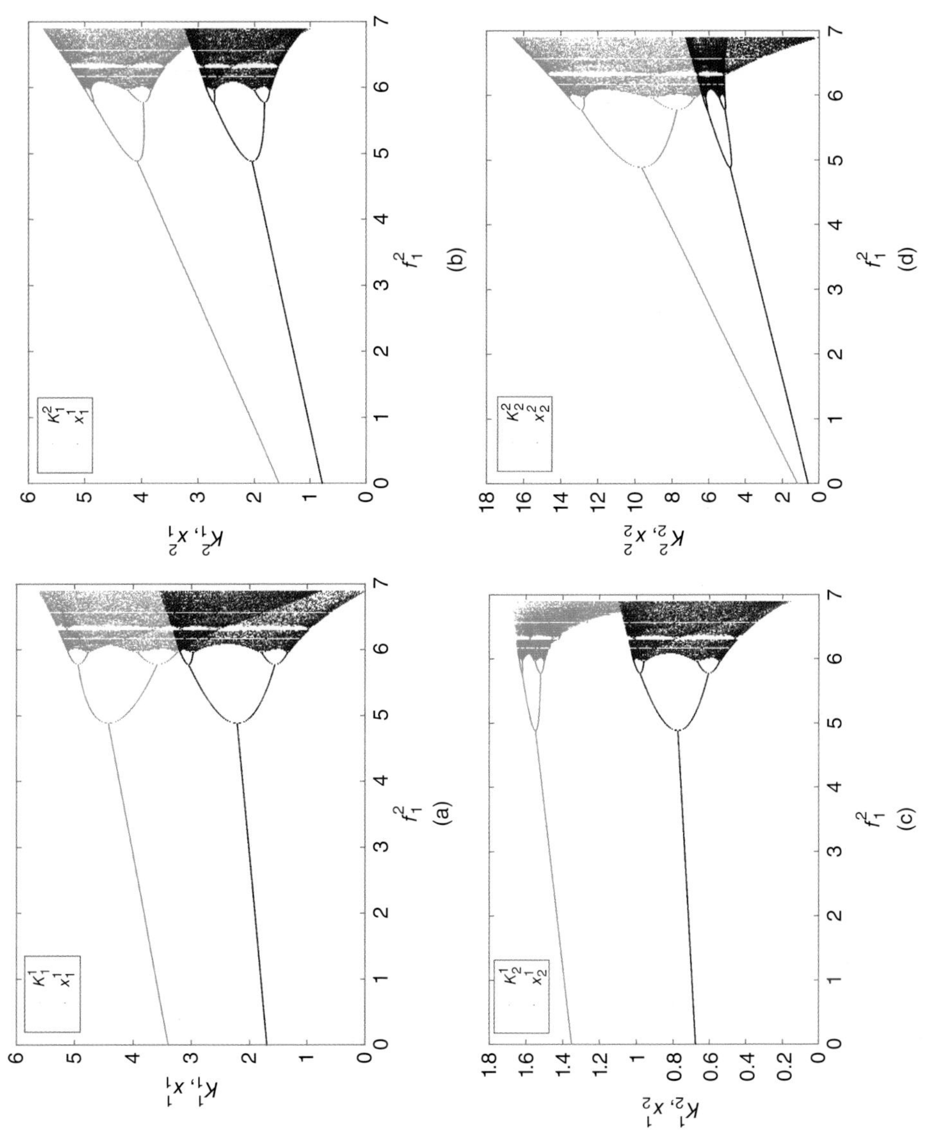

Figure 7.8 Two-country, two-sector, reversible model, production capacity (blue)/production (black) versus f_1^2. (a, b) Country 1; (c, d) country 2

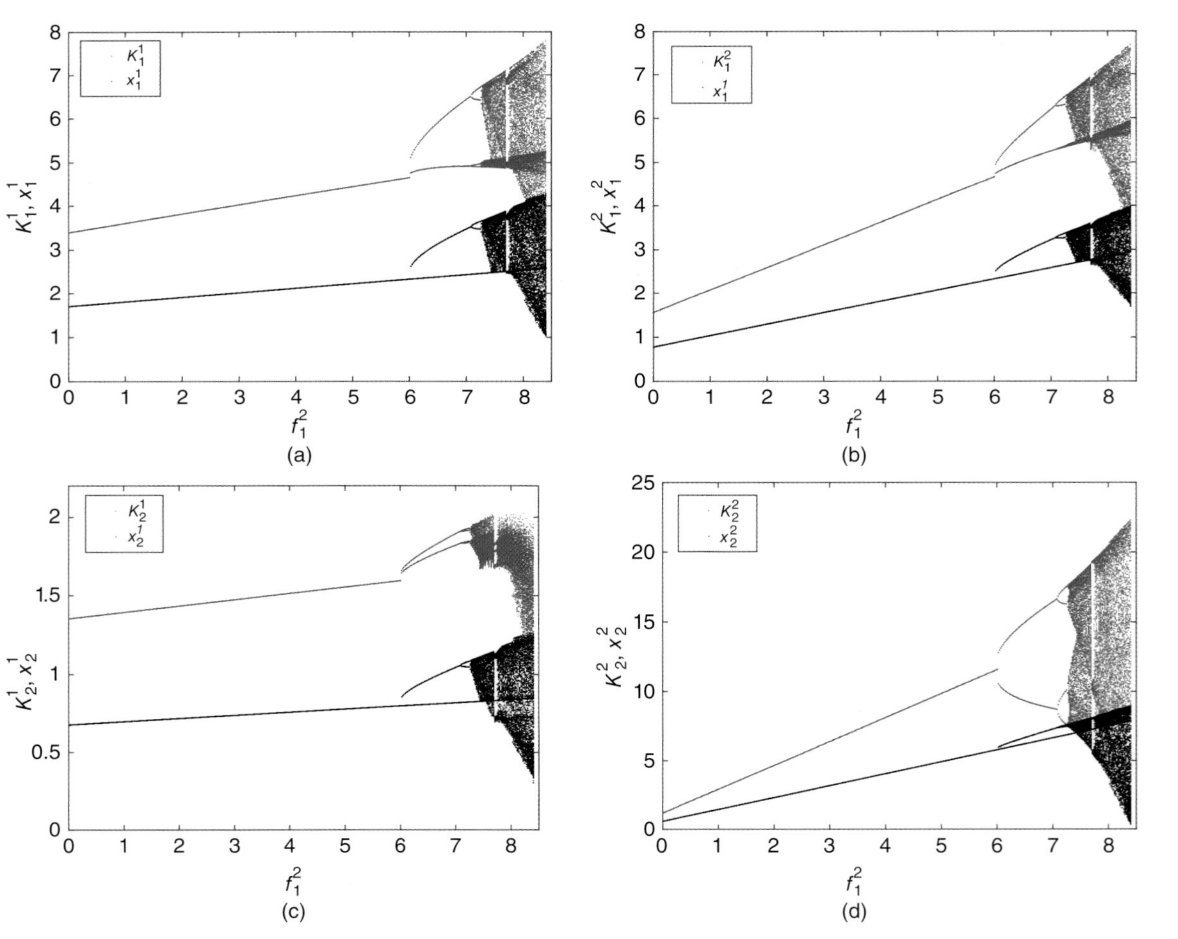

Figure 7.9 Two-country, two-sector non-reversible model, production capacity (blue)/production (black) versus f_1^2. (a, b) Country 1; (c, d) country 2

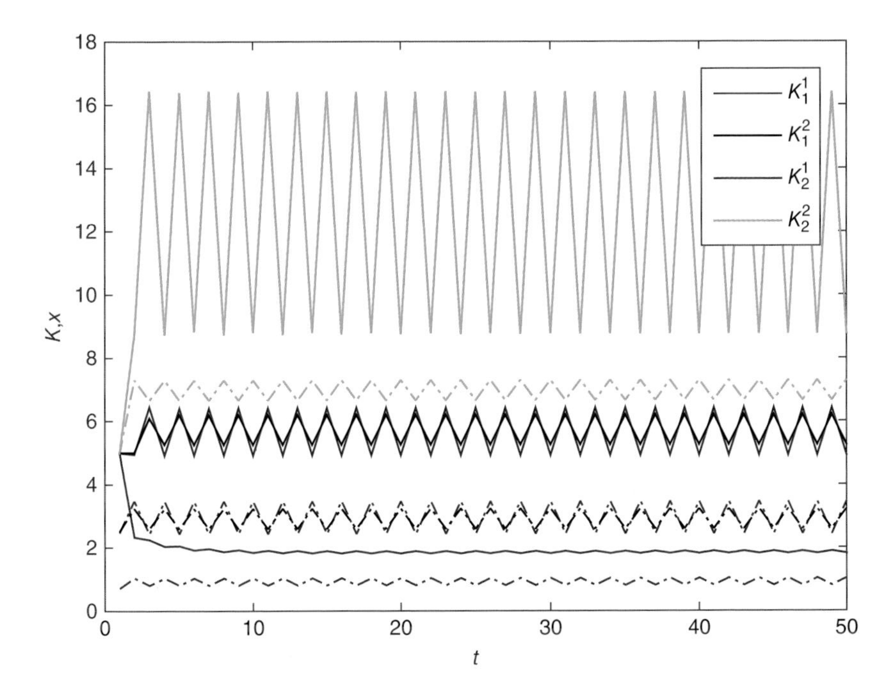

Figure 7.10 Two-country, two-sector non-reversible model: time series (period doubling) curves for production capacity (solid lines), production (dash-dot lines) for $f_1^2 = 7.0$ (same colours for x as for K in legend)

the two-country, two-sector (logistic) model of "Two-country, two-sector investment growth model with reversibility" section. In this model, the constraints $x, K \geq 0, x \leq K$ are satisfied, and supply>demand only well within the period doubling region, for these parameters. This model predicts non-linear logistic behaviour in production and its capacity for varying final demand of one sector six times more than any other, with otherwise linear behaviour (where supply = demand). Remarkably, the presence of the max function does not destroy the bifurcation behaviour as in the Duchin model.

Shown in Figure 7.11 are time series curves for production capacity, production and independent trade coefficients for this model, for $f_1^2 = 7.0$, showing period doubling. The initial conditions are $K(1) = 5.0$. This model is discrete, but for clarity the discrete values are joined up to display the graph more clearly.

7.5.5 *A Multi-region, Multi-sector, Investment Growth Model without Reversibility, with Variable Trade Coefficients*

Keeping the same supply and demand equation, equation (7.40) and the same fixed coefficients as in the "Two-country, two-sector investment growth model with reversibility" section, but allowing the trade coefficients t_{ij}^m to freely vary, one obtains production capacity and production curves shown in Figure 7.11. For this model, the constraints $x, K \geq 0, x \leq K$ are satisfied, and

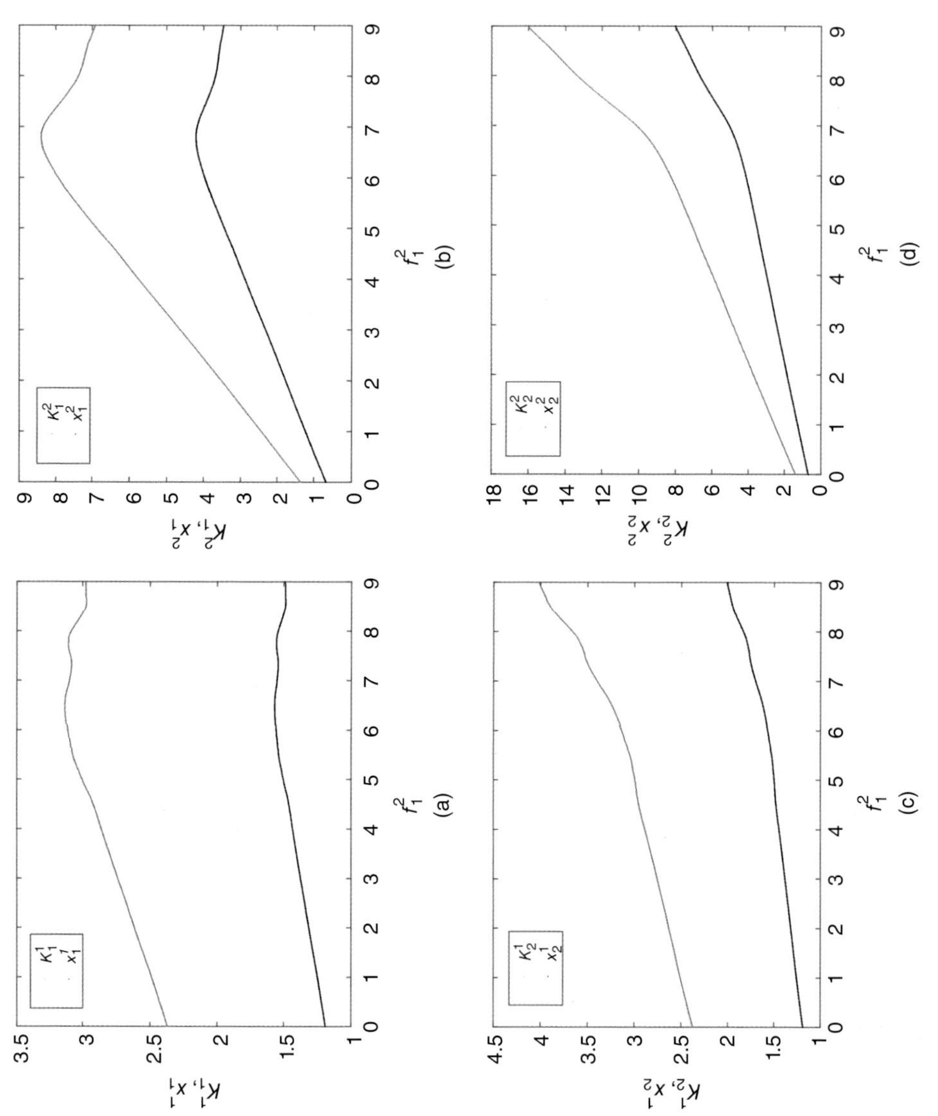

Figure 7.11 Two-country, two-sector non-reversible model, with variable trade coefficients, production capacity (blue)/production (black) versus f_1^2. (a, b) Country 1; (c, d) country 2

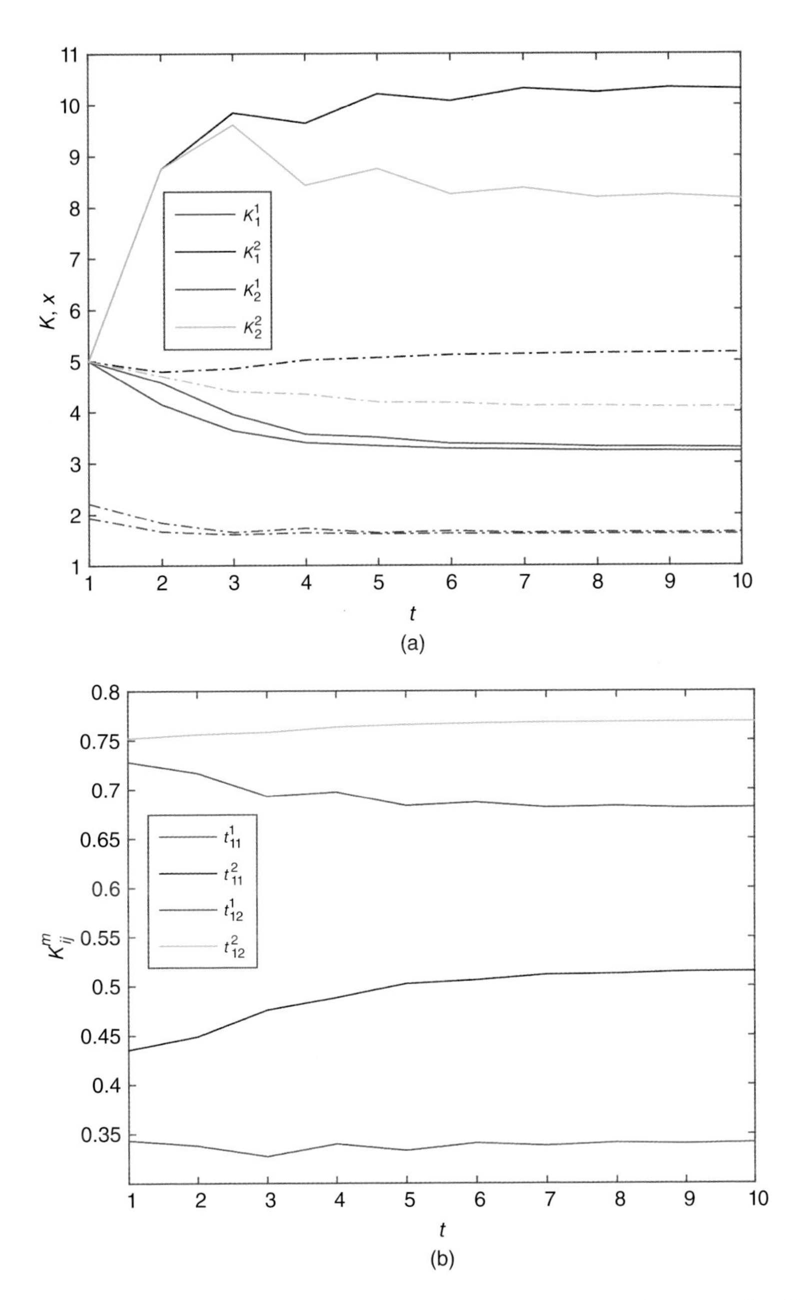

Figure 7.12 Two-country, two-sector model without reversibility: time series curves for (a) production capacity (solid lines), production (dash-dot lines) (same colours for x as for K in legend); (b) trade coefficients $t_{ij}^m, f_1^2 = 7.0$

supply = demand is maintained for all t. The initial conditions are $K(1) = 5.0$, and for T the initial conditions are the same as for the previous models.

Shown in Figure 7.12a are time series curves for production capacity, production and independent trade coefficients for this model, for $f_1^2 = 7.0$. This model is discrete in time, but for clarity the discrete values are joined up to display the graph more clearly.

Shown in Figure 7.12b is the variation of the trade coefficients t_{ij}^m with f_1^2. It can be seen that the model finds solutions such that $x = pK$, so that $(1 - TA)x = Tf$, but at the expense of varying T. The T coefficients are completely free to vary, subject only to the unit sum constraint $\sum_i t_{ij}^m = 1$, representing the ability of trading partners to freely choose who they trade with. Further results are shown in Figure 7.13 for the non-reversible case.

7.5.6 Dynamical Final Demand

The national account flows show that aid and remittances feed into the production process as general consumption processes. Assume that net aid receipts are given by $A(t) = \beta(1 - a)^2 x(t - 1)^2$, and remittances are $R(t) = \gamma(1 - a)x(t - 1)$, where population dependence is omitted, and where aid and remittances are dependent on production one time step earlier (as there is a time lag between assigning aid and its receipt). Using equations (7.8) and (7.9), one obtains for the one-country model, assuming $\tau = 1$:

$$f(t) = f_0 + g^C|(1 - a)(1 + g^W)(1 - g^T)x(t) + \gamma(1 - a)x(t - 1)| \\ + g^G \max(0, (1 - a)(g^T(1 + g^W) - g^W)x(t) + \beta(1 - a)^2 x(t - 1)^2). \quad (7.41)$$

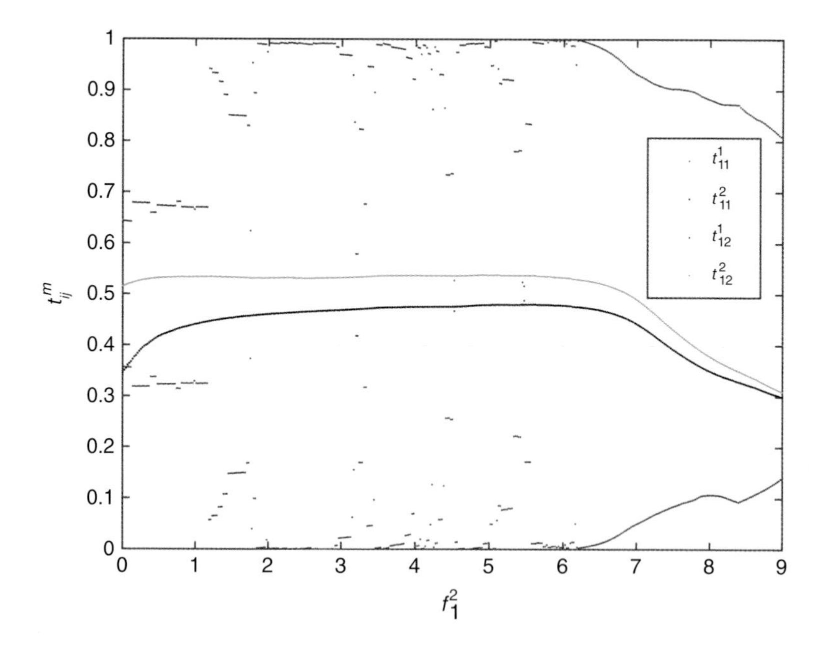

Figure 7.13 Two-country, two-sector non-reversible model, trade coefficients t_{ij}^m versus f_1^2

The max function disallows government borrowing, which can otherwise be implemented using an abs function. For household spending, with the parameters chosen, the abs is superfluous in this case, as each term is positive, although it would be required in the general case. The supply = demand (reversible) and capacity equations are given by, respectively:

$$x(t) = \max(0, a\, x(t) + f(t) + b\, k(t+1)), \tag{7.42}$$

$$k(t+1) \equiv K(t+1) - K(t) = rK(t)(x(t) - pK(t)). \tag{7.43}$$

This is shown in Figure 7.14a for 2% aid giving, varying the growth parameter r. Compare Figure 7.14b, which has just 1% for household and government spending, with the previous static final demand model in the "One-country, one-sector investment growth model with reversibility" section.

Taking $a = b = 0.1$, $p = 0.5$, $\gamma = 20/9$, $f_0 = 1$, $r = 0.55$, $g^T = 0.2$, $g^C = 0.2$, $g^W = 0.1$, $g^G = 0.2$, one obtains the results shown in Figure 7.14, varying the aid parameter β. $\beta < 0$ (i.e. for aid giving) leads to reduced government consumption, reducing final demand and keeping the dynamics away from the strange attractor as for the case of static final demand (recall that increasing final demand as a parameter drives behaviour towards the strange attractor). Remarkably, the non-linear period-2 behaviour survives, not leading to a linear branch of solutions. This is probably due to the value of the growth parameter $r = 0.55$, which for $\beta = -0.02$ was in the non-linear regime. On the other hand, aid receipts ($\beta > 0$) increase government consumption, which if too big leads to the strange attractor, as for the static final demand models. The lack of dependence on β is because the second argument of the max function is negative which occurs as tax receipts net of welfare transfers for these parameter values are less than the putative aid donations for this model where the government is prevented from borrowing by the max function. Figure 7.14a gives the corresponding results for $\gamma = 0.9$, showing the onset of period doubling.

Figure 7.14b shows the fractal set of the strange attractor for aid receipts corresponding to $\beta = 0.04, \gamma = 20/9$. Figure 7.15 shows the effect of varying β.

7.5.7 *Labour*

The value added in an industry is the expenditure to attract the resources (land, labour and capital) necessary for production, such as property, income, wages and salaries. Following Duchin and Szyld (1985), Takayama (1985) and Dietzenbacher (1988), labour requirements in I/O modelling can be obtained via labour coefficients:

$$\tilde{E}_i^m(t) = \sum_n L_i^{mn}(t) x_i^n(t) / p_i^n(t), \tag{7.44}$$

where $\tilde{E}_i^m$ is the labour demand of sector m in country i with labour coefficients $L_i^{mn}(t)$ giving the amount of labour of occupation m required to produce a unit of output of sector n during period t. Wages w_i^m are defined through $w_i^m E_i^m = L_i^m$, where factor payments to households L_i^m are given by equation (7.21), with x_i^m measured in dollar value, and E_i^m is the number of people employed in country i by sector m.

Equation (7.44) says that the movement of x_i^m determines the labour requirement, but there is no mechanism to eliminate the gap between the labour requirement and the supply of labour.

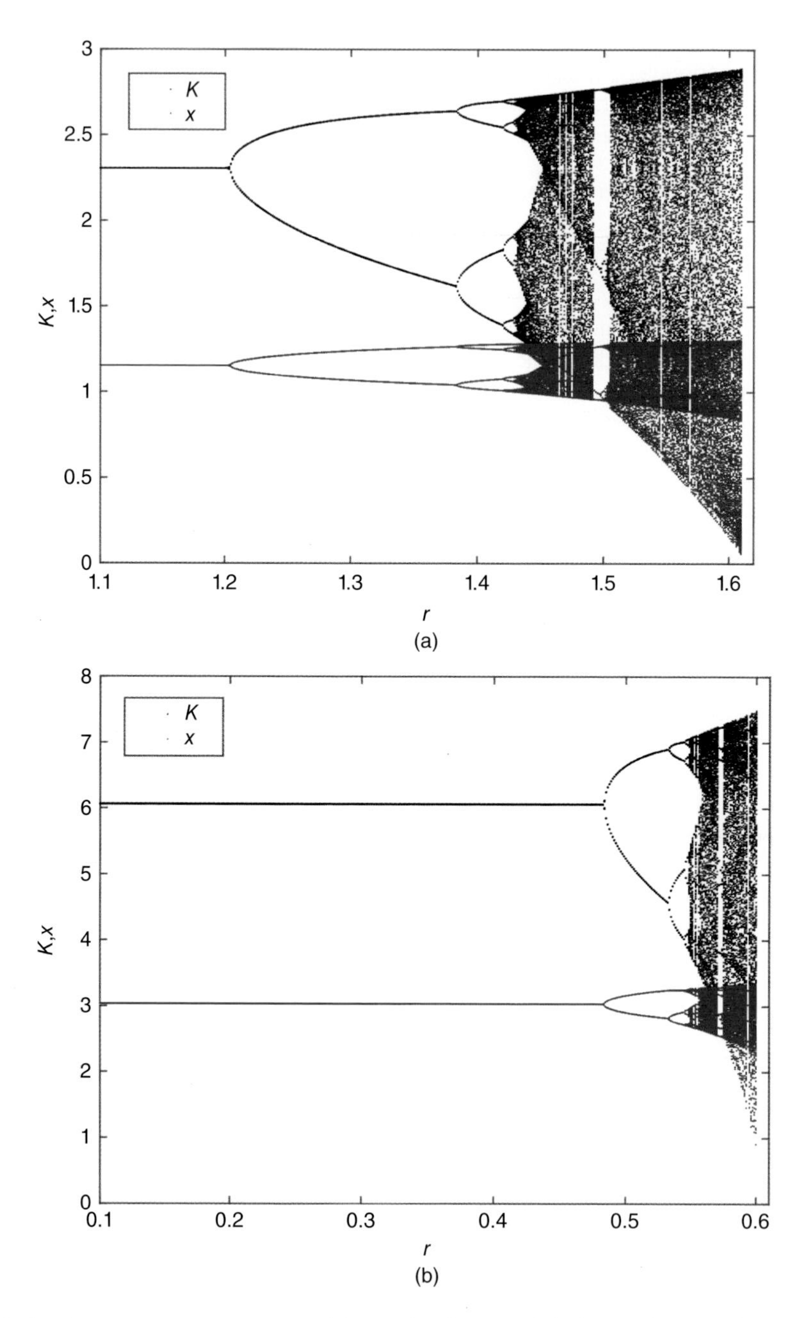

Figure 7.14 One-country, one-sector model, $a = b = 0.1$, $p = 0.5$, $\beta = -0.02$, $\gamma = 20/9$, $f_0 = 1$, (a) $g^C = 0.01$, $g^G = 0.01$, $g^T = 0.5$, $g^W = 1.8$, (b) $g^C = 0.2$, $g^G = 0.2$, $g^T = 0.2$, $g^W = 0.1$, including aid and remittances: production capacity (black)/production (red) versus r

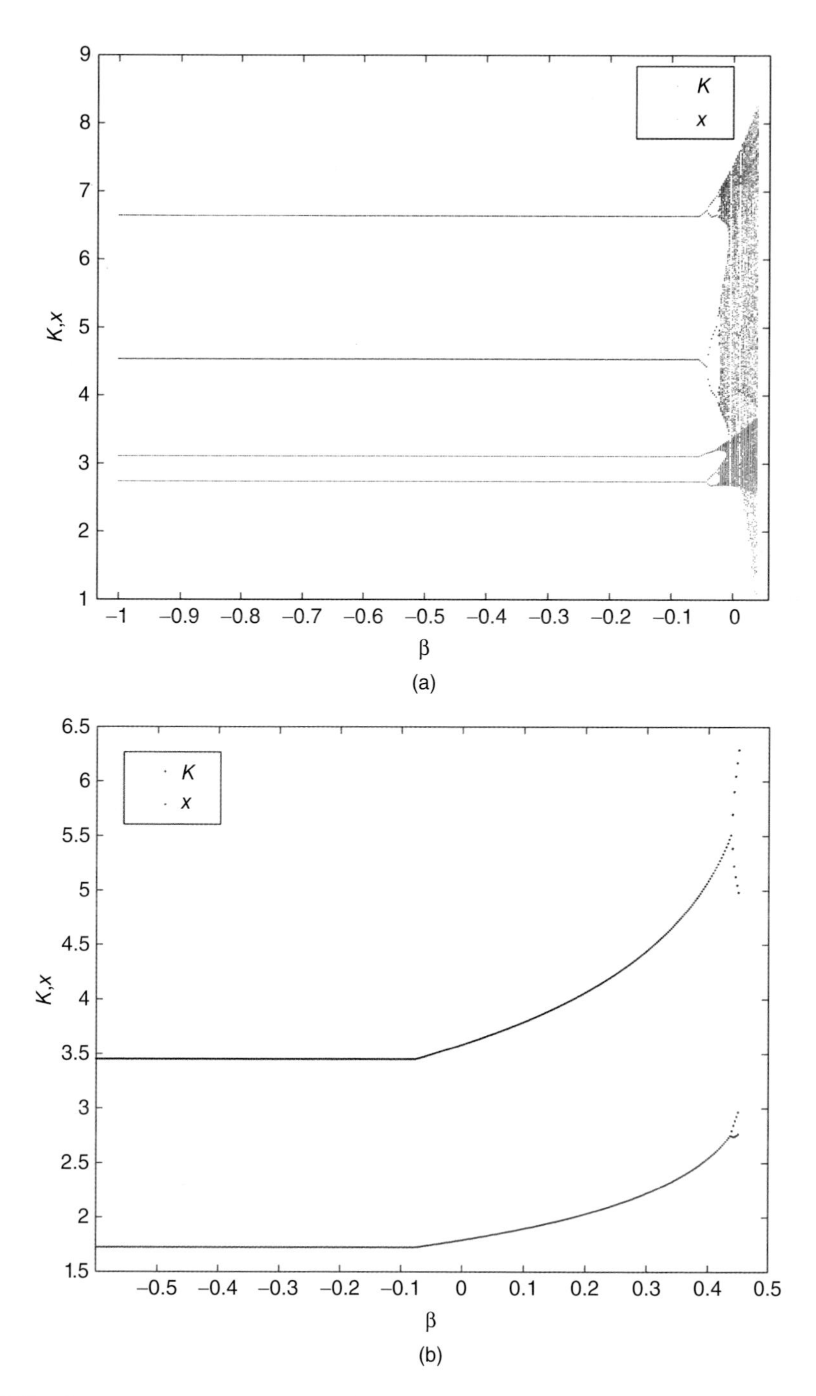

Figure 7.15 One-country, one-sector model, production capacity (black)/production (red) versus β. (Left) $\gamma = 20/9$; (right) $\gamma = 0.9$

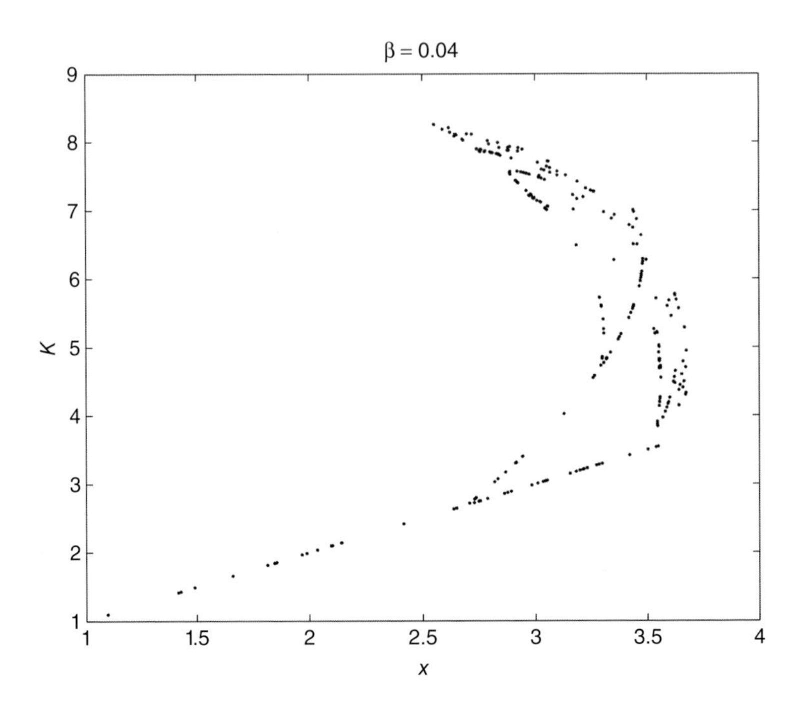

Figure 7.16 One-country, one-sector model, strange attractor, $\beta = 0.04$

This can be avoided by introducing labour in the mechanism of the producer's substitution. That is, if the labour supply grows faster than is required, then the price of labour will go down and encourage the use of labour in production which in turn increases the labour requirement. Thus, the labour coefficients are not fixed constants but functions of prices. Let a_{0i}^{m} be the amount of labour necessary to produce one unit of the mth good that is, $a_{0i}^{m} = \sum_{n} L_{i}^{nm}$, which (as well as wages) feeds into the price model discussed in the "The price model" section. The labour supply pool will depend on population parameters, with migration inducing a response in final demand (Hewings and Jensen, 1987). Migration can have a considerable impact on the labour market. Migrants tend to be of working age, and while the general effect is to increase the supply of labour at all wage rates, migration especially affects supply at lower wage rates. This is because migrants tend to come from low-wage economies. Minimising the difference between labour supply and demand leads to wages decreasing when there is an increase in the labour supply pool. As in the "A multi-region, multi-sector, investment growth model" section, the labour coefficients can be made to freely vary in the minimisation to allow equilibrium to be reached between labour supply and demand.

7.5.8 The Price Model

Most I/O studies value the entries in I/O accounts in producer's prices (i.e. prices at which the seller completes the transaction, sometimes called free-on-board (FOB) prices) (Miller and Blair, 2009; Wilson and Fry, 2012). The purchaser incurs the producer's price plus trade and transportation margins (and often excise taxes). The convention in most I/O studies is to assign the margins on all interindustry transactions in a column to the industry responsible for the margin. Since trade and transportation margins for all transactions into an industry are accumulated as single values for each industry, they in effect become service inputs to that industry. Thus, the sum of all inputs measured in producer's prices plus the value of all transportation and trade margins valued as service inputs are then the value of all inputs in consumer prices.

The money supply is typically taken to be the policy instrument of the central bank, and the interest rate adjusted to bring money supply and money demand into equilibrium. Here, we assume the central bank is assumed to set a target for the nominal interest rate. It then adjusts the money supply to whatever level is necessary to ensure that the equilibrium interest rate (which balances money supply and demand) hits the target. The main advantage of using the interest rate, rather than the money supply, as the policy instrument in the dynamic model is that it is more realistic. Today, most central banks, including the Federal Reserve, set a short-term target for the nominal interest rate. Hitting that target requires adjustments in the money supply. For this model, we do not need to specify the equilibrium condition for the money market, but one should keep in mind that when a central bank decides to change the interest rate, it is also committing itself to adjusting the money supply accordingly.

The formulation of the price system here follows Solow (1959). Consider an individual with money to invest who can either lend it (say, to a bank) at the interest rate r_i or invest it in the production of the mth good.

For establishing an increase in capacity of one unit in sector m in country i, the bundle of goods required is $\sum_{nj} t^n_{ji} \sum_{\theta=1}^{\tau} b_i^{nm\theta}$. In order for the capacity to be ready for use in period $(t+1)$, money would have to be invested $\sum_{nj} t^n_{ji} p^n_j(s) b_i^{nm\phi_s}$, at times s, where $s = t, t-1, \ldots\ t-\tau+1$, $\phi_s = t+1-s$, that is, the investment would be of total value $\sum_{nj} t^n_{ji} \sum_{\theta=1}^{\tau} p^n_j(t-\theta+1) b_i^{nm\theta}$, where $p^n_j(t)$ is the producer (i.e. seller) price of the mth good in country j. By instead lending the money during the same time periods, he can obtain $\sum_{nj} t^n_{ji} \sum_{\theta=1}^{\tau} p^n_j(t-\theta+1) b_i^{nm\theta} \prod_{\phi=0}^{\theta}(1 + r_i(t-\phi+1))$ at the beginning of the $(t+2)$th period.

Instead of lending his money, he can invest it in the production of one unit of the mth good. That is, he buys the configuration of capital equipment necessary to increase capacity by one unit in sector m. When the capacity is installed, it will produce one unit of output (Johansen, 1978). This requires him to tie up his money for at most the maximum gestation lag τ. Let capacity be installed at the beginning of time period $t+1$, and assume that wages and material cost are paid at the end of that period (i.e. in period $t+2$). Assume that labour is the only primary factor, and let $p^m_{0i}(t)$ be the price of labour in period t: $p^m_{0i}(t) \equiv w^m_i(t)$. Then the wage cost and material cost for the production of one unit of the mth good in country i in period $t+1$ are given, respectively, by $p^m_{0i}(t+2)a^m_{0i}$ and $\sum_n p^n_i(t+2) a_i^{nm}$. As the value of the capital equipment at the beginning of period $(t+2)$ is $\sum_{nj} p^n_j(t+2) t^n_{ji} \sum_{\theta=1}^{\tau} b_i^{nm\theta}$, then the basic price equation is:

$$\sum_{nj} t^n_{ji} \sum_{\theta=1}^{\tau} p^n_j(t-\theta+1) b_i^{nm\theta} \prod_{\phi=0}^{\theta} (1 + r_i(t-\phi+1)) = p^m_i(t+2) - p^m_{0i}(t+2)a^m_{0i}$$

$$- \sum_{nj} p^n_j(t+2) t^n_{ji} a_i^{nm} + \sum_{nj} p^n_j(t+2) t^n_{ji} \sum_{\theta=1}^{\tau} b_i^{nm\theta}. \tag{7.45}$$

We use per unit prices, but wish to use technical coefficients in notional dollar units. This is done by making the replacement $a_i^{mn} \rightarrow a_i^{mn} p_i^n / p_i^m$ (Miller and Blair, 2009), with the same for $b_i^{mn\theta}$. This amounts to changing from pre-multiplication by p to post-multiplication.

The c_{ij}^m in the trade coefficients in equation (7.16) can now be updated to include producer prices $c_{ij}^m = \sigma^m D_{ij} + p_i^m$ (cf. Wilson and Fry, 2012), with D_{ij} the distance from i to j, and transport costs σ^m to be calibrated within the modelling process.

In global security modelling, if T_{ij} is the threat that flows from nation i to nation j, which is the military capability of nation i that exists to counter the threat posed by nation j, and T_{ii} is the military capability of nation i that exists to counter the internal threats brought about by instabilities within that nation, proportional to the level of instability in nation i, then c_{ij}^m can be updated to include an additional cost based on the level of threat between two countries (i.e. $c_{ij}^m = \sigma^m D_{ij} + \psi^m T_{ij} + p_i^m$).

As before, equation (7.45) can be cast in matrix form. Assemble matrices $\hat{A}^{mn} = \text{diag}(a_1^{mn}, \ldots, a_P^{mn})$, $\tilde{A}^{mn} = \hat{A}^{mn}$ whereas $(T^m)_{ij} \equiv t_{ij}^m$, $\tilde{T} = \text{diag}(T^1, \ldots, T^m)$ (i.e. $\tilde{T}$ is as defined in the appendix to this chapter). The matrix $\tilde{B}^\theta$ is assembled in the same way as for $\tilde{A}$, but using coefficients $b_i^{mn\theta}$, and a matrix $\tilde{B}_r^\theta$ is assembled in the same way using coefficients $b_i^{mn\theta} \prod_{\phi=0}^{\theta}(1 + r_i(t - \phi + 1))$. Note that the matrices $A, \tilde{A}$ (and the same for B^θ, B_r^θ and their tilde versions) are related via a similarity transformation in a similar way to $T, \tilde{T}$ (for $T, \tilde{T}$, see the appendix): $\tilde{A} = \tilde{Q}^{-1} A \tilde{Q}$ where $\tilde{Q}_{ij} = \delta_{ij}, j = (i - (k-1)M - 1)P + k, k = \lceil i/M \rceil$, $i, j = 1, \ldots, MP$ (i.e. the roles of i and m are interchanged), and $\tilde{Q}^{-1} = \tilde{Q}^T$.

Then, for column vector $p = (p_1^1, \ldots, p_P^1, \ldots, p_1^N, \ldots, p_P^N)^T$, and a_0 a column vector of the same form:

$$\left(1 - \tilde{A}\tilde{T} + \sum_{\theta=1}^{\tau} \tilde{T}\tilde{B}^\theta\right) p(t+2) = \sum_{\theta=1}^{\tau} \tilde{T}\tilde{B}_r^\theta p(t - \theta + 1) + p_0(t+2) \cdot a_0. \tag{7.46}$$

The initial conditions for solving this equation are to specify $p_i^m(t)$ for $t = t_0 - \tau + 1, \ldots, t_0$ and $r_i(t)$ at $t = t_0 - \tau + 1, \ldots, t_0 + 1$. Once the prices are calculated, these can be used to update the trade coefficients for the next time period.

The interest rate $r_i(t)$ can be specified by modelling the banking system, demand for money, inflation and so on. Solow (1959) let the interest rate be an arbitrary function of time, then assumed it was constant, whereas Jorgenson (1960) assumed $r_i(t) = 0$ in proving his dual stability theorem.

As the global system is closed, one expects that an export surge, leading to an increase in interest rates, causes inflation to decrease and jobs in interest rate-sensitive sectors (e.g. construction) to decrease. Conversely, with an import surge, the central banks decrease interest rates, resulting in the loss of jobs due to imports being compensated by the gain in the number of jobs in other interest rate-sensitive sectors (e.g. construction) (Krugman, 1996).

An alternative to this price model would be to adjust prices in a similar way to Wilson and Fry (2012), but instead of trying to match production supply to production capacity (this is eschewed in Duchin and Szyld, 1985), adjust prices to minimise the gap between supply and demand.

7.6 Conclusions

We show that trade coefficients arise naturally from the entropy maximisation procedure associated with trade flows and also that the model of Levy *et al.* (2014) is equivalent to that of Wilson (1970). We also show the one-to-one relationship between trade coefficients in the literature (see e.g. Miller and Blair, 2009; Campisi *et al.*, 1991) and the import propensities and import ratios of Levy *et al.* (2014).

We develop non-linear dynamical global trade models. Capacity constraints are implemented and the reversibility of capital stock is considered, but with preference given to models not allowing reversibility, as in the literature (see e.g. Duchin and Szyld, 1985; Edler and Ribakova, 1993). Full capacity utilisation is not assumed, as guided by the literature (see e.g. Duchin and Szyld, 1985).

By contrast, for the Wilson investment growth model, fluctuations in capacity cannot be avoided or controlled at source. This creates an unavoidable gap between supply and demand, not present in the static model. When supply and demand are not equal, the difference is minimised. It is predicted that if the final demand of a sector is five or six times that of any other sector, non-linear bifurcation behaviour will occur in the trade model. Allowing trade coefficients to vary freely in a dynamical manner not only avoids sudden non-linear bifurcation behaviour but also allows further minimisation of the gap between supply and demand. This is equivalent to allowing trade relationships and prices to vary freely to minimise the gap between supply and demand within a trading period. Explicit expressions are given for the non-linear map, the location of the strange attractor and the onset of period doubling in terms of economic parameters in the one-dimensional case. This gives policy planners the ability to avoid period doubling by altering or influencing the economic parameters.

Additionally, non-constant final demand is modelled by considering how aid and remittance flows enter (or leave) an economy, as well as military spending flows by governments. These flows are therefore coupled to the trade flows as well as to the trade coefficients. Migration and population are coupled to labour which in turn is coupled to trade.

In future work, the mechanism for minimising the difference between supply and demand could be made smoother, such as by the introduction of a parameter in the cut-off function. The presence of a cut-off function, however, ensures that the dynamics converges exponentially quickly, making fixed point analysis useful for short-term forecasting. In the capacity planning approach of Duchin and Szyld (1985), the projected capacity function could be changed from its present geometric increase form to a predictor–corrector mechanism. This function can be freely chosen by the capacity planner or policy maker.

For the non-linear models, no contact has as yet been made with big data for a total number of countries and economic sectors in the world as has been done with the linear static model of Levy *et al.* (2014). These non-linear trade models can be coupled to models of aid and remittances, migration and security, again using big data for all the models. In addition, it is of interest to couple these models to city modelling in order to investigate the effect of coupling trade flows between countries and those of cities.

References

Baudains P, Davies T, Fry H and Wilson A. (2011). Introducing space to mathematical models of conflict. http://www.bartlett.ucl.ac.uk/casa/publications/working-paper-171

Bentley B. (2015). The welfare budget. https://docs.google.com/spreadsheets/d/1LKYgTGNeOJL-RrrgASkX526uLNagSeQUSMLpe_4jaSo/edit#gid=0

Breisinger C, Thomas M and Thurlow J. (2010). Social accounting matrices and multiplier analysis. http://www.ifpri.org/sites/default/files/publications/sp5.pdf

Burnside C and Dollar D. (1988). Aid, policies, and growth: revisiting the evidence. http://http://elibrary.worldbank.org/doi/book/10.1596/1813-9450-3251

Burress D, Eglinski M and Oslund P. (1988). A survey of static and dynamic state-level input-output models. http://ipsr.ku.edu/resrep/pdf/r4.pdf

Campisi D, Nastasi A, La Bella A and Schachter G. (1991). The dynamic behavior of multiregional multisectoral models: A biregional application to the Italian economy. *Applied Mathematical Modelling* 15: 525–533.

Campisi D and Gastaldi M. (1996). Decomposing growth in a multiregional I-O framework. *Annals of Regional Science* 30: 409–425.

Coyle D. (2014). *GDP: A Brief but Affectionate History*. Princeton, NJ: Princeton University Press.

Crowther K and Haimes Y. (2010). Development of the multiregional inoperability input-output model (MRIIM) for spatial explicitness in preparedness of interdependent regions. *Systems Engineering* 13(1): 28–46.

Danao R. (2007). *Mathematical Methods in Business*. Manila: University of the Philippines Press.

Dietzenbacher E. (1988). Trade imbalance in an input-output model. *Mathematical and Computer Modelling* 10(12): 883–889.

Doedel E and Oldeman B. (2012). Software for continuation and bifurcation problems in ordinary differential equations. http://indy.cs.concordia.ca/auto/

Duchin F. (2004). Input-output economics and material flows. http://www.economics.rpi.edu/workingpapers/rpi0424.pdf

Duchin F and Szyld D. (1985). A dynamic input-output model with assured positive output. *Metroeconomica* 37: 269–282.

Edler D and Ribakova T. (1993). The Leontief-Duchin-Szyld dynamic input-output model with reduction of idle capacity and modified decision function. *Structural Change and Economic Dynamics* 4(2): 279–297.

EconPort. (2006). Examples of calculating GDP. http://www.econport.org/content/handbook/NatIncAccount/CalculatingGDP/Examples.html

Hewings G and Jensen R. (1987). *Handbook of Regional and Urban Economics*, vol. 1. Amsterdam: Elsevier.

Johansen L. (1978). On the theory of dynamic input-output models with different time profiles of capital construction and finite life-time of capital equipment. *Journal of Economic Theory* 19: 513–533.

Jorgenson D. (1960). A dual stability theorem. *Econometrica* 28(4): 892–899.

Kaddar A and Alaoui HT. (2008). Fluctuations in a mixed IS-LM business cycle model. *Electronic Journal of Differential Equations* 134: 1–9.

Krugman P. (1996). A country is not a company. *Harvard Business Review*: 40–51.

Levy R, Evans TE and Wilson A. (2014). A global inter-country economic model based on linked input-output models. http://www.bartlett.ucl.ac.uk/casa/publications/working-paper-198

May R. (1976). Simple mathematical models with very complicated dynamics. *Nature* 261: 459–467.

Mankiw G and Taylor M. (2007). *Macroeconomics*. New York: Worth.

Miller R and Blair P. (2009). *Input-Output Analysis: Foundations and Extensions*. Cambridge: Cambridge University Press.

Mohapatra M. (2010). Taxing remittances is not a good idea. http://blogs.worldbank.org/peoplemove/taxing-remittances-is-not-a-good-idea

Pooler J. (1994). An extended family of spatial interaction models. *Progress in Human Geography* 18(1): 17–39.

Redwood A, El-Hodiri M and Burress D. (1987). Final Report Research Improvement Award for Economic Modeling. http://ipsr.ku.edu/resrep/pdf/m124.pdf

Richardson L. (1960). *Arms and Insecurity: A Mathematical Study of Causes and Origins of War*. Pittsburgh, PA: Boxwood Press.

Samuelson P. (1951). *Abstract of a Model Concerning Substitutability in Open Leontief Models*. New York: John Wiley & Sons Inc.

Solow RM. (1959). Competitive valuation in a dynamic input-output system. *Econometrica* 27(1): 30–53.

Takayama A. (1985). *Mathematical Economics*. Cambridge: Cambridge University Press.
Timmer M, Erumban A, Gouma R, Los B, Temurshoev U, de Vries G, *et al.* (2012). The World Input-Output Database (WIOD): contents, sources and methods. contents, sources and methods. http://www.wiod.org/publications/source\LY1\textbackslash_docs/WIOD\LY1\textbackslash_sources.pdf
Wagner D. (2003). Aid and trade: an empirical study. *Japanese and international Economies* 17: 153–173.
Wilson A. (1970). Inter-regional commodity flows: entropy maximizing approaches. *Geographical Analysis* 2(3): 254–282.
Wilson A. (2010). Urban and regional dynamics from the global to the local: 'hierarchies', 'DNA' and 'genetic planning'. *Environmental and Planning B, Planning and Design* 37(5): 823–837.
Wilson A. (2012). *The Science of Cities and Regions*. Berlin: Springer.
Wilson A. (2015). *Global Dynamics: Models from Complexity Science*. Chichester: John Wiley & Sons Ltd.
Wilson A, Coelho J, Macgill S and Williams H. (1981). *Optimisation in Locational and Transport Analysis*. Chichester: John Wiley & Sons Ltd.
Wilson A and Fry H. (2012). A dynamic global trade model with four sectors: food, natural resources, manufactured goods and labour. http://www.bartlett.ucl.ac.uk/casa/publications/working-paper-178

Appendix

A.1 Proof of Linearity of the Static Model and the Equivalence of Two Modelling Approaches

In Redwood *et al.* (1987), an import coefficient μ is defined, equal to $d/(1-d)$, with d the import ratio defined in Levy *et al.* (2014). The static model in Levy *et al.* (2014) defines the following equations for each country:

$$x_i^m = (1 - d_i^m)\left(\sum_n a_i^{mn} x_i^n + f_i^m\right) + e_i^m \tag{A.1.1}$$

$$i_i^m = d_i^m \left(\sum_n a_i^{mn} x_i^n + f_i^m\right), \tag{A.1.2}$$

for imports i_i^m and exports e_i^m.

The technical coefficients satisfy $0 \leq a_i^{mn} < 1$ and the Brauer–Solow condition $\sum_m a_i^{mn} < 1$ (see Danao, 2007). Intermediate demand z_i^{mn} is given by $z_i^{mn} = a_i^{mn} x_i^n$. Note that in the notation of Wilson (1970), x_{*i}^m is intermediate plus final demand of country i, $x_{i*}^m \equiv x_i^m$, and the first term on the right-hand side of equation (A.1.1) is x_{ii}^m. In the notation of Miller and Blair (2009), $\mathrm{Zd}_i = \sum_{m,n} z_i^{mn}$ and in the notation of Wilson (1970), $\mathrm{Xf}_i = \sum_m x_{ii}^m$.

These two equations say nothing more than that the production of a country supplies the exports of that country and a fraction $1-d$ of its own demand (intermediate plus final) and that a country's imports supply the remaining fraction d of its demand (intermediate plus final) requirement. For each sector, an import propensity matrix $(P_m)^{ij} \equiv p_m^{ij}$ is defined (with zero diagonal elements), $\sum_i p_m^{ij} = 1$, such that:

$$e_i^m = \sum_{j \neq i} p_m^{ij} i_j^m, \tag{A.1.3}$$

That is, the static model of Levy *et al.* (2014) can be written as the single equation:

$$x_i^m = \sum_j t_{ij}^m \left(\sum_n a_j^{mn} x_j^n + f_j^m \right) \quad \text{(A.1.4)}$$

where the trade coefficients are given by $t_{ij}^m = p_m^{ij} d_j^m, i \neq j$ and $t_{ii}^m = 1 - d_i^m$. The three-index variable x_{ij}^m of Wilson (1970) is then defined by $x_{ij}^m = t_{ij}^m \left(\sum_n a_j^{mn} x_j^n + f_j^m \right)$. One can easily show that $x_{ij}^m = p_m^{ij} \sum_{k \neq j} x_{kj}^m$ for $i \neq j$, i.e. x_{ij}^m is the same as the y_{ij}^m variable in Levy *et al.* (2014) for $i \neq j$. Using the t_{ij}^m notation instead of p_{ij}^m, d_i^m, one has that $x_{ij}^m = t_{ij}^m \sum_k x_{kj}^m$. In addition, x_{ij}^m is the same as $Y_{ij}^{(m)}$ in Wilson and Fry (2012) and z_i^{rs} in Levy *et al.* (2014), where a different index convention is used.

Defining T by $T_{ij} = \hat{t}_{ij}$ where $\hat{t}_{ij} = \mathrm{diag}(t_{ij}^1, \ldots, t_{ij}^N)$ and $A = \mathrm{diag}(A_1, \ldots, A_P)$, with $(A_i)^{mn} = a_i^{mn}$, then for all countries and all sectors, the static model in Levy *et al.* (2014) is given by the matrix equation:

$$x = T(Ax + f) \quad \text{(A.1.5)}$$

where vectors x, f are of the form $x^T = (x_1^1, \ldots, x_1^N, \ldots, x_P^1, \ldots, x_P^N)$. This matrix equation is clearly linear in x.

The above used the Levy *et al.* (2014) model as a starting point. One can instead start with the Leontief–Strout formulation of Wilson (1970). There one has by definition that $x_i^m = \sum_j x_{ij}^m$. The result of the "Supply and demand" section was that the entropy maximisation in Wilson (1970) naturally gives rise to trade coefficients t_{ij}^m defined by $x_{ij}^m = t_{ij}^m \sum_k x_{kj}^m$. Therefore one obtains $x_i^m = \sum_j t_{ij}^m \sum_k x_{kj}^m$. From the Leontief–Strout equation (Wilson, 1970), $\sum_k x_{ki}^m = \sum_n z_i^{mn} + f_i^m$, which says that the supply to country i equals the demand of country i, one obtains $x_i^m = \sum_j t_{ij}^m (\sum_n z_j^{mn} + f_j^m)$ which is equation (A.1.3). This paragraph shows that one can properly say that the model of Levy *et al.* (2014) can be derived directly (as they are equivalent) from the Leontief–Strout equation and entropy maximisation of Wilson (2010). The model of Levy *et al.* (2014) says that production from country i supplies every other country (including itself) but this is implied by, and implies, the formulation of Wilson (1970), which has that the supply to country i (including supply from itself) equals the demand of country i. This paragraph shows the two modelling approaches are equivalent because the flows occur in a closed global system and are circular, as well as the fact that the entropy maximisation of Wilson (1970) naturally gives rise to trade coefficients.

Note that equations (A.1.2) and (A.1.3) are in any case already given in Miller and Blair (2009) (equation (3.22)), where the trade coefficients are instead called interregional coefficients. They are also given in Campisi *et al.* (1991), except that there the model is 'closed', that is, non-investment final demand is considered as another sector (i.e. f_i^m is effectively set to zero). This model is also given explicitly in Crowther and Haimes (2010).

Equation (A.1.3) is easily solved for x. For computation of sparse matrices, it is useful to write T in terms of a matrix defined in a similar way to A, that is, define $\tilde{T} = \mathrm{diag}(T^1, \ldots, T^N)$, with the matrices T^m having components $(T^m)_{ij} \equiv t_{ij}^m$. Then, $T = Q^{-1} \tilde{T} Q$, where Q is given by $Q_{ij} = \delta_{ij'}, j = (i - (k-1)P - 1)N + k, k = \lceil i/P \rceil$, and $Q^{-1} = Q^T$.

Part Four

Space–Time Statistical Analysis

8

Space–Time Analysis of Point Patterns in Crime and Security Events

Toby P. Davies, Shane D. Johnson, Alex Braithwaite and Elio Marchione

8.1 Introduction

The analysis of point data plays a fundamental role in the study of any system in which events of interest occur at discrete locations in space and time. Examples can most readily be found in epidemiology, in which such events might represent cases of disease (see Pfeiffer *et al.*, 2008), but similar analysis has also been applied more recently in the context of crime, in which events correspond to offences (e.g. Johnson *et al.*, 2007), and international conflict, in which events correspond to militarised disputes between countries (e.g. Braithwaite, 2010). In either case, the identification of patterns within such data can offer insight into the underlying generative process, while also suggesting possible courses of preventative action.

8.1.1 Clustering

One of the primary phenomena of interest in point pattern analysis is that of clustering, whereby events tend to occur 'close' to each other, in some sense. This is most naturally seen as the disproportionate occurrence of events at some location in either time or space; that is, a density of points that is greater than would be expected if the distribution of risk was uniform. Clustering of this form may suggest that some property of those places or times is particularly conducive to the phenomenon in question. In terms of disease or crime, the presence of such clustering implies that certain locations are at particular risk, with practical implications for the prevention of the events in question.

Approaches to Geo-mathematical Modelling: New Tools for Complexity Science, First Edition. Edited by Alan G. Wilson.

Companion Website: www.wiley.com/go/wilson/ApproachestoGeo-mathematicalModelling

In addition to the analysis of clustering in either space or time, further insight can be gained by examining the interaction between the two dimensions. Of particular interest is the existence of dependence between the spatial and temporal separation of events; that is, whether events which are close in time tend also to be close in space (and *vice versa*), and whether such space–time clustering exceeds that which would be expected, given any observed (spatial or temporal) clustering. Such dependence is of particular significance because it implies that the spatial distribution of events at any given time depends upon where and how recently events have previously occurred. Spatial clusters of events may be stationary, but where space–time clustering exists they may be more fluid, tending to occur in the spatial vicinity of recent events (i.e. those that are near temporally). This implies that risk is communicable, in some sense, which has both theoretical and practical implications: it implies causality between the events in question, and also provides a basis upon which future events might be predicted.

Clustering of this form can be tested for, and quantified, in a number of ways, the majority of which involve comparison between the observed separation of events and that which would be expected if their spatial and temporal distributions were independent. This separation is most commonly measured in a pair-wise sense, in which closeness is determined in terms of either absolute spatio-temporal distance (Knox, 1964; Mantel, 1967) or a nearest-neighbour basis (Jacquez, 1996). These separations can then be compared against what would be expected under a suitable null hypothesis (e.g. independence in space and time).

One of the most prominent of these techniques is the Knox (1964) test, in which pairs of incidents are categorised according to their 'closeness'. This has been used widely, and is the basis for many recent developments in the analysis of crime. Empirical studies have shown that clustering is near-ubiquitous in urban crime, and these findings form the basis for a number of preventative strategies. In this chapter, we will outline recent work which has sought to extend this research in two key respects: the application of these methods to different types of event, and the extension of the technique in order to allow the identification of more specific fine-grained patterns.

In the first part of the chapter, we will describe the application of these methods to data concerning maritime piracy and insurgent activity. Both of these activities differ fundamentally from urban crimes, and there is little reason *prima facie* to suppose that the same targeting phenomena should also be observed. In fact, however, significant clustering is observed in both cases, suggesting that elements of more general offender spatial decision-making processes may be at play in these novel contexts.

The latter part of the chapter is focussed on technical development. The work is motivated by the observation that existing methods of clustering analysis, while capable of establishing its existence, are unable to discriminate between types of clustering; that is, the underlying patterns that contribute to it. In particular, patterns may be found to be equivalently clustered while displaying notable qualitative differences (of both theoretical and practical significance). By encoding the 'close pair' relationship as a network, we show that techniques from network science can be adapted to measure more nuanced patterns. In particular, we show how 'motifs' can be identified in event patterns: these correspond to targeting 'signatures', and we show that these differ considerably between crime types.

Before describing this work, we will first outline previous research concerning clustering in criminal phenomena, before formally defining the analytic setting and introducing the statistical techniques upon which this chapter is based.

8.1.2 Clustering of Urban Crime

The clustering of criminal phenomena has been studied extensively, and empirical research has consistently found it to be present in real-world data. Although the clustering in question can take several forms, the range of contexts in which it is observed suggests that it is a fundamental feature of criminality, and the desire to explain this has motivated a number of theoretical developments.

The earliest example of the spatial analysis of crime is the work of Shaw and McKay (1969), who demonstrated that the residences of offenders in Chicago tended to be located in close proximity to each other. Numerous studies subsequently showed that crime itself also tended to be concentrated in space: Sherman *et al.* (1989), for example, demonstrated that the majority of predatory crime in Minneapolis occurred in only a small number of locations. Similar findings for alcohol-related crime were also identified by Block and Block (1995), and the theme has latterly been developed by Weisburd *et al.* (2012), who have shown in particular that street segments can also display distinctive criminal character.

The temporal distribution of crime has also been the subject of empirical research. Several studies have shown that the concentration of crime varies with the time of day (Felson and Poulsen, 2003; Ratcliffe, 2002; Tompson and Bowers, 2013), with particular crime types displaying distinctive profiles. In addition to this, crime is also known to display seasonality (Farrell and Pease, 1994; Hipp *et al.*, 2004), with the variation in light and temperature levels cited as one explanation of this. More generally, this can be reconciled with routine activity theory (Cohen and Felson, 1979), which asserts that the distribution of crime is shaped by the regularity and tempo of daily life: the tendency of people to do certain things, in certain places, at certain times.

More recent work has sought to build on these foundations by considering spatial and temporal patterns in tandem. Interaction effects can be observed in a number of forms and at a number of scales. Lersch (2004), for example, showed that certain places are more prone to particular crimes at certain times of day. Over a longer scale, Weisburd *et al.* (2004) have suggested that the criminal character of micro-places can evolve over time, displaying distinctive trajectories. Ratcliffe (2004) has also argued that ‘hot spots’ of crime can be classified according to criteria such as their density and persistence over time.

One particular topic of research has focussed on the formal concept of spatio-temporal clustering. As stated here, this refers to the tendency of offences to occur close to other recent offences, and is most clearly manifested in the commonly observed phenomena of repeat (Johnson *et al.*, 1997; Pease, 1998) and near-repeat (Bowers and Johnson, 2005; Morgan, 2001) victimisation. The presence of this effect has been demonstrated in a number of studies (e.g. Bowers and Johnson, 2005; Johnson and Bowers, 2004a; Townsley *et al.*, 2003), the majority of which are concerned with burglary. Indeed, the clustering of burglary in particular has been shown to be near-ubiquitous in data from several countries (Johnson *et al.*, 2007). It is apparent, however, that the effect is not restricted to burglary only: similar techniques have also shown it to be present for a variety of urban crimes (for a review, see Johnson and Bowers, 2014), including shootings (Ratcliffe and Rengert, 2008), assault and robbery (Grubesic and Mack, 2008).

The desire to explain space–time clustering has led to a number of theoretical developments. Although a number of explanations have been offered (see Pease, 1998), examination of offender data suggests that the majority of instances of (near-)repeat victimisation are the work of the same offender (Bernasco, 2008; Johnson *et al.*, 2009b). In light of this, it has been suggested that such offenders are engaged in 'foraging' behaviour, in which they seek to maximise the rewards associated with their activity whilst minimising the effort and uncertainty associated with the target selection process (see Brantingham, 2013; Johnson and Bowers, 2004b; Johnson *et al.*, 2009b). The implication of this is that offenders tend to return to targets that they have recently victimised successfully, since the knowledge acquired during previous incidents (concerning both the likely risk and reward) renders them more appealing than alternative targets about which little is known. Targets near to those recently victimised are also thought to be favoured since, in line with the "first law of geography" (Tobler, 1970), they will have characteristics in common, including those that are likely to influence offender targeting decisions (see Johnson, 2014; Johnson and Bowers, 2004a).

The prevalence of space–time clustering is of particular significance in the study of crime, since it suggests immediate opportunities for crime prevention. The fact that offences occur near recent ones suggests that some crimes are, to some extent, predictable: certain areas are known to be at higher risk at certain points in time. Recent research has sought to take advantage of this by modelling risk dynamically (Bowers *et al.*, 2004; Johnson *et al.*, 2009a; Mohler, 2011), and such approaches have shown increased predictive performance when compared with methods that consider only the spatial heterogeneity of risk. Ideas and models such as these provide the foundation for much of the field of 'predictive policing' (see Perry *et al.*, 2013), which aims to efficiently direct preventative activity on the basis of dynamic risk predictions.

8.1.3 The Knox Test

Many of the studies described so far in this chapter employ statistical techniques based on an approach first introduced by Knox (1964). Since it is the basis for the sections which follow, we will outline the method here in general terms before discussing refinements in later sections.

The motivation for the Knox test, which was originally developed in epidemiology, is to test whether a set of event data is consistent with having been generated by a contagion-like process. The rationale is that, if such an effect is present, events will tend to be followed by other events in close spatial proximity more than would be expected if the timing and location of events were independent. The central principle of the Knox test is to measure proximity by classifying pairs of incidents in terms of their closeness in space and time.

Before describing the steps involved, it is useful to specify the analytical setting in formal terms (which apply generically, regardless of the type of event). It is assumed that there are N events, indexed by i, each of which has some spatial location x_i and a time of occurrence t_i. The spatial location can take any form (e.g. latitude/longitude or easting/northing), as long as a corresponding distance metric d can be applied to any pair of locations. The shorthand d_{ij} will be used to represent the distance between two events i and j, so that

$$d_{ij} = d(x_i, x_j) \geq 0. \tag{8.1}$$

Similarly, $t_{ij} = t_j - t_i$ is defined as the temporal separation between events, though in this case the value can be positive or negative, depending on whether j occurs after or before i, respectively. Although only the absolute value is relevant in some contexts (e.g. the basic Knox test), the distinction is significant in others.

The first step in the Knox test is to divide the ranges of possible spatial and temporal separations into a number of bands (in the original test, only two were used – near and far); that is, the spatial and temporal scales are partitioned. The choice of partition is arbitrary, but typically is informed by the anticipated radius of effect: for residential burglary, for example, bandwidths of 100 metres and 7 days are common. The resulting categories represent different degrees of 'closeness' in each dimension.

The next stage of the procedure involves comparing all possible pairs of events $\{i, j\}$ (for a dataset of size N, there will be $\frac{N(N-1)}{2}$ comparisons) and calculating the separations d_{ij} and t_{ij} in each case. For each of these, the band into which the measurement falls can then be identified. These results are then used to populate a contingency table: a two-way table in which the value of each cell represents the number of pairs $\{i, j\}$ which fall within the corresponding spatial and temporal bands. This encodes the observed separations.

Having computed the contingency table, its values must be compared with those that would be expected under the null hypothesis that the timing and location of events are independent. This can be achieved in a number of ways, such as by estimating the expected counts arithmetically under the assumption that they are Poisson distributed (see Knox, 1964). Here, we describe a Monte Carlo permutation approach in which alternative datasets are explicitly constructed under the null hypothesis (see Johnson *et al.*, 2007). This method is advantageous for a number of reasons. For example, it is a non-parametric approach and hence is not subject to the same distributional assumptions used in the original Knox test (and parametric approaches more generally). In addition, it controls for the underlying spatial and temporal distributions of the data, so that any significant pattern must be due to interaction effects.

With the observed event data as a starting point, permuted versions of the data are generated by repeatedly shuffling the timings of events while maintaining their spatial location. The motivation for this shuffling of event times is to break down any alignment with the spatial components: if the null hypothesis is true, the shuffling ought to make no significant difference to the number of pairs observed in each space–time band. Separations can be calculated for any pair of events under a given shuffling: denoting the permutation by σ, the comparison between two events i and j is therefore based on their true spatial locations x_i and x_j (as before) but their permuted time-points $t_{\sigma(i)}$ and $t_{\sigma(j)}$. In this way, a contingency table can be calculated for each permuted dataset in exactly the same way as for the observed data.

This process is repeated n times, each with a different permutation, and a contingency table produced in each case. This allows the significance of the observed values to be evaluated. For each cell of the contingency table, the observed value can be compared with the corresponding values in the n alternative contingency tables. These represent a null distribution, the deviation from which can be quantified: if r is the rank at which the observed value would appear in an ordered list of the n values, a pseudo-significance (see North *et al.*, 2002) is given by:

$$p = \frac{r}{n+1}. \tag{8.2}$$

Furthermore, the magnitude of the discrepancy can be quantified by computing either the z-score of the observed value, relative to the median, or its ratio with the median.

These tests can, of course, be carried out for every cell of the contingency table. Overall, therefore, we have an estimate of significance for each pair of spatial and temporal bands. These can be interpreted as measuring clustering at various scales: a typical finding, for example, would be that clustering is significant at a range of 1–199 metres and 1–13 days. Such findings are of value in estimating the effective range of clustering phenomena.

Sections 8.2 and 8.3 will describe the application of this technique to data for two types of events that have not commonly been the subject of spatial, let alone space–time, analysis: maritime piracy and insurgent activity.

8.2 Application in Novel Areas

8.2.1 Maritime Piracy

Maritime piracy (hereafter, piracy) is not a new phenomenon, but increases in the number of attacks over the last decade, and the associated cost to the global economy, have attracted attention from scholars, the military and the United Nations alike. However, there are relatively few quantitative analyses of patterns of piracy, and the majority of the published research has focussed on macro-level variation in the risk of piracy across the world's oceans. Such research clearly shows that the risk of piracy is not evenly distributed (e.g. Chalk and Hansen, 2012), and econometric approaches to analysis (e.g. Daxecker and Prins, 2013, 2015a, b; Hastings, 2009) suggest that the political stability and power of states – taken as indicators of a country's likely ability to deal with the problem – are associated with the rate of attacks around a country's coastlines.

Here, we consider patterns of piracy at much finer spatial and temporal scales and examine if, like urban crimes, there exist any regularities in the space–time distribution of pirate attacks. As noted, many different types of urban crime have been found to cluster in space and time. The precise theoretical explanations for anticipating such patterns vary, but one factor that is common to all of the crimes so far considered is that offenders (irrespective of the type of crime they commit) are typically affected by spatial and temporal constraints (Ratcliffe, 2006). That is, they are generally only capable of travelling finite (often short) distances, and can only travel so far so fast. This makes it not unreasonable to expect that they will commit sequential offences near to each other. Moreover, the conditions that make a location conducive for one type of criminal activity on one day may change over time, or may be perceived as so doing, meaning that offenders are perhaps more likely to return to the same locations, or those nearby, before the conditions that are favourable for offending change.

Such generic influences may play a role in explaining the spatial decision making of many types of offenders, including those involved in piracy (and insurgency). For example, if pirates evaluate (however crudely) the costs and benefits of targeting particular locations, it seems reasonable to suggest that – just like burglars – they would be more likely to select locations that are perceived to offer benefits that outweigh the associated risks and effort involved. Most pirates are unlikely to have constantly up-to-date knowledge of the seas, and the types of vessels they use (skiffs) have limited ranges, which reduces opportunities for reconnaissance. Thus, pirates may be expected to return to locations at which they have previously launched successful attacks or at which the conditions are currently perceived to be conducive to piracy.

Furthermore, pirates may perceive that the accuracy of their knowledge – including merchant (or other) vessels' use of particular shipping routes and their levels of protection – will decay over time, and so a quick return to previously targeted locations would be logical. However, targeting vessels at the exact same locations repeatedly is likely to lead to particular routes being avoided or pirates captured. Research suggests that pirates may not fear arrest *per se* (see Ploch, 2010) due to the difficulties associated with prosecuting them (discussed further in this chapter), but detection nevertheless incurs costs in terms of time and resources. Consequently, pirates would be expected to vary attack locations over time, but perhaps in a predictable manner – targeting locations not so far from previously attacked locations, at least some of the time. For all of these reasons (for an extended discussion, see Marchione and Johnson, 2013), it seems reasonable to suggest that incidents of piracy are likely to cluster in space and time.

To examine this, we exploit data concerning 5715 incidents of piracy obtained from the National Geospatial Intelligence Agency's (NGIA) Anti-Shipping Activity Messages database for the 24-year period 1978–2012. The NGIA data come from a variety of sources including, but not limited to, the International Maritime Bureau and the International Maritime Organization. Recorded incidents include successful and attempted attacks, as well as (for example) suspicious approaches. We consider all incidents here since they are all likely to inform understanding of pirate activity. For each incident, the location and date are recorded, along with a description of the event. Prior to analysis, duplicate records were identified and removed.

Figure 8.1 shows a kernel density map of the spatial distribution of all recorded pirate attacks for those parts of the world in which most attacks took place. Visual inspection of the map suggests that attacks are spatially clustered, and formal analyses (see Marchione and Johnson, 2013) confirm this to be the case.

The above patterns are, of course, aggregated over the study period and mask temporal variation. To illustrate, Figure 8.2 shows the monthly time series for two ocean regions – defined using NGIA boundaries – with high monthly counts of incidents. Off the coast of Somalia, changes in the rate of attacks (per unit time) are quite startling towards the end of the time series. In contrast, the peak in attacks off the coast of Malaysia occurs much earlier (between 2000 and 2006), and there appears to be less volatility in the monthly counts of attacks. The graph of monthly trends (see Figure 8.2, inset) suggests that the seasonal patterns for these two regions also differ. Overall, Figure 8.2 suggests that the risk of attack clusters in time but that patterns change over time, and vary for different regions of the ocean.

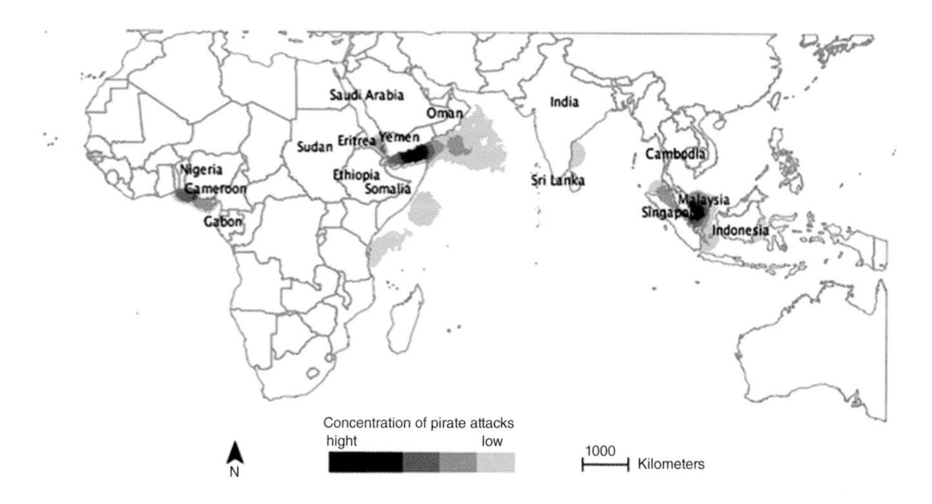

Figure 8.1 Kernel density map of incidents of piracy, January 1985 to January 2012

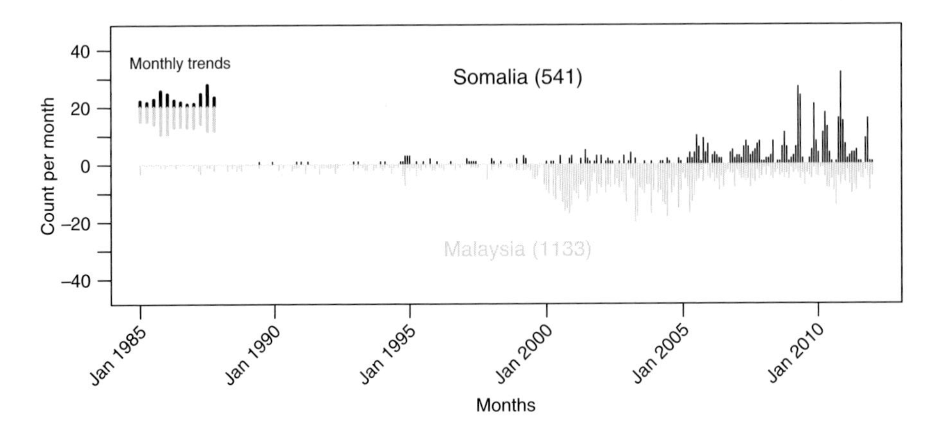

Figure 8.2 Monthly time series for two NGIA ocean subregions with high numbers of attacks (inset: mean monthly trends scaled by a factor of 2)

8.2.2 Space–Time Clustering of Piracy

To examine whether patterns of piracy cluster in space and time, we use a variant of the Knox test discussed in Section 8.1.3. Previous research on space–time clustering has used data for a city, or possibly a country. Here, we use data for the whole world. One issue with doing this is that generating the expected distribution by shuffling the dates in a completely random fashion may be unreasonable. This is because analyses of the time series for different regions of the ocean indicated that there were different seasonal patterns (for an example, see Figure 8.2). Ignoring this may lead to evidence of space–time clustering that is actually illusory, and can be explained by the aggregation of data for different regions of the ocean for which there is seasonal variation in the risk of attacks. For this reason, when shuffling the dates, we do this within NGIA subregions to preserve any (local) seasonal trends apparent in the permuted data.

As noted in this chapter, the selection of the bandwidths used to produce the Knox table is at the discretion of the researcher. In the current study, we use spatial intervals of 10 km and temporal intervals of one week. The spatial intervals may seem a little large, but the reader is invited to consider that at sea what would be a large distance in an urban area will be relatively small. Moreover, sensitivity analyses suggested that – within reason – alternative bandwidths generated qualitatively identical results.

Rather than presenting the results of the analyses for all years (which show similar patterns), Figure 8.3 shows the results for 2011. In the plot shown, each cell displays a Knox ratio for a particular space–time interval, and the shading indicates whether the observed number of event pairs is equal to (Knox ratios of 1), exceeds (Knox values above 1) or is below expectation (Knox values below 1). The cells shaded are those that are statistically significant at the 0.01 level. This more conservative approach to significance testing was adopted as the plot summarises so many test statistics. Like other crime types, it is clear that more events occur close to each other in space and time than would be expected if the timing and location of attacks were independent. That is, the Knox ratio in the bottom left of the plot is clearly much

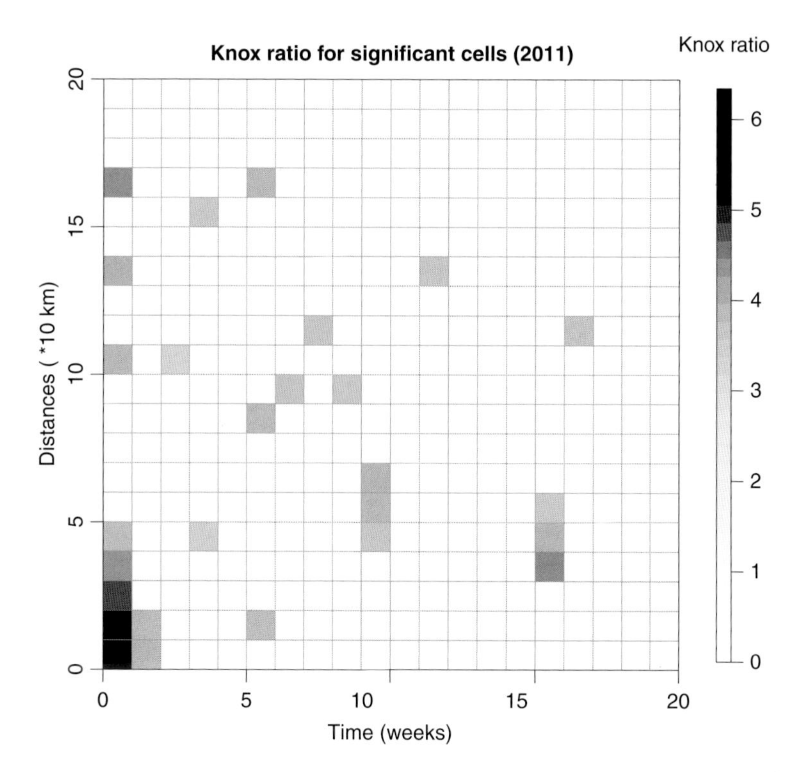

Figure 8.3 Knox ratios for attacks taking place in 2011 (shaded cells are statistically significant, $p < 0.01$)

larger than any of the others. Moreover, there is evidence of a pattern of space–time decay as the next largest Knox ratio is for events that occurred within one week and 20 km of each other. For events that occurred within 2 weeks of each other, the Knox ratios for incidents that occurred nearest to each other are smaller than those for events that occurred within one week of each other. The Knox ratios for some of the other space–time intervals are also statistically significant, but the ratios are smaller and there is no clearly discernible pattern to these values. This is quite typical for studies of this kind (see Johnson *et al.*, 2007).

The findings thus suggest that, just like urban crimes, piracy is more likely to occur in some places, and at some times, than others. They also suggest that when an incident occurs at one location, others are more likely to take place at the same location or nearby shortly afterwards. It is, however, important to acknowledge that the data analysed here are imperfect and suffer from the same kinds of issues associated with (for example) crimes recorded by the police. A more thorough discussion of these issues can be found in Marchione and Johnson (2013), but these include the fact that not all incidents will be reported by victims, and what is defined as an incident of piracy may vary over time or location. Nevertheless, the analyses suggest that this type of approach shows considerable promise in terms of informing practical solutions to the problem of piracy. Since these patterns are explored in more detail in Section 8.3, we conclude our discussion of the findings for now.

8.2.3 *Insurgency and Counterinsurgency in Iraq*

As for piracy, the majority of research concerned with spatial patterns of terrorist and insurgent activity has focussed on macro-level patterns. In this section, we examine patterns at the micro level. A few studies have done this for improvised explosive device (IED) attacks in Iraq (Johnson and Braithwaite, 2009; Townsley *et al.*, 2008), but here we consider the co-evolution of insurgent activity and counter insurgent (COIN) responses to it in space and time. To achieve this, we employ a novel variant of the Knox test to assess the sequential relationship between these two event types (see also Braithwaite and Johnson, 2012).

To many, the roadside bombings commonly reported during the Iraq insurgency may appear to be nothing more than random attacks. Consequently, before proceeding we briefly consider why we might expect patterns of insurgency to cluster in space and time, and in turn why we might expect COIN activity to also do so.

As discussed in this chapter, human mobility is subject to constraints that influence when and where people can engage in different forms of activity. This applies to the general public, offenders, and insurgents alike, albeit (perhaps) to differing degrees. On this basis alone, we might expect insurgent activity to cluster in space and time. There are, however, a number of other good reasons to expect such clustering to occur. These are discussed in some detail in Braithwaite and Johnson (2015), and so we will only briefly consider one example here. Terrorist or insurgent forces generally face a power asymmetry, having much smaller numbers than those they oppose. As a result, one strategy adopted by insurgents is that of exhaustion (see for example, Lapan and Sandler, 1993). This involves engaging in 'bursts' of attacks against small, vulnerable targets. These bursts of activity are intended to exhaust the state's resources or troop morale, resulting in insurgent victory. Such strategies will be undermined, however, if the state is able to replenish depleted troops, and so insurgents engaged in such manoeuvres need to strike a balance between continuing to fight and taking alternative courses of action when the benefits of further activity are eclipsed by the risks of so doing. This dynamic, where it occurs, is likely to lead to insurgent attacks clustering in space and time.

In line with the above, analyses of IED attacks in Iraq suggest that they do cluster in space and time (Johnson and Braithwaite, 2009; Townsley *et al.*, 2008). However, because such analyses have previously considered only where attacks take place, without reference to COIN activity, it is difficult to determine whether observed patterns reflect insurgent strategy or variation in the location of COIN forces – a possible key target of IED attacks – over time. Moreover, if insurgents engage in 'tit-for-tat' behaviour (e.g. Axelrod, 1984), whereby COIN action is met by retaliation from local insurgent forces, the observed patterns may be driven by the targeting strategy of COIN operatives.

Considering the space–time dynamics of COIN activity, this may be determined by both proactive strategies and those that are implemented in response to insurgent action. Both types of strategy can be expected to lead to the clustering of COIN activity in space and time due to constraints on troop mobility (see above). Furthermore, if troops are deployed in response to insurgent action, since the latter is found to cluster in both space and time (Johnson and Braithwaite, 2009; Townsley *et al.*, 2008), then so too perhaps should COIN activity. This leads to the expectation that COIN activity will cluster in space and time, and that COIN activity will occur shortly after and near to insurgent activity.

In what follows, we first examine the space–time distribution of IED explosions and COIN activity independently, using the Knox test. We then examine co-variation in these two types of events using a bivariate version of the Knox test (see Johnson *et al.*, 2009b). The latter involves enumerating how many IED–COIN event pairs occur at different spatial and temporal bandwidths, and then comparing this distribution with that expected, assuming the timing and location of the two types of events are independent. Instead of considering the simple joint distribution, because the ordering of events is of interest, we use a novel implementation of the test. To examine the space–time distribution of events where IED attacks (indexed by i) follow COIN activity (indexed by j), we populate a Knox table representing only pairings for which $t_i > t_j$ (t is the timing of events). To examine patterns for when COIN activity follows IED attacks, we populate a Knox table using event pairings for which $t_i < t_j$.

To test hypotheses, we analyse data for three different types of events drawn from Significant Activity (SIGACTS) Reports for Iraq. Recorded by coalition forces, the data cover the period January 1–June 30, 2005 (for a detailed discussion, see Johnson and Braithwaite, 2009). For insurgent activity, we examine the space–time distribution of IED attacks. IEDs take a variety of forms, including being vehicle borne, remote controlled, personnel borne and the more familiar form of a roadside bomb, triggered by moving vehicles. Their common characteristic is that they are 'booby traps'. IEDs have been employed in conflict zones ever since the first explosives were invented and are currently the most common tactic amongst insurgents in Iraq (as well as Afghanistan and elsewhere). For the 6-month period, there were a total of 3775 IED attacks. In the case of COIN activity, we examine patterns for a collection of proactive activities, cordon/searches, weapons cache discoveries and raids. Here we examine patterns for these types of events collectively, but in Braithwaite and Johnson (2012) we consider them individually. A second category of COIN activity that we examine are incidents that involved the identification and dismantling of unexploded IEDs. We differentiate between these two classes of events since the former are likely to reflect more strategic activity than the latter.

Figure 8.4 shows the results for both the univariate and bivariate Knox analyses. Considering IED attacks and COIN activity independently, it is clear that both cluster in space and time. This is particularly acute in the case of COIN activity, for which activities were significantly more likely to be observed near to each other in space and time, but significantly less likely to be observed to have occurred close in space but not in time. This is true for the three types of COIN activity considered collectively, and for the clearing of IEDs. Such a pattern suggests that COIN forces tended to be deployed near to each other but did not stay in the same places over time. Patterns of IED attacks also clustered in space and time, and showed some evidence of being less likely to occur in the same places over time, but the latter effect was much less pronounced than in the case of COIN activity.

Turning to the co-evolution of the two types of activity, Figure 8.4 suggests that IED explosions and COIN activity are associated with unexploded IEDs clustered in space and time. And, perhaps not surprisingly, it was particularly the case that COIN activity of this kind tended to follow IED explosions (which suggests the effectiveness of this COIN action). In general, however, after accounting for the fact that both types of activity were more likely at some locations than others and at certain times more than others, there was little evidence of COIN and insurgent activity clustering in space and time. This suggests that the space–time clustering observed for IED attacks and COIN activity independently cannot easily be explained by a process of tit-for-tat, or other similar dynamics. When considering these findings, of course, the reader should note that the analyses presented are for the whole of Iraq, and it is possible that different patterns would be evident for particular locations or cities within the country. Different patterns might also be observed for other types of COIN or insurgent activity.

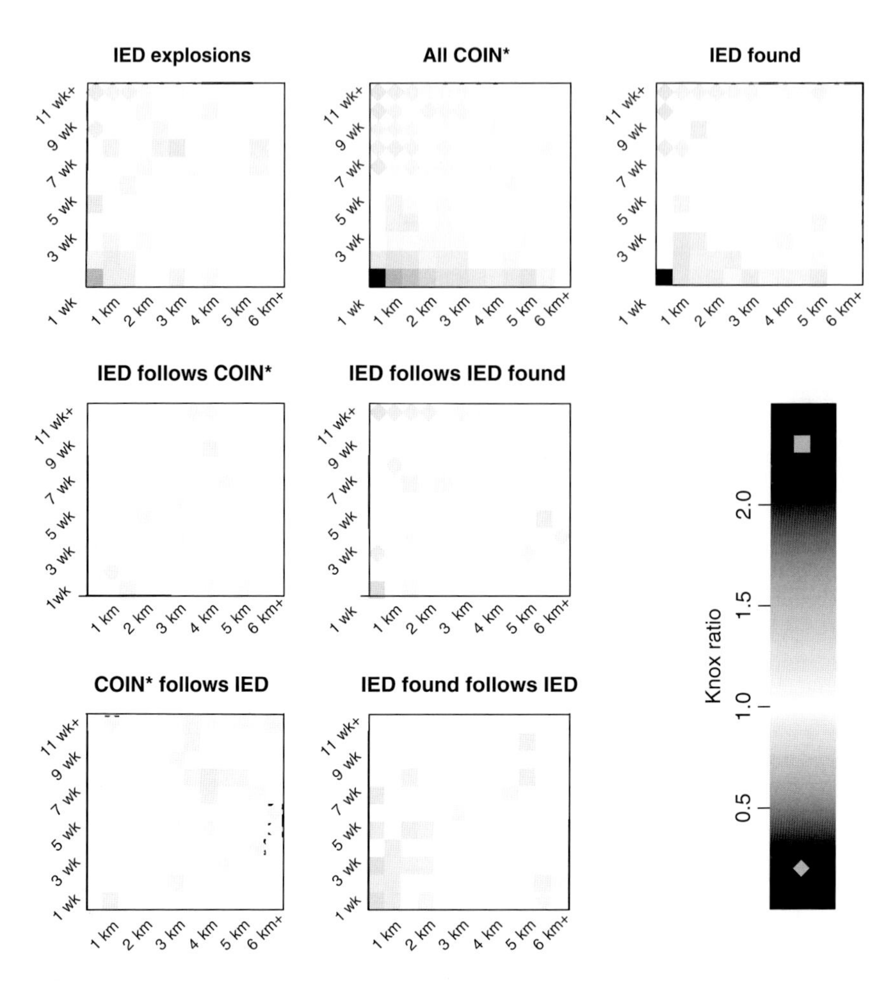

Figure 8.4 Univariate and bivariate Knox analyses for insurgent and COIN activity (COIN events are cordon/searches, weapons cache discoveries and raids)

8.3 Motif Analysis

8.3.1 Introduction

The results described in the Section 8.2 are typical examples of those produced by Knox-type analysis, in the sense that they show the existence of clustering in a general sense. In particular, they provide macro-level insight, in the sense that they show that certain levels of separation are over- or under-represented in the data as a whole, rather than identifying specific relationships. Such findings are informative and of material significance, as has been noted; nevertheless, there are some aspects of event patterns that such techniques are unable to capture.

In particular, techniques such as the Knox test are unable to identify the *type* of clustering present in any set of events; that is, the configurations in which events tend to co-occur. This is illustrated by the hypothetical examples shown in Figure 8.5, in which we consider a

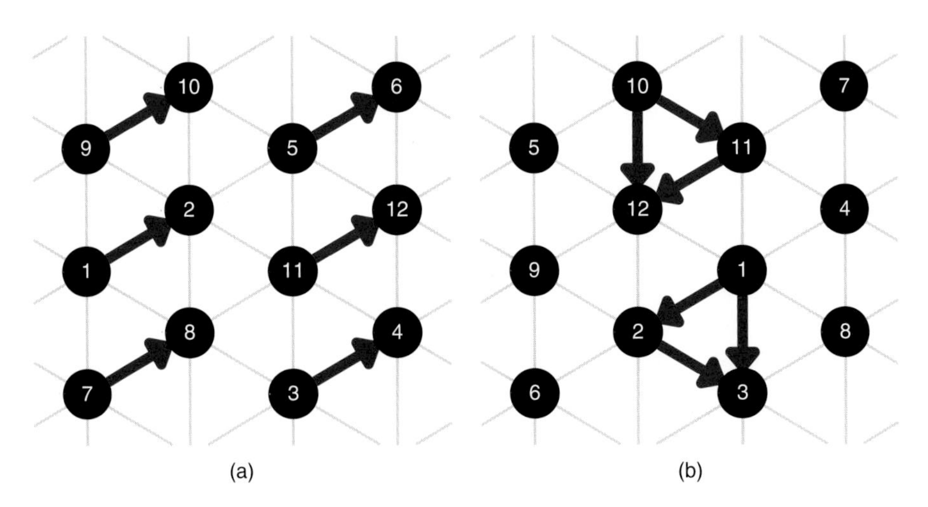

Figure 8.5 Close pair relationships for simple hypothetical sets of events. Vertex labels represent the time at which events occur, and their location is given by their position on the underlying grid, shown in grey. Red arrows represent close pairs, defined in this case as incidents occurring within two time units and one grid spacing

single-band version of the Knox test (i.e. one in which pairs are either close or not). Both sets contain the same number of close pairs, yet exhibit clear qualitative differences: Figure 8.5a shows a series of isolated close pairs, whereas two distinct groups are present in Figure 8.5b. Since the marginal spatial and temporal distributions are the same in each case, the Knox test would be unable to discriminate between them.

The fundamental difference between the cases shown in Figure 8.5 lies not in the number of close pairs, but rather their configuration. It is apparent, therefore, that in order to discriminate more precisely between patterns, the set of close pairs must be considered as a whole, and the relationships between them examined. The approach we describe here seeks to do this by using the close pair relationship to define an *event network*, in which vertices represent events and edges denote closeness. By examining the structural features of this network, insight can be gained into the nature of the underlying event data.

While the range of network features that can be analysed is large, we will focus here on the identification of 'motifs'. These are small subgraphs which occur disproportionately in a network (relative to a null model) and can be interpreted as the 'building blocks' of the network (Milo *et al.*, 2002). The relationships represented by these sub-structures can be reconciled with hypotheses concerning the formation of the network, and the approach has been applied extensively in ecology and biology (Alon, 2007; Mangan and Alon, 2003; Proulx *et al.*, 2005).

In the contexts of spatio-temporal events, motifs represent small groups of events which occur together in specific configurations. Their prevalence therefore reveals more than that of pairs only: it reveals whether events occur together in larger groups, and whether distinctive configurations can be identified. For criminal events, these can be interpreted as 'signatures' of targeting processes.

In order to identify these structures, however, it is necessary to control for the pair-wise clustering present in the data (which would otherwise bias the results). This is a non-trivial problem, and the spatio-temporal nature of the data means that methods typically used in motif analysis cannot be applied. We will describe here how this technical challenge can be overcome, and present the results of motif analysis for different crime types.

8.3.2 Event Networks

The basis for event network construction is similar to that of the Knox test, in that pairs of events are classified in terms of their closeness in space and time. Rather than defining several bands of closeness, however, it is treated as a binary concept: two events, i and j, are close in space if they lie within a spatial radius D, and close in time if they lie within a spatial radius T. These radii are parameters for network construction, and their variation will be explored later in this chapter.

These close pair relationships can be used to define two networks: one, G_d^D, on the basis of spatial proximity, and another, G_t^T, on the basis of temporal proximity. In both cases, the vertices represent the events in the dataset; vertex i corresponds to event i.

G_d^D is an undirected network, where any two vertices are connected by an edge if the events occurred within a distance D. Its edge set, E_d^D, is therefore

$$E_d^D = \{(i,j) \mid d_{ij} \leq D\}. \tag{8.3}$$

Since there is temporal ordering in the data, G_t^T is instead taken to be a directed network. In this case, if two events occurred within a time window T, an edge is added from the earlier event to the later, so that the edge-set E_t^T is

$$E_t^T = \{(i,j) \mid 0 < t_{ij} \leq T\}. \tag{8.4}$$

In terms of close pairs, G_d^D and G_t^T encode all information about the set of events (indeed, the basic Knox test could be expressed in these terms). Trivially, pairs of events which are close in both space *and* time can be identified by consulting the two networks: events i and j are close in space and time if they are adjacent in G_d^D and G_t^T. Formulating this explicitly, we also define G_{dt}^{DT} as the directed network of pairs which are close in space and time; we refer to this as the *event network* for the dataset. Its edge-set E_{dt}^{DT} is the intersection of those of the spatial and temporal networks:

$$E_{dt}^{DT} = \{(i,j) \mid (i,j) \in E_d^D \quad \textbf{and} \quad (i,j) \in E_t^T\}. \tag{8.5}$$

8.3.3 Network Motifs

The object of this analysis is to identify small subgraphs which are statistically prevalent within event networks, in order to characterise the spatio-temporal patterns that they represent. For subgraphs of small order, this analysis is particularly viable because the number of possible subgraphs is relatively small: there are, for example, only 13 connected directed networks of three vertices, up to isomorphism.

The number of possible motifs is further reduced for event networks, as a result of the structural constraints imposed by the nature of the data. Since the direction of edges is determined

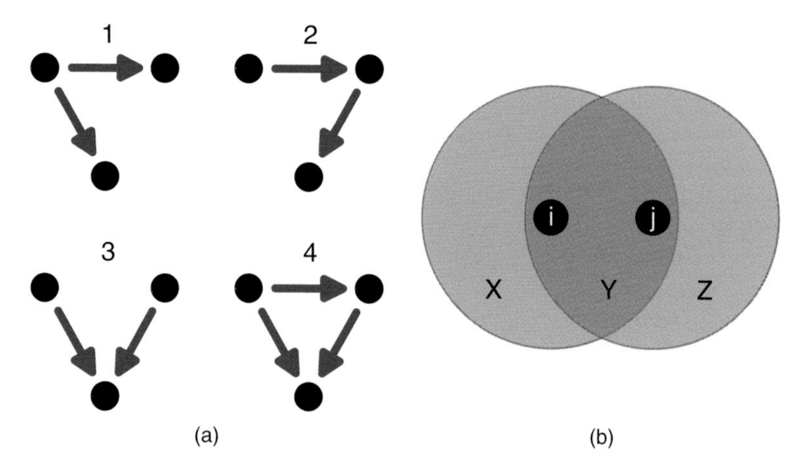

Figure 8.6 Motifs of three vertices and their real-world interpretation: (a) all three-vertex subgraphs which can arise in event networks, up to isomorphism; and (b) two example events, i and j, with a circular region of radius D indicated for each

by temporal information, for example, the fact that events can be ordered temporally means that any event network must be cycle-free. Such restrictions mean that only four motifs of order 3 (see Figure 8.6a), and 24 motifs of order 4, are possible for event networks.

The value of motif analysis can be understood by considering the spatio-temporal patterns represented by individual motifs. Each motif corresponds to a particular spatial configuration and temporal ordering among a small set of events, and it is the prevalence of these which we seek ultimately to characterise. The real-world patterns in question can be understood by considering the arrangements in graphical terms.

Figure 8.6 shows all possible three-vertex motifs, together with a stylised representation of a close pair of events, i and j. Since each motif contains at least one edge, representing a close pair of events, the existence of two such events can be taken as a common point of reference. Figure 8.6b shows these events, with regions of radius D indicated for each. It can then intuitively be seen that each motif corresponds to the occurrence of a third event in one of the regions **X**, **Y** or **Z** (with temporal precedence also known). Similar reasoning can also be applied to four-vertex structures: in either case, the motifs correspond to a distinctive pattern of events.

8.3.4 Statistical Analysis

Counting the occurrences of small subgraphs within a network is a task for which a number of well-documented methods are available. For sufficiently small networks, simple brute-force enumeration can be used (as it is here); in cases where this is computationally prohibitive, efficient sampling-based methods are also available (Kashtan *et al.*, 2004; Wernicke and Rasche, 2006). The main challenge in the statistical characterisation of motifs therefore concerns the choice of 'random' networks against which the observed data should be compared. Crucially, since the aim of the analysis is to gain insight beyond that offered by pair-counting methods, these reference networks should correspond to sets of events which exhibit the same level of pair-wise clustering as the observed data.

The need to generate appropriate reference networks is well-known in motif analysis: in particular, it is necessary to maintain a constant edge-count when generating randomised graphs, since otherwise changes in subgraph frequency may simply be an artefact of varying network density. This immediately excludes the possibility of simply permuting temporal information, as in the Knox test: for events that are pair-wise clustered, such a procedure would (by definition) generate networks with fewer edges.

The approach most commonly used in motif analysis is to generate the required randomised networks by simply re-wiring the edges of the observed network; that is, by re-assigning one or both of the end-points of some edges. In this way, no edges are created or destroyed, and other structural features of the network can also be preserved. This approach is not, however, applicable for event networks. As noted, the fact that such networks are derived from spatial and temporal data constrains the space of possible configurations; they must, for example, be cycle-free. Furthermore, since edges represent spatial proximity, their existence is not mutually independent, and some combinations cannot arise from any possible set of events (Penrose, 2003). The nine-vertex 'star' network is such a combination: in two-dimensional space, it is impossible for eight points to all lie within a radius D of a given point without at least two of those eight being within D of each other themselves.

An even more subtle issue concerns the fact that not all valid event networks are equally likely to arise from random event data. The mapping from datasets to their event networks is many-to-one, and the number of datasets which give rise to each event network is not equal; some networks will only be induced by very particular arrangements. Given that the analysis is concerned with the randomness (or otherwise) of the event data, this means that, even if the set of all valid event networks could be found, randomly sampling from it would still be inadequate.

8.3.5 Random Network Generation

The observations of Section 8.3.4 make clear that the generation of appropriate random networks cannot be done while remaining agnostic to the nature of the underlying data. Instead, it is necessary to devise a method capable of generating valid event networks in a way which represents a random sampling from the space of underlying event-sets. In order to provide a valid null comparison, these networks must also contain the same number of edges (i.e. close pairs) as the observed data.

The approach we take to this task is to explicitly simulate synthetic sets of events, constructed so as to have the required number of close pairs. Because these events are generated explicitly, the event networks derived from them are guaranteed to be valid, and the main question therefore concerns how to achieve the correct level of clustering.

The event-sets generated are, in fact, required to be clustered in a number of senses. Although the analysis is concerned with the closeness of events simultaneously in space and time, this is influenced by their closeness in each dimension independently. All else being equal, therefore, datasets containing more close spatial pairs will tend to contain more close pairs in space and time. This must be controlled for in order to ensure that any motifs detected are the result of spatio-temporal interactions, rather than simply spatial or temporal heterogeneity.

The implication of this is that randomly generated event sets, as well as containing the same number of close space–time pairs as the observed data, should also contain the same number of close pairs in space and time independently. Although the exact spatial and temporal distributions may differ, this ensures that the synthetic events will be 'as clustered', in at least a basic sense, as the observed data. Defining the spatial, temporal and spatio-temporal networks derived from the synthetic data as $\widetilde{G}_d^D$, $\widetilde{G}_t^T$ and $\widetilde{G}_{dt}^{DT}$, respectively, the requirement is that all of these should have the same edge-count as their counterparts (G_d^D, G_t^T and G_{dt}^{DT}) for the observed data.

The method used to achieve this involves taking a set of fully random events as a start point, and making iterative adjustments to it until the required conditions are satisfied. We begin by generating N events within the region in question, uniformly at random in both space and time. The number of close pairs in space, time and both space and time are calculated for these events, and denoted $\widetilde{M}_d$, $\widetilde{M}_t$ and $\widetilde{M}_{dt}$, respectively. We then define an 'energy', E, quantifying the extent to which these measurements differ from those required:

$$E = \frac{|\widetilde{M}_d - M_d|}{\widetilde{M}_d + M_d} + \frac{|\widetilde{M}_t - M_t|}{\widetilde{M}_t + M_t} + \frac{|\widetilde{M}_{dt} - M_{dt}|}{\widetilde{M}_{dt} + M_{dt}}. \tag{8.6}$$

This energy is zero only when the required counts are equal for the observed and synthetic data, and increases as the deviation increases. The aim of the following process is therefore to make adjustments to the event set which reduce E until it eventually reaches zero.

At each iteration, the current energy, E^{cur}, is calculated, before one of the events is then selected uniformly at random as a candidate for adjustment. A prospective new position and time are then generated for the selected event, chosen uniformly at random in the known spatio-temporal region. The value of the energy under this prospective change, E^{new}, is calculated and compared with E^{cur}. If $E^{new} < E^{cur}$, the prospective change is accepted and implemented; otherwise, it is rejected and the original positions are retained (see Figure 8.7). In this way, E decreases monotonically as the algorithm iterates, and the process eventually ends when it reaches zero. In order to avoid local minima, a small random error can be introduced whereby energy-increasing changes are accepted; this is similar to a simulated annealing approach (see Newman and Barkema, 1999).

Given a set of events, a spatial threshold D and a temporal threshold T, this method can be used to produce synthetic sets of events with the same level of clustering as the observed data. Since the synthetic events are matched in terms of pair-wise clustering, yet random in all other respects, the subgraph frequencies present in their event networks can be compared with those in the observed network G_{dt}^{DT}.

8.3.6 Results

The motif analysis technique is applied here to real-world event data concerning two distinct criminal phenomena: residential burglary and piracy. As described elsewhere in this chapter, both of these are known to display significant space–time clustering (measured according to the Knox test) but are otherwise of significantly different character.

The burglary data consist of 5690 incidents, representing all residential burglaries recorded in the city of Birmingham, UK, between March 2012 and February 2013, inclusive. For each

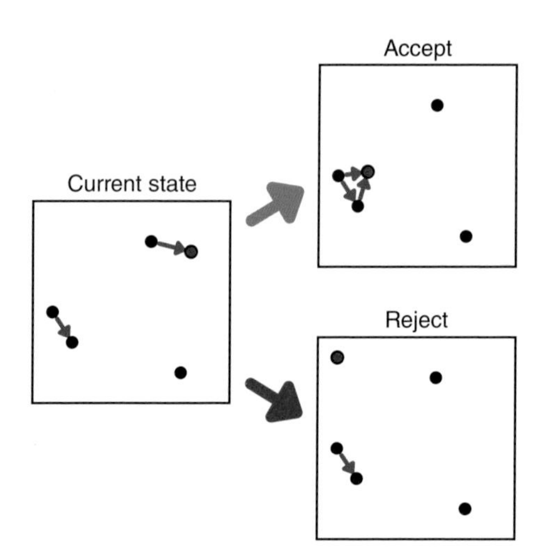

Figure 8.7 The process of randomised network generation involves iterative movement of points. The plots show the two possible outcomes at each iteration, for a hypothetical situation where four close pairs are required and it is assumed that all events are close temporally. The initial configuration, shown left, contains two close pairs. The point highlighted red is selected to move: one possible change gives three pairs and would be retained, and the other reduces the number to 1 and would be rejected

incident, the location of the victimised property is recorded in terms of a British National Grid co-ordinate reference, to an accuracy of 1 metre. The temporal data take the form of a window, representing the earliest and latest possible times at which the incident could have taken place; to establish a point estimate, we take the midpoint of these times.

The piracy dataset used is exactly the same as that introduced in Section 8.2.1. In this analysis, we consider all incidents taking place within 2010.

The analytical process is identical in both cases. For any given spatial threshold D and temporal threshold T, the first step is to construct the associated proximity networks: the spatial network G_d^D, temporal network G_t^T and event network G_{dt}^{DT}. The frequency of each possible motif in G_{dt}^{DT} is calculated, and the edge-counts of all three networks are noted.

Using the process described in Section 8.3.5, 99 synthetic sets of events are then produced, and the event network of each is derived. The frequencies of each possible motif across these 99 networks form a null distribution against which the observed frequencies can be compared. Since the networks are constructed so as to have the same degree of pairwise clustering as the observed data, any significant patterns must be due to variation in the type of clustering present, rather than its presence *per se*.

Considering three-vertex motifs first, Figure 8.8 shows the results of the analysis for a variety of spatial and temporal thresholds. An array is shown for each motif: for each cell of the array, a black outline indicates significance at the $p = 0.01$ level, and the colour represents the z-score for the effect (where blue and red represent under- and over-representation, respectively).

The most striking result for both event types is the highly significant over-representation of motif 4. That this motif – a fully connected triple – should be so prevalent suggests that the occurrence of edges in these networks is not independent, and that links have a significant

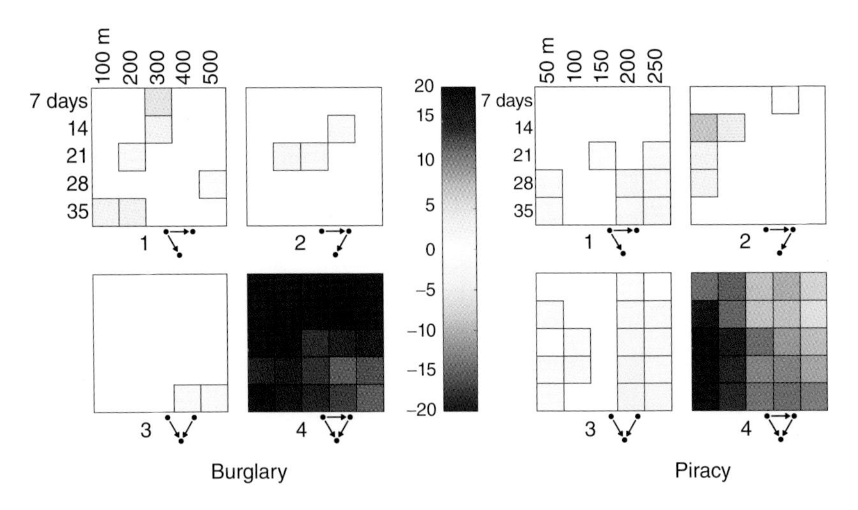

Figure 8.8 Statistical analysis of three-vertex motifs for a variety of spatio-temporal thresholds. Each cell corresponds to a particular combination of D and T, and the colours indicate the effect size

tendency to co-occur in this way. In spatio-temporal terms, the result suggests that events tend to occur not just in pairs, but in fact in larger groups within a restricted spatio-temporal domain.

The significance of motif 4 is ubiquitous across all close pair thresholds, suggesting that it is a fundamental signature of both processes. For piracy, the fact that its effect size is noticeably larger at smaller spatial scales but larger temporal scales can be interpreted as implying a characteristic scale, in some sense, for behaviour of this type. In the case of burglary, the largest effects are seen at the shortest temporal scales, which again may be instructive as to the time course of victimisation.

One other notable result is the under-representation of motif 3 in the piracy data. This motif corresponds to situations in which an event occurs in the common neighbourhood of two previous events which were not themselves a close pair. Though many explanations are possible, the pattern could be interpreted as corresponding to a situation in which the two initial events are committed by different offenders (otherwise, they imply a reversal of direction). Its absence would therefore be consistent with the hypothesis that the presence of multiple active offenders in some region is unlikely. This also accords with the under-representation of motif 1 – both of these correspond to spatially diverse offending, and their absence might be suggestive of localised territorial behaviour by a small number of offenders.

The corresponding results for four-vertex motifs are shown in Figure 8.9: again, several motifs are evident in each case. A number of these, such as motifs 11, 15, 21 and 24, show highly significant over-representation in both cases. Motif 24, the fully connected pattern, is analogous to the fully connected motif in the three-vertex case, and its significance here represents further evidence of particularly dense clustering in both datasets. The reasons for the appearance of each of the other three are not intuitively obvious, but all are close relations of the fully connected three-vertex motif, augmented in various ways by an additional event.

The number of subgraphs represented in Figure 8.9 (and their close similarity) means that there is little value in providing an exhaustive exploration of each. However, perhaps the most significant observation concerns the results as a whole, and in particular the difference in motifs

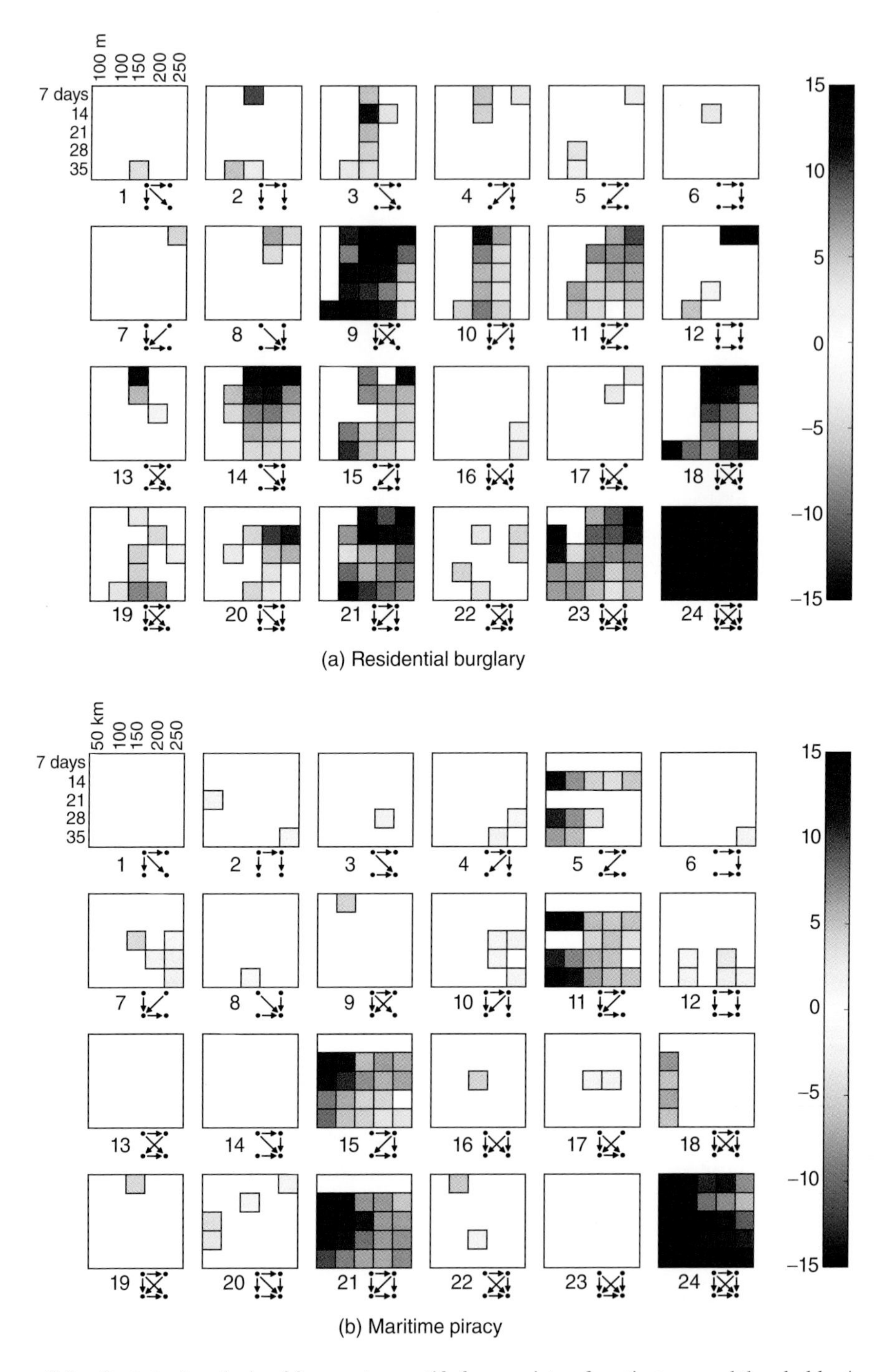

Figure 8.9 Statistical analysis of four-vertex motifs for a variety of spatio-temporal thresholds. Again, outlines and colours represent statistical significance and effect sizes, respectively

between event types. Although several motifs are common to both cases, a number of others display contrasting behaviour.

Particularly clear examples are motifs 9, 10, 14, 18 and 23, which are highly significant for burglary but barely so at all in the case of piracy. Indeed, several motifs are significantly under-represented for piracy, such as 7, 10 and 12. Though unpicking their precise meaning is difficult, those patterns do, in general, appear to correspond to spatially diverse patterns of events, the unlikeliness of which can be reconciled with the maritime setting and the known use of anchored 'motherships'.

The higher overall number of motifs in the burglary case also indicates a more dense, though still stylised, pattern of offending. Again, however, the most crucial observation is simply the difference in patterns between the two event types, which indicates that this analysis is capable of revealing differences in the patterning of offences at a deeper level than previously possible.

8.4 Discussion

In this chapter, we have summarised a programme of research that has sought to advance the understanding of spatio-temporal point patterns which arise in the context of criminal events. This sought to build on a substantial body of recent empirical work concerning urban crime, in which distinctive clustering patterns have been shown to be prevalent in incident data. Here, we applied similar analytical principles to events which have not been examined in this way before – maritime piracy and insurgent activity – finding that significant clustering was indeed present. In the latter part of the chapter, we also demonstrated how the encoding of pair-wise relationships as a network allows more complex features of clustering patterns to be characterised. Distinctive patterns, corresponding to 'signatures' of offending, were identified in incident data.

The results outlined have a number of implications for both theoretical development and the formulation of preventative strategies. Most immediately, the presence of clustering for the novel event types considered is a property which had not previously been established. The fact that it is observed so clearly suggests that those behavioural principles thought to apply to urban crime may also be relevant to the understanding of these event types. In particular, it suggests that the actors involved in these events are subject to spatial and temporal constraints, and accordingly engage in foraging-like behaviour. More specifically, the results also serve as partial validation of some of the speculative hypotheses, outlined in the chapter, concerning factors particular to these domains. In the case of piracy, for example, clustering can be reconciled with the logistical constraints associated with maritime travel. Similarly, the tendency of COIN activities to follow IED attacks confirms the expectation that these are, to some extent, responsive in nature.

The analogy with urban crime can also be extended to the consideration of preventative strategies. In the context of traditional crime, the prevalence of clustering has provided the foundation for novel policing strategies, based on the key observation that recent offending can be used to predict future incident locations. The fact that similar patterns are present for other event types raises the prospect that analogous strategies could be developed for these: in short, that prediction may also be viable for piracy and insurgency. Of course, the settings differ in numerous ways – the spatial scales and terrain, for example – and distinct preventative strategies would also be required, but the possibility of targeting preventative resources is one with considerable potential value.

The analysis of event motifs provides further insight into the nature of clustering patterns for both traditional and novel event types. The motifs identified in these cases represent 'signatures' of offending patterns, and offer qualitative insight into the configurations of events that occur at the micro level. In particular, they show that clustering is not simply pair-wise; instead, offences occur in larger groups of tightly clustered incidents. Perhaps more interesting, though, is the fact that different motifs are identified for events of different types. This suggests that, although both events are clustered, the patterns present differ at a deeper level: that is, although the overall clustering is equivalent, there are fundamental differences in the micro-level patterns that contribute to it. An avenue for future work is to establish the behavioural differences to which this can be attributed.

The presence of crime motifs also has potential practical implications, again in the context of prediction. The fact that higher-order motifs occur with differing prevalence implies that extra information can be leveraged when evaluating the prospective risk at any particular location. For example, if two close incidents have occurred, the ensuing risk in the various surrounding regions is different: the likelihood of the corresponding motifs suggests that some regions are more likely to be victimised than others. This differs from what can be said on the basis of pair-wise relationships alone: in that case, all regions would be treated as equal. Ultimately, this principle could be used as the basis for a predictive system in which areas are profiled differently according to the particular configuration (and number) of recent events present.

References

Alon U. (2007). Network motifs: theory and experimental approaches. *Nature Reviews Genetics* 8(6): 450–461.

Axelrod RM. (1984). *The Evolution of Cooperation*. New York: Basic Books.

Bernasco W. (2008). Them again? Same-offender involvement in repeat and near repeat burglaries. *European Journal of Criminology* 5(4): 411–431.

Block RL and Block CR. (1995). Space, place, and crime: hot spot areas and hot places of liquor-related crime. In Eck JE and Weisburd D (eds), *Crime and Place*, vol. 4 of *Crime Prevention Studies*. Monsey, NY: Criminal Justice Press, pp. 145–183.

Bowers KJ and Johnson SD. (2005). Domestic burglary repeats and space-time clusters. *European Journal of Criminology* 2(1): 67–92.

Bowers KJ, Johnson SD, and Pease K. (2004). Prospective hot-spotting: the future of crime mapping? *British Journal of Criminology* 44(5): 641–658.

Braithwaite A. (2010). *Conflict Hot Spots: Emergence, Causes, and Consequences*. Farnham, UK: Ashgate Press.

Braithwaite A and Johnson S. (2012). Space–time modeling of insurgency and counterinsurgency in Iraq. *Journal of Quantitative Criminology* 28(1): 31–48.

Braithwaite A and Johnson SD. (2015). The battle for Baghdad: testing hypotheses about insurgency from risk heterogeneity, repeat victimization, and denial policing approaches. *Terrorism and Political Violence* 27(1): 112–132.

Brantingham PJ. (2013). Prey selection among Los Angeles car thieves. *Crime Science* 2(1): 3.

Chalk P and Hansen SJ. (2012). Present day piracy: scope, dimensions, dangers, and causes. *Studies in Conflict and Terrorism* 35(7–8): 497–506.

Cohen LE and Felson M. (1979). Social change and crime rate trends: a routine activity approach. *American Sociological Review* 44(4): 588–608.

Daxecker U and Prins B. (2013). Insurgents of the sea: institutional and economic opportunities for maritime piracy. *Journal of Conflict Resolution* 57(6): 940–965.

Daxecker UE and Prins BC. (2015a). The new Barbary wars: forecasting maritime piracy. *Foreign Policy Analysis* 11(1): 23–44.

Daxecker UE and Prins BC. (2015b). Searching for sanctuary: government power and the location of maritime piracy. *International Interactions*, forthcoming.
Farrell G and Pease K. (1994). Crime seasonality: domestic disputes and residential burglary in Merseyside 1988–90. *British Journal of Criminology* 34(4): 487–498.
Felson M and Poulsen E. (2003). Simple indicators of crime by time of day. *International Journal of Forecasting* 19(4): 595–601.
Grubesic TH and Mack EA. (2008). Spatio-temporal interaction of urban crime. *Journal of Quantitative Criminology* 24(3): 285–306.
Hastings JV. (2009). Geographies of state failure and sophistication in maritime piracy hijackings. *Political Geography* 28(4): 213–223.
Hipp JR, Curran PJ, Bollen KA and Bauer DJ. (2004). Crimes of opportunity or crimes of emotion? Testing two explanations of seasonal change in crime. *Social Forces* 82(4): 1333–1372.
Jacquez GM. (1996). A *k* nearest neighbor test for space-time interaction. *Statistics in Medicine* 15(18): 1935–1949.
Johnson SD. (2014). How do offenders choose where to offend? Perspectives from animal foraging. *Legal and Criminological Psychology* 19(2): 193–210.
Johnson SD, Bernasco W, Bowers KJ, Elffers H, Ratcliffe J, Rengert G and Townsley M. (2007). Space–time patterns of risk: a cross national assessment of residential burglary victimization. *Journal of Quantitative Criminology* 23(3): 201–219.
Johnson SD, Bowers K and Hirschfield A. (1997). New insights into the spatial and temporal distribution of repeat victimization. *British Journal of Criminology* 37(2): 224–241.
Johnson SD and Bowers KJ. (2004a). The burglary as clue to the future: the beginnings of prospective hot-spotting. *European Journal of Criminology* 1(2): 237–255.
Johnson SD and Bowers KJ. (2004b). The stability of space-time clusters of burglary. *British Journal of Criminology* 44(1): 55–65.
Johnson SD and Bowers KJ. (2014). Near repeats and crime forecasting. In Bruinsma G and Weisburd D (eds.), *Encyclopedia of Criminology and Criminal Justice*. New York: Springer, pp. 3242–3254.
Johnson SD, Bowers KJ, Birks, DJ and Pease K. (2009a). Predictive mapping of crime by ProMap: accuracy, units of analysis, and the environmental backcloth. In Weisburd D, Bernasco W and Bruinsma GJ (eds), *Putting Crime in Its Place*. New York: Springer, pp. 171–198.
Johnson SD and Braithwaite A. (2009). Spatio-temporal modelling of insurgency in Iraq. In Freilich JD and Newman GR (eds), *Reducing Terrorism through Situational Crime Prevention*. New York: Criminal Justice Press, pp. 9–32.
Johnson SD, Summers L and Pease K. (2009b). Offender as forager? A direct test of the boost account of victimization. *Journal of Quantitative Criminology* 25: 181–200.
Kashtan N, Itzkovitz S, Milo R and Alon U. (2004). Efficient sampling algorithm for estimating subgraph concentrations and detecting network motifs. *Bioinformatics* 20(11): 1746–1758.
Knox G. (1964). Epidemiology of childhood leukaemia in Northumberland and Durham. *British Journal of Preventive and Social Medicine* 18(1): 17–24.
Lapan HE and Sandler T. (1993). Terrorism and signalling. *European Journal of Political Economy* 9(3): 383–397.
Lersch KM. (2004). *Space, Time, and Crime*. Durham, NC: Carolina Academic Press.
Mangan S and Alon U. (2003). Structure and function of the feed-forward loop network motif. *Proceedings of the National Academy of Sciences* 100(21): 11980–11985.
Mantel N. (1967). The detection of disease clustering and a generalized regression approach. *Cancer Research* 27(2): 209–220.
Marchione E and Johnson SD. (2013). Spatial, temporal and spatio-temporal patterns of maritime piracy. *Journal of Research in Crime and Delinquency* 50(4): 504–524.
Milo R, Shen-Orr S, Itzkovitz S, Kashtan N, Chklovskii D and Alon U. (2002). Network motifs: simple building blocks of complex networks. *Science* 298(5594): 824–827.
Mohler GO. (2011). Self-exciting point process modeling of crime. *Journal of the American Statistical Association* 106(493): 100–108.
Morgan F. (2001). Repeat burglary in a Perth suburb: indicator of short-term or long-term risk? In Farrell G and Pease K (eds.), *Repeat Victimisation*, vol. 12 of *Crime Prevention Studies*. Monsey, NY: Criminal Justice Press, pp. 83–118.
Newman MEJ and Barkema GT. (1999). *Monte Carlo Methods in Statistical Physics*. Oxford: Oxford University Press.

North BV, Curtis D and Sham PC. (2002). A note on the calculation of empirical *p* values from Monte Carlo procedures. *American Journal of Human Genetics* 71(2): 439–441.
Pease K. (1998). *Repeat Victimisation: Taking Stock*. London: Home Office Police Research Group.
Penrose M. (2003). *Random Geometric Graphs*. Oxford: Oxford University Press.
Perry WL, McInnis B, Price CC, Smith S and Hollywood JS. (2013). *Predictive Policing*. Santa Monica, CA: RAND.
Pfeiffer DU, Robinson TP, Stevenson M, Stevens KB, Rogers DJ and Clements ACA. (2008). *Spatial Analysis in Epidemiology*. Oxford: Oxford University Press.
Ploch L. (2010). *Piracy off the Horn of Africa*. Darby, PA: DIANE.
Proulx SR, Promislow DE and Phillips PC. (2005). Network thinking in ecology and evolution. *Trends in Ecology and Evolution* 20(6): 345–353.
Ratcliffe JH. (2002). Aoristic signatures and the spatio-temporal analysis of high volume crime patterns. *Journal of Quantitative Criminology* 18(1): 23–43.
Ratcliffe JH. (2004). The hotspot matrix: a framework for the spatio-temporal targeting of crime reduction. *Police Practice and Research* 5(1): 5–23.
Ratcliffe JH. (2006). A temporal constraint theory to explain opportunity-based spatial offending patterns. *Journal of Research in Crime and Delinquency* 43(3): 261–291.
Ratcliffe JH and Rengert GF. (2008). Near-repeat patterns in Philadelphia shootings. *Security Journal* 21(1–2): 58–76.
Shaw CR and McKay HD. (1969). *Juvenile Delinquency and Urban Areas: A Study of Rates of Delinquency in Relation to Differential Characteristics of Local Communities in American Cities*. Chicago: University of Chicago Press.
Sherman LW, Gartin PR and Buerger ME. (1989). Hot spots of predatory crime: routine activities and the criminology of place. *Criminology* 27(1): 27–56.
Tobler WR. (1970). A computer movie simulating urban growth in the Detroit region. *Economic Geography* 46: 234–240.
Tompson L and Bowers K. (2013). A stab in the dark? A research note on temporal patterns of street robbery. *Journal of Research in Crime and Delinquency* 50(4): 616–631.
Townsley M, Homel R and Chaseling J. (2003). Infectious burglaries: a test of the near repeat hypothesis. *British Journal of Criminology* 43(3): 615–633.
Townsley M, Johnson SD and Ratcliffe JH. (2008). Space time dynamics of insurgent activity in Iraq. *Security Journal* 21(3): 139–146.
Weisburd D, Bushway S, Lum C and Yang S-M. (2004). Trajectories of crime at places: a longitudinal study of street segments in the city of Seattle. *Criminology* 42(2): 283–322.
Weisburd D, Groff ER and Yang S-M. (2012). *The Criminology of Place*. Oxford: Oxford University Press.
Wernicke S and Rasche F. (2006). FANMOD: a tool for fast network motif detection. *Bioinformatics* 22(9): 1152–1153.

Part Five

Real-Time Response Models

9

The London Riots – 1: Epidemiology, Spatial Interaction and Probability of Arrest

Toby P. Davies, Hannah M. Fry, Alan G. Wilson and Steven R. Bishop

9.1 Introduction

The need for public policy to be informed by an evidence-based approach has been recognised for some time; however, this is often problematic, particularly in the context of rare events. The difficulty is especially evident in situations where quantitative recommendations are required, such as estimating the appropriate contingency for a certain scenario, since traditional hypothesis testing is undermined by the paucity of data. In such cases, mathematical modelling has much to offer, allowing rigorous quantitative analysis of the system in question and the testing of varying scenarios. Recent advances in the modelling of large-scale social systems using techniques of complexity science mean this is now a viable approach, and indeed its potential has been demonstrated in such fields as epidemic modelling (Balcan *et al.*, 2009; Colizza *et al.*, 2006), crowd control (Johansson *et al.*, 2012) and infrastructure resilience (Kinney *et al.*, 2005). Here, we employ such an approach in the context of the 2011 London riots and the policy questions subsequently arising.

The London riots occurred between 6 and 10 August 2011, as the United Kingdom experienced its most widespread and sustained period of civil unrest in at least 20 years. Repeated episodes of looting, rioting, arson and interpersonal violence took place in several cities, including London, Manchester and Birmingham. The consequences of the events include numerous instances of injury, including five deaths, and extensive property damage, for which liability has been estimated as £250 million (Metropolitan Police, 2011). Here, we focus on the disorder in London, the worst-affected city.

Approaches to Geo-mathematical Modelling: New Tools for Complexity Science, First Edition. Edited by Alan G. Wilson.

Companion Website: www.wiley.com/go/wilson/ApproachestoGeo-mathematicalModelling

The London riots have been the subject of much research in the academic (Gross, 2011), governmental (Riots Communities and Victims Panel, 2011) and journalistic (The Guardian and LSE, 2011) communities. The majority of this research, however, as with much of that considering previous episodes (Haddock and Polsby, 1994), has focussed on the psycho-social motives of individual rioters, ascribing willingness to participate to various social factors, including unemployment, poor police relations and endemic criminality. Our work is distinct from this; rather than consider how and why the riots began, we take their initiation as our starting point and instead consider their spatio-temporal development. This approach is informed by the policy question which motivates our work: how, once such an incident is in progress, the police might best respond in order to suppress disorder as quickly as possible.

Several questions relating to the response of the authorities were raised following the disorder. Although order was restored after five days, it has been variously claimed that the police were inadequately prepared and slow to react to developments, and that disorder might have been suppressed sooner; indeed, official inquiries have acknowledged such shortcomings (HMIC, 2011; Metropolitan Police, 2011, 2012). Alongside a need to anticipate better the disorder itself, these inquiries have emphasised the need to establish a level of policing resource, and a mode of response, commensurate with an outbreak of this magnitude. Mathematical modelling can contribute to this by allowing quantitative examination of the effect of varying police responses, and the investigation of a range of scenarios.

A crucial factor in determining police response to a riot, and indeed for policing in general (Yarwood, 2007), is an understanding of the spatio-temporal distribution of events. The London riots are notable for the fact that, despite being apparently catalysed by a specific incident – the fatal shooting by a police officer of a suspect in Tottenham, North London, and a subsequent peaceful protest – disorder escalated in a dramatic and unanticipated way, spreading widely across the city. Understanding why and how this spreading occurred, and why some areas were afflicted more than others, is therefore fundamental to the planning of responses.

The use of mathematical modelling to explore the mechanisms behind the spatial heterogeneity seen in the disorder can go some way towards explaining the events. In this respect, our work follows the 'generative' approach previously used in the modelling of, for example, residential segregation (Schelling, 1971), state formation (Cederman, 1997) and collective action in social networks (Siegel, 2009), in that our aim is simply to establish whether our hypothesised mechanisms are capable of giving rise to realistic patterns. Although our work is motivated by empirical data, we make no attempt to replicate the London disorder, but rather to imitate the general 'stylised facts' observed.

A significant body of theory exists concerning criminal involvement and target choice, which, in tandem with the empirical results presented in Section 10.2, informs the model. Although such theories are premised on the assumption that offenders act rationally during a riot, and so cannot necessarily be invoked *a priori*, previous research suggests that, in general, that assumption is defensible (Auyero and Moran, 2007; Mason, 1984; McPhail, 1991). Indeed, research specific to the London disorder suggests that criminological theory, such as social disorganisation theory (Shaw and McKay, 1969) and rational choice theory (Cornish and Clarke, 1986), is applicable (Baudains *et al.*, 2013) and that its implications are therefore a well-justified basis for a model.

The concept of the rational offender implies that their decisions are influenced by the relative merits of actions – for instance, the potential rewards available at different sites – and that they seek to minimise costs, choosing nearby targets and seeking to avoid capture, as appears to be the case during rioting (McPhail, 1994). From a modelling perspective, this implies that these factors should be included in any formulation of the utility associated with a particular location. Closely related to this, crime pattern theory (Brantingham and Brantingham, 1993) seeks to explain the location of crime in terms of the 'awareness spaces' of potential offenders (i.e. the locations of opportunities and offenders' knowledge of them). For a riot, this would again suggest that participants are more likely to target nearby locations, as has been demonstrated across a variety of crime types (Townsley and Sidebottom, 2010), and that these locations are likely to be those which are commonly perceived to offer large potential rewards. These ideas inform our choice of spatial system in the model.

Considering instead the initial decision to offend, theory can inform both the mechanism by which potential participants are influenced, and the differing effect this influence might have. Theories of environmental criminology state that environmental precipitators (in this case, knowledge of ongoing rioting) can serve to prompt, pressure, permit or provoke offending (Wortley, 2001); the 'safety in numbers' effect in riots, in terms of the risk of arrest (Epstein, 2002; Granovetter, 1978), is a particularly clear example of how this might modify the cost–benefit structure (Myers, 2000). In the case of London, there is a widely- held perception (Gross, 2011) that awareness of disorder provided a self-reinforcing stimulus to rioter involvement, facilitated in many cases by social media. These ideas are clearly fundamental to the evolution and spreading of disorder, and suggest a contagion-like mechanism for this. Such cues may not, however, act uniformly, and whether they lead to offending may depend on local circumstances. The notion that environment affects the propensity of residents to engage in crime is well developed in criminology: social disorganisation theory (Shaw and McKay, 1969) suggests that criminality is more likely to take hold in areas with weak social fabric, due to a lack of informal social control (Sampson *et al.*, 1997). Deprivation relates closely to these ideas: more deprived communities lack the resources and structure to regulate themselves in this way. For the purpose of model building, this implies a need to incorporate geo-demographic factors in the proposed mechanisms.

Previous attempts to model riots have employed both continuous and agent-based approaches. The former have generally been attempts to adapt models of crowd dynamics to the case of rioting whilst doing little to accommodate realistic human behaviour (Bhat and Maciejewski, 2006; Kirkland and Maciejewski, 2003). Agent-based models, following the civil violence model of Epstein (2002), have had some success in using game-theoretic concepts to inform agent behaviour (Yiu *et al.*, 2002; Goh *et al.*, 2006). These have generally, though, given little attention to geographical concerns, either treating movement as random or else in a fairly naïve sense. More recent approaches, however, have remedied this somewhat by incorporating real spatial data via GIS (Weidmann and Salehyan, 2013) and including sophisticated spatial decision making (Torrens and McDaniel, 2013). Our work is, as far as we are aware, the first to incorporate such behaviour outside an agent-based framework, and is also differentiated by its focus on the case of London, in both its incorporation of data and consideration of particular policy concerns.

After describing the general trends observed in the data, we describe a model which incorporates several phases of riot development: a contagious process of involvement, a target choice stage and an interaction between participants and police. We then demonstrate, via numerical simulation, that the model is capable of reproducing the general trends identified. With a realistic simulation established, this framework is used to explore policy issues, such as the effect of varying police response and the predisposition of certain areas to riot activity.

9.2 Characteristics of Disorder

In seeking insight into the behaviour of individuals during an episode of rioting, we consider both existing theoretical research into such incidents (Wilkinson, 2009) and criminal activity in general, and specific observations from the London disorder. The latter takes the form of analysis of data provided by the Metropolitan Police, which contains the details of all individuals arrested in relation to the riots and matches the home addresses and offence locations of suspects. Since it is typically argued that individuals act rationally during a riot (i.e. that their decisions are based on some cost–benefit analysis; (Auyero and Moran, 2007; Mason, 1984; McPhail, 1991), these observations can be used to inform a model of the actions of rioters.

A fundamental observation is the predominant targeting of retail sites, reflecting the acquisitive nature of much offending. Crimes against commercial premises, including both acquisitive crime and criminal damage, accounted for 51% of all offences in the United Kingdom as a whole (Home Office, 2011), and offences clustered in areas such as Clapham Junction, Croydon, Ealing and Brixton. This can be immediately reconciled with crime pattern theory (Brantingham and Brantingham, 1993); the richness of opportunity at retail premises is likely to be common knowledge amongst riot participants, and they therefore act as *crime attractors*. In line with this, for our model we adopt a system of retail centres as the sites of disorder.

We also consider the origins of offenders (i.e. the locations of their residences) and, therefore, the distances they travelled to the sites where they offended. As seen in Figure 9.1a, the flows of offenders follow a clear distance–decay relationship. Although statistical tests (Clauset *et al.*, 2009) find that the distribution does not correspond to most common forms, the best fit is provided by an exponential distribution with parameter 0.274. An offender's perception of distance does not necessarily aggregate to an exponential distance decay, since other factors, some of which are temporally varying, are likely to contribute, and we nevertheless incorporate an exponential distance decay within our model. Distributions such as these are reminiscent of those seen in the analysis of flows in retail systems (Wilson, 1970, 2012), and so, noting also the central role of commercial centres, we model the behaviour of rioters partly by analogy with this.

Analysing the riot locations further, we explore the relationship between deprivation and offending. Figure 9.1b shows that a disproportionately high number of offences occurred in more deprived areas (approximately 50% within the 20% most deprived), using the UK's Index of Multiple Deprivation (IMD) to rank census units. Looking instead at suspects' residences, Figure 9.1c shows the average proportion of riot suspects for groups of LSOAs ordered by deprivation, where in this case, anticipating its incorporation as a variable in the model, we use a deprivation score based on IMD ranking. A relationship between offending and deprivation has also been found elsewhere (Home Office, 2011), and youth unemployment and child poverty have also been identified (Ben-Galim and Gottfried, 2011). That the most deprived areas acted disproportionately as both origins and destinations will clearly influence the distance distribution, and *vice versa*, but analysis shows that both effects persist when controlling for the other (see Chapter 10).

With this notion in mind, we incorporate the deprivation score discussed here as a feature within our model, allowing for a higher probability of offending in deprived areas.

We also note distinctive temporal patterns in the riot data, as seen in Figure 9.1d. From the small initial disturbance, incidents escalated in volume and intensity on each successive day, with police response growing in line with this, from 3480 on Saturday evening to 16,000 by

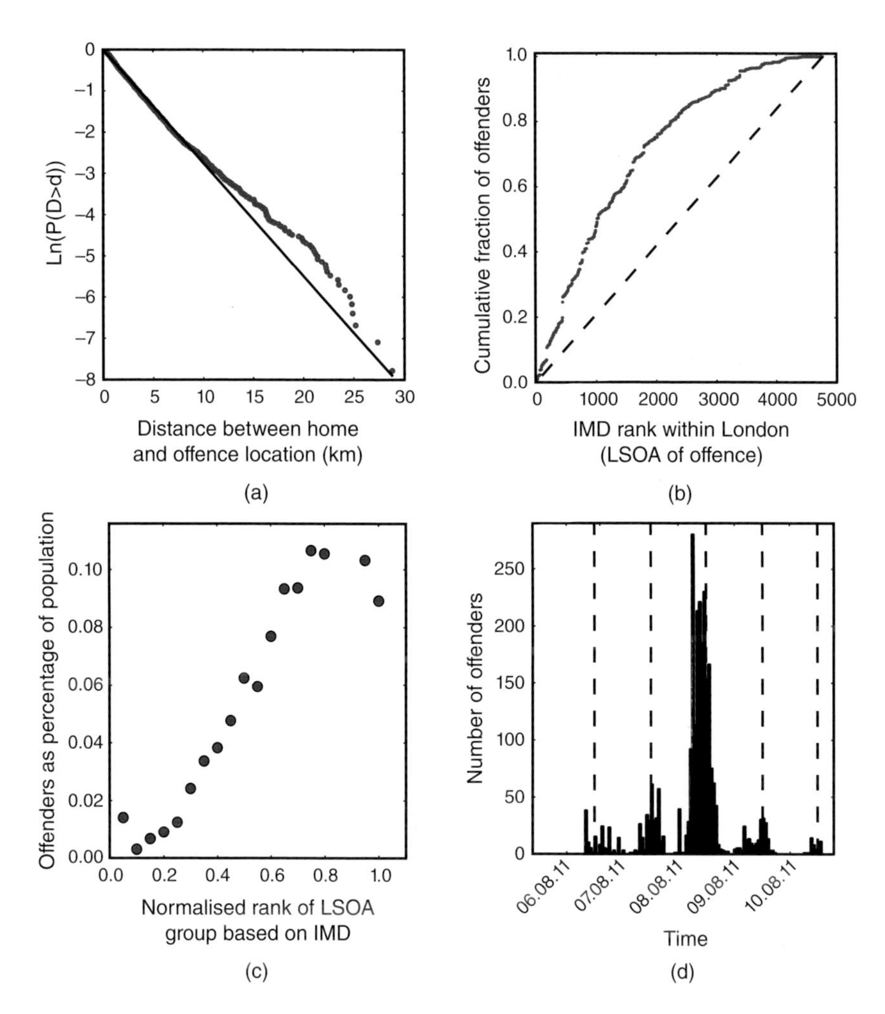

Figure 9.1 Observations from arrest data: (a) log-linear plot of the complementary cumulative distribution function of D, the distance between residential and offence locations. The straight line shows a hypothetical exponential distribution with parameter 0.274 (± 0.01 for a 95% confidence interval), for which the Kolmogorov–Smirnov distance statistic is 0.0246 (which compares with 0.332 for the equivalent fitted power-law). (b) Lorenz curve for the distribution of riot locations amongst Lower Super Output Areas (LSOAs; UK census units with an average population of approximately 1500) ranked according to deprivation (where 1 is the most deprived). The dashed line represents perfect equality. (c) Relationship between area-level deprivation and the proportion of residents involved in the disorder, where the horizontal axis represents a score derived from the UK Index of Multiple Deprivation (IMD) so that all values lie in [0, 1] and so that London's most deprived area is given a value of 1. (d) Temporal distribution of recorded crime

Tuesday (Metropolitan Police, 2011). This may also be seen at the scale of individual days, where the majority of criminality took place at night and built to a peak in the early hours. Whilst various explanations for this have been put forward, a particularly compelling one suggests that awareness of disorder provided a self-reinforcing stimulus to rioter involvement (Gross, 2011), and a contagion-based model is therefore appealing.

9.3 The Model

9.3.1 *Outline*

We develop a mathematical model with the aim of exploring the spatial and temporal patterns of the events in London. Recognising that non-linearities inherent in the system imply a significant dependence on initial conditions (which are unknown), and that numerous factors not considered here are likely to play a material role, we do not seek to replicate exactly the London events. Rather, we aim to produce a 'generative' type model which can give rise to realistic patterns and macro-level behaviour, so that insight might plausibly be gained through analysis of the underlying dynamics.

Our model draws on elements of several existing ones; our contribution is in their combination and adaptation to produce an integrated spatial model of disorder, and in the analysis of varying police strategies. The model can be divided into three components: an epidemiological model for riot participation, a spatial interaction model (SIM) (Wilson, 1971) for the spatial allocation of rioters and police and a model for interaction between rioters and police previously applied in the context of civil violence (Epstein, 2002).

9.3.2 *General Concepts*

The model is defined across a discrete system of two entities: residential areas and retail centres. These are indexed by i and j, respectively, and embedded in space, and we use LSOAs and defined 'retail cores' when considering London. Participating individuals are notionally tracked through the system via a logical sequence which involves a decision to participate taking place at their home, a choice of site at which to offend and possible removal due to arrest by police officers. These officers are active at all times but may move and be located according to different principles.

To model rioters' decisions, some concept of the attractiveness of a riot site is required. This is formulated using a 'cost–benefit' structure, as is normal for SIMs, where benefit represents the potential reward at a site and cost embodies both travel cost and the deterrent effect of police.

We assume that the *benefit* for site j is given by the logarithm of Z_j, a non-dimensional measure of its relative value (e.g. the ratio of j's floorspace to the mean across the system), as is a standard assumption in retail models of this type (Harris and Wilson, 1978; Wilson, 2008), reflecting diminishing returns to scale. For b_{ij}, the benefit of site j as perceived by an individual in i, we therefore have

$$b_{ij} \sim \log Z_j, \qquad \forall i. \tag{9.1}$$

Turning to *deterrence*, we suggest that the primary gauge by which an individual assesses whether the situation at a site is conducive to riot is the probability of arrest, determined by the relative numbers of rioters and police: a low perceived chance of capture encourages participation. Several such expressions for probability of arrest have been proposed; in this case, we take an adapted version of the formulation of Epstein (2002) as our starting point:

$$\mathbf{P}(\text{arrest in } j \text{ in one time unit}) = 1 - \exp\left(-\left\lfloor \frac{Q_j}{aD_j} \right\rfloor\right), \tag{9.2}$$

where Q_j is the number of police officers in j, D_j the number of rioters in j, and a the number of police officers required, on average, to 'contain' one rioter. The use of the floor function $\lfloor Q_j/aD_j \rfloor$ has empirical motivation; the Metropolitan Police review of the London disorder (Metropolitan Police, 2012) explicitly states that "decisions were made not to arrest due to the prioritisation of competing demands … specifically, the need to protect emergency services, prevent the spread of further disorder and hold ground until the arrival of more police resources". Accordingly, when the police are 'outnumbered' at a site (i.e. $Q_j < aD_j$), the situation is considered to be out of control and the police are unable to make any arrests without the addition of 'backup' (and thus the probability of arrest is 0). On the basis that increased probability corresponds to increased deterrence, we therefore express deterrence thus:

$$\text{deterrence} \sim \left\lfloor \frac{Q_j}{aD_j} \right\rfloor. \tag{9.3}$$

We also incorporate a linear function of the distance between residential areas and riot sites, as is typical for analogous retail systems. Taking this as proportional to d_{ij}, the distance between the centroids of i and j, we can then combine with equations (9.2) and (9.3) to obtain the full expression for *benefit–cost*:

$$w_1 \log Z_j - w_2 d_{ij} - w_3 \left\lfloor \frac{Q_j}{aD_j} \right\rfloor, \tag{9.4}$$

where the w_n are constants. The associated *attractiveness* term W_{ij} which appears in the terms of the spatial interaction model can, as described elsewhere (Wilson, 2008), then be written as follows:

$$W_{ij} = Z_j^{\alpha_r} \exp(-\beta_r d_{ij}) \exp\left(-\left\lfloor \frac{\gamma_r Q_j}{D_j} \right\rfloor\right), \tag{9.5}$$

where α_r, β_r and γ_r (which itself absorbs a) are parameters to be obtained in calibration with real-world data (the subscript r denoting reference to riot participants). It is through the form of equation (9.5) that an exponential distance decay, discussed in Section 10.2, features in the model.

9.3.3 *Riot Participation*

Motivated by the hypothesis, consistent with the temporal progression of the riots, that exposure to nearby disorder had the effect of inciting participation, we propose a Susceptible-Infected-Removed (SIR) model (Anderson and May, 1992); that is, a mechanism akin to infection by which individuals transfer to an active rioting state according to their level of exposure. Recalling the correlation between propensity to riot and deprivation, we also incorporate this, and the function we propose is therefore:

$$\mathbf{P}(\text{individual in } i \text{ chooses to offend}) = \rho_i^{\mu} \frac{\sum_j W_{ij}}{1 + \sum_j W_{ij}}, \tag{9.6}$$

where ρ_i is a measure of the deprivation in i (which we take to be based upon the IMD), and μ an exponent to be calibrated. A logistic function is used here to represent the existence of a threshold at which rioting becomes appealing; any transition is likely to be localised rather than gradual. Intuitively, this probability will be small when the overall attractiveness of potential riot areas is low, whereas, when the 'ambient' level of rioting is high, the probability of offending tends towards ρ_i^{μ}. From another perspective, where two areas were equally exposed to disorder, greater participation would arise in the more deprived of the two.

Translating this to the macro level for a residential area i, we therefore find an expression for $N_i(t)$, the rate at which individuals choose to participate at time t. Under the assumption that decisions are independent between individuals, this is given by the product of population size and decision probability,

$$N_i(t) = \eta I_i(t) \rho_i^{\mu} \frac{\sum_j W_{ij}(t)}{1 + \sum_j W_{ij}(t)}, \tag{9.7}$$

where η is an *infection rate*, and $I_i(t)$ the number of inactive individuals resident in area i. We can now formulate expressions for $I_i(t)$ and $R_i(t)$, the number of rioters whose residence is in a given zone i, as well as their change in a time period $[t, t + \delta t]$. These, along with their initial conditions ($I_i(0)$ is the residential population of i, and $R_i(0)$ a seed of participants, to be chosen), determine the numbers of individuals of each type, in each residential area, at all times. The choice to structure the model in this way is motivated by our focus on the residential origins of rioters, since it enables us to understand the composition of rioting groups in these terms. At this stage, we also include an extra term $C_i(t)$, to be fully defined later in this chapter, for the rate at which participants from i are arrested at time t:

$$R_i(t + \delta t) = R_i(t) + \delta t (N_i(t) - C_i(t)) \tag{9.8}$$

$$I_i(t + \delta t) = I_i(t) - \delta t N_i(t) \tag{9.9}$$

9.3.4 Spatial Assignment

We assign active rioters to sites of disorder using an entropy-maximising SIM; the purpose of these models is to estimate the most probable flows in a spatial system such as ours, given certain constraints (Wilson, 1971).

Rather than incorporating the attractiveness function W_{ij} directly into the spatial interaction equations, we use its moving average over a number of previous time steps, for several reasons: to account for factors such as travel time on the part of rioters, to represent 'lag' in the spread of information through the system and to dampen the effect of sudden fluctuations in attractiveness. The values used to determine the assignments at a given time, referred to as *effective attractiveness* and denoted W_{ij}^{e}, are therefore the average values of W_{ij} over the

L_r most recent time steps in our discretised temporal scheme (which has intervals δt; when $t < (L_r - 1)\delta t$, we 'pad' with the $t = 0$ value):

$$W^e_{ij}(t) = \frac{1}{L_r}\sum_{l=0}^{L_r-1} W_{ij}(t - l\delta t). \tag{9.10}$$

Following the standard entropy-maximising derivation of a SIM (Wilson, 2008), it can be shown that S_{ij}, an estimate of the number of rioters from i who are participating in disorder in j at time t, is given by:

$$S_{ij}(t) = \frac{R_i(t)W^e_{ij}(t)}{\sum_k W^e_{ik}(t)}. \tag{9.11}$$

An identical expression for S_{ij} may be formulated by using an alternative derivation: by considering equation (9.4) as a utility term in a conditional logit model (McFadden, 1984). In either case, summing over residential areas i yields the total number of rioters D_j in j:

$$D_j(t) = \sum_i \frac{R_i(t)W^e_{ij}(t)}{\sum_k W^e_{ik}(t)}. \tag{9.12}$$

It should be noted here that each time unit is therefore implicitly defined as the mean time taken for each participant to travel from a home location to a chosen riot site.

The assignment of police resources to areas of disorder is also realised via a SIM, as for riot participants; there are, however, noteworthy differences. First, police units have no 'home' location and are active and situated at potential sites of disorder at all times. The response lag L_p is also different to that for rioters (and intended to be higher), reflecting the delay in learning of the plans and movements of rioters, and conferring upon the rioters a degree of 'first-mover advantage'.

The main difference for police, however, is in the *attractiveness* function, analogous here to the requirement for officers at a given site. Following a similar argument to that of the rioters seen in equation (9.4), we assume the *benefit–cost* of police follows:

$$\bar{w}_1 \log Z_j + \bar{w}_2 D_j, \tag{9.13}$$

This expression includes no spatial decay term, reflecting the fact that the police do not prioritise incidents on the basis of proximity (HMIC, 2011) and can travel to incidents rapidly. In addition, the second term is a function of rioter numbers only: given that their aim is to eliminate all disorder, the number of police already at a site is likely to be immaterial to the police. As in equation (9.5) and described elsewhere (Wilson, 2008), the *attractiveness* function V_j representing police *requirement* is therefore:

$$V_j = Z_j^{\alpha_p} \exp(\gamma_p D_j), \tag{9.14}$$

where α_p and γ_p are, as before, parameters to be calibrated which encode the relative importance of the two factors. Following the identical process seen with equation (9.5) we may first calculate *effective requirement* to take into account time lags in the system,

$$V^e_j(t) = \frac{1}{L_p}\sum_{l=0}^{L_p-1} V_j(t - l\delta t), \tag{9.15}$$

and, in conjunction with a SIM, as in equations (9.11) and (9.12), can derive an expression for the total number of police officers in location j at time t:

$$Q_j(t) = P\frac{V_j^e(t)}{\sum_k V_k^e(t)}, \tag{9.16}$$

where P is the total number of police officers in the system.

9.3.5 *Interaction between Police and Rioters*

To model the interaction of police and rioters, we return to the mechanism of arrest and its associated probability described previously in this chapter. This gives the probability of capture for an individual rioter, and multiplying by the number of participants present therefore gives the expected number arrested. Since, for reasons explained previously, we classify participants by residential location, this is done separately for each area to give $C_i(t)$, the rate at which individuals who originated in i are arrested at time t:

$$C_i(t) = \tau \sum_j S_{ij}(t)\left(1 - \exp\left(-\left\lfloor\frac{Q_j}{D_j}\right\rfloor\right)\right). \tag{9.17}$$

where τ is an *arrest rate* parameter.

9.4 Demonstration Case

As a step towards verification of the model, and to establish a 'base case' for further investigation, a series of numerical simulations were run, representing the escalation of events during a typical evening. Individual simulations ran for 10 time units (where one unit is the time taken for a rioter to travel to their destination) and involved sequential iteration through the model equations in the order of (9.12), (9.15) and (9.8). The system was seeded with 100 riot participants, assigned to residential areas in proportion to population and allocated to sites of disorder according to the static component of attractiveness (*i.e.* $Z_j^{\alpha_r}$). Similarly, 5000 police officers (the approximate number deployed on each of the first 3 days in London) were initially placed at retail sites according to $Z_j^{\alpha_p}$. Given the high dimensionality of the model, many parameter sets were found to yield feasible results. To focus our discussion, an example parameter set was chosen (Table 9.1) which gives rise to outcomes broadly in agreement with the features observed in the data, both in terms of borough level participants (Figure 9.2) and distance decay (Figure 9.3).

Table 9.1 Parameters used in base case simulation.

Parameter	α_r	β	γ_r	α_p	γ_p	η	k	τ	L_r	L_p	δt
Value	0.6	0.5	0.11	0.65	0.012	0.006	6	0.75	30	60	0.0143

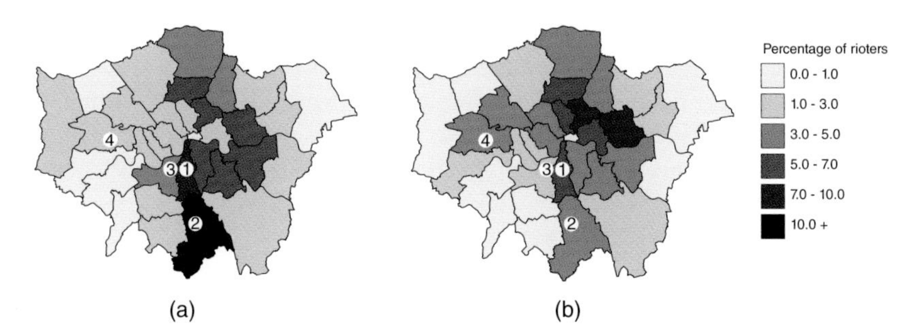

Figure 9.2 Borough-level choropleth of rioter residential locations from (a) data and (b) simulated results. Although the extreme dependence on initial conditions precludes our model from generating an exact replica of the observed incidents, the results show good qualitative agreement, with 26 of the 33 boroughs showing rioter percentages in the same or adjacent bands as the data. The remaining discrepancy may be accounted for by factors specific to the London disorder, such as communication between groups, other activity patterns occurring at the time, or social factors beyond the scope of this work. The labels 1, 2, 3 or 4 correspond to retail centres in Brixton, Croydon, Clapham Junction and Ealing, respectively, which are considered individually in our later simulations

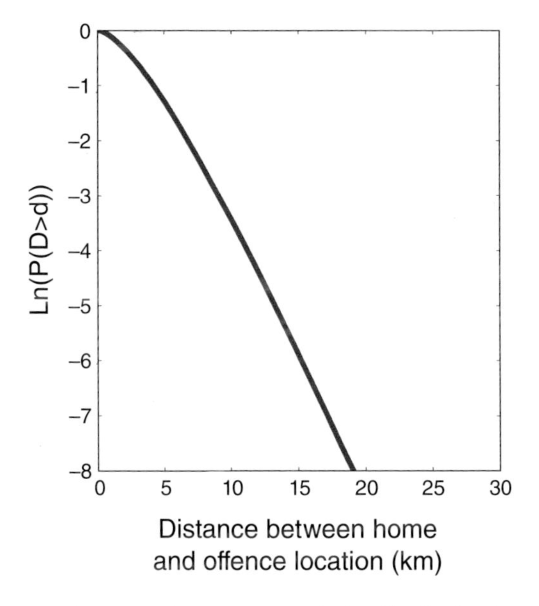

Figure 9.3 Log-linear plot of the complementary cumulative distribution function for d, the distances between residences and offence locations within the demonstration simulation

Since the riots occurred over 5 days, with incidents initialised in various locations across that period, these aggregate results offer little validation other than to confirm that the model is capable of replicating the general characteristics of the data. By instead initialising small incidents at just two locations, rather than simultaneously across the city, we may explore the susceptibility of retail centres. Such initialisation is also reflective of the way in which real incidents are thought to arise: many of the outbreaks began as small local gatherings of unrest (Metropolitan Police, 2012; Riots Communities and Victims Panel, 2011).

Our analysis considers retail sites which were worst affected: Brixton, Croydon, Clapham Junction and Ealing. We ran four such simulations in each case, pairing the site of interest with each of its closest geographical neighbours. In all simulations (Figure 9.4), the centres which

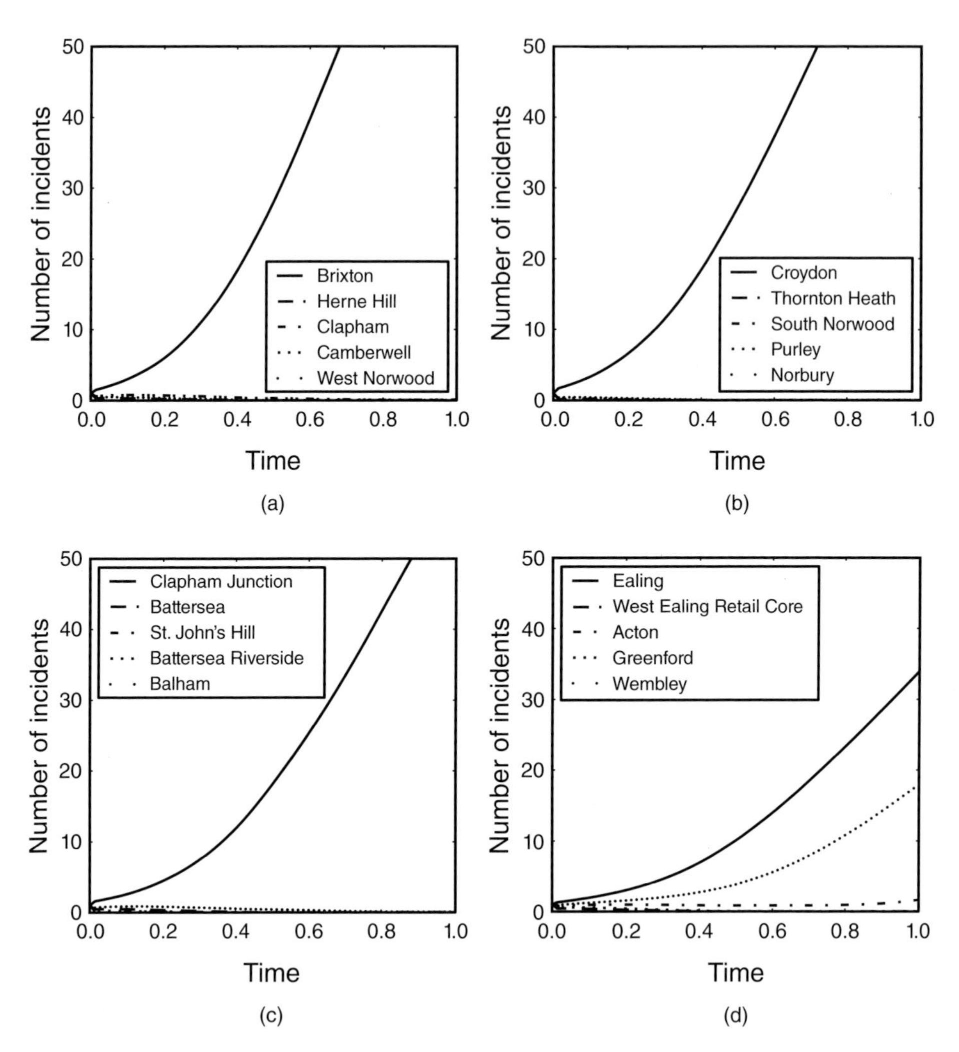

Figure 9.4 The susceptibility of retail sites. For each of the four centres worst affected in the riots – (a) Brixton, (b) Croydon, (c) Clapham Junction and (d) Ealing – we ran four separate simulations, pairing the site of interest with each of its closest geographical neighbours in turn. An initial disturbance of one rioter was included at both sites, and the model run to allow the incidents to evolve. Results shown for the sites of interest are the average of their four simulations, and in each case substantial growth is seen, particularly in comparison to the neighbouring centres

experienced widespread rioting in reality also saw substantial growth from the initial small disturbance in the model, while the vast majority of other retail locations saw incidents decay to zero. These results serve as further validation but also, given the structure of the model, offer insight into why some sites were more susceptible than others, since the dynamics are based on a combination of factors: proximity to populous areas of high deprivation, and the balance of centre size and police presence. These are important results, as such an approach might be applied as an indicator of future susceptibility.

Police Resources and Response

To gain quantitative insight into the level of police resources required to maintain control in a situation such as London's, we used the results of our demonstration case to analyse the effect of policing configuration on the development of disorder. To meaningfully compare realisations of the system, we define a quantity *severity* to summarise the cumulative disorder, given by the overall extent to which police are outnumbered by rioters:

$$\text{Severity} = \sum_j \sum_t^T \frac{D_j(t)}{Q_j(t)} \tag{9.18}$$

Two parameters were varied independently in our simulations – total police P and response lag L_p – with parameters as in Table 9.1 otherwise, and results are shown in Figure 9.5. Police numbers correspond to those seen in London, and reflect what was seen in data: numbers above approximately 10,000 appear sufficient to suppress disorder. In the case of speed of response, the difference in severity as L_p increases, relative to a base case of $L_p = 0$, is plotted. After a noisy stage at small values, the severity appears to increase with lag. Although the increase is

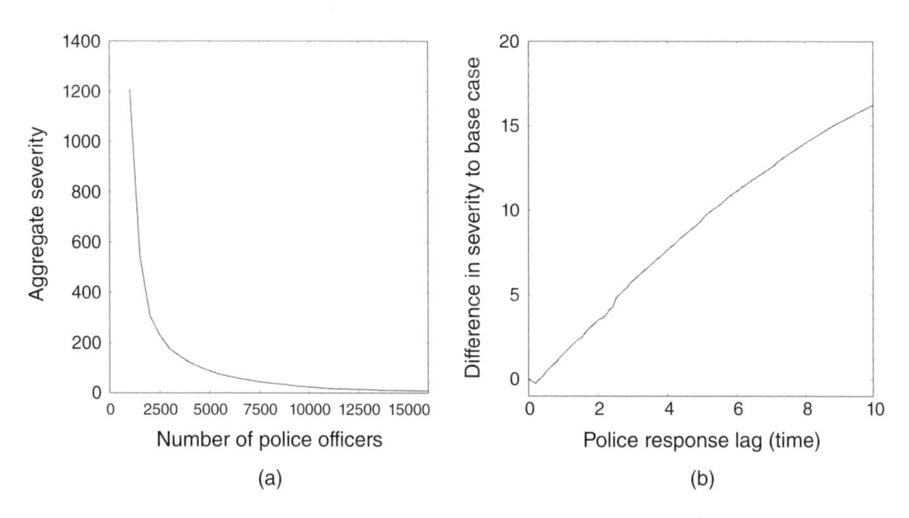

Figure 9.5 The effect on severity of modifying (a) the number of police officers, and (b) their response lag

small as a proportion of absolute value, it should be borne in mind that these simulations are run with parameter values chosen such that a certain level of severity is assumed. Any changes, therefore, are variations around a level which has been implied *a priori* by other factors, such as police and rioter numbers. As expected, the trend observed reflects the importance of delivering police to scenes of disorder before control is lost. The same simulations were also run for other police configurations – specifically where police are assigned to locations initially, either uniformly or proportionally with Z^{α_p}, and remain static throughout – but results differ only slightly from the dynamic case and are not shown.

9.5 Concluding Comments

Motivated by the events in London in 2011, we have presented a model of civil disorder which is able to replicate the general features of that outbreak. Whilst simple, the model incorporates the fundamental features of such an incident, and might be applicable to others of a similar nature. We have used the model to explore how the level of resource available to be deployed by the police might affect the outcome of such an incident, which is currently an open question in the United Kingdom and has clear and timely implications for policy. The availability and use of police intelligence as a means of deploying officers rapidly and efficiently – for which our concept of *lag* is a proxy – are also of particular interest. Both of these are found to have a material, and quantifiable, effect in our work. More sophisticated strategies on the part of both police and rioters could be incorporated in future iterations of the model, with the ultimate aim of making further quantitative recommendations.

References

Anderson RM and May RM. (1992). *Infectious Diseases of Humans: Dynamics and Control*. Oxford: Oxford University Press.

Auyero J and Moran TP. (2007). The dynamics of collective violence: dissecting food riots in contemporary Argentina. *Social Forces* 85: 1341–1367.

Balcan D, Colizza V, Gonçalves B, Hu H, Ramasco J and Vespignani A. (2009). Multiscale mobility networks and the spatial spreading of infectious diseases. *Proceedings of the National Academy of Sciences* 106: 21484–21489.

Baudains P, Braithwaite A and Johnson SD. (2013). Target choice during extreme events: a discrete spatial choice model of the 2011 London riots. *Criminology* 52: 251–285.

Ben-Galim D and Gottfried G. (2011). Exploring the relationship between riot areas and deprivation: an IPPR analysis. Available at http://www.ippr.org/articles/56/7857

Bhat SS and Maciejewski AA. (2006). An agent-based simulation of the L.A. 1992 riots. In *Proceedings of the 2006 International Conference on Artificial Intelligence*. Athens, GA: CSREA Press.

Brantingham PL and Brantingham PJ. (1993). Nodes, paths and edges: considerations on the complexity of crime and the physical environment. *Journal of Environmental Psychology* 13: 3–28.

Cederman LE. (1997). *Emergent Actors in World Politics: How States and Nations Develop and Dissolve*. Princeton, NJ: Princeton University Press.

Clauset A, Shalizi CR and Newman MEJ. (2009). Power-law distributions in empirical data. *SIAM Review* 51: 661–703.

Colizza V, Barrat A, Barthélemy M and Vespignani A. (2006). The role of the airline transportation network in the prediction and predictability of global epidemics. *Proceedings of the National Academy of Sciences of the United States of America* 103: 2015–2020.

Cornish DB and Clarke RV. (1986). *The Reasoning Criminal: Rational Choice Perspectives on Offending*. New York: Springer-Verlag.

Epstein JM. (2002). Modeling civil violence: an agent-based computational approach. *Proceedings of the National Academy of Science* 99: 7243–7250.

Goh CK, Quek HY, Tan KC and Abbass HA. (2006). Modeling civil violence: an evolutionary multi-agent, game theoretic approach. In *Proceedings of the IEEE Congress on Evolutionary Computation 2006*. Piscataway, NJ: IEEE, pp. 1624–1631.

Granovetter M. (1978). Threshold models of collective behavior. *American Journal of Sociology* 83: 1420–1443.

Gross M. (2011). Why do people riot? *Current Biology* 21: 673–676.

Haddock D and Polsby D. (1994). Understanding riots. *Cato Journal* 14: 147–157.

Harris B and Wilson AG. (1978). Equilibrium values and dynamics of attractiveness terms in production-constrained spatial-interaction models. *Environment and Planning A* 10: 371–388.

HMIC. (2011). The rules of engagement: a review of the August 2011 disorders. Available at http://www.hmic.gov.uk/media/a-review-of-the-august-2011-disorders-20111220.pdf

Home Office. (2011). An overview of recorded crimes and arrests resulting from disorder events in August 2011. Available at http://www.homeoffice.gov.uk/publications/scienceresearch-statistics/researchstatistics/crime-research/overview-disorder-aug2011/

Johansson A, Batty M, Hayashi K, Al Bar O, Marcozzi D and Memish ZA. (2012). Crowd and environmental management during mass gatherings. *The Lancet Infectious Diseases* 12: 150–156.

Kinney R, Crucitti P, Albert R and Latora V. (2005). Modeling cascading failures in the North American power grid. *The European Physical Journal B – Condensed Matter and Complex Systems* 46: 101–107.

Kirkland JA and Maciejewski AA. (2003). A simulation of attempts to influence crowd dynamics. In *Proceedings of the IEEE International Conference on Systems, Man, and Cybernetics 2003*, vol. 5. Piscataway, NJ: IEEE, pp. 4328–4333.

Mason TD. (1984). Individual participation in collective racial violence: a rational choice synthesis. *The American Political Science Review* 78: 1040–1056.

McFadden DL. (1984). Econometric analysis of qualitative response models. *Handbook of Econometrics* 2: 1395–1457.

McPhail C. (1991). *The Myth of the Madding Crowd*. New York: Aldine.

McPhail C. (1994). The dark side of purpose: individual and collective violence in riots. *The Sociological Quarterly* 35: 1–32.

Metropolitan Police. (2011). Operation Kirkin strategic review: interim report. Available at http://content.met.police.uk/News/August-Disorder–Police-Interim-Report/1400005002176/1257246745756

Metropolitan Police (2012). 4 days in August: strategic review into the disorder of August 2011. Available at http://content.met.police.uk/News/MPS-report-into-summer-disorder/1400007360193/1257246745756

Myers DJ. (2000). The diffusion of collective violence: infectiousness, susceptibility, and mass media networks. *American Journal of Sociology* 106: 173–208.

Riots Communities and Victims Panel. (2011). 5 days in August: an interim report on the 2011 English riots. Available at http://www.5daysinaugust.co.uk/PDF/downloads/Interim-Report-UK-Riots.pdf

Sampson RJ, Raudenbush SW and Earls F. (1997). Neighborhoods and violent crime: a multilevel study of collective efficacy. *Science* 277: 918–924.

Schelling TC. (1971). Dynamic models of segregation. *Journal of Mathematical Sociology* 1: 143–186.

Shaw C and McKay H. (1969). *Juvenile Delinquency and Urban Areas: A Study of Rates of Delinquency in Relation to Differential Characteristics of Local Communities in American Cities*. Chicago: University of Chicago Press.

Siegel DA. (2009). Social networks and collective action. *American Journal of Political Science* 53: 122–138.

The Guardian and LSE. (2011). Reading the riots: investigating England's summer of disorder. Available at http://www.guardian.co.uk/uk/series/reading-the-riots

Torrens PM and McDaniel AW. (2013). Modeling geographic behavior in riotous crowds. *Annals of the Association of American Geographers* 103(1): 20–46.

Townsley M and Sidebottom A. (2010). All offenders are equal, but some are more equal than others: variation in journeys to crime between offenders. *Criminology* 48: 897–917.

Weidmann AG and Salehyan I. (2013). Violence and ethnic segregation: a computational model applied to Baghdad. *International Studies Quarterly* 57: 52–64.

Wilkinson SI. (2009). Riots. *Annual Review of Political Science* 12: 329–343.

Wilson AG. (1970). *Entropy in Urban and Regional Modelling*. London: Pion.

Wilson AG. (1971). A family of spatial interaction models, and associated developments. *Environment and Planning* 3: 1–32.

Wilson AG. (2008). Boltzmann, Lotka and Volterra and spatial structural evolution: an integrated methodology for some dynamical systems. *Journal of the Royal Society Interface* 5: 865–871.
Wilson AG. (2012). *The Science of Cities and Regions: Lectures on Mathematical Model Design*. Berlin: Springer.
Wortley R. (2001). A classification of techniques for controlling situational precipitators of crime. *Security Journal* 14: 63–82.
Yarwood R. (2007). The geographies of policing. *Progress in Human Geography* 31: 447–465.
Yiu SY, Gill A and Shi P. (2002). Investigating strategies for managing civil violence using the MANA agent-based distillation. In *Proceedings of the Land Warfare Conference 2002*. Salisbury, Australia: Defence Science and Technology Organisation.

Appendix

A.1 Note on Methods: Data

The behaviour of rioters is studied via the analysis of recorded crime data, provided by the Metropolitan Police. This consists of all offences detected by the police in London in the period 6–11 August 2011 which have been classified as being associated with the riots. Although the usual caveats concerning recorded crime data – regarding completeness and representativeness – do apply, these data are the best available for the London riots. Each record corresponds to a single incident and specifies the area where the offence took place, the date and time at which the offence was estimated to have occurred, the residence of the offender and the age of the offender. It should be noted that no offender appears more than once in the data. This is preferable, since our focus is on involvement only, and the fact that a single instance of participation might comprise several crimes (perhaps determined subjectively) is liable to introduce bias to the data.

The dataset is composed of 3914 records; however, only 2299 of these contain entries for both residential and offence location. Since both of these elements are fundamental to the model, only those records were analysed. In using this information to determine the distribution of 'journeys to crime' (i.e. the distance between home and offence), the Euclidean distance between the centroids of the basic census units within which the two points fall is used. Although it is recognised that this is not necessarily a true representation of the cost associated with travel, it is the only metric which can be applied consistently to our data; furthermore, common alternatives incorporating travel time may well not apply in such extraordinary circumstances.

The geographical data we use in our analysis of residential areas are Lower and Medium Super Output Areas, a hierarchical geographical structure defined by the UK government for census purposes. The census itself is also used to provide the residential populations of each of these areas, as used in the model. Government statistics for deprivation are also available at the level of Lower Super Output Areas, in the form of the Index of Multiple Deprivation (produced by the Department for Communities and Local Government in 2011). This is a UK-wide indicator, by which areas are ranked according to a combination of employment, health, education, housing and other factors.

We also use data concerning London's retail centres, as defined by the Department for Communities and Local Government (see www.planningstatistics.org.uk). These are consistently defined 'areas of town centre activity', and measurements of the total area of retail floorspace are given for each; it is this quantity that is used as a proxy for the size of each centre in our model. To calculate the distances between residential areas and retail centres required by our model, the Euclidean distance between zone centroids is used.

A.2 Numerical Simulations

The data described in Section A.1 are used as the inputs for numerical simulations of the model described previously. The model is implemented as a discrete-time system, with each time step involving sequential iteration through the model equations given in the earlier text.

In order to obtain the configuration used for the 'demonstration case', a parameter search was carried out across the space of all tunable parameters. The process began by selecting plausible ranges for each parameter, informed by previous work with similar models. One simulation was carried out for each configuration in a coarse sampling of this space, with several observables computed for each completed run: the distribution of riot magnitude across all sites, the number of riot sites where the level of offending was of an order higher than the mean level, the temporal progression of the simulation and the distribution of flow-weighted travel distances. Using these observations, a smaller region of parameter space was identified for which all observations were of similar character to the riot data (in the sense that their relative difference was within a certain tolerance). The process was then repeated for the smaller parameter space, using a lower tolerance, and several further similar iterations followed.

10

The London Riots – 2: A Discrete Choice Model

Peter Baudains, Alex Braithwaite and Shane D. Johnson

10.1 Introduction

The literatures on offender decision making and collective behaviour contain a number of theories regarding the nature by which individuals make decisions during outbreaks of rioting. These theories are employed in this chapter to investigate offender spatial decision making for the 2011 London riots. A parametric statistical model is presented that evaluates the extent to which theories of offender behaviour offer explanations for the distinctive space–time patterns of the riots. This model is presented in more detail in Baudains et al. (2013). The individual behaviour modelled is the choice of spatial location selected by each offender, a key driver in the emergent spatio-temporal profile of the system. By considering the observed riots as just one realisation of a probabilistic model, a simulation is then constructed in an attempt to recreate the spatio-temporal profile of the riots. This simulation is considered from the perspective of its possible application in the policy domain, with the findings of the model used to calculate optimal police deployment strategies.

10.2 Model Setup

For any type of crime, the rational choice perspective (Cornish and Clarke, 1986) suggests that offenders undergo at least two basic decision-making processes. The first concerns their readiness to offend, and this involves a consideration of the offender's needs, experience and alternative courses of action. The second, which is the focus of this chapter, concerns event-level decision making and includes the choice of where to offend. Offenders are assumed to follow a spatially structured process, whereby (for example) they select an area to target, then

Approaches to Geo-mathematical Modelling: New Tools for Complexity Science, First Edition. Edited by Alan G. Wilson.

Companion Website: www.wiley.com/go/wilson/ApproachestoGeo-mathematicalModelling

a particular street and then a particular place. From this perspective, in the case of the riots, having decided to engage in the disorder, an offender must select a location at which to offend. They do so from a set of locations that represent a choice set of alternatives that are available to each rioter. Discrete location choice models (Bernasco and Niewbeerta, 2005; McFadden, 1974), the particular form of random utility model employed here, enable us to estimate the extent to which area-level factors affect offender spatial decision making. Using such a model, the first aim of this chapter is to determine what types of area level factors, measured using a variety of data sources, appear to influence where rioting broke out in London in 2011. This analysis informs the second aim of the chapter, which is to develop a model of what types of areas might be more at risk if a similar outbreak of disorder were to occur in the future.

In order to differentiate between the different choices, it is assumed that each choice that could be made has associated with it some intrinsic value, or utility, to the decision maker. If the decision maker were to choose a particular option, they would then obtain that level of utility. Utility is often thought of as the difference between the benefits and costs of selecting a particular option and, as well as tangible constituents such as financial gain, can also incorporate abstract concepts, such as the well-being or potential happiness of the decision maker. As will be demonstrated in the case of rioting, utility can consist of a number of factors, including the ease of accessing a particular target, or the value of potential goods that may be looted from a target. Here, we describe the choice model employed in more detail.

For decision maker i, suppose that alternative j has utility $U_{ij} \in \mathbb{R}$ associated with it, for alternatives $j = 1, \ldots, J$, and for decision makers $i = 1, 2, \ldots, N$. In other words, U_{ij} is the utility decision maker i obtains by selecting alternative j. The principal assumption of the model is that a decision maker will select the alternative that offers the maximum utility across all possible alternatives. That is, decision maker i will select the alternative j with $U_{ij} > U_{il}$ for all $l \neq j$.

A random utility model estimates U_{ij} for all i and j by supposing the perception of the utility to decision maker i is composed of two components given by

$$U_{ij} = V_{ij} + \epsilon_{ij}. \tag{10.1}$$

The first component, denoted by V_{ij}, is the observable component of the utility U_{ij}. That is, V_{ij} is the utility of alternative j according to decision maker i that is perceptible to an observer of that decision maker. This is the portion of utility that a researcher can attempt to model. The second component, denoted by ϵ_{ij}, corresponds to the unobserved utility. This corresponds to the desires and objectives of the decision maker that are unknown to an observer, and it can be used to incorporate idiosyncratic preferences across individual decision makers. The unobserved component of utility cannot be accurately obtained and thus is treated as a random error term.

The introduction of a random error term means that a random variable Z_i is introduced, defined as the choice that is made by decision maker i. In what follows, a model is derived for the probability distribution $Pr(Z_i = j)$ for each choice $j = 1, 2, \ldots, J$ and for each decision maker $i = 1, 2, \ldots, N$.

Different models can be specified by assuming different functional forms on the error term ϵ_{ij}. The model that is specified here is the multinomial logit model, and it is derived by assuming that the errors ϵ_{ij} for $j = 1, \ldots, J$ are independent and identically distributed according to an extreme value type I distribution (which is also known as a Gumbel distribution). It can be shown (e.g. Train, 2003) that under this assumption, the distribution of Z_i is given by

$$Pr(Z_i = j) = \frac{e^{V_{ij}}}{\sum_{l=1}^{J} e^{V_{il}}}. \quad (10.2)$$

The probability of each decision maker i selecting alternative j is therefore given by an expression that is dependent on only the observed component of utility for each choice, and does not depend on the unknown error ϵ_{ij}. Given the observed component of utility for each alternative and for each decision maker, this probability can be found by calculating the ratio of the exponential of the observed utility, compared against the sum of the exponentials for all alternatives. The model emphasises the comparative nature of the discrete choice model: the decision maker is more likely to select those alternatives which offer comparatively greater observed utility. An account of the history of this model and the range of different uses is given in McFadden (2001).

The model in equation (10.2) leads to what is known as independence of irrelevant alternatives. That is, the ratio of choice probabilities between any two alternatives is unaffected by the presence of other alternatives, which arises due to the following equation:

$$\frac{Pr(Z_i = j)}{Pr(Z_i = l)} = \frac{e^{V_{ij}}}{e^{V_{il}}}, \quad (10.3)$$

for two different alternatives j and l. Independence of irrelevant alternatives arises as a result of the assumption of independence over the error terms ϵ_{ij}. In many scenarios, particularly those related to spatial choice problems, this assumption is likely to be violated. In particular, it implies that an area is chosen as a result of just the features of that area, and not as a result of the features of areas nearby. Since it is conceivable that a rioter may offend in a particular area due to the characteristics of a neighbouring area, the errors are likely to be correlated between nearby targets. This mechanism would manifest as a spillover effect. To account for this limitation of the model, and the fact that there may well be correlated error terms over different alternatives, there have been a number of more complex models proposed that do not have a closed analytic form (Train, 2003). However, as Bernasco et al. (2013) explain, whilst such models allow for spatial dependence, they do not directly treat it as an active process, instead accounting for it indirectly as part of the error term. Accounting for spatial effects in the observed part of the model enables investigation into the spatial processes that might be at play (see also Beck et al., 2006). This is the approach taken in this study, and spillover effects are incorporated into the observed part of the model.

10.3 Modelling the Observed Utility

For each rioter, the choice set is taken to be the set of 4765 Lower Super Output Areas (LSOAs) in Greater London. LSOAs are a geographic partition of the United Kingdom for the purposes of reporting census data. Each LSOA is designed to contain around 1500 residents, and, consequently, the set of LSOAs vary in size according to the underlying population density.

The data used to calibrate the model of target choice was obtained from the Metropolitan Police Service and consists of all crimes associated with the 2011 London riots. For each offence, the data included identifiers for the LSOA within which the offence took place, the LSOA in which the offender was recorded as living, the date and time on which the offence was estimated to have occurred, and the age of the offender. Of the 3914 recorded events, complete information was available for 2299 offences, and it is these data that are analysed here. Unlike other studies of offending that employ the discrete choice approach (e.g. Johnson and Summers, 2015), no offender appears in the data more than once. This simplifies the analysis since the assumption of independence (e.g. that prolific offenders do not bias estimates obtained) is not violated.

The observed component of utility V_{ij} for offender i and target j can be modelled by considering the characteristics of the target j, its relationship to the offender i and how this might change based on the time at which the offender chooses to engage in the disorder. It is modelled as a linear combination of n variables, denoted by $W_{1ij}, W_{2ij}, \ldots, W_{nij}$, so that

$$V_{ij} = \beta_1 W_{1ij} + \beta_2 W_{2ij} + \ldots + \beta_n W_{nij}, \tag{10.4}$$

for parameters $\beta_1, \beta_2, \ldots, \beta_n$.

The construction of the model requires the specification of each of these variables for all values of i and j, and, in what follows, this is described for each W_{gij}, for $g = 1, 2, \ldots, n$. These variables are chosen in accordance with three criminological theories that have previously been used to explain the target choice of crime and rioting. These are: the theory of crowds, crime pattern theory, and the theory of social disorganisation. These theories, each of which is discussed in more detail below, describe, respectively, how the behaviour of a crowd, and therefore the presence of rioting at a particular location, influences the likelihood of selecting that area in which to riot; how decisions made with respect to rioting are influenced by the routine activities of rioters combined with the environment and urban form of the potential locations; and how rioting is more likely to occur in areas with weak social ties.

Modern treatments of crowds assert that rioters retain some level of control over their actions. This is in contrast to historical perspectives of rioting which viewed the actions of crowds as irrational or 'animal-like' (Freud, 1921; Le Bon, 1896). Ongoing rioting can, however, serve as a situational precipitator by prompting, pressuring, permitting or provoking others to engage in the disorder (Wortley, 2008). Indeed, recent treatments of crowds have incorporated individual incentives with some degree of rationality, but also allow individuals to be influenced by the actions of those around them. This perspective of crowd behaviour, in which individuals, or agents, influence the behaviour of others, each of whom has their own set of behaviours, attributes or objectives, all of which might vary widely over the population, has been considered in a range of models of rioting and civil disorder (Epstein, 2002; Granovetter, 1978; Midlarsky, 1978; Myers, 2000). It is important to emphasise that such an argument does not assume that offenders cease to act like rational agents, but that the decision to engage in a criminal event can be rather dynamic: actors are perhaps more willing to engage in disorder due to the precipitating influence of crowd behaviour after weighing up the costs and benefits of doing so.

In order to incorporate riot precipitators into the model of target choice, it is assumed that the utility of each target depends on the number of riot-related offences that have recently occurred at that target. It is hypothesised that, all other things being equal, areas in which riots have recently occurred will be more likely to be selected by rioters in which to offend. We therefore select the variable W_{1ij} to consist of the number of riot offences that occurred in area j in the 24 hours prior to the time at which rioter i makes their decision.

Crime pattern theory asserts that the spatial and temporal distribution of crime can be explained by the routine activities (Cohen and Felson, 1979) of potential offenders (Brantingham and Brantingham, 1993). Specifically, it considers how routine activity patterns shape offender awareness of criminal opportunities, and how this may lead to the emergence of spatial concentrations of crime. According to the theory, people create mental maps of their routine activity spaces, which typically consist of routine activity nodes (the locations individuals frequently visit, or at which they spend much of their time) and the routes they take to travel between them. It is within these awareness spaces that offenders gain knowledge of crime opportunities and the likely risks and rewards associated with them and, thus, are predicted to be more likely to engage in crime.

Prominent features of the urban environment are expected to influence the awareness spaces of a range of different people, including offenders. In particular, for much of the population of London and other urban areas around the world, routine activity nodes are likely to include local landmarks such as retail centres, transport hubs and schools. As such, these activity nodes may act as crime generators through their influence on the routine activities of offenders. Using data for the city of Chicago, Bernasco and Block (2009) provide evidence that this is indeed the case for the crime of robbery.

Consequently, on the basis of crime pattern theory, it is hypothesised that during rioting, with all other things being equal, offenders will be more likely to choose locations at which to offend that are near to schools, public transport hubs, retail centres and locations that are proximate to the city centre, as these represent locations which are likely to be prominent within the mental maps of a wide range of rioters. In addition, consistent with the literature on the 'journey to crime' (e.g. Rengert et al., 1999), rioters are predicted to be more likely to select locations that are closer to their residential location, since they should be rather familiar with the urban environment around their home. As well as increasing the likelihood that people will become familiar with the particular locations, features of the urban environment can also reduce it. Natural barriers, such as rivers, are one such example (e.g. Clare et al., 2009). In this case, the cost of traversal can reduce people's familiarity with those places that are located on the other side of a river from their routine activity nodes. Consequently, we predict that rioters will be less likely to cross the River Thames, all else being equal. Table 10.1 presents the variables included in the model of target choice used to capture arguments associated with crime pattern theory.

Table 10.1 The variables used to estimate the observed utility of each target associated with crime pattern theory

Variable	Description	Expected effect of higher values on attractiveness
W_{2j}	Number of schools (with interaction term)	Increase, particularly to juveniles
W_{3j}	Underground station indicator	Increase
W_{4j}	Retail floorspace	Increase
W_{5j}	Distance to city centre	Decrease
W_{6ij}	Distance between residence and target	Decrease, particularly to juveniles
W_{7ij}	Thames between residence and target	Decrease

Proponents of social disorganisation theory argue that the inability of a community to jointly identify common social values, and to subsequently exert effective informal social controls, substantially increases the crime and delinquency within an area. That is, for neighbourhoods in which there is a strong sense of community and mutual cooperation, residents are more likely to intervene to prevent crime. Tests for social disorganisation theory typically identify conditions that might lead to a lack of social cohesion, which can be affected by a number of different neighbourhood characteristics. For neighbourhoods with a transient population, for example, brought about by a large flux of inward and outward migration, it is asserted that there will be relatively fewer opportunities for the formation of stable social ties, leading to the lack of social cohesion which fosters inability to jointly act to prevent and mitigate crime. Other conditions identified as having an impact on the resulting crime and delinquency rates include ethnic heterogeneity (it is argued that diversity among individuals can act as a barrier to social cohesion as different communities can fail to share consensus), family disruption (close family is often viewed as a first opportunity to exert such informal social control) and deprivation (rather than having a direct result on levels of crime, it is argued that within disadvantaged neighbourhoods, communities may lack the resources and organisational base of their more affluent counterparts, and so are less likely to exert control) (Bernasco, 2006; Sampson and Groves, 1989; Sampson et al., 1997).

The level of social disorganisation of an area could influence the likelihood of rioting taking place in slightly different ways. First, cohesive neighbourhoods may exert control over their own residents to reduce the likelihood that they will engage in disorder or form a rioting crowd. Second, signs of cohesion within a neighbourhood might affect whether offenders, regardless of where they live, choose to engage in disorder within that neighbourhood. In this case, social cohesion might be seen as acting as a social barrier to deter rioters from targeting or coalescing in a given neighbourhood (Bernasco, 2006, argues this from the perspective of target choice for residential burglary).

We therefore hypothesise that there is an increased likelihood of areas being selected as targets for rioting if those areas have greater levels of social disorganisation. We measure social disorganisation using three variables. The first variable measures population churn of an area j, given by

$$W_{8j} = \left(\frac{D_j + O_j + M_j}{P_j} \right) \times 10, \tag{10.5}$$

where D_j is the in-migration to area j over a particular period of time, O_j is the out-migration from the area and M_j is the total migrants who relocate from one residence to another whilst remaining within the same area j over that time period. P_j is the population of area j.

The second variable measures the ethnic heterogeneity of each target area. The index of qualitative variation is used, and defined as

$$W_{9j} = \left(1 - \sum_{k=1}^{E} e_{kj}^2 \right) \times 10, \tag{10.6}$$

where E is the total number of distinct ethnic groups and e_{kj} is the proportion of individuals belonging to ethnic group k, who reside in area j. W_{9j} is interpreted to be a measure of the probability that two individuals selected at random from the population of zone j will be of different ethnicity.

Finally, a measure of deprivation is incorporated into the observed utility of each target. Denoted by W_{10j}, this is given by the Index of Multiple Deprivation, a measure used extensively in the United Kingdom to determine disadvantaged areas.

In addition to the variables discussed above, a measure of population density of each target area is also included in the model, denoted by W_{11j} to control for its potential effects.

A number of spatially lagged variables are also included in order to mitigate some of the unintended effects that arise due to independence of irrelevant alternatives. W_{12ij} denotes a spatially lagged variable to measure the number of offences that occur in neighbouring areas to j for the 24 hours prior to the offence i taking place. In addition, spatial spillover effects are included for: the average number of schools in neighbouring areas, denoted by W_{13j}; an indicator variable to determine whether the neighbouring area contained an underground station, denoted by W_{14j}; and the average retail floor space in neighbouring areas, denoted by W_{15j}. All neighbouring areas are defined with queen contiguity: areas need to just share a single point of a boundary in order to be classed as neighbours.

10.4 Results

Parameter estimation is achieved using a maximum likelihood procedure. For reasons of computational efficiency, parameter estimation is performed separately for each day of rioting. This provides the added advantage of comparing the parameters across the different days of rioting. If the parameter estimates are consistent over the different days tested, it would provide evidence that the conclusions that may be drawn from them are robust under the data for the different days (which involves different offenders since each offender appears in the dataset only once). The number of offences that occur for each day of rioting, and therefore the number of offences included in each optimisation procedure, together with the number of LSOAs which contain offences during each day are shown in Table 10.2. For purposes of clarity, and to avoid making conclusions from an insufficient amount of data, the results for the first and last days of rioting – the 6th and 10th of August – are excluded from the presentation in what follows. Therefore, only results for the 7th, 8th and 9th of August are reported, which included 93.7% (2155) of available records.

Models with and without spillover effects are presented separately due to the anticipated high levels of collinearity between the spatially lagged variables and the variables associated with each target area, which can lead to problems in the interpretation of parameter estimates. A model without neighbourhood effects would be preferred from the perspective of minimising collinearity in the explanatory variables, but a model with these effects would be preferred from the perspective of mitigating the impact from independence of irrelevant alternatives. If the inclusion of spatial spillover effects has little effect on the estimation of the other parameters, then there will be more confidence in the robustness of the parameter estimates.

In order to measure the extent to which the model is able to reproduce the observed data, a pseudo R^2 statistic is calculated for each day of rioting. Denoting the maximum likelihood estimator by $\bar{\boldsymbol{\beta}}$, this statistic compares the value of the log-likelihood function $\ln \mathcal{L}(\boldsymbol{\beta})$ with $\boldsymbol{\beta} = \bar{\boldsymbol{\beta}}$ and the value of the same function with $\boldsymbol{\beta} = \mathbf{0}$, and is defined as

$$R^2 = 1 - \frac{\ln \mathcal{L}(\bar{\boldsymbol{\beta}})}{\ln \mathcal{L}(\mathbf{0})}. \tag{10.7}$$

Table 10.2 The number of offences and the number of LSOAs affected by day of rioting. The total number number of LSOAs affected is the total number of LSOAs that experienced rioting over the 5 days

Dates	Number of arrest records	Number of LSOAs affected
6th August 2011	54	20
7th August 2011	232	42
8th August 2011	1477	247
9th August 2011	446	162
10th August 2011	90	55
Total	2299	436

When $\boldsymbol{\beta} = \mathbf{0}$, the observed component of the utility of each choice is equal to 0, and there is an equal probability that each site is chosen. In this case, there are no distinguishable features accounted for across the possible choices. Given that the log-likelihood $\ln\ \mathcal{L}(\boldsymbol{\beta})$ is a measure of the probability that the model with parameter $\boldsymbol{\beta}$ will result in the observed data, the value of R^2 indicates the extent to which the model estimated with the parameter $\bar{\boldsymbol{\beta}}$ increases this probability against a null model in which targets are selected uniformly randomly. It can be interpreted as the extent to which the model with the parameter $\bar{\boldsymbol{\beta}}$ explains the variance observed in the data.

The weighted-average value of R^2 (weighted according to the number of offences included in each day of analysis) without spillover effects is 0.34 and with spillover effects is 0.36. The first observation that can be made is that the model performs well in explaining the variation of target choice in the data: the likelihood function increases by around 35% when explanatory variables are included. In particular, McFadden (1979) states that values between 0.2 and 0.4 represent an excellent fit to the data. It should be noted that the R^2 values are typically much lower for maximum likelihood estimation than R^2 values that can be calculated in ordinary least-squares regression. This is because the model uses probabilities to estimate a binary choice (whether each area is chosen).

In Figure 10.1, the maximum likelihood estimates are presented without spillover effects. In Figure 10.2, spillover effects are included. A 95% normal confidence interval is also shown for each parameter. The exponentiated value of the parameter β_g is the multiplicative effect of a one-unit increase in attribute W_{gij} on the odds that decision maker i selects target j. The odds are defined as the probability that i selects j, divided by the probability that i does not select j. If $e^{\beta_g} = 1$, then there is no association between that variable and offender spatial decision making during the London riots. Values above 1 suggest that the likelihood of an area being chosen is positively associated with the variable considered, and values below 1 suggest a negative association. The value of each exponentiated parameter in Figures 10.1 and 10.2 can therefore enable the interpretation of each attribute in the model.

The estimates of the first parameter, which measures the effect that offences occurring in the previous 24 hours at each location have on the attractiveness of that target, are consistently positive and significant. Although the magnitude of the exponentiated variable is only slightly greater than 1, the relatively small confidence intervals for the parameter estimate suggest

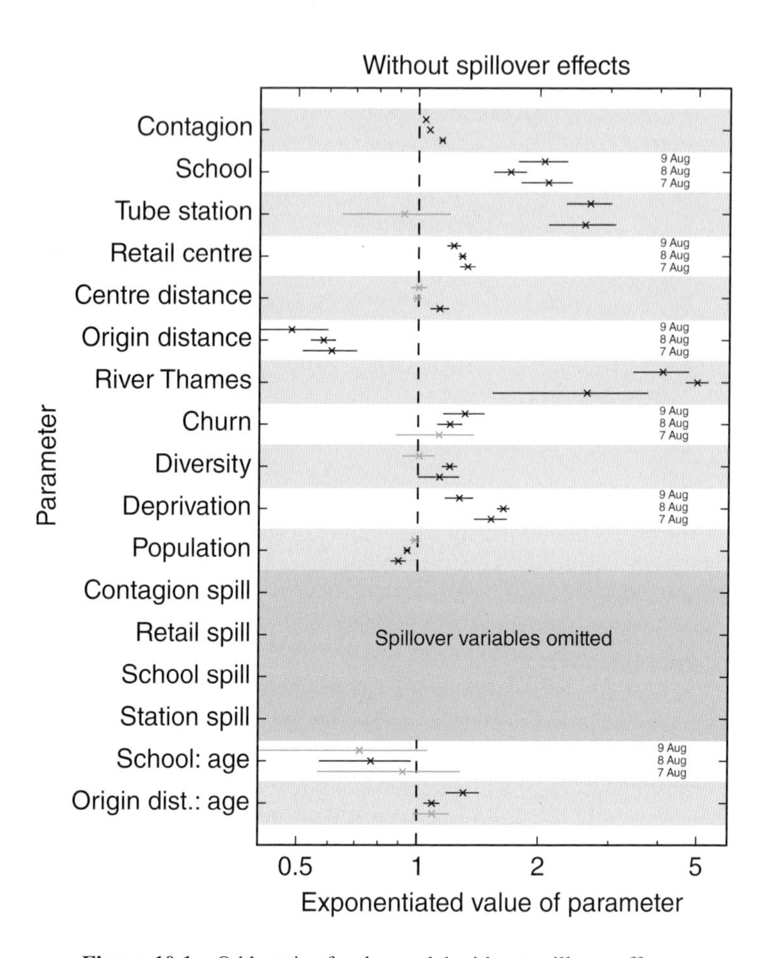

Figure 10.1 Odds ratios for the model without spillover effects

that this is a highly significant finding. Moreover, it is important to consider that rescaling the parameter (e.g. dividing the number of prior offences by 10) would produce a much larger estimated coefficient. Thus, we conclude that ongoing rioting can act as a situational precipitator, in which rioters are encouraged to engage in the disorder more so than they otherwise would. To illustrate, on the 7th of September, the odds of an area being targeted by an offender increased by a factor of 1.143 for every additional (detected) offence that occurred in that area in the previous 24 hours.

Similar statements can be made for each day of unrest; however, when considering how this variable changes over the three days of rioting – from 1.143 on the 7th, to 1.064 on the 8th, to 1.039 on the 9th – the temporal distribution of offences throughout the duration of rioting requires consideration. This is because the number of offences that occur within any 24-hour period prior to an offence changes significantly over time, and such a change may well affect the parameter estimates. Indeed, since the parameter estimates measure the increased attractiveness of each area due to a single extra offence with all other things being equal,

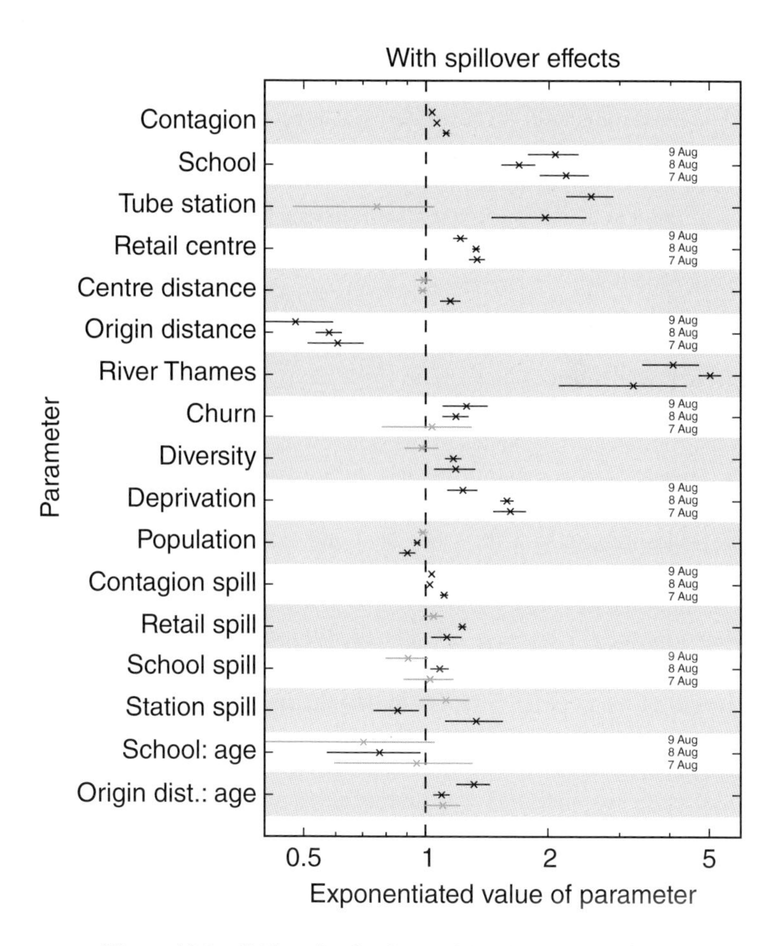

Figure 10.2 Odds ratios for the model with spillover effects

one might expect to experience diminishing returns on the extent to which attractiveness can increase as the number of rioters increases. That is, the increased attractiveness per rioter is likely to decrease with more rioters: it has been hypothesised elsewhere that the first rioter can be the most important in influencing the chance of a full-scale outburst (Granovetter, 1978). Since the parameter estimates considered here decrease with each passing day, and since the number of rioters increased from the 6th August to the 8th August (see Table 10.2), the period of time prior to events on the 9th August would, in all likelihood, include the greatest number of rioters. Thus, if there was a diminishing effect on the increased attractiveness as the number of rioters at each site increased, then the estimates of the parameters for this variable would be expected to decrease over time, which is indeed what is observed.

The parameter estimates for the presence of schools in the target area are all positively associated and significant at the 0.05 level with the choices made. Thus, rioters were more likely to offend in areas containing schools. In order to further test the hypothesis that the target choice of offenders was influenced by schools being prominent in the routine activity nodes of offenders, an interaction parameter was also estimated to determine the extra effect brought about by the offender being over the age of 18. Our hypothesis was that the location of schools would be most influential for juveniles, since they would be more likely to represent current or recent routine activity nodes for them. Although only significant at the 0.05 level for the 8th August, there is some indication that the effect of schools on the decision making of adults is less prominent than it is for juvenile offenders. For instance, on the 8th August, the point estimate is 0.769, meaning that the total impact of schools was estimated to be around 23% less for adult offenders.

The effect from the connectivity of an area, proxied by an indicator of the presence of an underground train station, was, for the 7th and 9th August, positively associated with the chance that the area was selected. In fact, on those days, the odds of an area being targeted by an offender more than doubled if it contained a station. This provides further support that those areas in the routine activity nodes of offenders were more likely to be targeted. On the 8th August, the estimate was not significant and, curiously, was in the opposite direction from the other two days. This might be explained by the fact that, on the 8th August, the rioting was much more widespread than on the other days, and so the ease of accessibility might have been less of a concern for rioters on this particular day.

The effect of retail centres was positively associated and significant with the likelihood of an area being selected for all days considered. Whilst this finding might be interpreted as evidence for the targeting of routine activity nodes, which may include retail centres, it may also be a result of many of the offences during the riots being associated with looting of high-value goods.

The effect on target attractiveness from its distance to the city centre did not appear to play a consistent role, as the estimate was only significant for one of the days tested. One reason for the apparent absence of the influence of this variable may be that, for a city like London, the city centre may be too crude to represent a routine activity node for all offenders.

The point estimates for the effect of distance between the offender's residence and their riot location are significantly negatively associated with the choices made by rioters for each day of unrest. This suggests that areas further away from a rioter's residence were less likely to be selected, which is entirely consistent with crime pattern theory. For interpretation, the odds of an offender selecting an area reduces by a factor of between 0.482 and 0.608 for each additional kilometre of distance between their residence and that target area, all other things being equal.

Considering the interaction term for distance and age, two of the three estimates are statistically significant at the 0.05 level, suggesting that the magnitude of the exponentiated parameter estimate for the journey to crime variable is closer to 1 for adults than it is for juveniles. This indicates that the effect of distance on the target choice of rioters is more pronounced for juvenile offenders, and adult offenders did indeed appear to travel further to commit their crimes. This likely reflects the extended awareness spaces of adults perhaps combined with their increased means to travel farther (see Townsley and Sidebottom, 2009).

The influence of the River Thames was significantly positively associated with rioter target choice, and consistent across all days. This supports the hypothesis derived from crime pattern theory, which states that the river acts as a natural barrier to the awareness spaces of offenders.

With respect to social disorganisation, as measured by population churn, ethnic diversity and deprivation, the parameter estimates were, in general, positively associated with target choice, although some results were not statistically significant at the 0.05 level. This finding lends some support to the argument that social disorganisation influenced the locations of rioting.

In Figure 10.2, in which estimates of the exponentiated coefficients for the spillover variables are also presented, it can be seen that the results are consistent with those discussed so far, with respect to both the direction in which the effect acts, as well as the significance of each variable. Thus, it can be concluded that the inclusion of spillover effects does not drastically alter the parameter estimates of the other variables. This demonstrates that the findings are robust, and implies that the substitution patterns captured by the spillover variables do not unduly impact the other estimates.

The spillover effect from prior offences is significant and positive for all days under consideration. The occurrence of offences in neighbouring areas therefore appears to increase the attractiveness of areas to rioters. Considering the spillover effects for the presence of schools, underground train stations and retail areas, the results are more mixed, with significant effects at the 0.05 level detected for schools on the 7th and 8th of August, for underground stations on the 8th of August and for retail areas on the 7th and 8th of August.

The interpretation of the individual spillover parameters is complicated due to high levels of collinearity with non-spillover variables that arise due to spatial autocorrelation of those variables. The importance of including the spillover effects is largely to determine whether the non-spillover parameters are consistent when the spillover effects are included. Since this appears to be the case, this provides evidence for the robustness of the parameter estimates and the model itself. In particular, variables associated with crowd theory, crime pattern theory and social disorganisation theory have been shown to provide robust estimates for influences on rioter target choice. Consistency of many of these estimates over the different days tested implies consistency in the decision making of rioters, providing some evidence for the presence of (bounded) rationality in rioter target choice.

10.5 Simulating the 2011 London Riots: Towards a Policy Tool

We now propose a microsimulation model of rioter target choice that is based upon the statistical model presented in this chapter. The objective of a microsimulation model is to generate realisations of individual decision making, based on aggregated empirical data, which might have applicability within a policy setting (Ballas et al., 2005). Microsimulation models typically consist of an empirical dataset for a particular population, which is used to specify the initial conditions, together with a series of probability distributions that may be conditional upon a range of factors. Pseudo-random number generators combined with these probability distributions are then used to simulate certain characteristics associated with each individual in the population, such as the decisions they make over a period of time.

The successful estimation of parameters in the statistical model of discrete choice suggests that an appropriate decision to be simulated is the target choice of rioters during the 2011 London riots. The probability distribution defined by the model is conditional upon the initial location of each rioter, the age of the rioter, the time at which the rioter decides to engage with the disorder, and the characteristics of the riot scenario up to that time. Thus, this model is well suited to being applied within a microsimulation framework.

The model is described as follows: suppose that each offender, indexed by i for $i = 1, 2, \ldots, N$, resides within an LSOA in Greater London, denoted by $s_i^{(o)}$, and is deemed to commit their offence at time t_i, corresponding to the hourly interval within which the offence occurred. Let $I_a(i)$ indicate whether offender i is an adult (in which case $I_{a(i)} = 1$) or under the age of 18 (in which case $I_{a(i)} = 0$), and let $s_i^{(d)}$ denote the LSOA that was chosen according to the empirical data. Suppose also that the offences are ordered so that $t_i < t_{i+1}$ for $i = 1, 2, \ldots, N-1$. The discrete choice model depends on the riot scenario up until each rioter makes their decision as to where to engage with the disorder. Thus, we define the history of the system at time t, denoted by $\mathcal{H}(t)$, by

$$\mathcal{H}(t) = \left\{ (t_i, s_i^{(d)}) | t_i < t \right\}. \tag{10.8}$$

The variable to be simulated is the target choice of each offender. Since there is uncertainty surrounding the choice that each offender makes, a random variable Z_i is modelled. Realisations of Z_i are required to correspond to the LSOA which offender i selects as a target within the simulation; thus, the set of possible values for Z_i is given by the set $\mathcal{D} = \{1, 2, 3, \ldots, 4765\}$, where each member of $\mathcal{D}$ corresponds to an LSOA. The probability mass function of Z_i prescribes the probability with which each member of the set $\mathcal{D}$ becomes a realisation of Z_i, and therefore determines the probability with which each LSOA is chosen by offender i in the model. We define

$$f_{Z_i}(j|s_i^{(o)}, I_a(i), \mathcal{H}(t_i)) = Pr(Z_i = j|s_i^{(o)}, I_a(i), \mathcal{H}(t_i)) = \frac{e^{V_i(j)}}{\sum_{l=1}^{J} e^{V_i(l)}}, \tag{10.9}$$

where $V_i(j)$ is the observed component of utility gained by offender i if they were to select option $j \in \mathcal{D}$.

A candidate for the function $V_i(j|s_i^{(o)}, I_a(i), \mathcal{H}(t_i))$ was constructed as a discrete choice model in Section 10.3, where it was written V_{ij}; however, not all of the components of the model were deemed to be significant predictors for the behaviour of rioters. As a consequence, the model taken in this section is chosen to only include the variables which provided the most predictive power, assessed by the corresponding confidence interval associated with each variable. Thus, the following function is defined:

$$\begin{aligned} V_i(j) = {} & \beta_1^{\delta t} W_{1ij} + \beta_2 W_{2j} + \beta_2^a I_a(i) W_{2j} + \beta_3 W_{3j} \\ & + \beta_4 W_{4j} + \beta_6 W_{6ij} + \beta_6^a I_a(i) W_{6j} + \beta_7 W_{7ij} + \beta_{10} W_{10j}, \end{aligned} \tag{10.10}$$

where the terms measure, respectively, the effect from offences occurring in target area j during the previous 24 hours to t_i; schools in target area j; underground train stations in target area j; retail floorspace in target area j; the distance between the offender's residence and target area j; whether or not target area j is on the same side of the River Thames as the offender's residence; and deprivation in target area j. As well as explaining a large amount of the variance in the data, these variables also capture the three theoretical perspectives – crowd theory, crime pattern theory and social disorganisation theory – discussed in the derivation of the model.

The values of the vector $\boldsymbol{\beta} = (\beta_1^{\delta t}, \beta_2, \beta_2^a, \beta_3, \beta_4, \beta_6, \beta_6^a, \beta_7$ and $\beta_{10})$ are selected in accordance with the estimation of these parameters above. For offender i, the corresponding parameter $\boldsymbol{\beta}$ is found by sampling independently from the joint normal distribution with mean given by the point estimates of $\boldsymbol{\beta}$ and standard deviation given by the corresponding standard errors. Recall that three sets of parameters were estimated: one for each day of rioting under consideration. The choice of distribution for each parameter therefore also depends upon the day on which the offence occurred. Random sampling of the parameter values is employed to better reflect the uncertainty associated with each parameter over the different decision makers.

The simulation then proceeds as follows:

1. Set $i = 1$.
2. At time t_i, offender i commits their offence at some location. Calculate the value of $f_{Z_i}(j|s_i^{(o)}, I_a(i), \mathcal{H}(t_i))$ for each $j \in \mathcal{D}$.
3. For each j, find the value of the function:

$$F_{Z_i}(j|s_i^{(o)}, I_a(i), \mathcal{H}(t_i)) = Pr(Z_i \leq j|s_i^{(o)}, I_a(i), \mathcal{H}(t_i))$$
$$= \sum_{l=1}^{j} f_{Z_i}(l|s_i^{(o)}, I_a(i), \mathcal{H}(t_i)), \tag{10.11}$$

which forms an increasing function on the set $\mathcal{D}$, taking values in the interval $[0, 1]$.
4. Generate a pseudo-random number between 0 and 1, denoted by $\mathcal{R}$.
5. Find a realisation of Z_i, given by $z_i = F_{Z_i}^{-1}(\mathcal{R})$.
6. If $i < N$, set $i \rightarrow i + 1$ and return to step 2; otherwise, stop.

This simulation produces a set of chosen target areas, $z_1, z_2, \ldots, z_N$, where the lower case notation is used to correspond to the realisation of the random variable Z_i for $i = 1, 2, \ldots, N$. The outputs of this simulation represent a riot scenario in which the rioters behave according to the discrete choice model. If the model is able to recreate the observed riot data, then it provides evidence to suggest that the theoretical perspectives discussed here are both necessary and sufficient to explain rioter target choice. Furthermore, if the model can provide accurate realisations of riots, then it may be possible for the model to be employed as a predictive tool. In what follows, a comparison between the model outputs and the empirical data is first made, before considering a potential policy application.

In order to assess the model, it is noted that each realisation is the result of a number of stochastic elements, and thus, to get a more complete understanding of the model outputs, a sample of 100 realisations is made, resulting in chosen target areas $z_1^{(g)}, z_2^{(g)}, \ldots, z_N^{(g)}$ for $g = 1, 2, \ldots, 100$. To determine whether the model is capable of producing similar output to the observed phenomenon, the average number of rioters who targeted LSOA j over the 100 simulations of the model, given by

$$\bar{C}_j = \frac{1}{100} \sum_{g=1}^{100} \sum_{i=1}^{N} \mathbf{1}(z_i^{(g)} = j), \tag{10.12}$$

where $\mathbf{1}(.)$ is an indicator function equal to 1 if the condition inside the bracket is satisfied, and equal to zero otherwise, is compared against the actual counts of events that occurred in LSOA j for $j = 1, 2, \ldots, J$.

Figure 10.3a shows the 30 LSOAs that were most targeted by the rioters according to the empirical data. The bars in the positive direction correspond to the empirical count of offences in each LSOA, and the bars in the negative direction correspond to the values of $\bar{C}_j$ that are obtained from the 100 iterations of the simulation for the corresponding LSOA j. Although a significant discrepancy between the model and the data is observed, there is some indication that the most targeted LSOAs were also those that were most targeted according to the

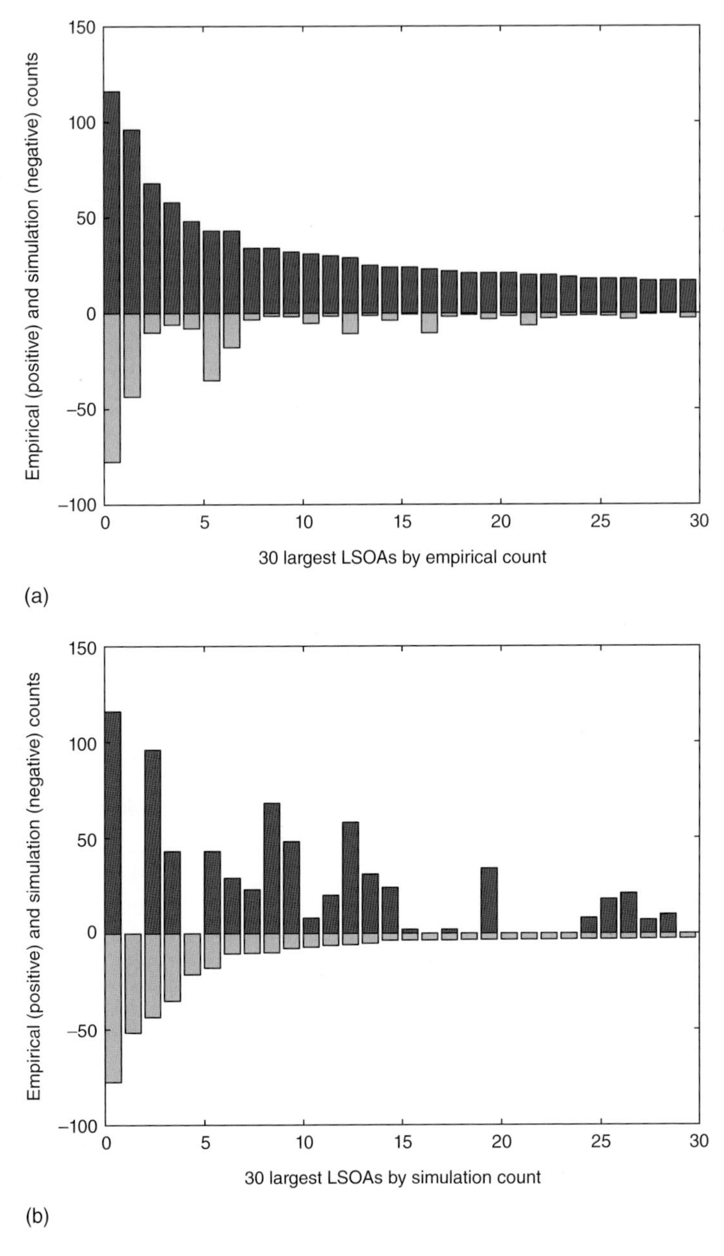

Figure 10.3 The 30 most targeted LSOAs sorted according to (a) the empirical data, and (b) the average of the simulations. The positive bar chart shows the empirical counts, and the negative bar chart shows the averaged simulated counts

simulation. This implies that the model might indeed be able to contribute to the prediction of riot locations; however, it doesn't necessarily capture all of the underlying behaviour of the rioters. In particular, the values of the counts in the empirical data are much larger than the counts resulting from the simulation, suggesting that there was greater clustering in some areas observed than is accounted for in the model.

If the model is to be used as a forecasting tool in a policy setting, one must also be wary of the false positive rate of the simulation, which might occur when the model erroneously predicts that a particular location will be targeted. Figure 10.3b shows a bar chart in which the x-axis represents the 30 LSOAs that were most targeted according to the averaged simulation counts, given by $\bar{C}_j$ for $j = 1, 2, \ldots, J$. The majority of locations most selected by rioters in the simulation were also those areas selected according to the empirical data. There are, however, two notable outliers that deserve attention.

The largest outlier, the second most selected as a target according to the simulation, experienced no offences according to the empirical data. This particular LSOA represents a region in North London containing five schools. Since the count of schools was used as an attractiveness factor in the model of target choice which, for this LSOA, would have been five times as strong, many simulated rioters selected it as a suitable location at which to commit their offence. Possible explanations for why this area was not selected by rioters according to the empirical data may be that the effect from schools is not additive, and that a school indicator function, taking values 0 or 1, would have been a more appropriate measure of the effect from schools, rather than the count. Another explanation may be that the effect from the presence of schools in each LSOA is, in reality, dependent upon a range of other area-level attributes such as retail floorspace. Nonlinear utility functions can be used to model such dependencies. A further possibility is that while we modelled variation in the parameters over time (i.e. for each day of the riots) we did not model variation over space. Thus, it is possible that the parameter estimates vary spatially too. A crude approach to remedy this would be to model the variation in the parameters for a small number of regions of London (e.g. the North and South). A more complicated approach would be to allow the parameters to vary randomly across areas.

Another outlier, representing the fifth most selected LSOA according to the simulation, contains part of London's largest retail centre in the area around Oxford Street. According to the empirical data, this LSOA experienced no riot offences. Since retail floorspace is an attractiveness factor within the simulation, the large retail floorspace of this particular LSOA in comparison to all other areas is likely to have attracted a greater proportion of rioters. Possible explanations for why rioters perceived the very centre of London's retail district as a poor target according to the empirical data may be the perception that, within the centre of London, there may be more law enforcement officers available to counter any riots, which may increase the chances that each rioter will be arrested. Furthermore, larger retail centres may also have higher levels of security, meaning that looting and other riot-related offences are difficult to commit. More generally, it is important to acknowledge that the models described so far (statistical and microsimulation) do not include data on police activity. Since this is unlikely to have been uniformly distributed across the capital, it may have had a significant effect on the space–time profile of the riots. The allocation model described in what follows incorporates some notion of police deployment, but future empirical work might seek to incorporate estimates of police presence in statistical models also.

Although each simulation of the riots has the same number of offences as in the empirical data by construction, the average variance of offence counts in each LSOA across the simulations is 3.29, compared to 11.90 for the empirical data. The offences are therefore more spread out over the LSOAs in the simulation of the riots than is actually observed. This suggests that, although the model goes someway to explaining the target choice of rioters, it does not incorporate all possible explanations as to why rioters selected certain locations over others. Nevertheless, although there are discrepancies between the model and the empirical data with respect to the counts of offences that occurred within each LSOA, the present model may still be of use in a policy setting if it is able to broadly reproduce the spatial patterning of the riots. To determine this, it is next considered whether the simulation broadly consistently highlights those areas that were most vulnerable to experiencing riots. For this purpose, another metric is employed: the ratio of the count of offences in the LSOA to its rank, where the LSOAs are ranked in ascending order according to the number of offences that occur within it. The inclusion of the rank of each LSOA is to reduce the dependency of the following tests on just the counts of offences, which have been shown to include notable discrepancies. If the model is able to broadly highlight the areas most at risk, then it may be of use for the prediction of the location of riots. Figure 10.4 plots the ratio of the count of offences to its rank for each

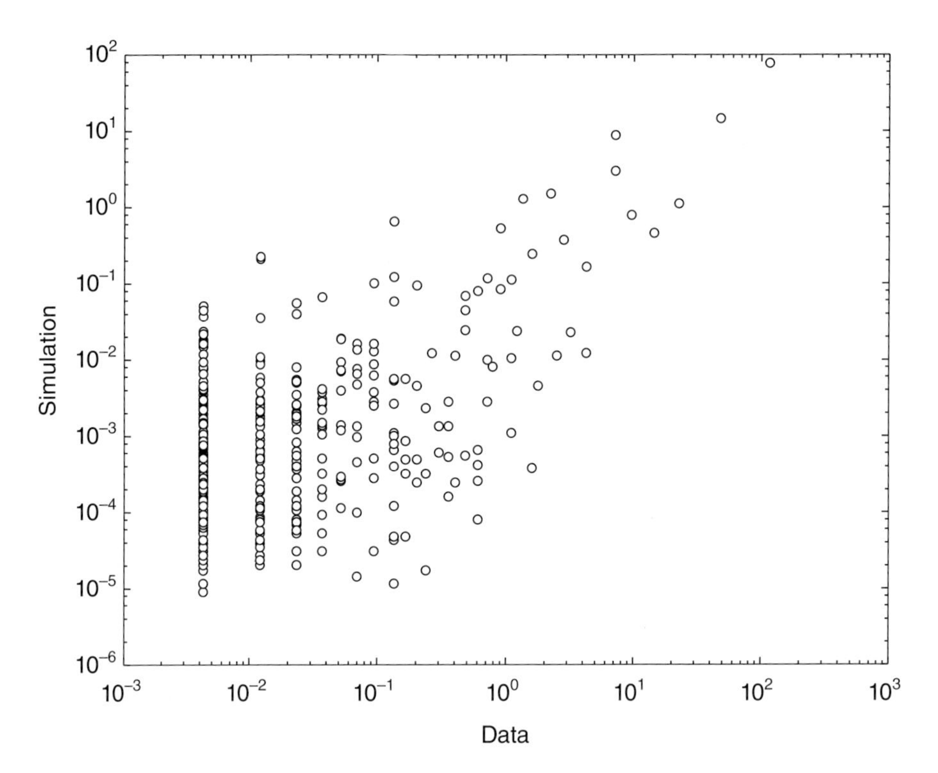

Figure 10.4 Ratio of count to rank for each LSOA for both the empirical data and the simulation. The ranks are obtained by sorting the LSOAs according to their count. The plot is shown on a log-scale for clarity

LSOA, comparing the empirical data to the averaged simulated data. If the model is a good fit to the data, a positive correlation would be expected. Although there is a significant amount of variation between the model and the simulated data, a positive correlation is also observed. The Pearson's product moment correlation coefficient is 0.906, confirming a strong positive correlation and suggesting that the simulation is indeed capable of reproducing some of the more general patterns observed in the empirical data.

10.6 Modelling Optimal Police Deployment

We now demonstrate how the microsimulation might be used within a policy context. The strategic deployment of police during a city-wide outbreak of rioting, as observed in London, is an important policy issue. Police allocations can be made in anticipation of rioting, and they are dynamic, meaning that law enforcement officers can move towards nearby sites at which rioting occurs. The objective for police commanders, therefore, is to optimally allocate law enforcement officers over different areas in the city so that the maximum number of police officers are within a short travel distance of anticipated riot locations, enabling police to arrive quickly once rioting occurs. We propose a dynamic allocation algorithm that uses the outputs of riot realisations generated from the model to produce potential deployment distributions in London.

As in Section 10.5, let the count of offences that occur in LSOA j (for $j = 1, 2, \ldots, J$) in the g-th riot realisation be denoted by $C_j^{(g)}$. Suppose also that the number of police officers available to be deployed prior to a potential riot outbreak is given by L. For scenarios in which the police are unable to be present over the entire region in which riots are anticipated, as was the case during the riots in London, it can be expected that $L \ll J$. When considering potential deployments, the police will consider the number of police officers that may be required for any given number of rioters at each location. We suppose that the number of police officers required to alleviate the threats posed by one rioter (in terms of the damage they may incur on property and danger to civilians) is given by the parameter ν. Thus, a riot of 50 rioters would require a deployment of 50ν police officers to be quelled.

The anticipated riot intensity at LSOA j is defined to be

$$\ln\left(\frac{1 + C_j^{(g)}}{1 + \nu L_j^{(g)}}\right), \tag{10.13}$$

where $L_j^{(g)}$ is the number of police officers deployed to area j in iteration g. This particular form is chosen partly because it increases logarithmically with increasing rioter count, meaning that whilst the intensity will be significantly increased when a single rioter decides to engage with a small disorder – indicating that the disorder is showing significant signs of growth – a rioter joining an already large disorder will increase the intensity by a smaller amount since it is of relatively less importance in comparison to the already large disorder. In addition, the measure decreases logarithmically with increasing $\nu L_j^{(g)}$, suggesting that a small number of police can drastically reduce the threats posed by a small disorder but that the allocation of additional police to larger disorders, at which there is already a significant police presence, does not have a similar reduction. The addition of 1 to both the numerator and denominator avoids the measure being undefined for all non-negative values of $C_j^{(g)}$ and $L_j^{(g)}$.

The allocation of police should also incorporate the time it takes for police to travel between expected rioter sites, since once rioting emerges in certain locations, police officers may wish to arrive quickly to alleviate its impact and to prevent the riot from growing. The proximity between two LSOAs l and j is defined to be

$$\exp(-\upsilon d_{lj}), \tag{10.14}$$

where d_{lj} is taken here to be the Euclidean distance between the centroids of LSOA j and LSOA l; and υ is a positive parameter. Other implementations might consider alternative distance metrics, such as travel time between two LSOAs. The form of this function is useful since it obtains a maximum value of 1 only for the LSOA in which police are already located, and decreases quickly for nearby LSOAs. Therefore, greater emphasis is placed on police being more inclined to remain where they are, rather than travelling too far and arriving at a location at which the presence of police is no longer required. The parameter υ controls the extent to which emphasis is placed upon nearby locations, rather than locations farther away.

Using the two components of riot intensity and proximity, a measure of deployment utility is next defined. The idea behind this measure is to determine the benefit of allocating a single additional police officer to a particular LSOA, whilst accounting for their ability to travel to nearby potential riot sites and, simultaneously, accounting for the police that might already be located nearby. Denoting deployment utility for LSOA l and iteration g by $\mathcal{Y}_l^{(g)}$, the measure is defined as

$$\mathcal{Y}_l^{(g)} = \sum_{j=1}^{J} \ln\left(\frac{1 + C_j^{(g)}}{1 + \nu L_j^{(g)}}\right) e^{-\upsilon d_{jl}}. \tag{10.15}$$

In order to produce a full allocation of the L available police, a dynamic allocation is required. After each police officer has been allocated to the LSOA with the maximum value of $\mathcal{Y}_l^{(g)}$, the value of $\mathcal{Y}_l^{(g)}$ requires recalculation, taking into account the effect of the previous deployment. Thus, a suitable algorithm for the allocation of police according to the microsimulation model is given by the following procedure:

1. Set $L_l^{(g)} = 0$ for $l = 1, 2, \ldots, 4765$.
2. Calculate $\mathcal{Y}_l^{(g)}$ as in equation (10.15) for each LSOA l.
3. Find the maximum value of $\mathcal{Y}_l^{(g)}$ over all LSOAs, and allocate one police unit to that location. Update the values of $L_l^{(g)}$ to reflect this deployment.
4. If there are still more police to allocate, return to step 2; otherwise, stop.

The average value of the deployment utility over $G = 100$ iterations, with $L_j^{(g)} = 0$ for each LSOA j and iteration g and with $\nu = \upsilon = 1$, is shown in Figure 10.5a as a heat map. LSOAs that are shaded darker have a higher initial deployment utility associated with them, and are therefore areas where rioting is predicted to occur. According to the simulation, there are two prominent areas that have the highest level of deployment utility: one above the River Thames and one below the River Thames (the river is indicated by the white line through the centre of Greater London). The value of the deployment utility of an LSOA j, in comparison to the other LSOAs, can be thought of as the relative importance of allocating police officers to that particular area; and this figure shows the spatial distribution of this measure.

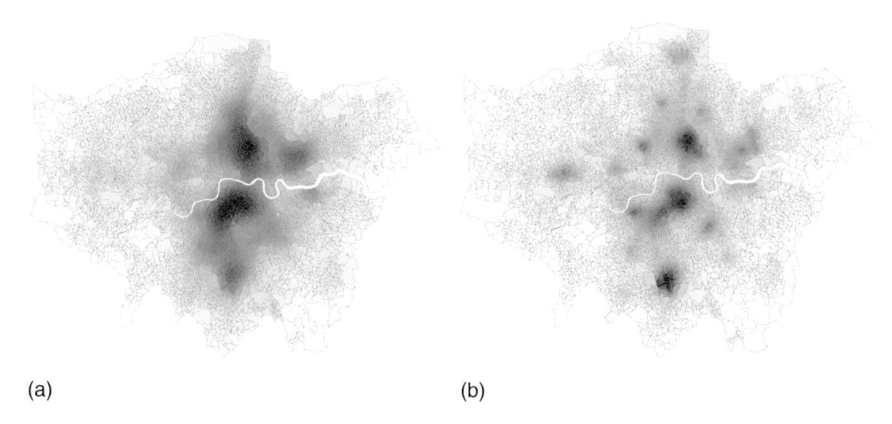

Figure 10.5 Deployment utility (a) for the simulation and (b) according to the empirical data. Darker shades indicate higher levels of deployment utility. The values for deployment utility are plotted assuming that no police officers have been deployed

As a final comparison, the equivalent value of the deployment utility calculated with the empirical counts of offences, rather than the simulated counts $C_j^{(g)}$, is shown in Figure 10.5b. Again, the darker the shading of the LSOA, the higher the deployment utility, and more value is assigned to that particular area. In this case, the darker areas of the map are more localised, with three or four prominent areas at which the deployment utility is highest.

Although there is some discrepancy between the model outputs and the empirical data, there is agreement in terms of the broad pattern. In particular, the model appears to generate a number of peaks of deployment utility in broadly similar areas to the deployment utility calculated with the empirical data. The distribution of the deployment utility with the actual data appears to be strongly clustered in certain locations in comparison to the model, where the clusters are much larger. Potential explanations for this effect include the fact that the deployment utility for the model is calculated using the average of a number of simulations, and thus may become more smooth. Another explanation may be due to the fact that the variance of the counts of offences is much greater for the empirical data than was observed in the simulation. The simulation led to the occurrence of offences that were more spread out in space, and the results presented here reflect this.

A limitation of this model as a policy tool is its conditional dependence on certain features of the empirical data. The model is conditional on the age and residential locations of each offender, the time at which each offender chooses to commit their offence, and the history of the riot up until one hour prior to the point at which the model is used. To ensure the model can be used in a policy context, each of the factors upon which the model is conditional requires the development and implementation of separate sub-models. Models for rioter age and residential location might be developed by exploring further the characteristics of the rioters who have previously engaged in rioting combined with demographic statistics from London. Models for the timing of rioter offences might be used to explore further mechanisms of contagion. In particular, such models will be required to specify precisely how a rioter chooses to engage in the disorder, and not just where they choose to do so. The history of the system can be provided by real-time police recording during a riot. Despite this restriction, the presentation of a policy model of target choice in this chapter has provided a proof of concept that statistical models of this type might be usefully incorporated into predictive policy models.

Another limitation of this model is that it does not account for the effect that the deployment of police officers may have on the target choice of rioters. Further development of the model might incorporate such effects, although this was not possible in the present study due to a lack of available data on law enforcement activity. Davies et al. (2013) model the interaction between rioters and police more explicitly. In addition, game theory might also usefully contribute to the models developed here since both rioters and police might aim to strategically position themselves in an attempt to maximise their utility (see e.g. Oléron Evans and Bishop, 2013; and Chapters 18 and 19 and Appendix D in this volume).

Finally, rioting is a dynamic phenomenon that occurs over relatively short time periods (the London riots occurred over five days in total). The optimal allocation of policing is therefore likely to vary on shorter timescales than is accounted for here, and further work might also account for this.

References

Ballas D, Rossiter D, Thomas B, Clarke G and Dorling D. (2005). *Geography Matters*. York: Joseph Rowntree Foundation.

Baudains P, Braithwaite A and Johnson SD. (2013). Target choice during extreme events: a discrete spatial choice model of the 2011 London riots. *Criminology* 51(2): 251–285.

Beck N, Gleditsch KS and Beardsley K. (2006). Space is more than geography: using spatial econometrics in the study of political economy. *International Studies Quarterly* 50: 27–44.

Bernasco W. (2006). Co-offending and the choice of target area in burglary. *Journal of Investigative Psychology and Offender Profiling* 3: 139–155.

Bernasco W and Block R. (2009). Where offenders choose to attack: a discrete choice model of robberies in Chicago. *Criminology* 47(1): 93–130.

Bernasco W, Block R and Ruiter S. (2013). Go where the money is: modeling street robbers' location choices. *Journal of Economic Geography* 13(1): 119–143.

Bernasco W and Nieuwbeerta P. (2005). How do residential burglars select target areas? A new approach to the analysis of criminal location choice. *British Journal of Criminology* 44(3): 296–315.

Brantingham PJ and Brantingham PL. (1993). Environment, routine and situation: toward and pattern theory of crime. In *Routine Activity and Rational Choice: Advances in Criminological Theory*, vol. 5. New Brunswick, NJ: Transaction, pp. 259–294.

Clare J, Fernandez J and Morgan F. (2009). Formal evaluation of the impact of barriers and connectors on residential burglars' macro-level offending location choices. *Australian and New Zealand Journal of Criminology* 42(2): 139–158.

Cohen LE and Felson M. (1979). Social change and crime rate trends: a routine activity approach. *American Sociological Review* 44(4): 588–608.

Cornish DB and Clarke RV. (1986). *The Reasoning Criminal: Rational Choice Perspectives on Offending*. New York: Springer-Verlag.

Davies TP, Fry HM, Wilson AG and Bishop SR. (2013). A mathematical model of the London riots and their policing. *Scientific Reports* 3: 1303.

Epstein JM. (2002). Modeling civil violence: an agent-based computational approach. *Proceedings of the National Academy of Sciences of the United States of America* 99 Suppl. 3(2): 7243–7250.

Freud S. (1921). *Group Psychology and Analysis of Ego*. London: International Psychoanalytic Press.

Granovetter M. (1978). Threshold models of collective behavior. *American Journal of Sociology* 83(6): 1420–1442.

Johnson SD and Summers L. (2015). Testing ecological theories of offender spatial decision making using a discrete choice model. *Crime and Delinquency* 61(3): 454–480.

Le Bon G. (1896/1960). *The Crowd: A Study of the Popular Mind*. New York: Viking Press.

McFadden D. (1974). Conditional logit analysis of qualitative choice behavior. In Zarembka P (ed), *Frontiers in Econometrics*, vol. 1 of Economic Theory and Mathematical Economics. New York: Academic Press, pp. 105–142.

McFadden D. (1979). Quantitative methods for analysing travel behaviour of individuals: some recent developments. *Behavioural Travel Modelling* 13: 279–319.

McFadden D. (2001). Economic choices. *The American Economic Review* 91(3): 351–378.

Midlarsky, M. (1978). Analyzing diffusion and contagion effects: the urban disorders of the 1960s. *The American Political Science Review* 72(3): 996–1008.

Myers DJ. (2000). The diffusion of collective violence: infectiousness, susceptibility, and mass media networks. *American Journal of Sociology* 106(1): 173–208.

Oléron Evans TP and Bishop SR. (2013). Static search games played over graphs and general metric spaces. *European Journal of Operational Research* 231(3): 667–689.

Rengert GF, Piquero AR and Jones PR. (1999). Distance decay reexamined. *Criminology* 37(2): 427–446.

Sampson RJ and Groves WB. (1989). Community structure and crime: testing social-disorganization theory. *American Journal of Sociology* 94: 774–802.

Sampson RJ, Raudenbush SW and Earls F. (1997). Neighborhoods and violent crime: a multilevel study of collective efficacy. *Science* 277: 918–924.

Townsley M and Sidebottom A. (2010). All offenders are equal, but some are more equal than others: Variation in journeys to crime between offenders. *Criminology* 48: 897–917.

Train K. (2003). *Discrete Choice Methods with Simulation*. Cambridge: Cambridge University Press.

Wortley R. (2008). Situational precipitators of crime. In Wortley R and Mazerolle L (eds), *Environmental Criminology and Crime Analysis*. London: Routledge, pp. 48–69.

Part Six

The Mathematics of War

11

Richardson Models with Space

Peter Baudains*

11.1 Introduction

Defining a set of differential equations (DEs) to model the evolution of some variable of interest is one way of representing social systems' dynamics. In contrast to agent-based simulations (another popular method, examples of which are presented in Part 7 of this volume), in which the interest is often focussed on individuals, the dependent variable in a DE-based model of a social system is often taken to be some attribute associated with a *group* of individuals. DEs are therefore typically used for more aggregate scenarios than agent-based models (although there are exceptions: DEs are employed with individual perspectives in Liebovitch et al. (2008) and Curtis and Smith (2008), and agent-based models are employed with aggregated perspectives in Cederman (2003)).

There are many examples of DEs being applied to conflict and warfare. The dependent variable of such models is often taken to be the number of individuals on each side of a conflict. For example, Lanchester (1916) uses DEs to model different types of attrition warfare between two adversaries. A number of studies have built upon Lanchester's work by modelling the change in the population size of adversaries with DEs (e.g. Atkinson et al., 2011; Deitchman, 1962; Intriligator and Brito, 1988; Kress and MacKay, 2014).

Another class of DEs in the conflict literature stems from the work of Richardson (1960) on the actions of nations during the lead-up to war. In this case, the dependent variable is not the population size of opponents, but the net level of threat on each adversary. Although an abstract measure, Richardson argues that this threat can be proxied by the yearly military expenditure of different nations. It is then assumed that a nation with more threats on them will spend proportionally more on their military in order to alleviate those threats. Richardson proposes a number of models, and explores how arms races might emerge from relatively simple competitive interactions between adversaries. One of his compelling conclusions emphasises

* The author would like to acknowledge the contributions of Steven Bishop, Alex Braithwaite, Toby Davies, Hannah Fry, Shane Johnson and Alan Wilson, each of whom contributed with a number of insightful discussions and points.

Approaches to Geo-mathematical Modelling: New Tools for Complexity Science, First Edition. Edited by Alan G. Wilson.

Companion Website: www.wiley.com/go/wilson/ApproachestoGeo-mathematicalModelling

how two nations might end up competing in an arms race merely by aiming to defend themselves against an adversary, without any explicit desire to exert threat over that adversary. Policies that result in such behaviour consequently lead to a very unstable international system, even when there are no overtly aggressive participants. Richardson contends that framing discussions of international arms accumulation in a mathematical way is advantageous in a policy setting since its "definiteness and brevity lead to a speeding up of discussions over disputable points, so that obscurities can be cleared away, errors refuted and truth found and expressed more quickly than they could have been, had a more cumbrous method of discussion been pursued" (Richardson, 1919).

Since Richardson initially proposed his models of arms races, a number of studies have attempted to fit the equations to empirical data, largely with limited success. Despite its appeal to policy discussions, it appears that Richardson's model is not complex enough to capture real-world processes. This is hardly surprising, given the simplicity with which the models are specified. Richardson himself stated that his models were never meant to capture all of the idiosyncrasies and externalities of international policy decisions. Nevertheless, in an effort to help remedy this, in this chapter, we demonstrate how Richardson's model can be disaggregated and made spatially explicit in order to account for more intricacy in the dynamics of international conflict.

Spatially varying dynamics of conflict processes are rarely incorporated into DE-based conflict models, and yet spatial processes are well known to play a important role in determining the nature and outcome of different types of conflict. The model presented in this chapter goes some way to addressing this discrepancy. It also enables a more general relationship between adversaries to be established, one that is not limited to geographic distance but can contain any number of factors that might influence the likelihood of nations being drawn into conflict, such as levels of trade or political similarity. Furthermore, we will demonstrate that the model presented in this chapter moves from considering the purely dyadic relationship between pairs of countries to incorporate the international security environment in which nations act. Thus, the model is more suited to representing real-world decision processes concerning military expenditure.

In what follows, we give an overview of the linear Richardson model of arms accumulation and present its analytical implications. We explain in more detail why the real-world application of Richardson's model has seen limited success. We then present a spatially explicit version of the model, which we argue captures the complexity of international security processes with more intricacy than Richardson's simple linear model, yet retains its essential dynamics. We present an example of an analytical result that highlights how the model is able to capture this increased level of intricacy. This result is presented in more detail in Baudains et al. (2016). In the final part of this chapter, we consider the application of the model to global military expenditure, building on two recent studies that aim to identify the determinants of military expenditure and to subsequently predict how it might change over time.

11.2 The Richardson Model

The key assumption in Richardson's model is that the defences of a nation (which we take as a measure of the steps taken to alleviate a certain level of threat), denoted by p, reacts to the military defences of their adversary, given by q, at a rate proportional to q. The adversary behaves similarly and reacts to the defence p. This reciprocal action–reaction process can result in an escalating arms race. A nation may react to the military defences of its rival both as a defensive measure, in order to provide protection from the threat posed by their opponent, as well as an aggressive measure, to exert threat over their rival.

Richardson included two further factors which he believed influenced military defences. The first was included in order to prevent unchecked growth of defences of a nation. It was hypothesised that economic fatigue, the cost of maintaining defences, and a natural inclination of nations to restrain growth would decrease defences at a rate proportional to their current size. The second comprised a constant term which was included to represent what Richardson termed 'external grievances', but which might incorporate a wide range of prejudices, aggressive or even pacifistic intentions. The model was first presented as a two-dimensional linear system of ordinary different equations given by

$$\frac{dp}{dt} = \dot{p} = -\sigma_1 p + \rho_1 q + \epsilon_1$$

$$\frac{dq}{dt} = \dot{q} = \rho_2 p - \sigma_2 q + \epsilon_2, \tag{11.1}$$

where σ_1 and σ_2 are parameters that specify the strength of inhibition associated with growing military defences, ρ_1 and ρ_2 are parameters that specify the rate of interaction between adversaries and ϵ_1 and ϵ_2 are parameters that specify the external grievances of each adversary. Typically, ρ_1 and ρ_2 will be positive, as military defences of one side will cause increasing defences of the other. σ_1 and σ_2 are also typically positive due to inhibition associated with increasing defences (which may arise due to deterioration and upkeep costs or even through pressures placed upon the government of each nation by the electorate).

Of particular interest when faced with an ordinary differential equation is not just on its analytical solutions (i.e. writing p and q as a function of their initial values and t), which are easy to find for the linear model in equation 11.1, but on the identification of different *families* of solutions. There may, for instance, be solutions arising from different initial conditions which eventually result in the same long-term behaviour. When this is the case, it is instructive to identify the set of all initial conditions that result in this behaviour. This is known as the basin of attraction for that particular long-term system state.

For general systems of differential equations, there are a range of different types of long-term behaviours, but just two are considered initially: divergence to infinity, and convergence to a single equilibrium point. Both of these behaviours will be shown to be present in the Richardson system, meaning that, according to the model, defence levels of both nations will either tend to a constant, or continue escalating (or de-escalating, depending on the sign of infinity). If $\sigma_1\sigma_2 \neq \rho_1\rho_2$, then there is a unique equilibrium[1] given by

$$(p_e, q_e) = \left(\frac{\sigma_2\epsilon_1 + \rho_1\epsilon_2}{\sigma_1\sigma_2 - \rho_1\rho_2}, \frac{\sigma_1\epsilon_2 + \rho_2\epsilon_1}{\sigma_1\sigma_2 - \rho_1\rho_2} \right). \tag{11.2}$$

Solutions that start at this location will remain there indefinitely, meaning that defence levels will not change. What largely determines the behaviour of the system is whether this equilibrium is attractive, in which case all solution curves converge towards this point, or whether it is repelling, in which case all solutions that do not start exactly on this point diverge away

[1] If $\sigma_1\sigma_2 = \rho_1\rho_2$, then if $\sigma_2\epsilon_1 + \rho_1\epsilon_2 = 0$ and $\sigma_1\epsilon_2 + \rho_2\epsilon_1 = 0$, there are infinitely many equilibria in the p–q plane; otherwise, there are none.

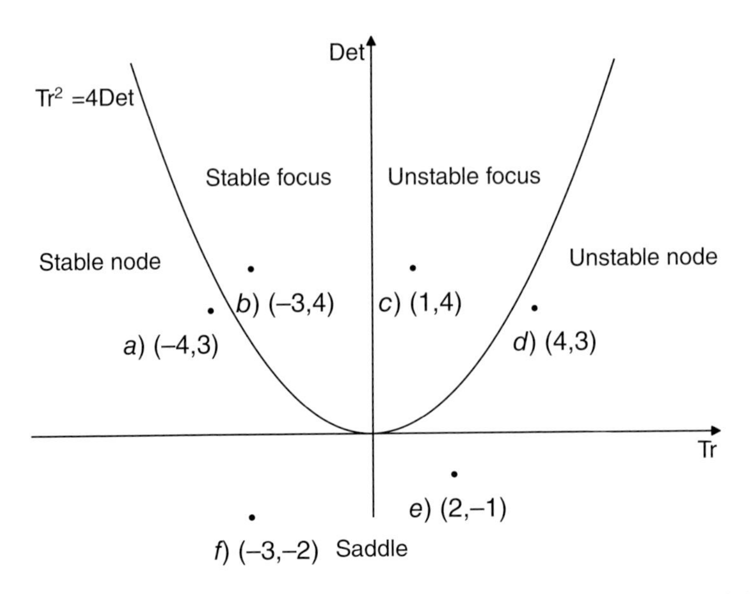

Figure 11.1 The trace–determinant diagram for linear planar systems. The location of the matrix P on this diagram determines the dynamics near to the equilibrium. The points (a–f) correspond to the subfigures in Figure 11.2, which show the qualitative dynamics for each case

from the point to infinity. The stability of the equilibrium, and consequently the behaviour of nearby solutions, is determined by the eigenvalues λ_1 and λ_2 of the matrix

$$P = \begin{pmatrix} -\sigma_1 & \rho_1 \\ \rho_2 & -\sigma_2 \end{pmatrix}. \quad (11.3)$$

A number of authors (e.g. Hirsch et al., 2004: p 63; Strogatz, 1994: p 137) have sought to demonstrate the range of behaviour for linear systems by presenting the trace–determinant diagram in Figure 11.1. This arises because the eigenvalues λ_i for $i = 1, 2$ of the two-dimensional matrix P are defined by its trace, Tr(P), and determinant, Det(P), according to

$$\lambda_i^2 - \mathrm{Tr}(P)\lambda_i + \mathrm{Det}(P) = 0. \quad (11.4)$$

In particular, the relationship between the trace and determinant determines the type of equilibrium. For the Richardson system, the trace, $\mathrm{Tr}(P) = -(\sigma_1 + \sigma_2)$, is the negative of the sum of the inhibition parameters, while the determinant, $\mathrm{Det}(P) = \sigma_1\sigma_2 - \rho_1\rho_2$, is a measure of the size of inhibition parameters in comparison to the action–reaction parameters. In Figure 11.2, the dynamics locally to the equilibrium of the system in equation 11.1 are shown for a range of parameter values. The parameters used for each of these cases are chosen to correspond with the points in Figure 11.1.

A stable node or focus (seen in Figures 11.2a and 11.2b, respectively) arises when $\mathrm{Tr}(P) < 0$ and $\mathrm{Det}(P) > 0$. Thus, two nations will typically only cease changing their defence levels if the sum of the inhibition parameters is positive, so that there is some damping in the system, and if those inhibition terms outweigh the escalation parameters. In this case, nations are being restrained by their internal dynamics – perhaps through pressures placed upon them by the

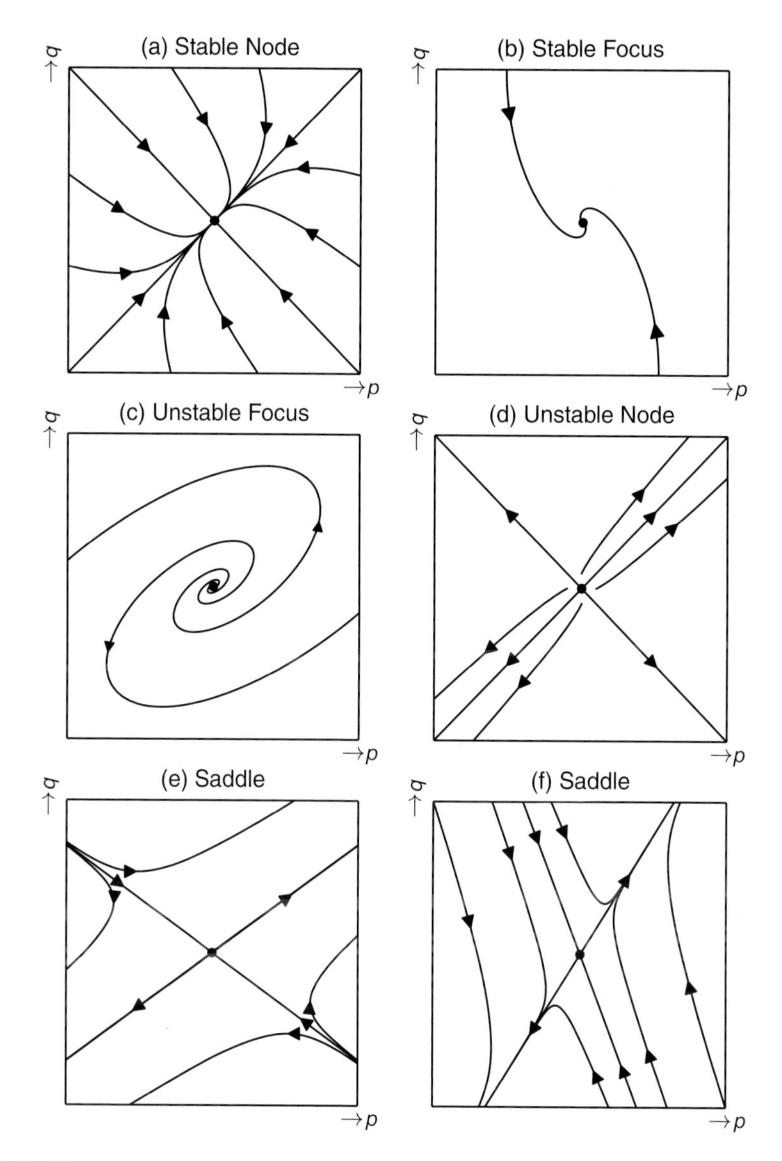

Figure 11.2 The dynamics around the equilibrium value of the Richardson model in equation 11.1 for different parameter values. The parameters are chosen to coincide with each point in Figure 11.1

electorate – rather than reacting to the threatening actions of their adversary. This heuristic result agrees with common sense, and begins to hint at how Richardson's model might be applied to real-world scenarios.

In the case of an unstable focus or node (seen in Figures 11.2c and 11.2d, respectively), nations are reacting to their own defence levels, as well as to the levels of their adversary, without any inhibition. There is no damping in the system, and it is consequently very unstable, with interactions compounding the escalation effect. Richardson argued that σ_1 and σ_2 are typically positive (which is to say there is some inherent damping behaviour), and it will therefore be assumed that such scenarios do not arise in the international system.

Saddles (shown in Figures 11.2e and 11.2f) provide further insights of real-world escalation processes. They occur when $\text{Det}(P) = \sigma_1\sigma_2 - \rho_1\rho_2 < 0$, which implies that the action–reaction parameters outweigh the inhibition parameters. In this scenario, the system can be susceptible to arms races. More insight into this scenario can be obtained by considering the eigenvectors of the matrix P. Assuming, without loss of generality, that $\rho_1 \neq 0$ holds in all cases of interest[2], then the eigenvectors are given by

$$\mathbf{v}_1 = \begin{pmatrix} \sigma_1 + \lambda_1 \\ \rho_1 \end{pmatrix}, \tag{11.5}$$

$$\mathbf{v}_2 = \begin{pmatrix} \sigma_1 + \lambda_2 \\ \rho_1 \end{pmatrix}. \tag{11.6}$$

Saddles occur when both eigenvalues are real numbers, with one being positive and one being negative. Denoting the positive eigenvalue by λ_1, then, if $\sigma_1 > 0$ and $\rho_1 > 0$, which Richardson argued occurs in most cases of interest, the eigenvector associated with the positive eigenvalue points in the direction of the positive quadrant in the plane. Almost all solution curves then either diverge to (∞, ∞) or $(-\infty, -\infty)$ (since the positive eigenvalue corresponds to divergence, rather than convergence). As Richardson (1960) stated, there is either a 'drift toward war' or a 'drift toward closer cooperation'[3]. For a given parameter set, the condition determining which of these occurs depends on the initial conditions. If the initial condition lies above the line defined by the eigenvector associated with the negative eigenvalue, then solutions diverge to positive infinity, whilst if the initial conditions lie below this line, then solutions diverge to negative infinity. In either case, the system is unstable, and the state of each individual nation is largely determined more by international dynamics than by internal processes.

According to Richardson's model, a nation hoping to avoid an escalating arms race with an adversary has several ways in which they can increase the stability of the system. The impacts of these strategies on the system parameters are summarised in Figure 11.3. They can, for instance, attempt to enforce a stable equilibrium by increasing the value of $\text{Det}(P) = \sigma_1\sigma_2 - \rho_1\rho_2$. They can do this either by decreasing their escalation parameter, as shown in Figure 11.3a or by increasing their inhibition parameter, as shown in Figure 11.3b.

As another strategy, if they perceive the system to be unstable, and in the form of a saddle, they can attempt to change the location of the system on the p–q plane so that any initial conditions will lie below the eigenvector associated with the negative eigenvalue. This could be done by decreasing the level of defences, or increasing the level of cooperation with their adversary.

Alternatively, they could attempt to alter the direction of the eigenvector associated with the negative eigenvalue so that the current state of the system falls below this line, and the system will result in an escalating process of cooperation. This could be done by altering the

[2] If both $\rho_1 = \rho_2 = 0$, then there are no action–reaction dynamics, and if $\rho_1 = 0$ but $\rho_2 \neq 0$, then the equations are relabelled.

[3] Cooperation between two nations in Richardson's framework is considered to be the opposite of threats. Net threat, as described by the dependent variables, is conceived to consist of threats minus cooperations. Richardson posits that cooperation between two nations might be measured by trade.

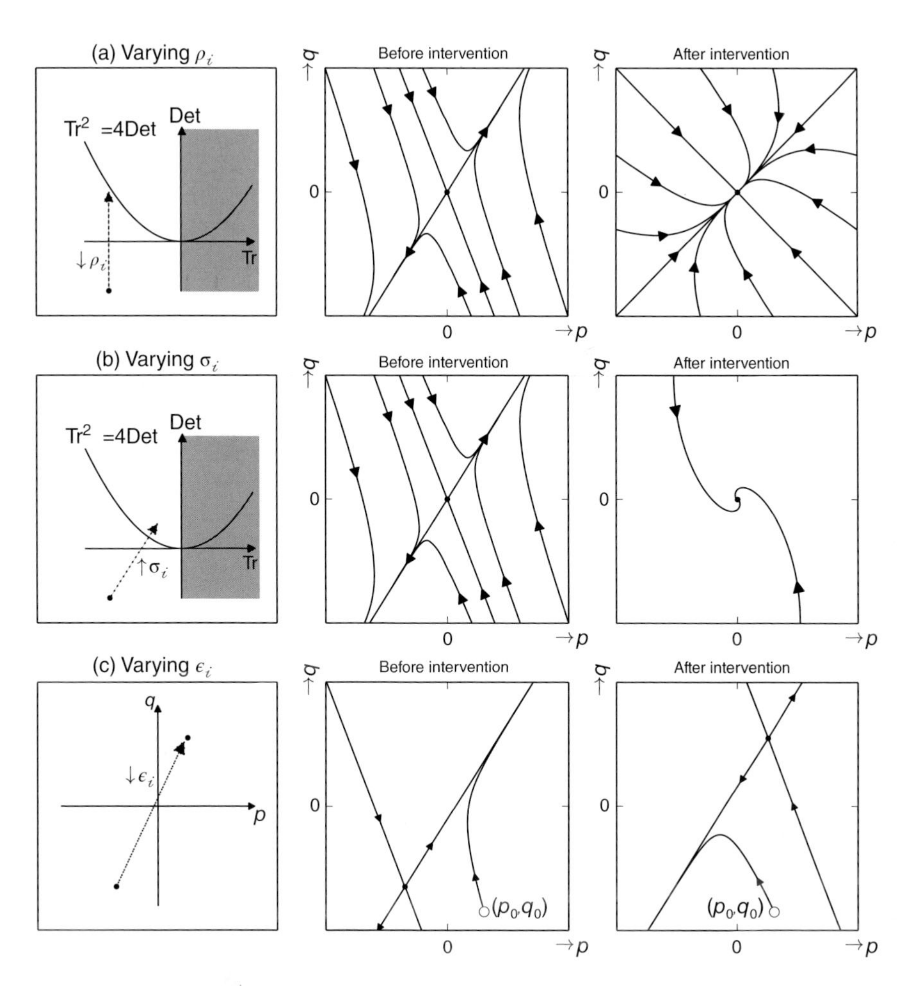

Figure 11.3 Unilateral policy options available to nation i. (a) The change in the system according to the trace–determinant diagram from Figure 11.1 when ρ_i is decreased; and (b) the change when σ_i is increased. Note that the half plane with Tr(P) > 0 is shaded since if $\sigma_1, \sigma_2 > 0$, as Richardson hypothesised, the system will not lie in this portion of the plane. (c) The impact of nation i reducing external grievances when the system is a saddle. The equilibrium point will move towards the positive quadrant, meaning that for initial conditions given by (p_0, q_0), the system will tend towards greater cooperation, rather than greater threats

parameters in the system in order to minimise the difference between $\mathbf{v}_1$ and $\mathbf{v}_2$ by ensuring that λ_1 and λ_2 are as close as possible. Given that the eigenvalues are equal when $\text{Tr}(P)^2 = 4\text{Det}(P)$, this again involves increasing the value of the determinant.

Finally, the position of the equilibrium can be changed by varying the level of grievances determined by ϵ, as shown in Figure 11.3c. If $\text{Det}(P) < 0$, an objective might be to minimise grievances so that the equilibrium point is as close to the origin as possible, thereby increasing the possibility that the state of the system will lie in the half of the plane which results in an escalation of cooperation.

Of course, even if a nation were to make these changes, there is no guarantee that their adversary will not change their dynamics in order to put the system back on a course to an escalating arms race. The Richardson model is useful in highlighting the possible consequences of a 'mechanical' arms race. As Richardson described, the model is "merely a description of what people would do if they did not stop to think" (Richardson, 1960: p 12). This implies a view of international conflict whose consequences, once set in motion, cannot be escaped. Subsequently, authors have considered various ways of extending the Richardson model in order to incorporate some notion of decision making on the part of the adversaries. Some, for example, have considered the Richardson model from the perspective of control theory and game theory, in which nations act according to a set of predefined objectives (Bennett, 1987; Gillespie, 1977; Intriligator and Brito, 1976). Although more closely considering the decision making of individuals that leads to the system outcomes, such approaches can lose some of the generality that a more descriptive model can sometimes afford. In what follows, we investigate some of the applications of Richardson's model to real-world arms races and examine the extent to which the model proposed by Richardson corresponds to real-world processes.

11.3 Empirical Applications of Richardson's Model

Measuring the threats on a nation – the dependent variable considered in Richardson's model – is difficult to achieve empirically. Richardson initially operationalised the variables p and q in equation 11.1 by considering annual military expenditures of two adversaries. Specifically, he showed how the increase in military expenditure of four nations – Russia, Germany, France and Austria-Hungary – on two sides of a conflict in the years prior to the First World War very closely follows a pattern that would have been predicted by the model. A figure from Richardson (1960) is reproduced in Figure 11.4, which shows the straight line expected from the model, against the data Richardson gathers for the years shown. The equation for the straight line is obtained by summing the two equations in 11.1 and assuming that both sides of the conflict react to their own defences and to the defences of their opponent at the same rate, so that $\sigma_1 = \sigma_2$ and $\rho_1 = \rho_2$.

Perhaps as a consequence of the very close fit between the model and the small dataset in Figure 11.4, Richardson's arms race model has been applied to various scenarios around the world which have been considered to exhibit 'arms race'–type behaviour. In many of these cases, however, when using modern estimation techniques with large datasets, the model has been unable to reproduce the empirical data to such a close extent. In fact, much of the time, the model prediction is found to be a poor fit to the data. Dunne and Smith (2007) give an overview of some of the econometric applications of Richardson's arms race model. They discuss the mixed results when the model is applied to the India–Pakistan arms race from 1960. In particular, using vector autoregression methods, they apply the Richardson model to expenditure data for India and Pakistan for the period 1960–2003. They find that, for some time periods, action–reaction-type dynamics present in the Richardson model can be observed; however, for other time periods, no such consistencies can be found.

Mixed results in the application of Richardson's model are common, and there are a number of possible explanations for this. In general, it is difficult to disentangle causal factors in the decision to increase military expenditure, which might be made as a direct response to an opponent's spending, but which might also be due to a wide range of geopolitical or socio-economic

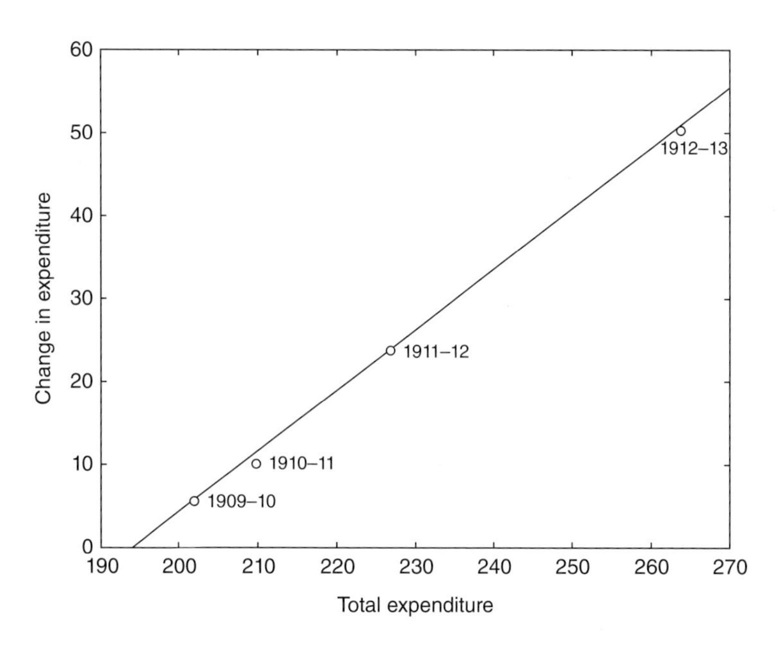

Figure 11.4 The change in the sum of defence budgets against the sum of defence budgets for four nations during the four years prior to the First World War. The four nations are Russia, Germany, France and Austria-Hungary, and the values plotted represent the sum of defence budgets over these nations. Defence expenditure data were gathered from various sources by Richardson, and the line represents the best fit that would be expected from the model in equation 11.1, assuming that $\sigma_1 = \sigma_2 = \sigma$ and $\rho_1 = \rho_2 = \rho$. This figure is reproduced from Richardson (1960). The gradient is given by $\rho - \sigma$ and is estimated by Richardson to be 0.73. An ordinary least squares regression produces the same output

reasons. One conclusion of Dunne and Smith (2007) is that arms races should not be treated in isolation using bivariate models, but instead should account for the global context and the dynamics of other nations.

Brauer (2002) reviews models applied to the Greco-Turkish arms race, and points out several issues associated with fitting them to data on military expenditure. These issues resonate with those encountered when applying DE-based models to social systems more generally. For instance, problems are regularly encountered with data availability, leading to complications in defining appropriate dependent variables. Furthermore, in the case of arms expenditure, decisions regarding whether to take the dependent variable as the absolute expenditure on defence for each nation, or the relative amount of expenditure on defence as a proportion of that nation's GDP (the so-called defence burden), can lead to varying levels of success with the fit of the model[4].

Parameter estimation can also be compromised as, in social systems in particular, parameter values can change very quickly. As Saperstein (2007) points out, the parameters of the original Richardson model in equation 11.1 are assumed to remain constant for time-scales over which the dependent variables change. Since decisions regarding military expenditure can be made

[4] Brauer (2002) identifies two studies of the Greco-Turkish arms race which use different definitions of the dependent variable and come to opposite conclusions: Georgiou et al. (1996) and Kollias and Makrydakis (1997).

by reacting to a single event that can occur on very short time-scales, there may be many scenarios in which this assumption is not valid. Saperstein (2007) goes on to define non-linear extensions of the model in which the parameters of the system change according to the strategic aims of each nation.

Finally, it should be borne in mind that Richardson's initial derivation of the model was not concerned primarily with military expenditure, but with threats faced by a nation. Although military expenditure is a plausible proxy for this effect, it does not necessarily capture all the threats that might be faced by a nation. Other factors such as strategic troop deployment or stocks of armaments might also contribute.

Despite these difficulties, interest in Richardson's model endures due to its attractive simplicity: it can be easily analysed, understood and used to explain the outcomes of different scenarios. Moreover, it is a useful tool to describe the possible states of an international system, and how the system might transition between these states. It was not intended by Richardson to be used as a predictive tool to forecast defence budgets (a point emphasised by Zinnes and Muncaster, 1984).

In what follows, we extend the Richardson model, first by considering how others have extended the simple model of two interacting adversaries into a model of N nations. This addresses one of the issues described above, that instead of focussing just on two nations in isolation, we should also incorporate the actions of other nations. We explain, however, how this model does not fully account for the international security environment since it neglects the important role of collective security. We then introduce a novel extension in which both space and collective security are incorporated. In this model, the intensity with which any two countries interact is able to depend on the distance between them and decisions regarding military spending are moderated by considering the international environment. We go on to consider an empirical application, in which it is not just the geographic distance between the two nations that determines the intensity of interactions but also the political dissimilarity. In this context, a nation will be less inclined to adjust their military expenditure due to the actions of a nation with which they share similar political goals since these nations will be less of a threat. The broad conclusion to draw from this chapter is that Richardson's model provides a stimulating framework to which refinements can be added. We offer one such contribution by adding non-linear interdependencies between nations via a spatial interaction model.

11.4 A Global Arms Race Model

Richardson himself extended the model in equation 11.1 to consider N interacting nations. This model is given by

$$\dot{p}_i = -\sigma_i p_i + \sum_{\substack{j=1 \\ j \neq i}}^{N} \rho_{ij} p_j + \epsilon_i, \tag{11.7}$$

for nations $i = 1, 2, \ldots, N$. The parameter ρ_{ij} is the rate at which nation i increases their spending due to the expenditure of j. The model in equation 11.7 requires the specification of $N(N+1)$ parameters detailing the dyadic relationship between every pair of nations.

Although this model incorporates the dynamics of an arbitrary number of nations, it does not fully overcome the issue of how the original Richardson model neglects the international security environment. This is because the dynamics of each nation i are governed by $N - 1$

dyadic interactions between nation i and every other nation. In practice, when considering the threats arising from a nation j, nation i will not only consider its relationship with j and adjust its spending accordingly, but also look at the international security environment and judge, based on the relationships between j and all other countries, how to adjust its expenditure. If j increases their expenditure rapidly, and i is an historic adversary of j, then i will be likely to also increase their spending. However, if j is known to have an adversarial relationship with another country k, then i might decide they do not need to adjust their spending by quite so much, since k is likely to adjust their spending on behalf of i (particularly if k is in an alliance with i), negating the threat posed by j.

Collective security refers to the phenomenon by which the threat from a nation is offset not just by your spending but also by the expenditure of others who would come to your aid should that nation attack. Epstein (1997) presents a non-linear variant of Richardson's model which accounts for collective security between three nations. However, this model results in yet more parameters to be specified. In this section, we propose a model that accounts for collective security in a different way, by incorporating within the model an entropy-maximising spatial interaction framework.

We proceed in a manner similar to the derivation of the spatial model of threat described in Chapter 9 of *Global Dynamics* and in Baudains et al. (2016). In these derivations, there are two adversaries, each of whom may be distributed over some spatial domain. In this chapter, the model is different in that there are N adversaries – our N nations – each of which we take to be located at just one point.

In this model, space arises not strictly due to geographical distance between adversaries (although that will play a role), but instead due to a measure of dyadic alliance between two nations. We therefore define C_{ij} to be the extent to which i considers j an ally (or, rather, the extent to which i considers j not to be a threat). C_{ij} plays an inverse role to ρ_{ij} in equation 11.7. Low values of C_{ij} indicate that i will be likely to increase their spending when j increases theirs. High values mean that i does not consider j a threat, and therefore will not react to the expenditure of j.

We then define T_{ji} to be the threat that nation j exerts on nation i. We assume that the total threat that can come from nation j is constrained by their current expenditure, so that

$$\sum_i T_{ji} = p_j. \tag{11.8}$$

In a similar way to Chapter 9 of *Global Dynamics* (see also Wilson, 2008), we also impose the following constraints:

$$\sum_{i,j} T_{ji} C_{ji} = c_1 \qquad \sum_{i,j} T_{ji} \log p_i = c_2, \tag{11.9}$$

for constants c_1 and c_2. Then, maximising the entropy measure

$$S = -\sum_{i,j} T_{ji} \log T_{ji}, \tag{11.10}$$

subject to these constraints leads to

$$T_{ji} = \frac{p_j p_i^{\alpha} \exp(-\beta C_{ji})}{\sum_k p_k^{\alpha} \exp(-\beta C_{jk})}, \tag{11.11}$$

for parameters α and β. Thus, the threat from j exerted on i is proportional to the expenditure of j, p_j, but weighted according to the extent to which p_i is a threat to j (as measured by the spending of i, p_i, and the alliance measure C_{ji}) when compared to all other nations that could be a threat to j. In this way, the international context is incorporated into the threat felt between i and j.

Summing T_{ji} over all possible sources of threat generates a measure of the total level of threat on nation i, given by

$$\tau_i = \sum_{j \neq i} T_{ji} = \sum_{j \neq i} \frac{p_j p_i^\alpha \exp(-\beta C_{ji})}{\sum_k p_k^\alpha \exp(-\beta C_{jk})}. \tag{11.12}$$

The threats on i can then be incorporated into the extended Richardson model of equation 11.7 according to

$$\dot{p}_i = -\sigma_i p_i + \rho_i \tau_i + \epsilon_i, \tag{11.13}$$

for $i = 1, 2, \ldots, N$, or, written in full as

$$\dot{p}_i = -\sigma_i p_i + \rho_i \sum_{j \neq i} \frac{p_j p_i^\alpha \exp(-\beta C_{ji})}{\sum_k p_k^\alpha \exp(-\beta C_{jk})} + \epsilon_i, \tag{11.14}$$

for $i = 1, 2, \ldots, N$.

11.5 Relationship to a Spatial Conflict Model

In Baudains et al. (2016), the model in equation 11.14 is explored but with two adversaries located over geographic space. One can arrive at this model from equation 11.14 by assuming alliances are absolute within groups and are dictated by geographic distance across groups. That is, when i and j are in the same group, $C_{ij} = C_{ji} = \infty$, and when i and j are in opposing groups, C_{ij} is equal to the geographic distance between i and j. In addition, within each group, the parameters ρ_i, σ_i and ϵ_i are taken to be constant. The resulting model can be written as

$$\dot{p}_j = -\sigma_1 p_j + \rho_1 \sum_{l=1}^{M} q_l \frac{p_j^\alpha e^{-\beta d(\mathbf{x}_j, \mathbf{y}_l)}}{\sum_{j'=1}^{N} p_{j'}^\alpha e^{-\beta d(\mathbf{x}_{j'}, \mathbf{y}_l)}} + \frac{\epsilon_1}{N} \tag{11.15}$$

$$\dot{q}_l = -\sigma_2 q_l + \rho_2 \sum_{j=1}^{N} p_j \frac{q_l^\alpha e^{-\beta d(\mathbf{y}_l, \mathbf{x}_j)}}{\sum_{l'=1}^{M} q_{l'}^\alpha e^{-\beta d(\mathbf{y}_{l'}, \mathbf{x}_j)}} + \frac{\epsilon_2}{M}, \tag{11.16}$$

for $j = 1, 2, \ldots, N$ and $l = 1, 2, \ldots, M$. The notation of the model in equations 11.15 and 11.16 is slightly different from the notation of equation 11.14. The variables p_j represent the net threats on one adversary located at $\mathbf{x}_j$, and q_l represents the net threats of their opponents at $\mathbf{y}_l$. The metric d defines the geography over which the conflict takes place.

Summing equation 11.15 over j and equation 11.16 over l generates a model for the aggregate change in threat between the two adversaries. For this model, it can be shown that the evolution of $p = \sum_j p_j$ and $q = \sum_l q_l$ obeys the original Richardson system in equation 11.1. Thus, the aggregate system either tends to an equilibrium or diverges to positive or negative infinity, depending on the location of the aggregated linear system on the trace–determinant diagram in Figure 11.1.

In this linear Richardson system, increasing the parameters ρ_1 and ρ_2 increases the total level of aggression in the system, and the equilibrium can undergo a *bifurcation* from a stable node, where the inhibition parameters outweigh the competition parameters, to an unstable node, where the opposite is true. In Baudains et al. (2016), a further bifurcation of the full non-linear model in equations 11.15 and 11.16 is identified, in which increasing levels of aggression between the two adversaries can cause spatial instabilities before the aggregate equilibrium turns unstable. That is, a transition can suddenly occur from a scenario in which threats are distributed relatively evenly over the locations of adversaries to one in which the vast majority of threats are felt within just a small number of locations.

The detection of spatial instabilities in the extended model in equations 11.15 and 11.16 demonstrates how non-linear effects can be incorporated into the Richardson model, and highlights the richness of the resulting behaviour. In particular, non-linear dynamics might be more capable of representing the complex decision processes of the real world such as simultaneously assessing threats from a number of sources and accounting for different forms of alliance. We next consider how the global threat model in equation 11.14 might be employed to model global military expenditure, building on two recent studies.

11.6 An Empirical Application

In this section, we present an example in which the spatially extended global Richardson model is applied to military expenditures. We note from the outset that the example described here is employed as a proof of concept of the spatial Richardson model. We build on the work of two previous studies but emphasise the utility of the spatial Richardson model in equation 11.14 as an avenue for further research.

11.6.1 Two Models of Global Military Expenditure

Two recent studies, Nordhaus et al. (2012) and Böhmelt and Bove (2014), have applied regression analyses to global yearly military expenditure data. We refer to these two studies in the remainder of this article as NOR and BB for brevity.

In NOR, a logistic regression is first specified that models the probability that any two pairs of states in the international system will go into conflict with each other. This model is calibrated using a number of independent variables which have been identified as correlates of conflict in previous literature investigating the causes of interstate war (e.g. Bremer, 1992). These include a measure of democracy of the two states, the geographic distance between them, relative GDP, and the level of bilateral trade. The outputs of this model are then used to estimate the probability that each individual state will engage in a new conflict at any given point in time, by considering the joint probability over the dyads involving each state.

The probabilities of each state going into conflict, as obtained by this procedure, are then assumed to represent the security threat on each country and are used as an independent variable in a model of military expenditure. Controlling for a range of factors that might further influence military spending, the authors find support that the external security environment, as measured by the modelled probability of conflict, provides some explanatory power in the variation of spending.

One of the control variables used is an estimation of the military expenditure of potential foes, which is included to account for the effects of arms races. This variable provides a further measure of the international security environment and is shown to have a significant positive influence on military expenditure. We explain in the following section how this variable is constructed.

Critiquing NOR, BB argue that more emphasis should be given to the ability of the variables to predict changes in the levels of military spending (see also Ward et al., 2010). They point out that using the modelled risk of an interstate conflict to predict military expenditure is problematic, since any variation between the covariates in the conflict model cannot be used to account for variation in military expenditure. Instead, the aggregation of the variables disguises the effect that any individual variable might have. As a consequence, in BB, the measure of conflict is disaggregated into monadic variables, and their ability to explain and predict military expenditure is examined individually.

BB consider the performance of each of the variables by assessing their ability to predict change in military expenditure in-sample, out-of-sample, and via a Bayesian model averaging procedure. They conclude that a number of the variables contained in the model of NOR do little to improve model fit. In fact, they recalibrate a model with just six predictor variables: a lagged dependent variable, a measure of democracy in a state, its GDP, the number of years since a state's last interstate conflict, the number of states in the world system and, pertaining to the international security environment, the military spending of foes. They show that this more parsimonious model is a more accurate predictor of global military expenditure than the full model proposed by NOR.

It is interesting to note that the final model of BB does not include any spatial or geographic variables. In addition, the only variables that reflect the international security environment are the number of states in the world system and the military spending of potential foes. In what follows, we recalibrate the model of BB using the spatially extended Richardson model in equation 11.14 to determine whether this model is capable of capturing the international security environment with respect to its effect on military expenditure. We first explain how potential foes are calculated in the work of NOR before employing a similar method to estimate the alliance measure C_{ij} in equation 11.14. This measure is taken in our work to be a combination of alliance portfolio similarity and the geographic distance between countries i and j. The incorporation of geographic distance is a departure from the models of NOR and BB. We then explain how the parameters α and β can be chosen via a brute force evaluation of models over feasible parameter regions. We then present our results before offering a conclusion and extensions for further research.

11.6.2 The Alliance Measure C_{ij}

In order to fully specify the global spatial Richardson model in equation 11.14, the measure C_{ij} is required. C_{ij} measures the extent to which state i considers state j *not* to be a threat. Small values of C_{ij} indicate potentially hostile relationships with a high risk of conflict between i and j. High values indicate that i has little concern about the actions of j. We use the term *alliance measure* to describe C_{ij}, but it should be borne in mind that two nations that are not necessarily in a formal alliance may also not consider each other to be a significant threat and so might produce a large value of C_{ij}.

In this section, we construct such a measure. Inspired by the approach of NOR, we suppose that the alliance measure can be constructed by considering the similarity of alliance portfolios between any two nations i and j. We also argue that geographic distance should play a role in the alliance measure and thus incorporate this.

Following Signorino and Ritter (1999) and NOR, we use data on formal alliances (version 4.1) from the Correlates of War project (Gibler, 2009) to construct, for each nation, a vector detailing its alliance status with every other nation. This is done by first rating the type of alliance that exists between any two states on an integer scale from 0 (for dyads with no formal alliance) to 3 (for a mutual defence pact, by which either nation will come to the other's aid should they be attacked). Denoting the strength of alliance between i and j by a_{ij}, the vector

$$A_i = (a_{i1}, a_{i2}, \dots, a_{iN}), \tag{11.17}$$

details the alliance portfolio of nation i.

A measure of vector similarity called the S-measure is then used to determine how every pair of nations differ in the way they form alliances with all other countries in the world system. We use a uniform weighting scheme of the S-measure (meaning that every potential alliance has the same weight when determining how similar the two alliance portfolios are), and thus the similarity measure between nations i and j is:

$$S_{ij} = 1 - \frac{2}{3N}\sum_{k=1}^{N} |a_{ik} - a_{jk}|. \tag{11.18}$$

To determine the foes of nation i, NOR rank all other nations according to the similarity measure of Signorino and Ritter. Those countries j whose similarity measure falls below the median are considered foes.

We depart from the formulation of NOR (and, indeed, of BB) by recognising the importance of geography in determining how alliances are formed. Proximity has long been shown to play a crucial role in determining countries' involvement in militarised interstate disputes (e.g. Bremer, 1992) and their military spending (Goldsmith, 2007). More broadly, regional geographies of countries' international environments play an important role in their actions (Gleditsch, 2002). In the present case, relying solely on formal alliances means that a dissimilar alliance portfolio between two countries will always generate an increased risk of conflict, regardless of their location. We assume that two countries with a dissimilar alliance portfolio will exert more threat on each other if they are geographically proximate than if they are farther apart. We construct C_{ij} accordingly by weighting the S-measure in equation 11.18 by the normalised distance between countries i and j. C_{ij} is then defined as

$$C_{ij} = (1 + S_{ij})(1 + \hat{d}_{ij}), \tag{11.19}$$

where $\hat{d}_{ij}$ is the (normalised) minimum geographic distance between countries i and j (obtained using the CShapes dataset; see Weidmann et al., 2010). The normalisation ensures that values of $\hat{d}_{ij}$ lie between 0 and 1.

The measure in equation 11.19 is simple to interpret: higher values of alliance portfolio similarity will produce higher values of C_{ij}, in which case i will consider j to be of little threat. C_{ij} is greater still if i and j are geographically distant due to the distance weighting employed.

Conversely, lower values of alliance portfolio similarity will decrease C_{ij}, producing higher levels of threat between i and j. This threat will be more pronounced if i and j are near to each other. The addition of 1 in the first bracket is to ensure that values of C_{ij} are greater than 0 (the S-measure in equation 11.18 generates values between -1 and 1), and the addition of 1 in the second bracket ensures that the distance measure does not dominate C_{ij} by forcing C_{ij} to be 0 between every two neighbouring countries regardless of their alliance similarity.

Next, we consider the final model for military expenditure proposed by BB, but adapt it to incorporate the threat measure of equation 11.12, operationalised using equation 11.19. The data used in this study are of panel form with variables specified for each year from 1952 to 2000. The measure for C_{ij} in equation 11.19 therefore varies, depending on the alliances that exist between countries in any given year (as well as any changing boundaries which may influence geographic proximity). The resulting regression model bears a close resemblance to the model of equation 11.14. In Sections 12.6.3 and 12.6.4, we first describe how the model is operationalised in more detail, including the specification of the parameters α and β in equation 11.12, before presenting the results of the regression analysis.

11.6.3 A Spatial Richardson Model of Global Military Expenditure

Using the dataset provided by BB, the variable τ_i is calculated for each state i for every year that state i appears in the dataset. τ_i varies in time due to variation in the military expenditure of nations and due to variation in C_{ij} as alliances and country borders change. Time-lagged military expenditure variables p_j are used in calculating τ_i to avoid issues with simultaneity.

We replace the military spending of foes in the final model of BB with τ_i. The military expenditure of foes is conceived to represent a similar measure of the international security environment as τ_i. Importantly, we suggest that τ_i captures the military spending of adversaries (and, in turn, the international security environment) more appropriately than the measure proposed by NOR due to the way it incorporates geography and collective security (as described in Section 11.4). Following BB and NOR, logged values of p_i and τ_i are used in the regression in order to retain comparability (the variable for the military spending of foes is logged in both of these studies).

To highlight the relationship between this regression model and the extended Richardson model in equation 11.14, we write the regression model to be calibrated as

$$p_i(t+1) = (1-\sigma)p_i(t) + \rho\tau_i(t,\alpha,\beta) + \epsilon + \boldsymbol{\theta}\cdot\mathbf{x}_i, \tag{11.20}$$

where the parameters σ, ρ and ϵ are as in equation 11.14, but do not vary over the different countries. It is instead assumed that the effects of each of the explanatory variables is consistent across these countries.

The extra term in equation 11.20 consists of the dot product of a vector of parameters $\theta = (\theta_1, \theta_2, \theta_3, \theta_4)$ with a vector of covariates $\mathbf{x}_i$. This term incorporates the remaining factors identified as predictors of military expenditure by BB. These are: the level of democracy in a country, its GDP, the number of years since the country's last interstate conflict, and the number of states in the world system at each point in time.

We use the same dataset as BB (omitting the military expenditure of foes, which is replaced with τ_i) and regress the log-transformed military expenditure data on the covariates. The dependent variable consists of military expenditure data collected from both Correlates of War and the Stockholm International Peace Research Institute (SIPRI) for the years 1952–2000,

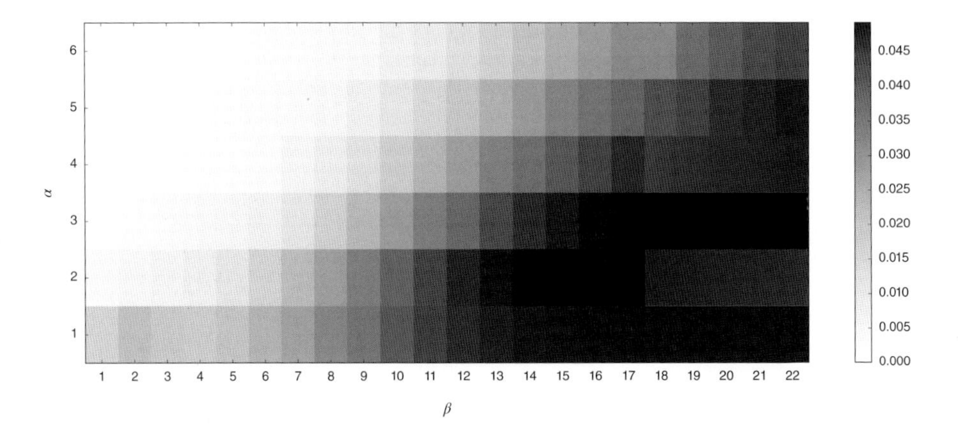

Figure 11.5 The increase in R^2 from the minimum of those tested, multiplied by 10^3, for different parameters α and β

adjusted by purchasing power parity. The reader can refer to NOR and BB for more detail on the specific data used.

In order to select the most appropriate values for α and β, which cannot be obtained via a simple linear regression, a brute force optimisation procedure is performed. To do this, the model is calibrated over a range of values for both α and β. The R^2 value of a model that includes a lagged dependent variable is calculated across all of these models, and the values of α and β that lead to the highest R^2 are those that are used in the analysis.

To perform this procedure, we let α take integer values between 1 and 6 and β take integer values between 1 and 22[5]. The increase in R^2 value from the minimum of the values found, for each pair of parameter values, is shown in Figure 11.5. The parameters that led to the maximum R^2 value were $\alpha = 3$, $\beta = 21$. These are the values that are used when presenting the results in Section 12.6.4.

11.6.4 Results

Table 11.1 presents the parameter values and standard errors (in parentheses) of three regression models calibrated with the threat measure τ_i. Following BB, the first model contains no lagged dependent variable, so it is assumed $\sigma = 1$ in equation 11.20. The second model includes the lagged dependent variable, and the third model clusters standard errors on each country in order to account for correlation between errors within countries. All estimates with the exception of the intercept in models 2 and 3 have $p < .05$, suggesting conventional statistical significance.

Our interest is principally on the threat variable τ_i. In all three models, this measure is positively associated with military expenditure. This confirms our hypothesis that a spatial Richardson model is capable of capturing interesting dynamics at the global level.

[5] A finer grained approach or the evaluation of a greater range of values may have further improved model fit; however, in the present chapter, in which a proof of concept of the model is all that is considered, we do not pursue a more extensive search of the parameter space.

Table 11.1 Regression results for three model specifications. Model 1 has no lagged dependent variable, model 2 includes this variable and model 3 incorporates clustered standard errors on each country. All parameters except the intercept in models 2 and 3 have $p < .05$

	Model 1	Model 2	Model 3
(Intercept)	3.7026	.0250	.0250
	(.1306)	(.0615)	(.0671)
Lagged dependent variable		.9131	.9131
		(.0146)	(.0208)
Peace years	−.0020	−.0005	−.0005
	(.0003)	(.0001)	(.0002)
Democracy	−.0100	−.0034	−.0034
	(.0013)	(.0007)	(.0009)
Size of world system	−.0041	−.0007	−.0007
	(.0003)	(.0001)	(.0001)
ln GDP	.4973	.0724	.0724
	(.0098)	(.0110)	(.0152)
Threat	.1718	.0057	.0057
	(.0025)	(.0020)	(.0023)
Number of observations	5684	5684	5684
R^2	.9124	.9809	.9809

Each of the models in Table 11.1 had a greater R^2 value than compared to a model in which τ_i was replaced with the military spending of foes as defined by NOR and BB. For models 2 and 3, however, the differences in the R^2 values were very small (a finding which might be explained by the fact that the values are already very close to 1). In particular, R^2 increased by 11% in model 1 when τ_i replaced the NOR specification of the military spending of foes but by just .002% in models 2 and 3, when the lagged dependent variable is incorporated.

We end this example by emphasising the encouraging performance of a spatial interaction measure of threat in a model of global military expenditure. The model in equation 11.14 has been operationalised by noting its similarity to two recent studies. In particular, the model of BB has been replicated with the variable measuring the military expenditure of foes replaced with our definition of threat in the international security environment. In what follows, we offer some concluding remarks and discuss some areas for further research that the work in this chapter has stimulated.

11.7 Conclusion

Richardson's model of an arms race provides a framework for investigating the implications of simplified interactions between two nations. We have explored the dynamics of this model and have considered the resulting policy insights. When applying the model to real-world data, as has been attempted in a number of previous studies, it is evident that the simplification of the Richardson model comes at a cost to its ability to account for a wider range of effects that might influence the spending decisions of nations.

In an attempt to remedy this, a number of efforts have extended the original model, often incorporating non-linear dynamics. In this chapter, we have proposed one such extension, which incorporates parameters on the strength of alliance between every two states as well as aspects of collective security. The strength of alliance measure determines the strength of a flow of threat between every two adversaries, whilst the spatial interaction framework moderates this flow so that a country considers the state of the world system, and how it fits into that system, before adjusting their level of military spending accordingly in order to alleviate the threat that is on them.

In Baudains et al. (2016), the dynamical properties of a special case of this model are explored, and rich nonlinear behaviour is observed including the onset of spatial instability as system-wide aggression increases.

As a proof of concept of the full global model, we have built on two previous studies and demonstrated that the proposed measure of threat is capable of explaining variance in the military spending of nations that is due to considerations of the external security environment. In order to operationalise this model, we defined a measure of alliance strength between every two countries, which incorporates both alliance portfolio similarity and geographic distance between every two nations.

Further research should consider the nonlinear dynamical properties of the global model when the alliance metric is not in the form of the special case investigated by Baudains et al. (2016). This might generate more insights into the way in which nations respond to the threats upon them. In addition, further investigation of how the model is applied to global military expenditure is warranted. Such work might apply a finer resolution exploration of the (α, β) parameter space, investigate the variance across countries of both military spending and the corresponding threat measure, and apply some of the more rigorous tests performed by BB to assess the predictive capability of the model (and, in particular, the threat variable). In addition, a number of studies have considered alternative measures of foreign policy similarity which might be explored further in the context of the model presented here (D'Orazio, 2012).

In summary, this chapter contributes to the challenge of modelling global interactions between countries by introducing a novel approach to handling relationships between countries and collective security. Further empirical analysis will shed more light on this model's ability to account for the threats that nations face, but work presented here has demonstrated the potential of such a model.

References

Atkinson MP, Gutfraind A and Kress M. (2011). When do armed revolts succeed: lessons from Lanchester theory. *Journal of the Operational Research Society* 63 (10): 1363–1373.

Baudains P, Fry HM, Davies TP, Wilson AG and Bishop SR. (2016). A dynamic spatial model of conflict escalation. European Journal of Applied Mathematics 27(3):530–553.

Bennett, P. (1987). Beyond game theory – where? In Bennett P (ed), *Analysing Conflict and Its Resolution: Some Mathematical Contributions*. Oxford: Clarendon Press, pp. 43–69.

Böhmelt T and Bove V. (2014). Forecasting military expenditure. *Research and Politics* 1(1): 1–8.

Brauer J. (2002). Survey and review of the defense economics literature on Greece and Turkey: what have we learned? *Defense and Peace Economics* 13(2): 85–107.

Bremer SA. (1992). Dangerous dyads: conditions affecting the likelihood of Interstate War, 1816–1965. *Journal of Conflict Resolution* 36(2): 309–341.

Cederman L-E. (2003). Modeling the size of wars: from billiard balls to sandpiles. *American Political Science Review* 97(1): 135.

Curtis JP and Smith FT. (2008). The dynamics of persuasion. *International Journal of Mathematical Models and Methods in Applied Sciences* 2(1): 115–122.

Deitchman SJ. (1962). A Lanchester model of guerrilla warfare. *Operations Research* 10(6): 818–827.

D'Orazio V. (2012). Advancing measurement of foreign policy similarity. Paper presented at the APSA 2012 annual meeting, December 3. Available at http://ssrn.com/abstract=2105547

Dunne PJ and Smith RP. (2007). The econometrics of military arms races. In Sandler T and Hartley K (eds), *Handbook of Defense Economics*, vol. 2. Amsterdam: North-Holland, pp. 913–940.

Epstein JM. (1997). *Nonlinear Dynamics, Mathematical Biology, and Social Science*. Reading, MA: Addison-Wesley.

Georgiou GM, Kapopoulos PT and Lazaretou S. (1996). Modelling Greek-Turkish rivalry: an empirical investigation of defence spending dynamics. *Journal of Peace Research* 33(2): 229–239.

Gibler DM. (2009). *International Military Alliances, 1648–2008*. Washington, DC: CQ Press.

Gillespie JV, Zinnes DA, Tahim G, Schrodt PA and Rubison RM. (1977). An optimal control model of arms races. *American Political Science Review* 71(1): 226–244.

Gleditsch KS. (2002). *All International Politics Is Local: The Diffusion of Conflict, Integration, and Democratization*. Ann Arbor: University of Michigan Press.

Goldsmith BE. (2007). Arms racing in 'space': spatial modelling of military spending around the world. *Australian Journal of Political Science* 42(3): 419–440.

Hirsch MW, Smale S and Devaney RL. (2004). *Differential Equations, Dynamical Systems and an Introduction to Chaos*, 2nd ed. San Diego, CA: Elsevier.

Intriligator MD and Brito DL. (1976). Formal models of arms races. *Conflict Management and Peace Science* 2(1): 77–88.

Intriligator MD and Brito DL. (1988). A predator-prey model of guerrilla warfare. *Synthese* 2: 235–244.

Kollias C and Makrydakis S. (1997). Is there a Greek-Turkish arms race? Evidence from cointegration and causality tests. *Defence and Peace Economics* 8(4): 355–379.

Kress M and MacKay NJ. (2014). Bits or shots in combat? The generalized Deitchman model of guerilla warfare. *Operations Research Letters* 42: 102–108.

Lanchester FW. (1916). *Aircraft in Warfare: The Dawn of the Fourth Arm*. London: Constable.

Liebovitch LS, Naudot V, Vallacher R, Nowak A, Bui-Wrzosinska L and Coleman P. (2008). Dynamics of two-actor cooperation-competition conflict models. *Physica A: Statistical Mechanics and Its Applications* 387(25): 6360–6378.

Nordhaus W, Oneal JR and Russett B. (2012). The effects of the international security environment on national military expenditures: a multicountry study. *International Organization* 66: 491–513.

Richardson LF. (1919). Mathematical psychology of war. In Sutherland I, Ashford OM, Charnock H, Drazin PG, Hunt JCR and Smoker P (eds), *The Collected Papers of Lewis Fry Richardson*, vol. 2, Quantitative Psychology and Studies of Conflict. Cambridge: Cambridge University Press, pp. 61–100.

Richardson LF. (1960). *Arms and Insecurity*. Pittsburgh, PA: Boxwood Press.

Saperstein AM. (2007). Chaos in models of arms races and the initiation of war. *Complexity* 12(3).

Signorino CS and Ritter JM. (1999). Tau-b or not tau-b: measuring the similarity of foreign policy positions. *International Studies Quarterly* 43(1): 115–144.

Strogatz SH. (1994). *Nonlinear Dynamics and Chaos with Applications to Physics, Biology, Chemistry and Engineering*. Reading, MA: Perseus Books.

Ward MD, Greenhill, BD and Bakke KM. (2010). The perils of policy by p-value: predicting civil conflicts. *Journal of Peace Research* 47: 363–375.

Weidmann NB, Kuse D and Gleditsch KS. (2010). The geography of the international system: the CShapes dataset. *International Interactions* 36(1): 86–106.

Wilson AG. (2008). Boltzmann, Lotka and Volterra and spatial structural evolution: An integrated methodology for some dynamical systems. *Journal of the Royal Society, Interface/the Royal Society* 5(25): 865–871.

Zinnes DA and Muncaster RG. (1984). The dynamics of hostile activity and the prediction of war. *Journal of Conflict Resolution* 28(2): 187–229.

Part Seven
Agent-Based Models

12

Agent-based Models of Piracy

Elio Marchione, Shane D. Johnson and Alan G. Wilson

12.1 Introduction

Maritime piracy has increased dramatically in recent years, and it is estimated that this type of crime costs the global economy over USD $7 billion per year (Ploch, 2010). Not surprisingly, this has attracted attention from the military and from international institutions such as the United Nations. It is claimed that the primary motivation for pirate attacks is financial gain obtained either through hijacking and theft of cargo or through ransoms collected following the kidnapping of the vessel and crew (Hastings, 2009).

A number of approaches to fight maritime piracy have been implemented (e.g. Rengelink, 2012). For example, a UN Security Council resolution (October 2008) provides the legal basis for the pursuit of pirates into Somali territorial waters. The UN sanction in 2008 and a US presidential order in 2010 both banned the payment of ransoms to a list of individuals known to be involved in piracy. In August 2008, the Maritime Security Patrol Area (MSPA) was set up in the Gulf of Aden by the Western Navies Combined Task Force, and in February 2009 a transit lane off the Gulf of Aden through which ships are advised to sail was established (the International Recommended Transit Corridor). Despite efforts to reduce piracy, a recent empirical analysis of daily counts of piracy in Somalia and the Gulf of Aden provided no evidence as to the effectiveness of such policies to date (Shortland and Vothknecht, 2010). However, fighting piracy is both an urgent and difficult task. To illustrate, consider that in 2010, 30 ships were allocated to the fight against piracy, but these are required to patrol a region of two million square miles (Chalk, 2010). Thus, understanding maritime piracy and optimising strategies aimed at reducing it are of clear importance.

Several scholars have attempted to investigate the phenomenon from qualitative (O'Meara, 2007) and descriptive (Abbot and Renwick, 1999; Vagg, 1995) perspectives, or have focussed on how to apply international law to this type of crime (Treves, 2009). The majority of the quantitative research has been carried out by political scientists, and in such studies, macro-level

Approaches to Geo-mathematical Modelling: New Tools for Complexity Science, First Edition. Edited by Alan G. Wilson.

Companion Website: www.wiley.com/go/wilson/ApproachestoGeo-mathematicalModelling

features of different states are used to explain the variation in the frequency of attacks. Perhaps the most relevant quantitative findings are twofold: first, the risk of piracy is not evenly distributed across the world's oceans (Marchione and Johnson, 2016); and, second, maritime piracy is thought to be more prevalent off the coasts of failed or weak states (Hastings, 2009).

To explain observed patterns, Daxecker and Prins (2012) propose a general explanation of maritime piracy that emphasises the importance of institutional and economic opportunities. They use a monodic-year unit of analysis that includes all states in the international system and cases from 1991 to 2007. By defining the country-year as the unit of analysis, they provide descriptive statistics on spatial and temporal trends. The authors conclude by showing a positive and statistically significant relationship between piracy attacks and both state failure and economic opportunity.

Such findings offer insight into how the problem may be addressed through political or other approaches – approaches that may be targeted towards the restoration of law and order in weak states – but offer only the most general intelligence as to where attacks are most likely to occur in the future while state fragility remains. Spatial analyses of maritime piracy that may assist in the latter endeavour are proposed by Caplan *et al.* (2011). Using *risk terrain modelling* (RTM), they define a way to standardise risk factors to common geographic units over a continuous surface. To generate such a map, separate layers representing the intensity of each risk factor are combined to produce a composite "risk terrain" map. In their study, to try to explain spatial patterns of maritime piracy, the authors put forward three risk factors: state status, shipping routes and maritime choke-points. In output, RTM generates a GIS map that highlights high-risk areas. Their results are promising, but the maps are time invariant and provide no insight into how the interactions between pirates, vessels (potential targets) and naval forces might lead to observed patterns of attacks.

Temporal patterns of maritime piracy are discussed by Shortland and Vothknecht (2010), who evaluate the effectiveness of the international naval mission in the Gulf of Aden between 2008 and 2010. The authors claim that pirates' actions can be modelled according to a rational choice framework, just as Becker (1968) and Sandler and Enders (2004) did for criminal activities and terrorism, respectively. Surprisingly, their evidence suggests that the incidence of piracy in the Gulf of Aden has increased since the inception of the naval counter-piracy measures discussed here.

The methods so far discussed are top-down approaches for which data are collected in the real world; associations are examined using statistical methods, and the data-generating process is inferred. However, spatial and temporal patterns of pirate attacks have also been explored using mathematical and agent-based models. Two examples are the work of Fu *et al.* (2010) and Jakob *et al.* (2010). Fu *et al.* attempt to explain changes in economic losses experienced by the global shipping industry over time in terms of the costs potentially incurred as a consequence of maritime piracy. To do so, they use a simulation model to investigate how maritime piracy might affect losses through the increased cost of insurance, and the potential increased costs associated with ships being forced to take (longer) alternate routes to avoid the risk of piracy. To do this, they use data on shipping movements between 2003 and 2008 and identify the primary routes used and those that may have to be used to avoid the risk of piracy. Focusing on the effect of piracy on the Far East–European markets, the authors conclude that their model is able to reproduce the losses observed in the real world, and hence suggest that these are likely to be due to piracy.

Jakob *et al.* propose a spatially explicit agent-based (Gilbert, 2008) platform for analysing, reasoning and visualising information in the maritime domain with the emphasis on detecting, anticipating and preventing pirate attacks. The platform is designed for the development and evaluation of counter-piracy methods; however, the authors neither explicitly state the methodologies used to calibrate and evaluate the model, nor describe the techniques they use to assess model outcomes.

The contribution of the current study is twofold. First, a method of calibrating the behaviours of a set of agents that, through their activity (or lack of it) likely contribute to patterns of maritime piracy, is introduced. As with other direct-contact predatory crimes (see Cohen and Felson, 1978), we hypothesise that incidents of piracy occur when motivated offenders (in this case, pirates) and suitable targets (in this case, vessels) converge in space and time in the absence of capable guardians. Capable guardians could be anyone who, through their presence, can prevent piracy. In the context of piracy, such guardians will typically be either naval vessels or private security teams – here, we focus on the former. Thus, we consider three classes of agents – pirates, vessels and naval forces – and see whether a simple model of their activity generates maps of piracy that resemble the empirical record. The second contribution is the use of techniques to visualise and compare simulation-model outputs with observed empirical patterns. This is of great importance for assessing model performance and (for example) considering how simulated strategies may affect maritime piracy.

In the work presented, an agent-based model is employed to produce dynamic maps of maritime piracy attacks in 2010 in the Gulf of Aden. This model is informed by a series of data sets, which are described in Section 12.2. A description of the model is in Section 12.3, and model calibration is discussed in Section 12.4. Section 12.5 concludes the chapter.

12.2 Data

Five sources of data are used in the proposed model. First, data concerning incidents of maritime piracy for the calendar year 2010 around the Gulf of Aden were obtained from the National Geospatial Intelligence Agency's[1] (NGIA) Anti-Shipping Activity Messages database. The NGIA collect and collate incidents of maritime piracy, and issue warnings to mariners in real time. Their data come from a variety of sources including, but not limited to, the International Maritime Bureau and the International Maritime Organization. For each attack, the location is recorded and represented using the latitude–longitude coordinate system. The date on which attacks took place is also recorded, along with information about the type of vessel attacked and a free-text description of the event.

We analyse all incidents of maritime piracy. Figure 12.1a shows each of the 56 incidents of piracy (red dots) that were recorded in 2010 around the Gulf of Aden.

Second are data concerned with the movement of ships through the area of interest. These positions are difficult to collect: one can either buy this information or collect Surface Synoptic Observation (SYNOP) data. We do the latter, collecting a sample of data for the period from 15 November to 31 December 2010.[2] SYNOP is a numerical code used for reporting weather observations. The data include meteorological measures along with the longitude and latitude of the location at which the observation was taken, and when it was taken. Individual

[1] NGIA: https://www1.nga.mil/Pages/default.aspx.

[2] See http://www.mundomanz.com/meteo_p/byind?l=1.

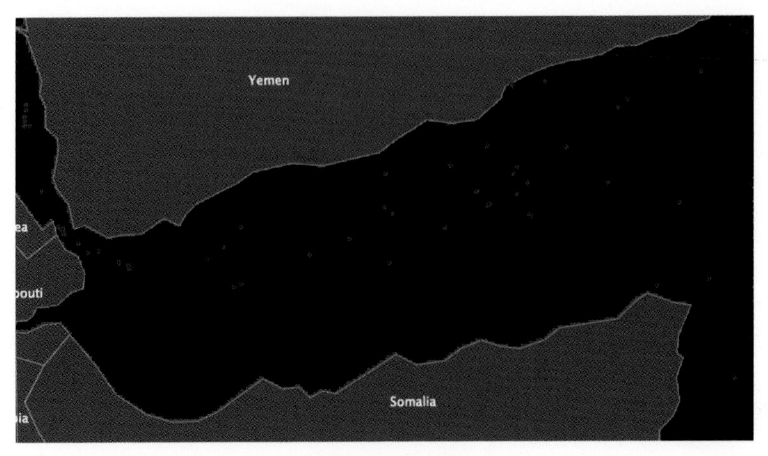

(a) Pirate attacks

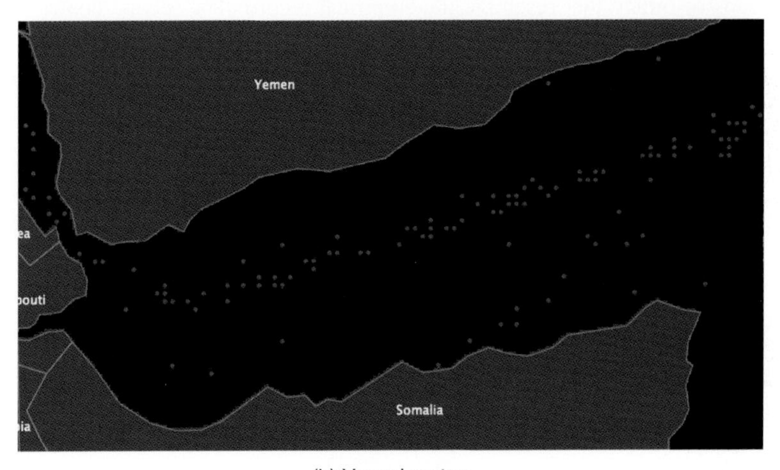

(b) Vessel routes

Figure 12.1 Pirate attacks and vessel routes in 2010

vessels routinely transmit SYNOP data, thereby allowing their location to be tracked over time. Figure 12.1b shows the entire dataset collected through SYNOP in 2010 for the Gulf of Aden. Each blue dot represents a SYNOP numerical code.

Third, the number of vessels passing through the Suez Canal in 2010, both from north to south and from south to north, is obtained from the Canal Suez Traffic Statistics.[3] These data are used to simulate vessel volumes in the Gulf of Aden. It is assumed that all the vessels that went through the Suez Canal also went through the Gulf of Aden. Table 12.1 shows figures for both flows: north to south and south to north.

Fourth, information on the number of naval units operating in the Gulf of Aden in 2010 is taken from the work of Chalk (2010), who identified 14 international naval units operating

[3] See http://www.suezcanal.gov.eg/TRstat.aspx?reportId=3.

Table 12.1 Vessels transiting the Gulf of Aden

Year	North/South	South/North
2010	9032	8961

in the Gulf. Apart from NATO, a number of other states sent frigates to protect and/or escort shipping off the Horn of Africa, including India, China, Russia, Pakistan, Saudi Arabia, the Netherlands and Malaysia.

And, finally, the flag (the country to which the vessel is registered) and type of all the vessels passing through the Suez Canal in the year 2010 were obtained from monthly reports issued by Canal Suez Traffic Statistics.[4]

12.3 An Agent-based Model

The model is implemented on and generates a dynamic map. It consists of a 286×160 grid for which each cell is represented by 3×3 pixels. Each pixel edge represents 1.4 miles, and therefore each cell is 17.64 square miles. Other dimensions could be used, but this scale is more than adequate to examine the phenomenon of interest and provides a tractable model that uses a relatively fine temporal resolution (see below). Land is coloured grey, and the sea black. Country borders are shown as green lines. There are four type of agents that are free to move across the sea: pirates coloured in magenta, successful pirate attacks coloured in red (they do not move), and vessels and navy units in cyan and green, respectively (see Figure 12.5a). The condition action rules that govern agent behaviour (see below) allow the testing of hypotheses and a method of generating dynamic maritime piracy risk maps.

12.3.1 Defining Maritime Piracy Maps

Before discussing the model further, we describe how the spatial pattern that we seek to simulate is defined. Let pirate attack *density* be a cell variable. It is based on the count of pirate attacks in that cell during the year 2010. As illustrated in Figure 12.1a, pirate attacks are rather scattered and, consequently for the cell dimensions used, rarely occur in the same cells. For this reason, a transformation is used to obtain a sensitive density value for the risk of piracy in each cell. One solution would be to increase the cell size. However, problems with using coarse grids for this type of analysis are well known and include the ecological fallacy (Robinson, 1950) of assuming that within a (large) cell the risk of attack is equal at all locations. A related and more important issue is the modifiable areal unit problem (Openshaw, 1984). In this case, the problem is that the distribution of a particular point pattern may look quite different if it is summarised by aggregating events to different sets of boundaries. This may be the case even if the different boundaries only differ in subtle ways, and the problem is particularly acute if events occur near to the boundaries. For these reasons, a different approach is proposed.

[4] Reports can be downloaded by changing the number from 45 to 56 at the end of http://www.suezcanal.gov.eg/Files/Publications/45.pdf.

Figure 12.2 Density map of observed pirate attacks in 2010

Instead of simply counting the number of attacks that occur in each cell, we use a kernel density estimator. Conceptually, for each attack, one can think of the approach as providing an estimate of how accurate it is to assign that attack to a particular cell. The density is computed using the method as described in Tarn (2007). As with other estimators, the algorithm used estimates density values for a grid of regular-sized cells by identifying events within a given bandwidth of each cell and then computing a weighted sum of events. Events are weighted by their proximity to the centre of the cell of interest, and so cells with large numbers of events clustered around the centroid of the cell will have the highest density values. This particular method (Tarn, 2007) is used as it not only computes a density estimate in two dimensions but also selects the optimal bandwidth to describe a particular point pattern.

Pirate attacks that occurred during the first six months of the year 2010 are transformed into the density map shown in Figure 12.2. Density is visualised in shades of magenta: the lighter the higher the density is, and the darker the lower. The transformation proposed here shows two areas with high concentrations of incidents of piracy. The one with the highest density is between Yemen and Somalia. The choke-point south-west of Yemen is also an area with a high density of attacks.

12.3.2 Defining Vessel Route Maps

Maritime piracy cannot occur in the absence of target vessels, and so estimates of the typical location of such ships comprise a key element of a model of maritime piracy. Figure 12.1b shows vessel locations, as recorded in the SYNOP data from 15 November 2010 to 31 December 2010, off the Gulf of Aden. It is possible to detect a main route starting from the Suez Canal, going through the Gulf of Aden and then forking north-east towards the open sea.

The vessel route map used here is derived from the SYNOP data and is defined as the likelihood of finding a vessel on a particular cell (vessel density) in the year 2010. This map has to be consistent with the shape of the land, and therefore around choke-points (where there is less room for manoeuvre) the density must increase, whereas the further a vessel moves into the open sea the lower the density will be. The vessel route map is generated through a two-step process. First, the main route is identified as a polyline as shown in Figure 12.3a.[5] The cells that intersect the polyline are then used to define the main route in the model. Ships can, of course, also move through other cells, but the probability of them doing so decreases the further a cell is from the main route. The second step is therefore to establish these probabilities. To do this, for each cell on the main route, we estimate the distance of that cell from the coastline. This value, labelled *breadth*, approximates the maximum distance a vessel could realistically be found between it and the coastline. For each cell on the main route, let *dist* be a vector of *n* random-normal values generated within the interval [0 *breadth*]. The probability

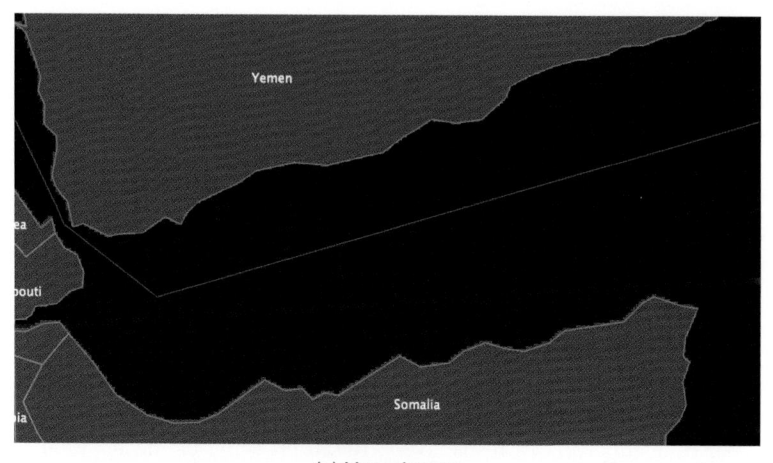

(a) Vessel route

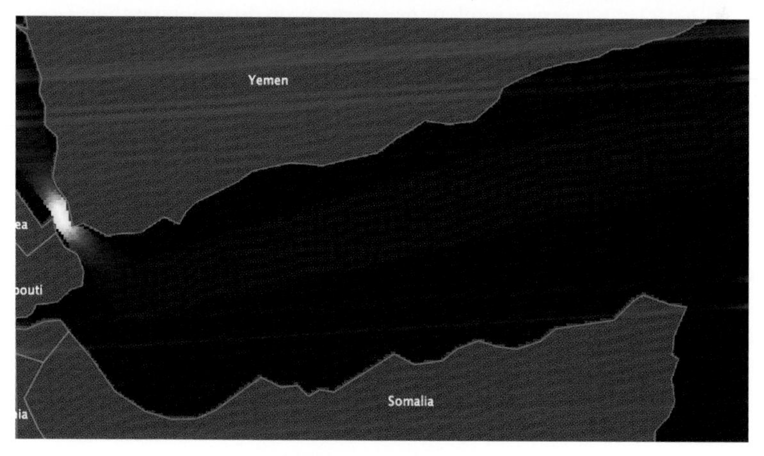

(b) Vessel density map

Figure 12.3 Vessel route map

[5] The main route was calculated by finding the polyline that returned the smallest sum of the squared distance from each point.

of a vessel travelling through a cell that is *d* cells from the main route is then estimated by counting the number of values of *dist* ∈[0 *breadth*] that are equal to *d*. Figure 12.3b shows the route map for vessels. Vessels sailing from south to north follow the same route as vessels sailing from north to south. Cell vessel density is visualised in shades of blue: the lighter the colour the higher the density, and the darker the colour the lower.

12.3.3 Defining Pirates', Naval Units' and Vessels' Behaviours

Pirates, naval units and vessel agents are free to move across the sea. A set of parameters is defined in order to simulate their behaviours:

na: naval unit's radius area of action
I: naval unit's radius of influence
np: number of pirates
pa: pirates' radius of action
pit: pirate inactivity time
npz: number of zones in which pirates operate
pI: pirates' radius of influence.

The parameter *npz* defines the number of (circular) zones in which pirates can operate, whereas *np* determines the number of pirates (where $np \geq npz$). Where the latter is greater than the former, the number of pirates allocated to each zone is the result of a stochastic process that ensures that similar numbers of pirates are allocated to each zone. The parameters *pa* and *na* define the zone of operation of pirates and naval units by identifying the radius within which these two types of agents can act, respectively. Based on empirical observation (see Section 12.2), the number of naval units is fixed to 14. However, as the number of pirates that operated in the region in 2010 is unknown, simulations are run with different numbers of pirates, which allows us to see how outcomes vary for different settings. Multiple pirate agents can operate within each pirate zone of action, but only one naval unit can operate in each naval zone. The centroids of the naval zones are allocated by dividing the main route into 14+1 equally spaced segments, and placing the centroid of a zone at the end of each of the first 14 segments. Pirate zones of action are allocated in the same way, but in this case the main route is divided into *npz+1* segments. An example, shown for the purposes of illustration only, for which there are five pirate zones each with a radius of 63 nautical miles is shown in Figure 12.4. Thus, the zones of action for both the pirate and naval agents will vary as a function of the number of zones used and the radius of action selected. Where there are few (many) zones of activity for any class of agent, there will be little or no (much) overlap across zones.

At regular intervals determined using a uniform random number generator, vessels begin journeys either from the north-west or east of the map. Given the geography of the area, those starting from the north-west can go only east, and those starting from east can move only north-west. Vessels follow a predefined route through the Gulf of Aden, as suggested by the handbook *Best Management Practices to Deter Piracy off the Coast of Somalia and in the Arabian Sea Area.*[6] The exact starting location of each vessel is determined by randomly selecting one of the cells at the margin of the route. As vessels move, they follow a route for which the

[6] http://www.secure-marine.com/bmp3/bpm3_pdf.pdf

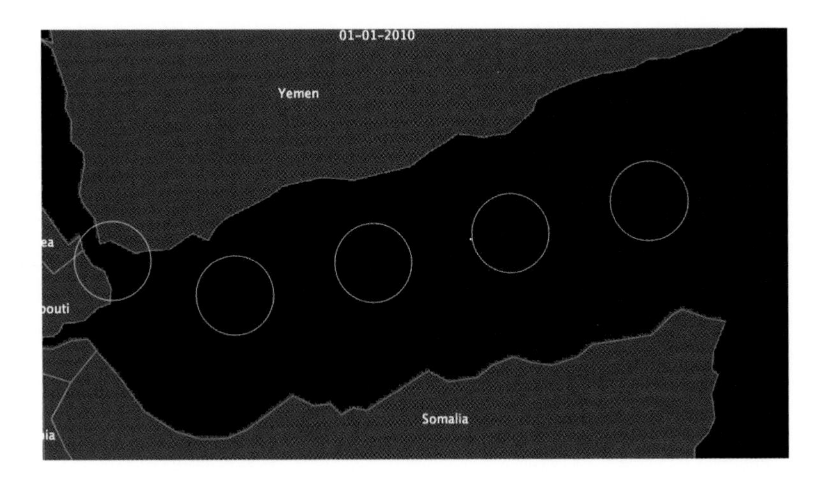

Figure 12.4 An example of the pirates' range of action

cells are of the same density as the starting cell. Thus, most (but not all) vessels will follow a route that is close to the centre of the shipping lanes shown in Figure 12.3a. On the basis of an empirical study conducted by the Office of Naval Intelligence (ONI) Piracy Team in 2009,[7] the mean vessel speed is set at 15 knots. To add some variability, the actual speed is sampled from a random normal distribution with a mean of 15 knots, and a standard deviation of 2 knots.

Insufficient literature is available to define the behaviours of the pirate and naval units using existing data. Consequently, for the purposes of this model, both pirate and naval agents are allocated an area of action, within which they conduct a biased random walk. The number of areas allocated to each class of agent, and the size of those areas, is estimated through model calibration. Both pirates and naval units follow the same rules for sailing. For each simulation step, they move 4.2 miles (which equates to 15 knots) and adjust their compass h as follows: $h = h \pm \text{random}\ (20)$. If they hit the boundary of their area of activity, their compass is set as $h = h\text{-}180$.

Figure 12.5a illustrates a snapshot of one simulation step (the simulated date is shown at the top of the figure). Every seven simulation steps correspond to an interval of two hours, and consequently a simulated year corresponds to 30,660 simulation steps.

Parameter I specifies the range over which the presence of a naval unit repels pirates, whereas pI specifies the range over which pirates can detect the presence of potential target vessels. Parameter pit indicates the number of days that pirates are inactive after a successful attack and is used to model the post-attack activity associated with collecting a ransom or transporting hijacked cargo.

Pirate attacks occur when three conditions are met:

1. The distance between a pirate and a potential target vessel is less than pI.
2. The distance from the closest navy unit is greater than that over which naval forces influence pirate decision making (I).

[7] http://www.marad.dot.gov/documents/Factors_Affecting_Pirate_Success_HOA.pdf.

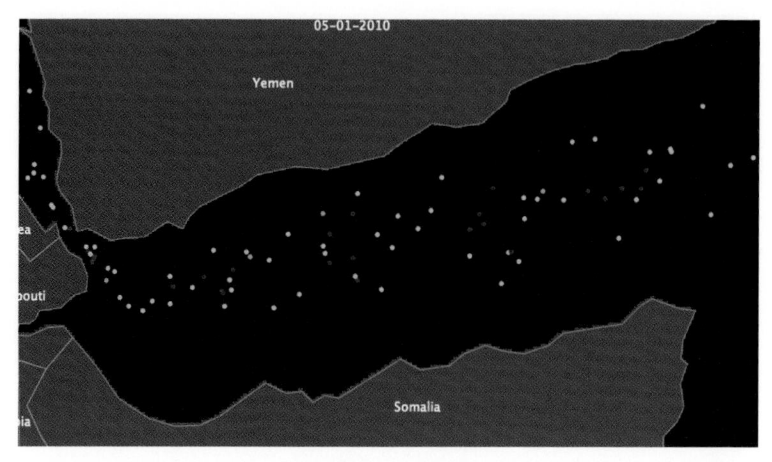

(a) Run snapshot

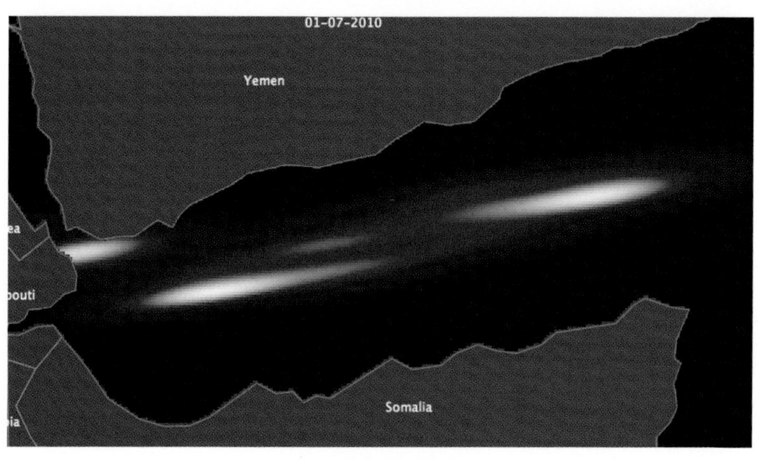

(b) Simulated map

Figure 12.5 Visualisation of the simulation (red dots, pirates; blue dots, vessels; green dots, naval units)

3. If conditions 1 and 2 are met, a pirate agent's decision as to whether to attack is influenced by three factors:
   - the (simulated) season
   - the type of vessel encountered (e.g. bulk cargo or oil tanker – see Table 12.3)
   - the flag (the country to which it is registered) of the vessel.

Conditions on the open seas vary by season, and research indicates clear seasonal patterns in the risk of maritime piracy (e.g. Marchione and Johnson, 2016). Consequently, account is taken of this in the model, and the monthly probability of a successful attack is estimated using NGIA data for the 10-year period (1999–2009) prior to the period for which we simulate attacks (see Figure 12.6). According to these data, attacks are more likely in the spring and autumn months.

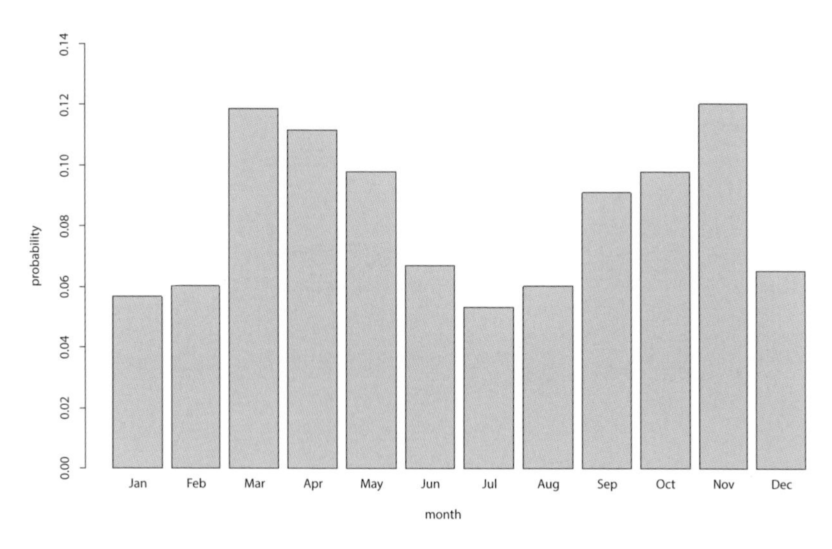

Figure 12.6 Probability of pirates attacks around the Gulf of Aden for the period 1999–2009. Source: National Geospatial Intelligence Agency's Anti-Shipping Activity Messages database

The type of vessel and where it comes from are also known to have a significant influence on the risk (number of vessels attacked divided by those transiting) of victimisation (Wolff *et al.*, 2009). Consequently, estimates of the risk associated with different types of vessels and those registered to particular states are taken from Wolff *et al.* (2009). Empirical data (shown in Tables 12.2 and 12.3, respectively) are used to replicate the flow of different types of vessels, and those registered to different states, through the Gulf of Aden. In the model, we include the 10 types of vessels and the 113 different flags that are considered in Wolff *et al.*

To model the probability of attack, let $s()$, $t()$ and $f()$ be season, vessels' type and vessels' flag attack probability density functions, respectively. Both time conditions 1 and 2 (see above) are met, a random number c is sampled from a uniform distribution between 0 and 1, and if $s()*t()*f() > c$, then a pirate attack takes place.

12.3.4 Comparing Risk Maps

The map shown in Figure 12.5b shows the full extent (the area shown in Video 2 is a much larger area) of a simulated maritime piracy map, generated using one set of the parameter values discussed above. It should be clear from the description of the model that data concerning the locations of attacks were not used to inform the condition action rules that govern agent behaviour, and so the simulated outcomes are not simply the product of presupposed emergence.

Visual inspection of the simulated map is insufficient to assess whether or not the model reproduces the patterns observed in the real data. To do this more systematically, a method to assess the goodness of fit is proposed. Each simulation is compared to the observed data with respect to the location, time and overall number of attacks that occurred during the period of interest. Specifically, to compare model performance, we use a kernel density–based

Table 12.2 Vessel type passing through the Gulf of Aden in 2010. Source: Canal Suez Traffic Statistics

	Ship. Type	Jan	Feb	Mar	Apr	May	Jun	Jul	Aug	Sep	Oct	Nov	Dec
1	Bulk Carrier	225.0	183.0	221.0	236.0	244.0	234.0	239.0	280.0	228.0	261.0	223	207.0
2	General Cargo	147.0	116.0	144.0	94.0	158.0	133.0	143.0	147.0	122.0	145.0	131	138.0
3	Container Ships	523.0	479.0	530.0	531.0	592.0	585.0	622.0	622.0	603.0	608.0	568	589.0
4	Oil Tankers	279.0	257.0	309.0	259.0	295.0	293.0	302.0	337.0	311.0	295.0	282	331.0
5	Liquefied Natual Gass Carrier	76.0	56.0	66.0	64.0	65.0	62.0	62.0	79.0	78.0	73.0	82	92.0
6	Combined Carrier	3.0	1.0	2.0	5.0	3.0	4.0	3.0	2.0	2.0	2.0	1	0.0
7	Wheeled Cargo	24.0	18.0	28.0	3.0	25.0	23.0	35.0	17.0	24.0	29.0	19	25.0
8	Car Carrier	75.0	75.0	80.0	86.0	98.0	87.0	83.0	88.0	82.0	81.0	88	81.0
9	Passenger Ships	5.0	5.0	11.0	23.0	14.0	3.0	1.0	2.0	0.0	7.0	20	9.0
10	Others	61.0	66.0	76.0	165.0	68.0	58.0	64.0	85.0	63.0	71.0	86	72.0
11	Total	1418.0	1256.0	1467.0	1466.0	1562.0	1482.0	1554.0	1659.0	1513.0	1572.0	1500	1544.0
12	Daily Average	45.7	44.9	47.3	48.9	50.4	49.4	50.1	53.5	50.4	50.7	50	49.8

Table 12.3 Vessel flag passing through the Gulf of Aden in 2010. Source: Canal Suez Traffic Statistics

	Jan	Feb	Mar	Apr	May	Jun	Jul	Aug	Sep	Oct	Nov	Dec
Algeria	0	0	0	0	0	0	0	0	0	1	1	0
Antigua And Barbuda	35	30	41	25	38	32	38	37	32	35	33	43
Argentina	0	0	0	0	0	0	0	0	0	0	0	0
Australia	0	1	0	0	0	0	0	0	0	0	0	1
Austria	0	0	0	0	7	1	0	1	0	2	4	3
Bahamas	67	62	74	68	76	68	73	66	75	81	84	70
Bahrain	4	3	3	2	2	2	6	1	6	2	4	5
Bangladesh	3	1	3	1	2	0	5	2	1	1	0	1
Barbados	4	1	1	0	0	0	0	1	0	1	1	4
Belgium	3	0	7	1	2	2	1	4	2	3	2	0
Belize	1	3	1	4	2	4	4	2	2	4	4	3
Bermuda	0	0	0	0	0	2	7	6	1	3	4	3
Bolivia	0	1	0	0	0	0	0	0	0	0	1	0
Bosnia And Herzegovi	0	0	0	0	2	0	0	0	0	0	0	0
Brazil	0	0	0	0	0	0	0	0	0	1	0	1
Bulgaria	2	1	1	5	5	3	3	0	1	2	1	1
Cambodia	3	0	2	6	2	2	2	3	0	4	0	1
Canada	0	0	0	1	0	0	0	0	0	0	0	0
Cayman Islands (Br)	0	0	0	0	2	9	12	10	3	18	14	9
China	19	15	26	28	25	18	21	19	20	23	19	18
Colombia	1	0	0	0	0	0	0	0	0	0	0	0
Comoros	0	0	0	0	4	0	3	2	0	2	0	2
Comoros Islands	2	1	0	3	0	2	0	0	0	0	0	0
Cook Islands	0	0	0	0	0	0	1	0	1	0	2	0
Cote De Ivoire	0	0	0	0	1	0	0	0	0	0	0	0
Croatia	1	1	1	1	2	1	0	1	0	0	2	0
Cyprus	27	27	32	30	32	23	33	28	25	27	15	25
Denmark	45	39	40	40	43	44	49	50	48	32	45	47
Dominica	0	0	0	0	0	0	0	0	1	3	1	1
Dominican Republic	1	5	2	6	0	0	0	0	2	0	0	0
Egypt	8	3	3	13	10	6	3	11	7	9	9	12
El Salvador	0	0	0	0	4	0	0	0	1	0	0	0
Eritrea	0	0	1	0	0	0	0	0	0	0	0	0
Faroe Islands	0	0	0	0	0	0	0	0	0	0	0	2
Finland	1	0	1	0	0	0	0	0	0	0	1	0
France	18	16	19	21	19	15	21	21	16	17	18	24
Georgia	0	0	0	0	0	0	1	0	0	4	0	2
Germany	50	45	54	46	63	57	52	52	61	47	48	53
Gibraltar	0	0	0	0	1	2	5	6	1	4	5	5
Goergia	0	0	0	0	5	1	0	2	3	0	3	0
Great Britain (UK)	103	88	94	109	90	78	75	74	96	83	67	77
Greece	33	34	39	38	40	45	40	49	43	47	40	38
Guinea	0	0	0	1	0	0	0	0	0	0	0	0
Honduras	0	1	0	1	1	0	1	0	0	2	0	0
Hong Kong	73	77	86	67	85	88	100	95	88	88	95	93

(*continued*)

Table 12.3 (*continued*)

	Jan	Feb	Mar	Apr	May	Jun	Jul	Aug	Sep	Oct	Nov	Dec
Iceland	0	0	0	0	1	0	1	0	2	2	1	1
India	16	11	12	10	8	12	10	10	8	12	14	9
Indonesia	0	1	0	0	2	0	0	0	1	3	0	0
Iran	12	12	9	10	11	10	11	12	5	7	8	6
Iraq	2	0	0	0	0	0	0	0	0	1	0	0
Ireland	0	0	0	0	0	0	0	0	0	1	0	0
Isle Of Man	0	0	0	0	0	2	1	1	3	1	0	0
Israel	2	2	2	1	1	1	1	2	1	0	1	1
Italy	15	23	31	22	30	23	23	37	28	28	25	30
Jamaica	1	0	1	1	0	1	0	0	0	2	1	1
Japan	1	5	4	7	4	4	5	1	6	5	3	4
Jordan	0	0	0	0	1	0	0	0	0	0	1	1
Kuwait	4	2	4	2	2	3	3	3	3	3	4	3
Latvia	0	0	0	0	0	0	0	0	0	0	0	0
Lebanon	1	1	1	0	1	0	1	0	1	1	0	0
Liberia	187	146	181	170	198	212	212	253	215	214	192	226
Libya	0	2	1	3	0	1	3	0	3	2	2	2
Lithuania	0	1	0	0	0	0	0	0	0	0	0	0
Luxembourg	1	2	3	2	3	3	2	4	1	2	2	5
Madeira Islands	0	0	0	0	0	0	0	0	0	1	0	0
Malaysia	4	3	5	5	0	7	2	6	5	6	2	4
Maldives	4	2	1	0	2	3	2	1	1	0	1	1
Malta	73	66	101	100	76	79	93	92	78	83	67	77
Malysia	0	0	0	0	8	0	0	0	0	0	0	0
Mangolia	2	2	0	1	1	0	0	1	1	1	1	0
Marshall Islands	89	68	69	75	67	74	75	93	82	90	103	103
Mauritius	0	0	0	0	0	0	0	0	0	0	0	0
Moldova	0	0	0	3	0	0	0	0	0	0	2	0
Morocco	0	0	0	0	6	0	1	4	4	1	0	0
Netherland Antilles	0	0	0	0	3	1	2	3	4	0	0	1
Netherlands	14	10	17	19	13	12	13	16	17	13	10	8
North Korea	0	0	0	0	1	0	0	0	0	0	0	0
Norway	27	24	27	25	37	17	25	35	25	24	31	21
Oman	0	0	0	0	1	1	0	0	0	1	0	0
Pakistan	0	0	2	0	3	0	0	0	0	0	1	0
Panama	254	250	259	278	313	287	283	303	283	294	268	264
Philippines	8	3	6	6	7	7	7	8	8	10	9	7
Poland	0	0	0	1	1	0	0	0	0	0	0	0
Portugal	2	3	5	6	0	1	1	0	0	0	1	3
Qatar	4	3	3	1	7	6	3	8	5	3	4	3
Romania	0	0	0	0	0	0	0	0	0	0	0	0
Russia	8	4	11	11	8	4	7	6	4	5	6	0
Saint Kitts And Nevi	0	0	0	0	0	6	5	2	2	2	6	2
Saint Vincent & Grenad	17	8	17	14	11	7	17	25	13	12	14	8
Saudi Arabia	10	11	9	11	8	12	8	12	7	13	9	13

Table 12.3 (*continued*)

	Jan	Feb	Mar	Apr	May	Jun	Jul	Aug	Sep	Oct	Nov	Dec
Seychelles	0	0	0	0	0	0	0	0	0	0	0	0
Sierra Leone	7	5	2	9	3	0	2	1	2	3	6	5
Singapore	54	52	53	55	69	67	71	67	72	84	77	87
Slovenia	0	0	0	0	1	0	0	0	0	0	0	0
South Korea	4	7	7	9	10	15	7	16	8	12	9	8
Spain	3	3	6	7	4	6	1	3	2	0	2	5
Sri Lanka	1	0	0	1	0	1	0	0	1	0	0	0
Sudan	0	0	0	0	1	0	0	1	0	0	0	0
Sweden	2	3	4	1	0	3	4	2	4	2	6	4
Switzerland	1	0	1	1	2	0	0	1	0	0	2	0
Syria	0	0	2	2	1	2	1	1	2	4	1	3
Tanzania	1	1	0	1	0	2	2	2	1	2	3	3
Thailand	5	6	5	3	2	1	3	4	1	2	3	2
Togo	0	1	0	1	2	1	1	1	2	0	0	0
Tonga	0	0	1	0	0	1	0	0	0	0	0	1
Tunisia	1	1	0	1	2	2	2	1	1	2	1	2
Turkey	11	14	17	18	14	23	26	27	17	21	29	16
Tuvalu	0	0	0	1	0	1	2	1	1	1	3	2
Ukraine	1	2	0	0	0	0	1	0	0	1	0	1
United Arab Emirates	3	2	4	4	3	4	8	5	5	4	4	5
United States	61	39	50	50	48	51	50	43	39	43	41	50
Vanuatu	0	0	2	1	0	1	0	2	3	2	1	1
Vietnam	1	0	1	0	0	0	0	0	0	0	0	1

two-sample comparison test for multidimensional data originally developed in biology to examine cell morphology after a given manipulation (Duong *et al.*, 2012).

The discrepancy measure *T* is:

$$T = \int [f(x_1(long, lat, time), H) - f(x_2(long, lat, time), H)]^2 dx \qquad (12.1)$$

where *f* is the kernel density estimate. In order to measure model performance with respect to the overall number of attacks produced, the bandwidth *H* is optimised only for the observed attacks (for a detailed description of the algorithm, see Chacon and Duong, 2010) and kept constant when estimating the density of the simulated attacks. *x1* and *x2* are the 3-variate (longitude, latitude and time) samples of observed and simulated pirates attacks, respectively. A *z*-score of the discrepancy measure is computed as follows:

$$z = \frac{T - \mu}{\sigma\sqrt{1/n_1 + 1/n_2}} \qquad (12.2)$$

where μ and σ are the mean and standard deviation of *T*, respectively (for the estimation of μ and σ, please see Duong *et al.*, 2012); and $n1$ and $n2$ are the number of observations for observed and simulated pirate attacks. The *z*-score is used to compare model performance, and high values suggest that two distributions are drawn from different populations.

12.4 Model Calibration

In the absence of data to calibrate some of the model variables, to examine the sufficiency of the model to generate patterns similar to those observed in the real world, we use a genetic algorithm to estimate parameters. Specifically, we use the Mebane *et al.* (2011) *genetic optimisation using derivatives* technique. This approach combines a derivative-based (Newton or quasi-Newton) method in the creation of new individuals (in the vocabulary of genetic algorithms) in the genetic algorithm with the typical mutation and crossover methods.

The fitness of each individual is the mean of the z-score (equation (12.2)) of 15 runs of the model for the first six months of 2010. Twenty generations of 20 individuals were used by the genetic algorithm to estimate parameters. Optimisation results are shown in Table 12.4.

Considering the influence of the parameter values on model outcomes, insight can be gained from examining which combinations of parameter values generate outcomes that most closely approximate the empirical record. The best results are observed when there are 60 pirates (np) operating in 42 zones (npz), with a range of action of (pa =) 147 miles, 16 days of pirate inactivity and a radius influence of 218.4 miles. For the naval forces, it seems that simulation outcomes most closely replicate the real-world patterns sampled when the range of influence is (I =) 37.8 miles and the range of action is (na =) 200 miles. The mean z-score (computed across 15 runs) for the parameter settings shown in Table 12.2 of 1.44 (standard deviation = 0.4) is less than the critical value of 1.96, suggesting that the model output does not differ significantly to that observed. The mean number of attacks is 57 with a standard deviation of 6.6, which compares well with the 56 attacks observed.

Figure 12.7 compares the observed map (Figure 12.7a) and a map generated during one run of the model that used the best parameter settings (Figure 12.7b). The latter has a z-score of 1.28 and 55 attacks. Visual inspection of the map generated indicates that the difference between the two distributions is due to the simulation model generating two high-density clusters too close to each other. This is partly caused by the restricted mobility of the vessel agents transiting the Gulf of Aden. For example, in the calibrated model, no vessel navigated off the south-east coast of the Yemen or off the north-east coast of Somalia.

12.5 Discussion

In this chapter, an agent-based approach is employed to test a simple model of maritime piracy. The model proposed is calibrated using empirical observations concerning the volume of vessels sailing through the Gulf of Aden, a sample of SYNOP data to estimate the

Table 12.4 Model calibration

Parameter	Value
na: naval units radius area of action	200 miles
I: naval units radius influence	37.8 miles
np: number of pirates	60
pa: pirates radius area of action	147 miles
pit: pirate inactivity time	16 days
npz: number of zones in which pirates operate	42
pl: pirates radius influence	218.4 miles

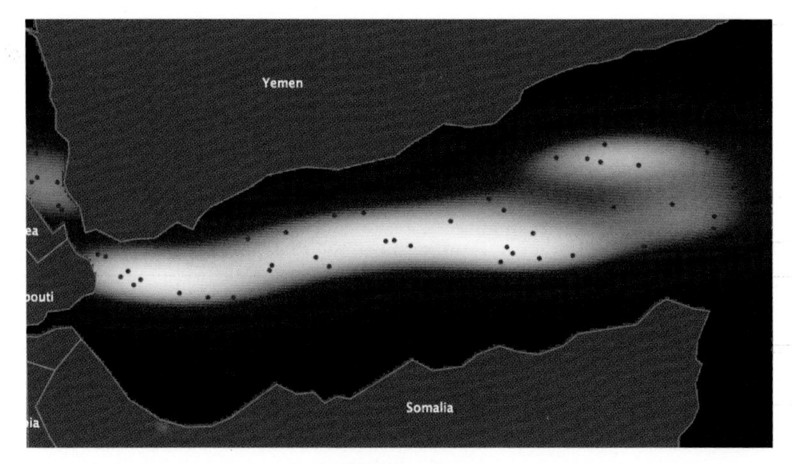

(a) Observed map

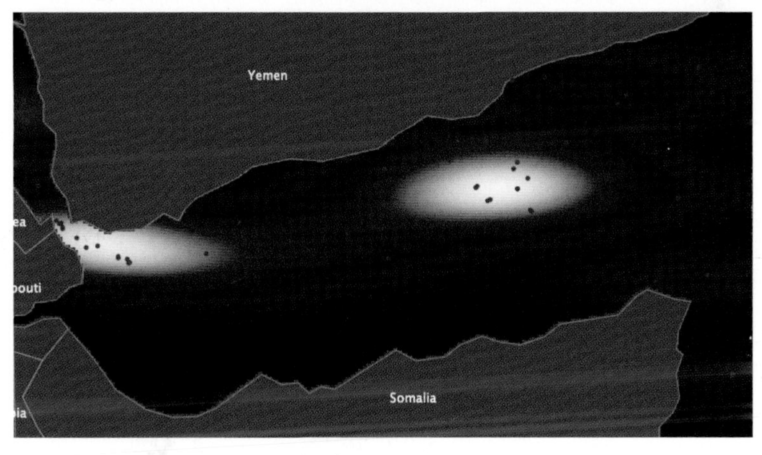

(b) Simulated map

Figure 12.7 (a) The observed map and (b) a map generated during one run of the model that used the best parameter settings.

geometry of shipping routes, estimates of seasonal variation in the risk of attack, and estimates of how the risk of attack differs by vessel type and state of registration. Model outputs are tested using empirical data (not used in the calibration of the model) concerning the location of pirate attacks around the Gulf of Aden during the first six months of 2010. A method for measuring the goodness of fit of the space–time distribution of attacks is suggested, and results are presented. Despite the simplicity of the model, it appears to approximate the observed spatial distribution of pirate attacks rather well. Consequently, we suggest that the model may prove useful as a way of providing insight into attack locations and for use as a test-bed for examining the potential impact of simulated policy interventions designed to reduce maritime piracy.

Due to a lack of information on pirate and naval strategies, pirate numbers and so on, we examine how varying model parameters affect model outcomes. The aim of so doing was to see if particular parameter combinations provide outcomes that are more similar to observed patterns, which in turn allows us to make some cautious inferences about the possible number of pirate vessels, their range of behaviour and that of the naval units. Considering model outcomes for the different parameter settings, we find that the model that most closely replicates the observed data is that for which there are 60 pirates operating in 42 zones, with a range of action of 147 nautical miles, a radius of influence of 218.4 miles, and 16 days of inactivity between attacks. For the 14 naval forces (this parameter is based upon empirical data), these were estimated to have an influence and range of action of 200 and 37.8 nautical miles, respectively. This is not to say that this is how pirates and naval forces do behave but that – of all of those tested – this model is the most able to explain observed patterns.

The pirates' radii of action and influence – as estimated by the model – are very large. However, these may not be unrealistic estimates. First, pirates have substantial mobility and can patrol large areas, either under power or by allowing their boats to drift in the open sea. Moreover, pirates operating off the Coast of Somalia are known to use *motherships* to sail for long periods in open water, which can substantially extend their range of operations (Shortland and Vothknecht, 2010). Pirates are frequently equipped with GPS navigation systems, mobile telephony and fast boats (skiffs) powered by two 100 bhp outboard motors (Shortland and Vothknecht, 2010) that can easily reach 30 knots. This technology allows them to monitor very large areas and to be able to attempt an attack on vessels that are far away from their position. For example, if a vessel is sailing at 15 knots, a pirate skiff 200 miles away can reach the vessel in 4 hours if the two are sailing towards each other, and in 13 hours if the pirate boat is chasing the target vessel. Therefore, a very large radius of influence, as found in the proposed model, may not be unrealistic and may support the idea that pirates use sophisticated technological tools, or coordinate their activity to plan their attacks.

In contrast, the best model parameter for the naval units suggests that they operate across large areas (a radius of 200) but have a relatively small range of influence (37.8 miles). This would suggest that naval units can only detect or deter the activity of pirate vessels that are in their vicinity. This difference in the area of influence of the pirates and naval vessels is not unreasonable given the asymmetry associated with the challenge of detecting targets for the two types of actor. To elaborate, for naval forces, the task of identification involves the detection of small vessels against a backdrop of waves encountered in the open seas, which can be challenging using conventional sensors such as radar. In contrast, pirates will often search for large vessels that actually transmit their position in real time, allowing their location to be more easily detected. Apropos the apparently limited effect of the naval forces in the model more generally, this is consistent with the research reviewed in the introduction regarding the impact of international policy against piracy off the coast of Somalia in early 2010 (Chalk, 2010; Shortland and Vothknecht, 2010).

In the current model, pirates choose their targets according to a probability function based on the current season, and the type and flag of potential target vessels. Vessels move following a predefined route across the Gulf of Aden, and naval units patrol the sea following a biased random walk. Further development of the model could explore more sophisticated agent design as well as the impact of exogenous factors such as political and economic influences. Doing so may lead to simulation outcomes that are able to approximate extended time series observed in the empirical record. A further refinement of the model would be to incorporate adaptation behaviours for the vessel agents so that, for example, with some constraints, vessels can change the routes they take as a consequence of recent pirate activity. Additionally, the model might be used to estimate the potential impacts of different naval strategies. For instance, what is the effect of increasing naval strength or different patrol patterns? Such questions could be examined using agent-based models such as the one proposed here, which explicitly models the complex interactions between actors, but they are beyond alternative approaches such as econometric analysis.

The work reported here is an initial demonstration, and although agent rules and interactions are simple, it provides the basis for developing and testing more accurate and realistic agent behaviour, and the simulated impact of specific types of policy intervention. Despite the limitations of the model, the results are encouraging insofar as they suggest that this approach to modelling may provide useful insight into the phenomenon of maritime piracy and interventions designed to reduce it.

References

Abbot J and Renwick N. (1999). Pirates? Maritime piracy and societal security in southeast Asia. *Global Change, Peace & Security* 11(1): 7–24.

Becker G. (1968). Crime and punishment: an economic approach. *Journal of Political Economy* 76(2): 169–217.

Caplan J, Moreto WD and Kennedy LW. (2011). Forecasting global maritime piracy utilizing the risk terrain modelling approach to spatial risk assessment. In *Crime and Terrorism Risk*. Boca Raton, FL: Taylor and Francis.

Chacon JE and Duong T. (2010). Multivariate plug-in bandwidth selection with unconstrained pilot matrices. *Test* 19: 375–398.

Chalk P. (2010). Piracy off the horn of Africa: scope, dimensions, causes and responses. *Brown Journal of World Affairs* 16(2): 89–108.

Daxecker U and Prins B. (2012). Insurgents of the sea: institutional and economic opportunities for maritime piracy. *Journal of Conflict Resolution* August 16.

Duong T, Goud B and Schauer K. (2012). Closed-form density-based framework for automatic detection of cellular morphology changes. *Proceedings of the National Academy of Science* 109: 8382–8387.

Fu X, Ng AK and Lau YY. (2010). The impacts of maritime piracy on global economic development: the case of Somalia. *Maritime Policy and Management* 37(7): 677–697.

Gilbert G. (2008). *Agent-based Models*. Thousand Oaks, CA: Sage.

Hastings J. (2009). Geographies of state failure and sophistication in maritime piracy hijackings. *Political Geography* 28(4): 213–223.

Jakob M, Vaněk O, Urban Š, Benda P and Pěchouček M. (2010). Agentc: agent-based testbed for adversarial modeling and reasoning in the maritime domain. In *Proceedings of the 9th International Conference on Autonomous Agents and Multiagent Systems*. vol. 1, pp. 1641–1642.

Marchione E and Johnson SD. (2016). Space-time dynamics of maritime piracy. *Journal of Research in Crime and Delinquency* in press.

Mebane WR Jr, and Sekhon JS. (2011). Genetic optimization using derivatives: the rgenoud package for R. *Journal of Statistical Software* 42(11): 1–26. http://www.jstatsoft.org/v42/i11/

O'Meara R. (2007). Maritime piracy in the 21 ST century: a short course for US policy makers. *Journal of Global Change and Governance* 1(1).

Openshaw S. (1984). The modifiable areal unit problem. *Concepts and Techniques in Modern Geography* 38: 41.

Ploch L. (2010). *Piracy off the Horn of Africa*. Darby, PA: DIANE Publishing.

Robinson WS. (1950). Ecological correlations and the behavior of individuals. *American Sociological Review* 15(3): 351–357.

Rengelink H. (2012). *Trends in Organized Crime*. Berlin: Springer.

Sandler T and Enders W. (2004). An economic perspective on transnational terrorism. *European Journal of Political Economy* 20(2): 301–316.

Shortland A and Vothknecht M. (2010). *Combating Maritime Terrorism off the Coast of Somalia*. Berlin: German Institute for Economic Research.

Tarn D. (2007). ks: kernel density estimation and kernel discriminant analysis for multivariate data in R. *Journal of Statistical Software* 21(7).

Treves T. (2009). Piracy, law of the sea, and use of force: developments off the coast of Somalia. *European Journal of International Law* 20(2): 399–414.

Vagg, J. (1995). Rough seas? *British Journal of Criminology* 35(1): 63.

Wolff F-C, Mejia M and Cariou P. (2009). Is maritime piracy random? *Applied Economics Letters* 16: 891–895.

13

A Simple Approach for the Prediction of Extinction Events in Multi-agent Models

Thomas P. Oléron Evans, Steven R. Bishop and Frank T. Smith

13.1 Introduction

The prediction of rare and extreme events in complicated real-world systems is a mathematical challenge for which the stakes are extremely high. Analysing seismological data to predict large earthquakes or meteorological data to predict hurricanes is of vital importance to preserving human life and taking measures to protect infrastructure. Predicting extreme events in complex systems of interacting agents, such as damaging crashes in financial markets or the extinction a certain species in an ecosystem, is an equally important goal and potentially an even more difficult one.

A common method of prediction, applied to a wide variety of such problems, is to model the system of interest with an agent-based model (ABM) or a cellular automaton (CA). The model can be calibrated with data observed from the real system and run multiple times to produce a range of possible outcomes. The probability of a given real-world event can then be inferred from the proportion of runs of the model in which it is observed. This variability of outcomes in such models may arise from in-built stochasticity or from the use of an ensemble forecast, in which differing initial conditions are applied to each run, reflecting uncertainty about the true state of the world (Smith, 2001).

However, comparatively little attention has been paid to the problem of predicting the behaviour of these ABMs and CA models themselves, without resorting to explicit simulation. On the one hand, this may seem like an inadvisable direction for research, since such models are often employed precisely because the scenarios being modelled involve such complex interactions that direct prediction is not possible. However, though long-term prediction may

Approaches to Geo-mathematical Modelling: New Tools for Complexity Science, First Edition. Edited by Alan G. Wilson.

Companion Website: www.wiley.com/go/wilson/ApproachestoGeo-mathematicalModelling

indeed be impossible, it does not seem unreasonable to investigate the prediction of ABM and CA models to some more limited future horizon, at least for some features of their behaviour.

In this chapter, we consider a simple individual-based model called the NANIA predator–prey model, which runs over a finite rectangular grid of cells. We observe that this model will eventually evolve to one of two fixed points, the extinction of both species or the extinction of the predators and the reproduction of the prey to fill the entire region. For different values of the parameters, evolution to one of these states may be rapid, or the system may settle into a regime of quasi-stable oscillations in which extinction can be seen as a rare event. If we desire to preserve both species, then it would be beneficial to be able to predict the moment at which the quasi-stability breaks down through analysing the current state of the model. This is the main aim of the chapter.

13.2 Key Concepts

13.2.1 Binary Classification

Binary classification is a general term for the analysis of data points that belong to one of two distinct categories, which are generally labelled −1 and 1. For example, we may be interested in analysing photographs that may or may not contain human faces, or examining documents whose content may or may not be related to a certain topic. In such scenarios, the goal of binary classification is to use data whose categories are known to deduce the categories of data whose categories are unknown.

More formally, given a data sample $\boldsymbol{x}_i, i \in \{1, \dots, D\}$, drawn from some set X, for which the categories $y_i \in \{-1, 1\}, i \in \{1, \dots, D\}$, are known, we suppose that these data and their categories have been drawn from some fixed but unknown probability distribution. The categorised data points $(\boldsymbol{x}_i, y_i)$ are thus considered to be realisations of a random variable $\Theta \in X \times \{-1, 1\}$ drawn from this unknown distribution.

The aim is to use the categorised data to create a **binary classifier** f, which may assign categories to points in X whose category is unknown:

$$\begin{aligned} f : X &\to \{-1, 1\} \\ \boldsymbol{x} &\mapsto f(\boldsymbol{x}) \end{aligned} \tag{13.1}$$

Ideally, given any data point $(\boldsymbol{x}, y)$, a classifier would assign the correct category to $\boldsymbol{x}$, such that $f(\boldsymbol{x}) = y$. However, if the underlying distribution is such that a point may belong to either category with non-zero probability, a classifier will not generally be able to categorise all points correctly. In such cases, measures of performance are required, both to compare the quality of different proposed classifiers and to establish the degree of confidence that we should have in a particular classifier's predictions.

For further details on the concepts and notation of binary classifiers, see Steinwart and Christmann (2008) and Gretton (2013).

13.2.2 Measures of Classifier Performance

There are several measures of the performance of a classifier. Which of these are the most appropriate in any given scenario depends on the purpose of the classifier and the priorities of those who designed it.

Suppose that we have a particular data set $x_1, \dots, x_D \in X$ and a corresponding set of estimated categories $\hat{y}_1, \dots, \hat{y}_D$ assigned to the data points by some binary classifier f. The true categories $y_1, \dots, y_D \in \{-1, 1\}$ of the observations may or may not be known.

For each observation x_i, there are four possible cases:

- **True positive:** $\hat{y}_i = y_i = 1$
- **False positive:** $\hat{y}_i = 1, y_i = -1$
- **True negative:** $\hat{y}_i = y_i = -1$
- **False negative:** $\hat{y}_i = -1, y_i = 1$.

Letting *TP*, *FP*, *FN* and *TN* represent the numbers of observations in each of these classes and letting $P = TP + FN$ and $N = FP + TN$, this information may be collected in a **confusion matrix**:

		True category (y_i)	
		1	−1
Estimated	1	*TP*	*FP*
category ($\hat{y}_i$)	−1	*FN*	*TN*
Column totals:		*P*	*N*

From the confusion matrix, we see that no single number could fully describe the performance of a classifier. Given that the number of observations D is known, a full description of classifier performance requires knowledge of any three of the values *TP*, *FP*, *FN* and *TN*. Several different scalar measures of classifier performance may therefore be derived from these quantities. For example:

- **Accuracy:** The overall rate of correct predictions; $(TP + TN)/D$.
- **True positive rate (or recall):** The proportion of truly positive observations that are correctly identified; TP/P.
- **False positive rate:** The proportion of truly negative observations that are incorrectly identified; FP/N.
- **Precision:** The proportion of observations estimated to be positive, which are indeed positive; $TP/(TP + FP)$.
- **Negative predictive value:** The proportion of observations estimated to be negative, which are indeed negative; $TN/(TN + FN)$.

Strictly speaking, these terms should refer to the expected values of the given expressions under a particular stochastic data-generating process. The expressions provided here are estimates of the true measures derived from the data.

Note that the true positive rate and the false positive rate are independent of the relative frequencies of positive and negative observations in the data. Whether the two categories are observed equally frequently or whether one category is more common than the other, these measures (considered as expectations) remain constant. This is not true of the accuracy, which is dominated by the performance of the classifier on observations in the more frequently occurring category, nor of the precision or the negative predictive value, which drop as the relative rarity of observations in their corresponding categories increases.

Because of the independence of the true positive rate and the false positive rate in relation to the relative frequencies of the categories, these two measures are often used in combination to assess classifier performance. If the true relative frequency of the categories is known, these two measures provide complete information about the performance of a classifier. For these reasons, classifier performances are often visualised and compared using a **receiver operating characteristics** (ROC) graph, in which the true and false positive rates of various proposed classifiers are plotted against each other for the purposes of comparison.

Alternatively, classifier performances may be visualised using a **precision-recall** graph, although, as noted above, such graphs are not invariant under changes to the relative frequency of the two classes. However, since we are interested in predicting rare extinctions, the quantities measured by the precision and recall (true positive rate) – specifically, the probability of an extinction when an extinction is predicted and the probability that a particular extinction will be successfully predicted – are of particular interest to us.

For further details on the various measures of classifier performance and on the ways of visualising this information, see Fawcett (2006), from which much of the terminology presented here was drawn.

13.2.3 Stochastic Processes

The material on stochastic processes presented here is derived from a number of sources, including Ross (1996), Grimmet and Stirzaker (2001), Stirzaker (2005), Cotar (2012) and Stroock (2014).

A **stochastic process** is a collection of random variables $\{Z(t) : t \in T\}$ taking values in a particular set $\mathfrak{Z}$ (the **state space**), each corresponding to a particular time $t \in T$, where T is generally a subset of $\mathbb{Z}$ or $\mathbb{R}$, depending on whether the process is discrete or continuous. A particular set of realisations of these random variables, $\{z_t \in \mathfrak{Z} : t \in T\}$ is called a **sample path** of the process, with z_t referred to as the **state** of the process at time t.

In this chapter, we are concerned exclusively with discrete time stochastic processes, with $T = \mathbb{N} \cup \{0\}$. We will also restrict our focus to consideration of stochastic processes for which $\mathfrak{Z}$ is a countable set.

The following definition specifies a particular family of stochastic processes:

Definition 13.1. *Given a stochastic process* $\{Z(t) : t \in T\}$ *with state space* $\mathfrak{Z}$ *and* $T = \mathbb{N} \cup \{0\}$, *we say that the process fulfils the* ***Markov property*** *if and only if, for all* $t \in T$, $y \in \mathfrak{Z}$ *and* $z_0, \ldots, z_t \in \mathfrak{Z}$:

$$\mathrm{P}[Z(t+1) = y \mid Z(0) = z_0, \ldots, Z(t) = z_t] = \mathrm{P}[Z(t+1) = y \mid Z(t) = z_t] \tag{13.2}$$

A stochastic process that fulfils the Markov property is called a **Markov process**. A discrete time Markov process is called a **Markov chain**. In this chapter, we will exclusively consider Markov chains that also fulfil the following property:

Definition 13.2. *A stochastic process* $\{Z(t) : t \in T\}$, *with state space* $\mathfrak{Z}$ *and* $T = \mathbb{N} \cup \{0\}$, *is described as* ***time homogeneous*** *if and only if, for all* $t \in T$ *and* $y_0, y_1 \in \mathfrak{Z}$:

$$\mathrm{P}[Z(t+1) = y_1 \mid Z(t) = y_0] = \mathrm{P}[Z(1) = y_1 \mid Z(0) = y_0] \tag{13.3}$$

Combining these two definitions, we see that a time-homogeneous Markov chain is a discrete time stochastic process in which, at all times, future states depend exclusively on the current state. Past states essentially have no influence on future behaviour in such processes, beyond their role in having determined the current state.

13.3 The NANIA Predator–prey Model

13.3.1 Background

Novel Approaches to Networks of Interacting Autonomes (NANIA) was a complexity science project funded by the Engineering and Physical Sciences Research Council (EPSRC), which ran from 2004 to 2009 (Marion, 2004; NANIA, 2009). One page of the NANIA website was dedicated to an interactive simulation of a predator–prey model (created by Ackland, 2009), described as "[an] autonome model[1] which simulates the behaviour of [a Lotka–Volterra] system and introduces space, with either periodic or fixed boundaries". The page has since been removed, but it can still be accessed at the Internet Archive (1996).

We refer to this model simply as the *NANIA model* (although many other models were also created as part of the NANIA project). Owing to the simplicity of its entities and interactions, we classify the NANIA model as an individual-based model (IBM) rather than an ABM, and we accordingly refer to *individuals* rather than *agents* throughout.

Although we focus specifically on the NANIA model, many predator–prey models have been proposed based on similar or identical mechanisms. Since these models are often presented in informal contexts, such as in lecture notes (Ackland, 2008; Forrest, 2014; He, 2013) or as interactive web simulations (Idkowiak and Komosinski, 2014; Sayama, 2015; Tyl, 2009), it is not possible to definitively identify any particular example as the original version. However, despite its simplicity, the fact that versions of the NANIA model may be found in such a diverse range of sources, and that it is often employed as an exemplar for the teaching and visualisation of IBMs, underlines its importance and relevance as an object of study.

The NANIA model is presented here using the ODD protocol (overview, design concepts and details), a standardised framework for the presentation of ABMs in scientific publications, created by an extensive group of authors working in the field (Grimm et al., 2006) and subsequently revised following an extensive review of its use (Grimm et al., 2010).

13.3.2 An ODD Description of the NANIA Model

13.3.2.1 Purpose

No specific aim was stated for the NANIA model in the original source (Ackland, 2009) beyond demonstrating that an 'autonome' model was capable of simulating the dynamics of the well-known Lotka–Volterra equations (see Section 13.3.3). However, it may be assumed that the model was intended to feed into the overall objectives of the NANIA project. On the project website (NANIA, 2009), the overall aim of the project is stated as:

[1] The NANIA website (NANIA, 2009) defines an *autonome* as "any interacting multi-state system which can encompass cellular automata, agent, organisms or species". Although this definition is a little ambiguous, in context, the word appears to be intended to serve as a generic term to cover all forms of interacting entity in complex models (rather than the models themselves).

> the search for overarching principles which apply in complex natural systems such as geophysical and ecological systems, and how such principles might be exploited for novel computation.

This general statement is then broken down into six "overarching scientific questions". However, it is unclear which of these, if any, the predator–prey model was intended to address.

In the context of this chapter, the purpose of the NANIA model is to serve as a minimally complex predator–prey IBM in which a simple method for the prediction of extinction events may be illustrated.

13.3.2.2 Entities, State Variables and Scales

The model involves two different types of entity: 'biological' entities (the individuals) and 'geographical' entities (comprising particular cells and the network that links them).

The individuals of the model are divided into two species, described as rabbits (the prey) and foxes (the predators), represented by the letters 'R' and 'F', respectively. The space inhabited by the individuals is a rectangular lattice of cells $L = \{C_{xy} : x, y \in \{0, \ldots, n-1\}\}$ of size $n \times n$. This lattice forms the vertex set of a graph whose edges are defined as linking either orthogonally adjacent cells (Von Neumann adjacency) or orthogonally and diagonally adjacent cells (Moore adjacency; see Figure 13.1). The boundaries of the lattice graph may be fixed, forming a square region, or periodic, forming a toroidal region.

At time t, a particular cell $C_{xy} \in L$ may be empty $C_{xy}[t] =$ 'E', inhabited by a rabbit $C_{xy}[t] =$ 'R' or inhabited by a fox $C_{xy}[t] =$ 'F'. A cell may never be occupied by more than one individual. The only variables associated with a particular cell are therefore its fixed x and y coordinates in the lattice and its current occupancy status. For a particular cell C_{xy}, the expression $N(C_{xy})$ refers to its neighbourhood[2], the set of all cells adjacent to C_{xy} in L.

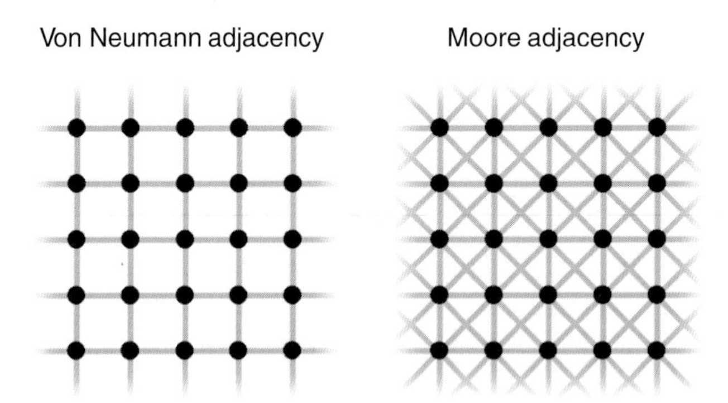

Figure 13.1 Comparison of the Von Neumann and Moore adjacency schemes in graphs over rectangular lattices. Source: Hogeweg 1988. Reproduced with permission of Elsevier

[2] More specifically, $N(C_{xy})$ is the **open neighbourhood** of C_{xy}. The **closed neighbourhood** of C_{xy}, denoted by $N[C_{xy}]$, includes the cell C_{xy} itself: $N[C_{xy}] = N(C_{xy}) \cup \{C_{xy}\}$.

The only state variable associated with an individual of a particular species is the current cell that it occupies. Individuals of a particular species are therefore indistinguishable and effectively interchangeable. Since cells may not contain more than one individual, the state of the model (and its behaviour) may be described entirely in terms of the cells and their occupancies. In this formulation of the model, the individuals are therefore disregarded as independent entities, considered to exist only implicitly through the occupancy states of the cells.

Note that, in common with many cellular automata (CAs), this model is based on a rectangular lattice of cells, each of which can take on one of a finite set of distinct states at each discrete time step. However, the model differs from typical CAs in the respect that cells are not updated simultaneously, but in randomly chosen pairs at each time step. Thus, it is more sensible to consider it as a temporally and spatially discrete IBM (even if individuals are only modelled indirectly), rather than as a CA.

Since the model is highly abstract, the spatial and temporal increments are not intended to correspond to any particular units in the real world.

13.3.2.3 Process Overview and Scheduling

Beginning at $t = 0$, time advances in discrete increments of size δt. For convenience, we suppose that $\delta t = 1$. At each time step:

1. A cell C_{xy} is randomly selected by means of a discrete uniform distribution over L;
2. One of its neighbours $C_{x'y'}$ is randomly selected according to a discrete uniform distribution over $N(C_{xy})$;
3. The occupancy states of these cells are updated according to a particular local transition rule $\mathfrak{T}_{\text{local}} : (C_{xy}[t], C_{x'y'}[t]) \rightarrow (C_{xy}[t+1], C_{x'y'}[t+1])$.

The details of the local transition rule $\mathfrak{T}_{\text{local}}$ are described in the "Submodels" Section (Section 13.3.2.7).

13.3.2.4 Design Concepts

Basic Principles:
The model is intended to represent the interaction of the populations of a prey species (called *rabbits*) and a predator species (called *foxes*) in a shared habitat, albeit in a highly abstract form. The fundamental implicit hypothesis underlying the model is that its behaviour in some sense matches that described by the Lotka–Volterra equations (see Section 13.3.3).

Emergence:
If the behaviour of the model *does* match that of the Lotka–Volterra equations, then we would expect to observe the emergence of cycles in the predator and prey populations similar to those of the equations. However, since the model also involves a spatial element, we may also expect to observe some form of emergent spatial waves in the densities of the two populations.

Interaction:
The only direct interactions between individuals in the model involve the predation of a rabbit by a fox, possibly resulting in the creation of a new fox individual. However, since cells cannot

be occupied by more than one individual, there is also *indirect* interaction, through individuals blocking space and preventing movement and reproduction.

Reproduction in the NANIA model is essentially asexual, since it does not require any interaction between individuals of the same species. The processes governing the model are discussed in detail in Section 13.3.2.7.

Stochasticity:

The model is governed almost entirely by stochastic processes. As described in Section 13.3.2.3, individuals may only 'act' at a particular time step if they are randomly selected. Reproduction and mortality are also modelled stochastically, as described in Section 13.3.2.7.

It has been argued that asynchronous stochastic updating, such as is employed in this model, may provide a good representation of the behaviour of systems of individuals in the real world (Caron-Lormier et al., 2008).

Observation:

Since the model is extremely simple, provided that the size of the lattice L is not too great, it is not computationally unreasonable to store a complete description of the system at each iteration. However, owing to the fact that only two cells are updated at each time step, for all but fairly small lattices, the model dynamics tend to evolve extremely slowly, such that meaningful temporal patterns can only be observed over many thousands or tens of thousands of iterations.

In all simulations of the model performed in this chapter, complete descriptions of the system state are recorded at fixed intervals, starting with the state at $t = 0$. However, while complete descriptions of the system are recorded, summary measures of the system state will generally be used, particularly the global population density of each species.

13.3.2.5 Initialisation

The model is initialised through assigning occupancy statuses to all cells. This may be achieved by means of any stochastic or deterministic process. The particular means of initialisation employed here will be described in Section 13.4.

13.3.2.6 Input Data

The model does not use any external data inputs during the course of a simulation.

13.3.2.7 Submodels

The local transition rule $\mathfrak{T}_{\text{local}} : (C_{xy}[t], C_{x'y'}[t]) \rightarrow (C_{xy}[t+1], C_{x'y'}[t+1])$ for the model is defined as follows:

1. If the first cell contains a rabbit and its neighbour is empty, $C_{xy}[t]$ = 'R', $C_{x'y'}[t]$ = 'E', then the rabbit reproduces with probability κ, $C_{xy}[t+1]$ = 'R', $C_{x'y'}[t+1]$ = 'R'. Otherwise there is no change, $C_{xy}[t+1]$ = 'R', $C_{x'y'}[t+1]$ = 'E'.

2. If the first cell contains a fox and its neighbour is a rabbit, $C_{xy}[t] =$ 'F', $C_{x'y'}[t] =$ 'R', then the fox eats the rabbit and reproduces with probability ς, $C_{xy}[t+1] =$ 'F', $C_{x'y'}[t+1] =$ 'F'. Otherwise the fox eats the rabbit but does not reproduce, $C_{xy}[t+1] =$ 'F', $C_{x'y'}[t+1] =$ 'E'.
3. If the first cell contains a fox and its neighbour is not a rabbit, $C_{xy}[t] =$ 'F', $C_{x'y'}[t] =$ 'F' or 'E', then the fox dies with probability τ, $C_{xy}[t+1] =$ 'E', $C_{x'y'}[t+1] = C_{x'y'}[t]$. Otherwise there is no change, $C_{xy}[t+1] =$ 'F', $C_{x'y'}[t+1] = C_{x'y'}[t]$.
4. If the first cell is empty, $C_{xy}[t] =$ 'E', then any adjacent individual moves into the cell $C_{xy}[t+1] = C_{x'y'}[t]$, $C_{x'y'}[t+1] = C_{xy}[t]$.
5. In any other case, there is no change, $C_{xy}[t+1] = C_{xy}[t]$, $C_{x'y'}[t+1] = C_{x'y'}[t]$.

Transitions that result in no change will be referred to as *null transitions*.

13.3.3 Behaviour of the NANIA Model

The NANIA model may be formulated as a stochastic process (see Section 13.2.3). This approach is in line with McKane and Newman's discussion of the closely related ecological patch models as stochastic processes (McKane and Newman, 2004, Sect. I).

Consider the NANIA model on an $n \times n$ square or toroidal grid, with either Von Neumann or Moore adjacency. Relabel the cells C_{xy}, $x, y \in \{0, \dots, n-1\}$ with a single index C_i, $i \in \{0, \dots, n^2-1\}$ by means of the transformation $i = nx + y$, and relabel the states taken by particular cells with a pair of binary variables $(r_i, f_i) \in \{0, 1\}^2$, constrained by the inequality $r_i + f_i \leq 1$, such that the states 'R', 'F' and 'E' correspond respectively to the pairs $(1, 0)$, $(0, 1)$ and $(0, 0)$.

The model may now be described as a stochastic process $\{X(t) : t \in T\}$, where $T = \mathbb{N} \cup \{0\}$ and where the $X(t)$ are discrete vector random variables taking values in the state space:

$$\Omega = \{(r_0 , \dots , r_{n^2-1} , f_0 , \dots , f_{n^2-1}) \in \{0, 1\}^{2n^2} : r_i + f_i \leq 1 , \forall i \in \{0, \dots , n^2 - 1\}\} \tag{13.4}$$

where Ω consists of the 3^{n^2} microstates representing all possible arrangements of foxes and rabbits across the lattice and is a subset of the $2n^2$-dimensional space $\mathbb{R}^{2n^2}$. If we specify a probability distribution for $X(0)$, then the family of joint probability density functions:

$$p_t(\boldsymbol{x}_0, \dots , \boldsymbol{x}_t) = \mathrm{P}[X(0) = \boldsymbol{x}_0, \dots , X(t) = \boldsymbol{x}_t] , \ t \in T , \ \boldsymbol{x}_0, \dots , \boldsymbol{x}_t \in \Omega$$

is unambiguously defined by the local transition rule $\mathfrak{T}_{\text{local}}$, which was described in Section 13.3.2. (assuming that the values of the parameters κ, ς and τ have been specified). Moreover, since $\mathfrak{T}_{\text{local}}$ determines the microstate at time $t+1$ with reference only to the microstate at time t and not to any previous microstates or to the value of t itself, we see that the NANIA model is a time-homogeneous Markov chain.

Since it is clear that every microstate in the model communicates with one or other of the two absorbing microstates (the General Extinction microstate, in which every cell is empty, and the microstate in which every cell is occupied by a rabbit), we conclude that all but these two microstates are transient: when occupying such a microstate, the probability of returning to it at some point in the future is strictly less than 1. Since the total number of possible microstates is finite, one consequence is that any initial condition will eventually necessarily evolve to one of the two absorbing states. Therefore, in any such model, extinction of the fox population is certain in the long term, leading either to extinction of both species (if the rabbit population is already extinct) or to the expansion of the rabbit population to fill the lattice.

However, while these observations are true, they are not necessarily very helpful in practice, since it may be the case that, for certain models and from certain initial microstates, the expected time required to evolve to one of the two absorbing states is so large that we can, in fact, meaningfully talk about situations of quasi-stable coexistence between the two species. This would be the case if, for example, evolving from a particular microstate to either of the absorbing microstates would require sequences of transitions that only occur with very low probability.[3]

Ackland (2009) asserts that the NANIA model has dynamics similar to those of the widely studied Lotka–Volterra equations (Lotka, 1920; Volterra, 1926; see also Murray, 1993: p 63.):

$$\dot{u} = u(a - bv)$$
$$\dot{v} = -v(c - du) \tag{13.5}$$

These equations display regular oscillations of the predator and prey population densities (u and v) about the fixed point $(u^*, v^*) = (c/d, a/b)$, but we have shown that the NANIA model necessarily evolves to one of its two absorbing states. It is therefore through this concept of quasi-stable coexistence in the NANIA model that the claims of Ackland should be understood. While it cannot truly replicate the permanent oscillatory behaviour of the Lotka–Volterra equations, over long time periods, the NANIA model can nevertheless exhibit quasi-stable behaviour that does display similar oscillatory dynamics.

13.3.4 Extinctions in the NANIA Model

Consider the following subsets of the state space Ω of the NANIA model:

$$\mathcal{E}_{\mathrm{R},0} = \{(r_1, \ldots, r_{n^2}, f_1, \ldots, f_{n^2}) \in \{0,1\}^{2n^2} : r_i = 0 \,, \forall i \in \{1, \ldots, n^2\}\} \tag{13.6}$$

$$\mathcal{E}_{\mathrm{F},0} = \{(r_1, \ldots, r_{n^2}, f_1, \ldots, f_{n^2}) \in \{0,1\}^{2n^2} : f_i = 0 \,, \forall i \in \{1, \ldots, n^2\}\} \tag{13.7}$$

$\mathcal{E}_{\mathrm{R},0}$ and $\mathcal{E}_{\mathrm{F},0}$ are respectively the sets of microstates in Ω for which the rabbits and the foxes are extinct, so $\mathcal{E}_{\mathrm{R},0} \cup \mathcal{E}_{\mathrm{F},0}$ is the set of microstates in Ω in which at least one of the two species is extinct.

Suppose that we wish to accurately predict the extinction of the fox population a certain number of time steps in advance of its occurrence, without consideration of the state of the rabbit population. Given a prediction lead time $t^* \in \mathbb{N}$, we wish to partition Ω into two subsets: a subset $\mathcal{E}_{\mathrm{F},t^*}$ in which we conclude that it is likely that the fox population will be extinct within t^* time steps and a subset $\mathcal{E}'_{\mathrm{F},t^*}$ in which we do not conclude that extinction in this time interval is likely (at this stage, we do not precisely define the word *likely*). We may then predict that extinction will occur within t^* time steps whenever the microstate of the system lies in the set

[3] A discussion of the concept of quasi-stable population cycles is provided by Pineda-Krch et al. (2007).

$\mathcal{E}_{F,t^*}$. Clearly, to predict extinctions of the rabbits, a partition of Ω into $\mathcal{E}_{R,t^*}$ and $\mathcal{E}'_{R,t^*}$ could be defined in a similar way.

We naturally expect that:

$$\mathcal{E}_{R,0} \subseteq \mathcal{E}_{R,1} \subseteq \mathcal{E}_{R,2} \subseteq \dots \tag{13.8}$$

$$\mathcal{E}_{F,0} \subseteq \mathcal{E}_{F,1} \subseteq \mathcal{E}_{F,2} \subseteq \dots \tag{13.9}$$

Also note that, following the discussion of Section 13.3.3, for sufficiently large t^*, we expect that $\mathcal{E}_{F,t^*} = \Omega$, since both absorbing states of the model (all cells equal to 'R' and all cells equal to 'E') lie in $\mathcal{E}_{F,0}$ and all other states are transient.

Rather than focussing exclusively on either foxes or rabbits, it may be of more interest to predict all extinctions. While a similar approach to that described could certainly be applied to this problem, it may be that the shape and location of the region of Ω from which imminent extinction of the rabbits is likely would be different from the shape and location of the corresponding region relating to imminent extinction of the foxes. Such situations should therefore be treated with care, since methods that can be effectively deployed to predict the extinction of a particular species may prove to be less effective if used to predict all extinctions.

13.4 Computer Simulation

13.4.1 Data Generation

Simulations of the NANIA model were performed over a 120×120 toroidal lattice $L = \{C_{xy} : x, y \in \{1, \dots, 120\}\}$. Parameters κ, ς and T were chosen such that simulations displayed quasi-stable oscillations in which extinctions were fairly rare, but not so rare as to make collecting relevant data prohibitively time-consuming.

The values chosen were:

$$\kappa = 0.02 \tag{13.10}$$

$$\varsigma = 1.00 \tag{13.11}$$

$$\mathrm{T} = 0.15 \tag{13.12}$$

Given the aim of studying destabilisation of the model dynamics, it was necessary to find a distribution of initial conditions from which the model would move quickly to the quasi-stable state, with rapid extinction of one or both species after initialisation an unlikely outcome. Following an investigation of various possible initialisations, the distribution chosen was the *Twin Ribbon Initialisation*, as described below and visualised in Figure 13.2.

Twin Ribbon Initialisation

- For all values of x, for $y \in \{1, 2, \dots, 6\} \cup \{61, 62, \dots, 66\}$, $C_{xy}[0] =$ 'F' with probability ν and 'E' otherwise (**A** in Figure 13.2).
- For all values of x, for $y \in \{7, 8, \dots, 12\} \cup \{67, 68, \dots, 72\}$, $C_{xy}[0] =$ 'F' with probability $\nu/2$, 'R' with probability $\nu/2$ and 'E' otherwise (**B** in Figure 13.2).

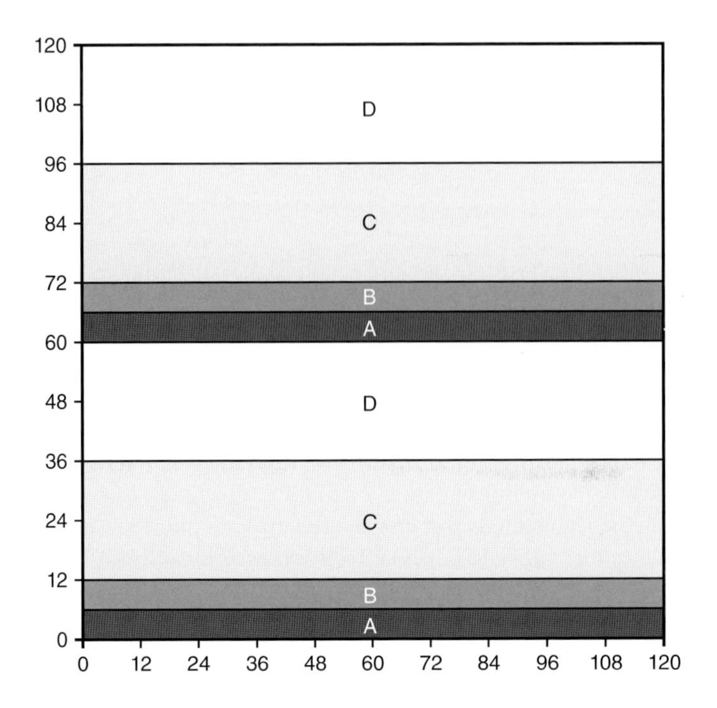

Figure 13.2 Visualisation of the Twin Ribbon Initialisation

- For all values of x, for $y \in \{13, 14, \ldots, 36\} \cup \{73, 74, \ldots, 96\}$, $C_{xy}[0] =$ 'R' with probability v and 'E' otherwise (**C** in Figure 13.2).
- For all other values of x and y, $C_{xy}[0] =$ 'E' (**D** in Figure 13.2).

The value of v was set at $v = 0.2$, following tests to determine the most suitable value to ensure that the model moved quickly to a state of quasi-stable oscillations.

Following these specifications, a simulation of 6,400,000,000 iterations of the NANIA model was performed, with the complete state of the system after iteration t given by $\omega_t \in \Omega$. These states were recorded every $\delta t = 16{,}000$ iterations, producing a sequence of 400,000 observations:

$$x_k = \omega_{k\delta t} \in \Omega \,, k \in \{0, \ldots, 399{,}999\}$$

For the sake of convenience, the word *step* is used to describe the interval between consecutive observations (δt iterations).

Each time an extinction $x_k \in \mathcal{E}_{\mathrm{R},0} \cup \mathcal{E}_{\mathrm{F},0}$ was observed for some $k \in \mathbb{N}$, the system was reinitialised using the Twin Ribbon Initialisation, such that the next iteration followed on from this reinitialisation rather than from the observed state x_k.

Also recorded was the total size of the rabbit population $x_{1,k}$ and the fox population $x_{2,k}$ for each observation x_k, creating a sequence of total population observations:

$$\hat{\boldsymbol{x}}_k = (\, x_{1,k} \,, x_{2,k} \,) \;\in\; \mathbb{N} \times \mathbb{N} \;\;,\;\; k \in \{0, \ldots, 399{,}999\}$$

In total, 934 fox extinctions and no rabbit extinctions were observed, with the final extinction occurring at observation 399,600. The mean number of steps from initialisation to extinction was therefore 428 (to the nearest whole number), corresponding to approximately 6.8 million iterations.

13.4.2 Categorisation of the Data

Based on test simulations, a decision was made to focus solely on the prediction of fox extinctions.

To study the prediction of such extinctions within a certain number of steps k^*, each observation $\boldsymbol{x}_k$ may be assigned a category $y_k \in \{-1, 1\}$, defined by:

$$y_k = \begin{cases} 1 \ , \ \text{if } x_{k+k^\dagger} \in \mathcal{E}_{\text{F},0} \text{ for some } k^\dagger \in \{0, \dots, k^*\} \\ -1 \ , \ \text{otherwise.} \end{cases}$$

In other words, all observations for which the fox population is observed to be extinct within k^* further steps are assigned the category 1, while other data points are assigned the category -1. We describe the value k^* as the **lead time** of the prediction.

It should be noted that, since the behaviour of the model was only simulated up to the 6.4 billionth iteration, it is impossible to assign categories to the final k^* observations. These observations must therefore be discarded. Thus, for a given lead time k^*, we have the following categorised observations:

$$(x_k, y_k) \ \in \ \Omega \times \{-1, 1\} \quad , \quad k \in \{0, \dots, 399{,}999 - k^*\} \tag{13.13}$$

and the corresponding categorised observations of population totals:

$$(\hat{\boldsymbol{x}}_k, y_k) \ \in \ \mathbb{N}^2 \times \{-1, 1\} \quad , \quad k \in \{0, \dots, 399{,}999 - k^*\} \tag{13.14}$$

13.5 Period Detection

In order to investigate the prediction of extinctions in this model in an informed way, it will be necessary to better understand the periodicity of the system. In Figure 13.3, which shows 501 consecutive observations of the predator and prey population densities, the noisy oscillatory character of the system dynamics is readily apparent. Across these observations, the time between successive peaks of the predator and prey populations appears to be fairly constant at roughly 20 steps (or 320,000 iterations), suggesting a period of approximately this size.

A better estimate of the period of the irregular oscillations displayed by the system may be obtained through consideration of the empirical autocorrelations of the time series of predator and prey population densities.

Since the system is reinitialised after each extinction, care must be taken when calculating autocorrelations, as it is meaningless to make calculations across such extinction events. Also, since we naturally expect the behaviour of the system to be different in the periods immediately before an extinction and immediately after an initialisation (i.e. the series does not display second-order stationarity[4] if considered from an initialisation onward), including observations

[4] Given a discrete stochastic process $(X_i)_{i \in \{0, \dots, n\}}$, second-order stationarity is the property that the joint distribution of X_i and X_{i+k} is independent of i (see Upton and Cook, 2011: pp 21, 90–93, 371).

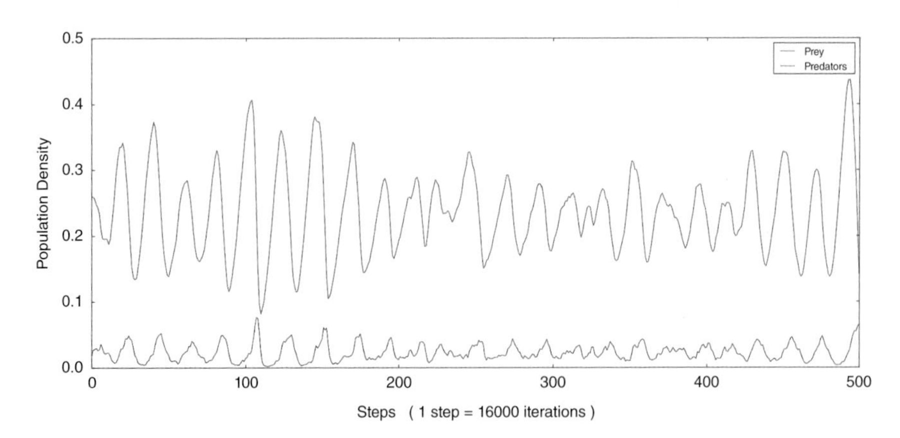

Figure 13.3 Sample dynamics of the predator and prey populations. The horizontal axis displays the number of steps since a particular chosen observation ($\hat{\boldsymbol{x}}_{936}$). This starting point was chosen since it provides a good example of the population cycles of the system, with no extinctions observed in the following 500 steps

from these periods in the calculations may reduce the strength of the autocorrelations relevant to the oscillatory dynamics. However, since extinctions are rare and since it is not clear how many observations should be removed, we disregard this issue in order to use the maximum possible amount of data in the calculations.

To calculate autocorrelations, we use a method that accounts both for the fact that the data is separated into blocks by extinctions and also, to some extent, for the fact that the behaviour of the system is expected to differ shortly before extinctions and after initialisations. For a sequence of observations $(X_i)_{i\in\{0,\ldots,n\}}$ from a (second-order) stationary time series with known mean μ and variance σ^2, the empirical autocorrelation of lag k^*, $\hat{R}(k^*)$, is defined as:

$$\hat{R}(k^*) = \frac{1}{(n-k^*)\sigma^2}\sum_{i=k^*}^{n}(X_i-\mu)(X_{i-k^*}-\mu)$$

However, while the behaviour of the predator and prey population density time series resembles that of a stationary stochastic process when displaying the irregular oscillations observed in Figure 13.3, it is not truly stationary since its behaviour is dependent on the time elapsed since the last initialisation, meaning that μ and σ^2 are not well defined.

Specifically, given that $(x_{1,i})_{i\in\{0,\ldots,n\}}$ is a sequence of observations of the prey population density (naturally, equivalent calculations could also be performed for the predator population), where $x_{1,0}$ represents an observation made one step after initialisation and $x_{1,n}$ represents the observation made one step prior to the next observed extinction (of either species), we would not expect the means and variances of the subsequences $(x_{1,i})_{i\in\{k^*,\ldots,n\}}$ and $(x_{1,i-k^*})_{i\in\{0,\ldots,n-k^*\}}$ to be the same, despite the fact that they are drawn from the same set of observations and will generally include many of the same values. This is because more of the $x_{1,i-k^*}$ are drawn from periods immediately after the initialisation, while more of the $x_{1,i}$ are drawn from periods immediately before the observed extinction, an issue that becomes increasingly acute for larger values of the lag k^*.

To get around this potential problem, we treat $(x_{1,i})_{i\in\{k^*,\dots,n\}}$ and $(x_{1,i-k^*})_{i\in\{0,\dots,n-k^*\}}$ as if they were separate sequences and find the cross-correlation of the two:

$$\frac{1}{(n-k^*)}\left(\frac{1}{\sigma_1\sigma_2}\right)\sum_{i=k^*}^{n}(x_{1,i}-\mu_1)(x_{1,i-k^*}-\mu_2)$$

where μ_1 and σ_1 are calculated from the set $\{x_{1,i}\}_{i\in\{k^*,\dots,n\}}$, and μ_2 and σ_2 are calculated from the set $\{x_{i-k^*}\}_{i\in\{0,\dots,n-k^*\}}$.

It only remains to take account of the fact that the full data set consists of many separate blocks of consecutive observations, each recording the trajectory of a particular run of the model from initialisation to extinction. To do this, we introduce the set of indices $I_{k^*} \subset \{1,\dots,400{,}000\}$, for which the two observations $x_{1,i}$ and $x_{1,i-k^*}$ with $i \in I_{k^*}$ are both drawn from the same run, with extinctions being excluded:

$$I_{k^*} = \{\ i \in \{1,\dots,400{,}000\}\ :\ \{x_{i-k^*},\dots,x_i\} \cap \mathcal{E}_{\mathrm{F},0} = \emptyset\ \}$$

and consider an adapted measure of autocorrelation $\tilde{R}(k^*)$, defined as follows:

$$\tilde{R}(k^*) = \frac{1}{(n-k^*)}\left(\frac{1}{\sigma_1\sigma_2}\right)\sum_{i\in I_{k^*}}(x_i-\mu_1)(x_{i-k^*}-\mu_2) \tag{13.15}$$

where μ_1 and σ_1 are calculated from the set $\{x_i\}_{i\in I}$, and μ_2 and σ_2 are calculated from the set $\{x_{i-k^*}\}_{i\in I}$.

We would expect that, as greater lags are considered, the autocorrelation of each series should drop off from an automatic maximum value of 1 (for the autocorrelation of lag 0), before peaking again at the lag that represents a best estimate for the period of the irregular oscillations. At lags that are multiples of this value, we would expect to observe further peaks of decreasing size, as the stochastic nature of the system causes a gradual loss of information over longer durations.

The values of $\tilde{R}(k^*)$ for $k^* \in \{0,\dots,500\}$ are presented in Figure 13.4a. Although the number of paired observations used to calculate each value of $\tilde{R}(k^*)$ necessarily decreases as k^* increases, all calculations performed to produce the plot used over 100,000 pairs, as shown in Figure 13.4b.

Figure 13.4a shows results exactly as were predicted. For both the predator and prey dynamics, the first peak of the autocorrelation occurs at the lag $k^* = 22$, close to the approximate period of 20 steps that was estimated from Figure 13.3. However, since each step represents 16,000 iterations, and since the inter-peak distance is almost uniform, with every such interval depicted in Figure 13.4a equal to between 20 and 23 steps, a more reliable estimate of the true peak of the autocorrelation may be found by finding the mean inter-peak distance across the complete figure and expressing the result in iterations.

For both species, the final peak shown in the figure, the 23rd such peak of the autocorrelation for positive lags, occurs where $k^* = 490$. This suggests an estimated period of the irregular oscillations of approximately 21.3 steps, or 340,870 iterations.

Categorising the complete set of 400,000 observations based on a prediction lead time of $k^* = 21$ (see Section 13.4.2) yields a total of 20,196 positively categorised observations (predator extinction within 21 steps), 379,783 negatively categorised observations (no predator extinction within 21 steps) and 21 observations to which categories cannot be assigned

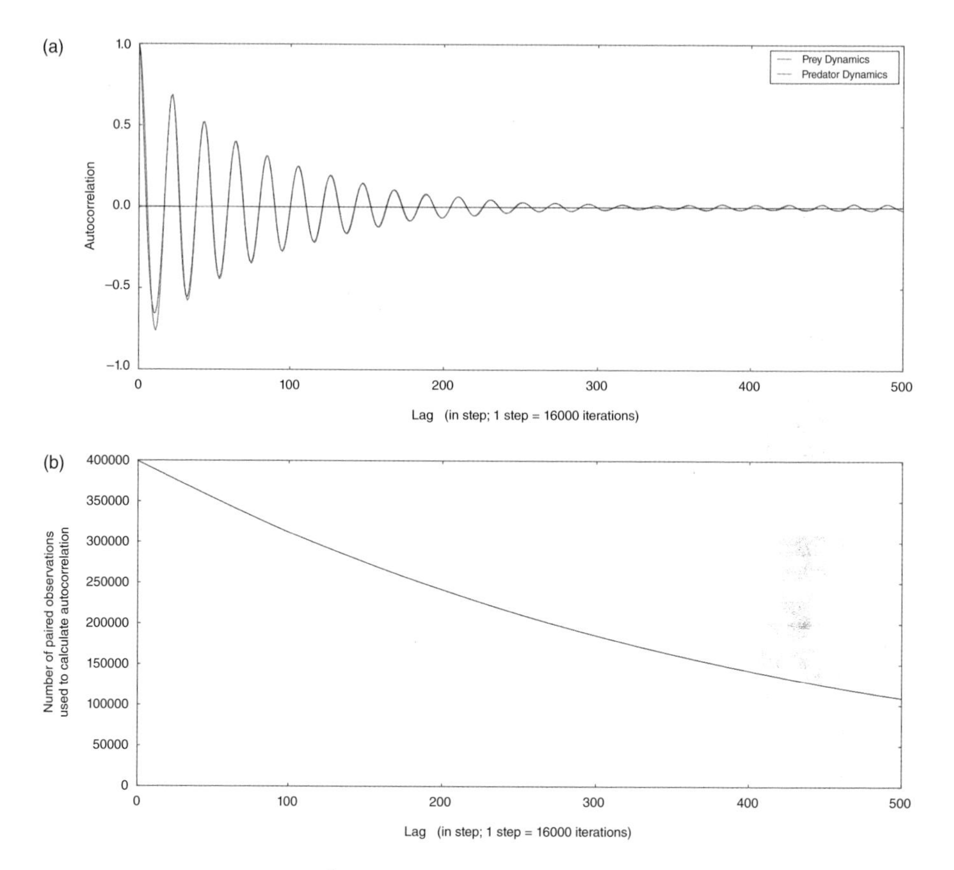

Figure 13.4 (a) Autocorrelations $\tilde{R}(k^*)$ of the predator and prey population density time series calculated from the full set of 400,000 observations. Note that the number of paired observations used in each calculation varies, because fewer high-lag autocorrelations may be calculated from a time series of given length than low-lag autocorrelations. For all lags in the range 0 to 500, the number of pairs used exceeds 100,000. (b) The number of pairs used in each of the 501 calculations

since they were collected within 21 steps of the end of the simulation. Positively categorised observations therefore make up approximately 5.05% of those that can be categorised, for a ratio of approximately 18.80 negatively categorised observations for every positively categorised observation.

13.6 A Monte Carlo Approach to Prediction

13.6.1 Binned Data

We will predict (predator) extinctions in the NANIA model by partitioning the phase space into a lattice of disjoint regions and using the observed data to calculate Monte Carlo probabilities of extinction in each region for a certain lead time k^*.

Based on the results of Section 13.5, we choose the lead time $k^* = 21$, meaning that we are considering extinctions that occur within approximately one complete period of a given observation. For ease of visualisation, we focus on the phase space of the two-dimensional data points $\hat{\boldsymbol{x}}_k$.

The complete set of 399,979 categorised observations $(\hat{\boldsymbol{x}}_k, y_k)$, $k \in \{0, \dots, 399{,}978\}$ for this lead time is shown in Figure 13.5. Those observations for which $y_k = -1$ (no extinction in the next 21 observations) are coloured green, while those observations for which $y_k = 1$ are coloured red (extinction in the next 21 observations).

Observations lying in the set $\mathcal{E}_{R,0} \cup \mathcal{E}_{F,0}$ are discarded, and the phase space $\{(x_{1,k}, x_{2,k}) \in \mathbb{N}^2 : x_{1,k}, x_{2,k} > 0\}$ of the remaining observations is partitioned into subsets S_{ij}, for $i, j \in \mathbb{N} \cup \{0\}$, called bins:

$$S_{ij} = \{(x_1, x_2) : x_1 \in \{500i + 1, \dots, 500(i + 1)\}, x_2 \in \{100j + 1, \dots, 100(j + 1)\}\} \tag{13.16}$$

Figure 13.6 shows the number of observations lying in each of the bins S_{ij}.

A Monte Carlo approach for using this partitioned data to determine the probability of imminent extinction proceeds as follows.

Consider a particular bin S_{ij} containing n_{ij} categorised observations $(\hat{\boldsymbol{x}}_k, y_k)$, of which z_{ij} have category $y_k = 1$. Suppose that, for a particular run of the NANIA model, given that the true state of the system at a particular time is $\omega \in \Omega$, a new observation $\hat{\boldsymbol{x}}$ of the system is made and found to lie in S_{ij}. Suppose that the simulation is allowed to continue, and let the random variable $Y_{ij} \in \{-1, 1\}$ represent the state of the system k^* steps after this observation, where the values 1 and -1 correspond to the events that the predator population is and is not extinct, respectively.

If we let $Y^{\dagger}_{ij} = (Y_{ij} + 1)/2$, it is natural to suppose that $Y^{\dagger}_{ij}$ follows a Bernoulli distribution with parameter $p_{ij} \in [0, 1]$, such that:

$$\mathrm{P}[Y_{ij} = -1] = 1 - p_{ij} \tag{13.17}$$

$$\mathrm{P}[Y_{ij} = 1] = p_{ij} \tag{13.18}$$

In this model, p_{ij} represents the probability that, given a new observation $\hat{\boldsymbol{x}} \in S_{ij}$ of the true state of the system $\omega \in \Omega$ at a particular time, the system will evolve to a state in which the predator population is extinct within k^* steps.

The categories y_k of the n_{ij} observations $\hat{\boldsymbol{x}}_k$ lying in S_{ij} are assumed to be independent realisations of the random variable Y_{ij}. Under this assumption, the number z_{ij} of these observations with category $y_k = 1$ can be considered to be a realisation of the random variable $Z_{ij} \sim \mathrm{Binomial}(n_{ij}, p_{ij})$. The maximum likelihood estimator (MLE) $\hat{p}_{ij}$ for the parameter p_{ij} is thus given by $\hat{p}_{ij} = z_{ij}/n_{ij}$.

These estimated probabilities of extinction $\hat{p}_{ij}$, for a lead time of $k^* = 21$, for observations lying in each of the bins S_{ij}, are summarised in Figure 13.7. Binary predictions of whether or not an extinction will occur within 21 steps could clearly be made from these probabilities simply by inspecting the estimated probability associated with the bin in which the system is currently located. However, such predictions would not take account of the uncertainty associated with these estimates.

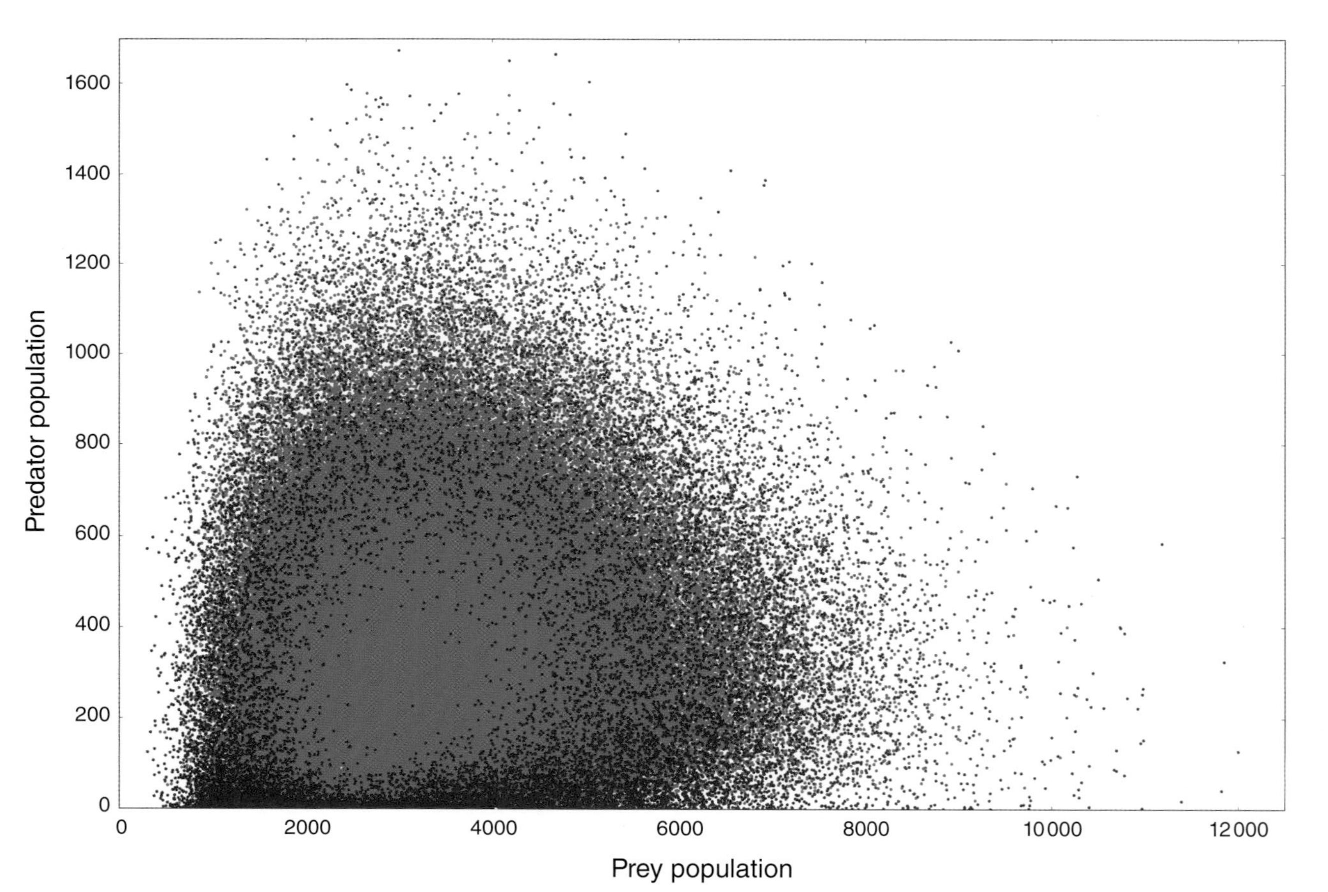

Figure 13.5 399,979 observations of the NANIA model using the specifications of Section 13.4.1. Observations are classified by extinction of the foxes with a lead time of $k^* = 21$. Observations for which the fox population was extinct within 21 steps are coloured red; otherwise, they are coloured green. Note that the red points are plotted over the green points, increasing their apparent prominence.

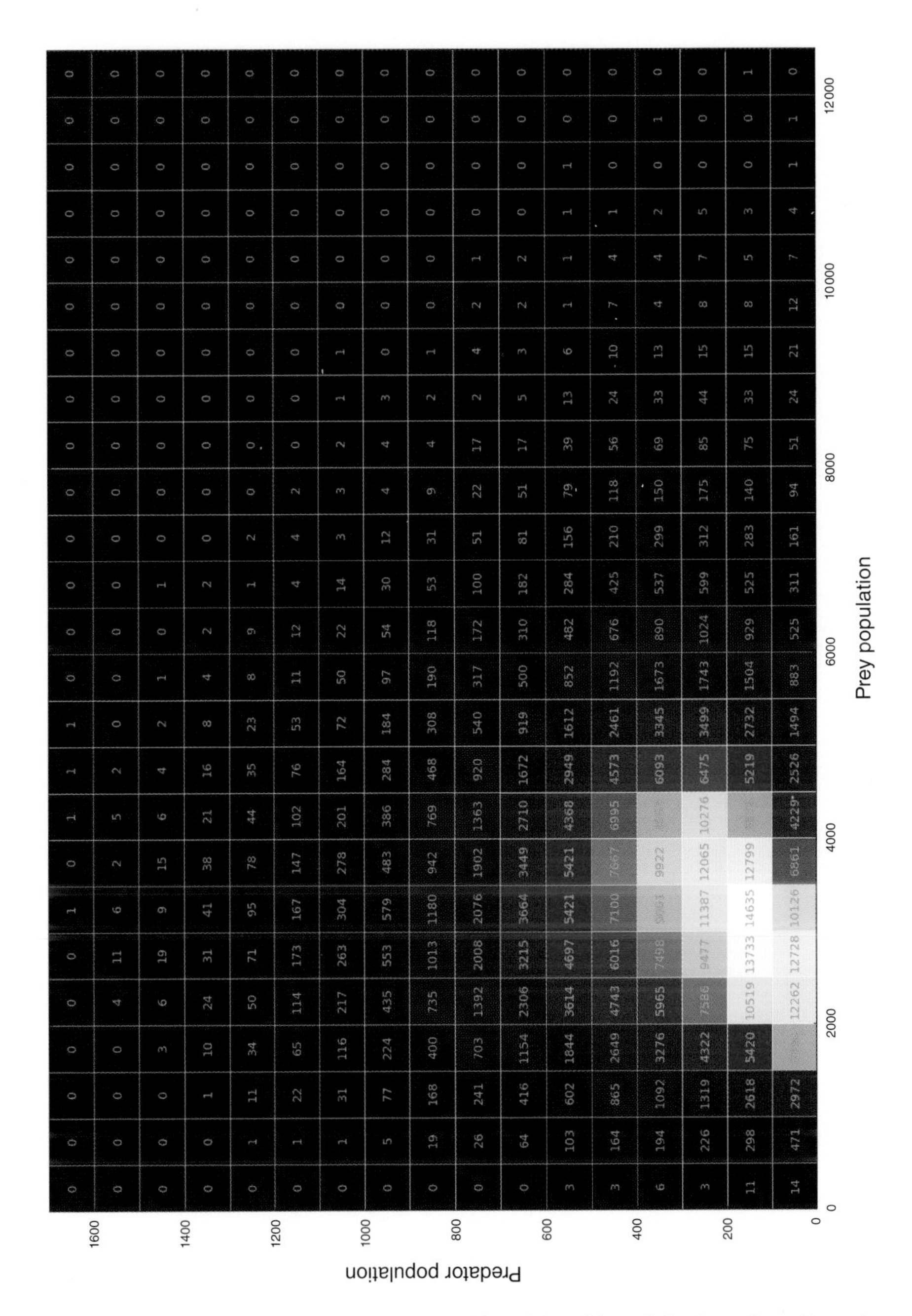

Figure 13.6 The observations of Figure 13.5 partitioned into bins of the form $S_{ij} = \{(x_1, x_2) : x_1 \in \{500i, \ldots, 500(i+1)\}\}$ for $i, j \in \mathbb{N}$. Observations for which the fox population was extinct are excluded. Brighter coloured cells indicate higher numbers of observations.

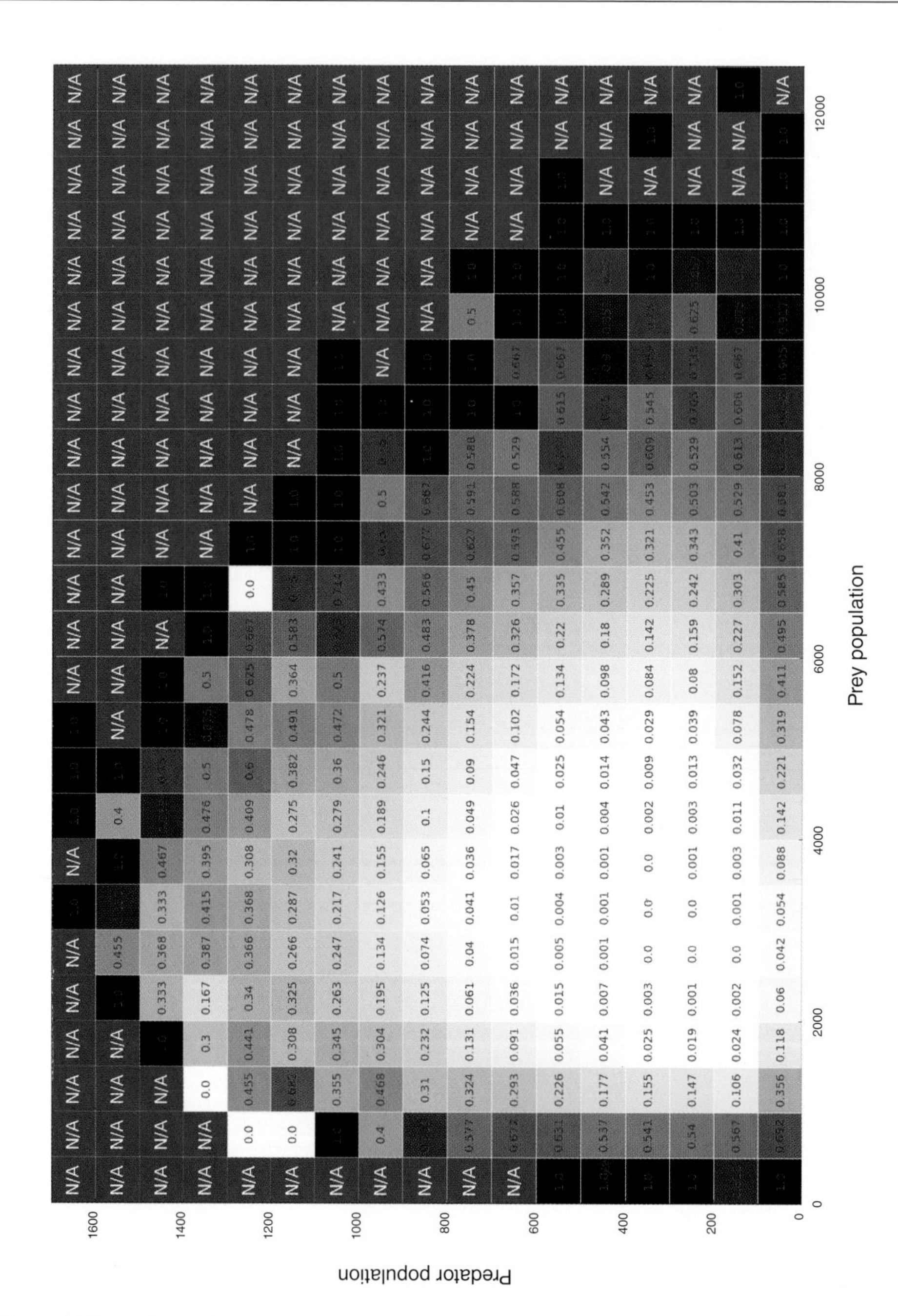

Figure 13.7 Estimated probabilities of fox extinction for observations in each of the bins of Figure 13.6 for a lead time of 21. Paler coloured cells indicate lower estimated probabilities. Cells marked N/A contain no observations, so no extinction probabilities could be estimated for these cells.

13.6.2 Confidence Intervals

To address the issue of uncertainty, confidence intervals must be determined for the estimated probabilities $\hat{p}_{ij}$. Confidence intervals are particularly important in this context, since the nature of the way in which the data were collected (observations made from simulations of the model) means that there are huge variations in the number of observations in each of the bins S_{ij} and therefore correspondingly large differences in the reliability of the MLEs $\hat{p}_{ij}$.

Given a particular realisation z of a random variable $Z \sim \text{Binomial}(n, p)$, where n is known and p is unknown, several methods exist for determining confidence intervals on the MLE $\hat{p}$ of p for a given confidence level $1 - \alpha$, for $\alpha \in [0, 1)$. The most intuitive is the Clopper–Pearson technique (Clopper and Pearson, 1934), which creates an interval that includes all values of p for which both of the probabilities $\text{P}[Z \geq z]$ and $\text{P}[Z \leq z]$ equal or exceed $\alpha/2$.

In this particular case, we have that the confidence interval associated with the bin S_{ij} is given by:

$$I_{ij,\alpha} = \{\ p : \text{P}[Z \geq z_{ij}] \geq \alpha/2 \ \text{ and } \ \text{P}[Z \leq z_{ij}] \geq \alpha/2\ , \ \text{if } \ Z \sim \text{Binomial}(n_{ij}, p)\ \} \quad (13.19)$$

The interpretation of this interval should be carefully stated, since its role may be easily misinterpreted. In particular, it cannot be stated that, given a particular realisation z_{ij} of the random variable $Z_{ij} \sim \text{Binomial}(n_{ij}, p_{ij})$, the probability that the true, unknown value p_{ij} lies in the interval $I_{ij,\alpha}$ is at least $1 - \alpha$. This is an epistemic probability, which cannot be determined without assuming some particular prior distribution for p_{ij} and applying the appropriate Bayesian methods.

It should rather be stated that, if the model of independent identically distributed Bernoulli variables is valid, then, whatever the true, unknown value of p_{ij}, the probability that an interval $I_{ij,\alpha}$, derived from a random realisation z_{ij} of Z_{ij}, will contain p_{ij} is at least $1 - \alpha$.

The sizes $[\sup(I_{ij,\alpha}) - \inf(I_{ij,\alpha})]$ of these confidence intervals, for a confidence level of 95% ($\alpha = 0.05$), are summarised in Figure 13.8. It should be noted that, owing to computational limitations, the method used to calculate the boundaries of these intervals was only accurate to the nearest 0.001. The bounds calculated were not, therefore, the precise greatest lower bound and the least upper bound of each interval, meaning that the sizes given in Figure 13.8 may each be overestimated by up to 0.002. For similar reasons, 0.001 is the smallest size of interval that could be calculated, and the true size of the confidence intervals shown as having this size in Figure 13.8 may in fact be much smaller.

13.6.3 Predicting Extinctions using Binned Population Data

The upper bounds calculated for each confidence interval $I_{ij,\alpha}$ are summarised in Figure 13.9. These values should be interpreted as follows. If the true probability of extinction within 21 steps for a single randomly generated observation lying in the corresponding bin were equal to or greater than the stated value, the probability of observing the number of extinctions (within 21 steps) that were indeed observed, or fewer, given the total number of observations in the bin, would be less than 0.025 (half the 5% significance level). For probabilities equal to or less than the stated value minus 0.001, the opposite is true, and this probability would be greater than or equal to 0.025. In other words, true probabilities of extinction (within 21 steps) greater than or equal to the values seen would not be consistent with the observed data.

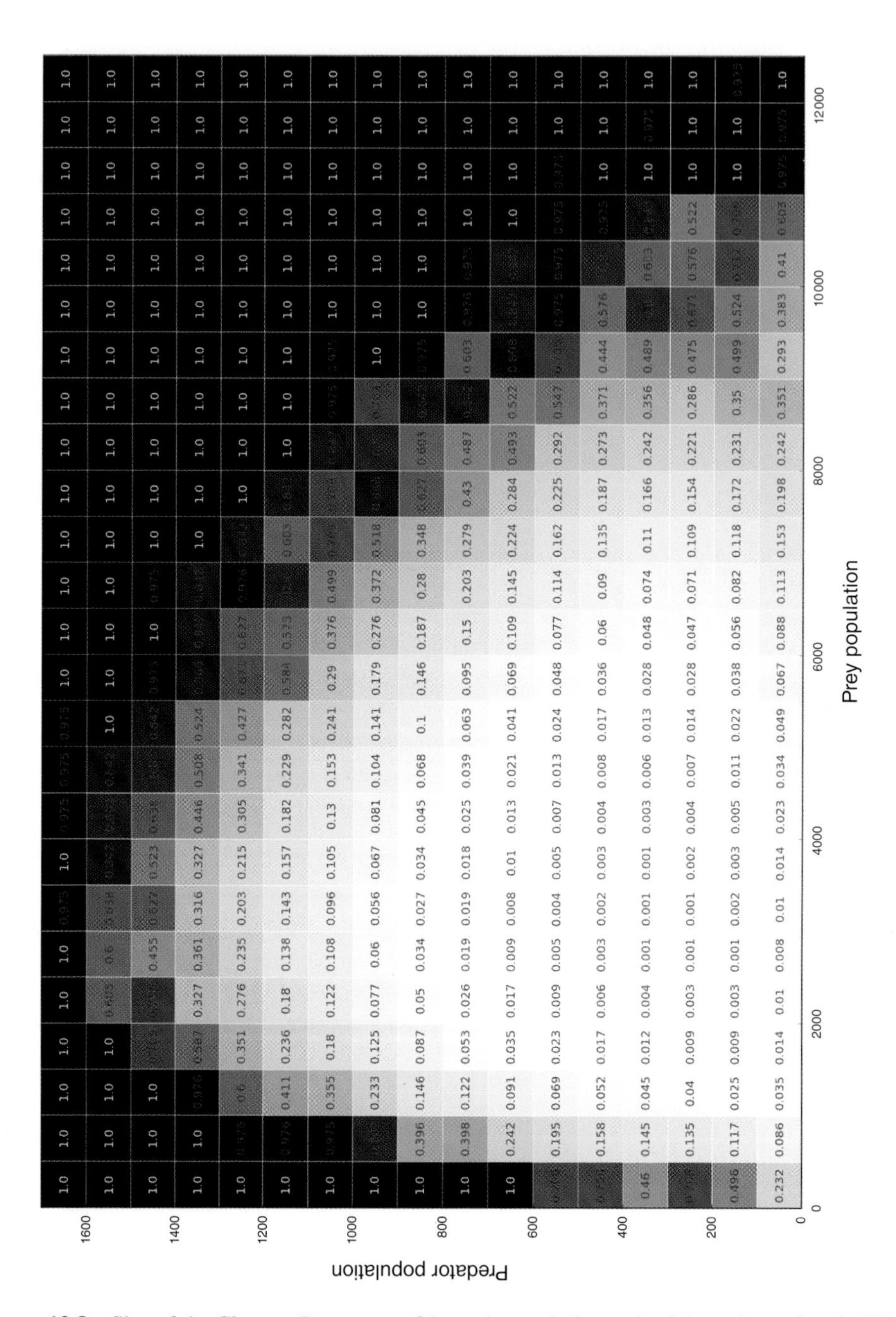

Figure 13.8 Size of the Clopper–Pearson confidence intervals for each of the estimated probabilities of Figure 13.7 for a confidence level of 95%. Paler coloured cells indicate smaller intervals and therefore less uncertainty over the true extinction probability for observations in that cell. Cells in which no observations were recorded necessarily have a confidence interval of size 1. Such cells are distinguished by their white text.

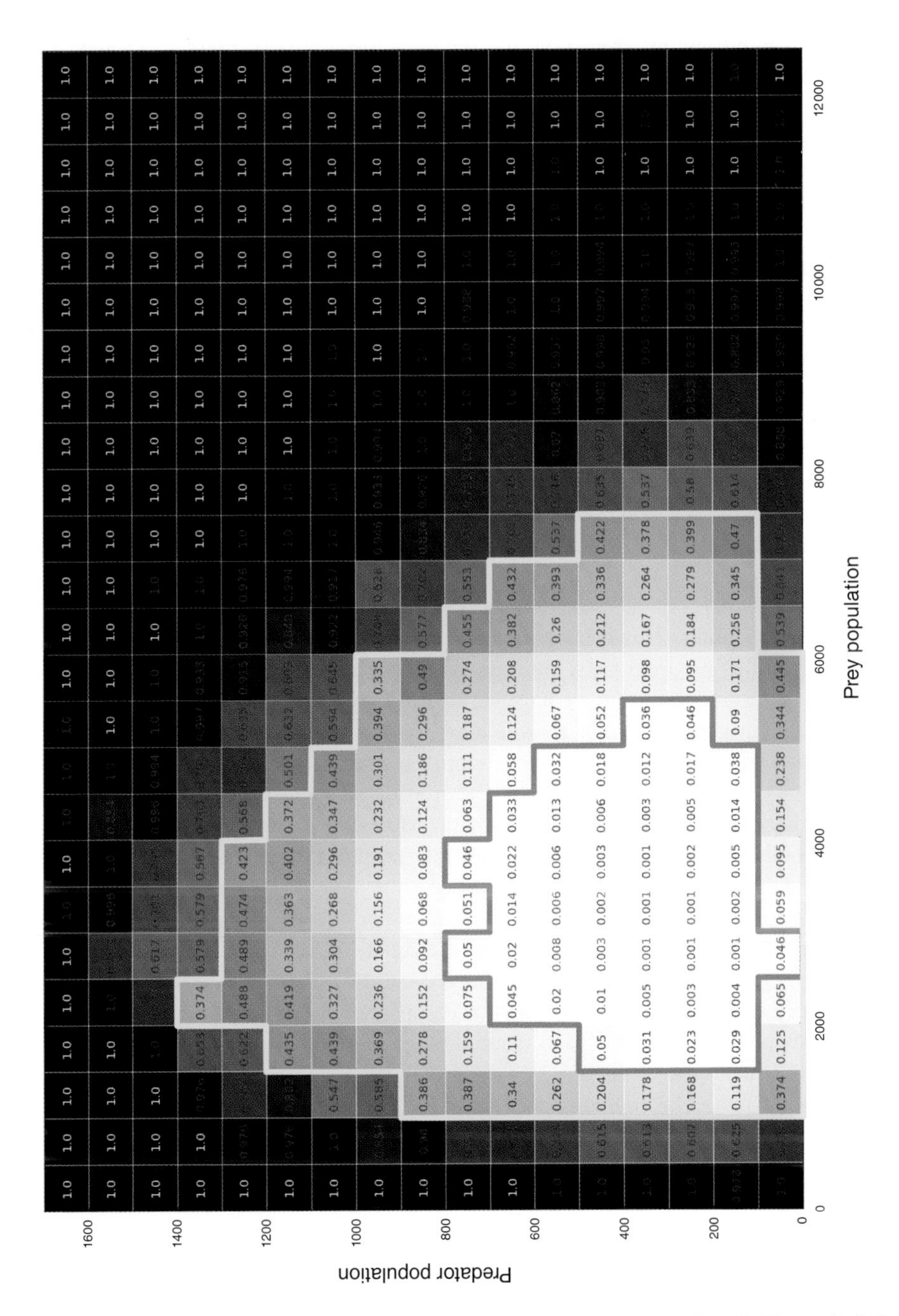

Figure 13.9 Upper bounds of the Clopper–Pearson confidence intervals visualised in Figure 13.8. Paler coloured cells indicate lower upper bounds on the estimated probability. Cells in which no observations were recorded necessarily have an upper bound of 1. Such cells are distinguished by their white text. The area enclosed by the orange (yellow) boundary indicates the region of phase space in which we may be 'confident' that the probability of extinction within 21 steps does not exceed 5% (50%).

Bearing in mind this interpretation, the inner enclosed area of Figure 13.9 indicates the region of phase space in which we may be confident that the probability of extinction (of the predators) within 21 steps (approximately one period of the irregular oscillation displayed by the model, by the results of Section 13.5) is less than 5%, in the sense that probabilities of 5% or greater would not be consistent with the observations made in this region. The outer enclosed area indicates the corresponding region for probabilities of 50% or greater. Obviously, the accuracy with which the boundaries of these regions are calculated is limited by the way in which the phase space has been partitioned into bins.

The identification of this 'safe' region (the inner area) offers a crude means of preventing extinctions of the predator population. This method would involve close monitoring of the sizes of the predator and prey populations and noting the moment at which the system leaves the region. At this point, predator or prey agents would be selectively introduced or removed to artificially return the system to the region.

However, it is worth remembering that all of the results given here do not refer to microstates selected by means of a random uniform distribution across the entire state space of the model Ω, but rather randomly selected according to the expected time that the system spends in each microstate given an infinite number of repetitions of the experimental runs described in Section 13.4.1. This means that an artificial manipulation of the system, as described here, would not necessarily be successful without a more thorough understanding of the system dynamics. This is because, if the microstate occupied by the system following an intervention lies in a region of the state space rarely visited by the natural dynamics of the system, the experiments that we have performed may give us no useful information about how the system might be expected to evolve.

One of the clearest problems with this approach, as it stands, is that the assumption that the categories y_k of observations in S_{ij} can be considered to be independent realisations of a Bernoulli random variable is clearly problematic. Most obviously, if consecutive observations $\hat{\boldsymbol{x}}_k$ and $\hat{\boldsymbol{x}}_{k+1}$ both lie in S_{ij}, then there will be a strong effective correlation between their respective categories y_k and y_{k+1}. This problem can be alleviated to some extent through a careful choice of the partition of the phase space into bins S_{ij}, or through sampling of the data at sufficiently large intervals, but long-range temporal correlations between category and position in the phase space cannot be ruled out.

In any case, Figures 13.5–13.9 clearly demonstrate the existence of a region of the predator–prey population phase space, centred roughly around the point (3000,300), in which we can be confident that extinction of the predator population within 21 steps is unlikely. Figure 13.6, in particular, demonstrates that this region very roughly coincides with the area of the phase space in which the system spends the most time, under the experimental conditions described in Section 13.4.1.

13.6.4 ROC and Precision-recall Curves for Monte Carlo Prediction of Predator Extinctions

In this section, we construct collections of predictors of varying sensitivity from the binned data discussed in Section 13.6. The predictors will be based largely on the upper bounds of the Clopper–Pearson confidence intervals on the probabilities of extinction within 21 steps for each of the bins, as visualised in Figure 13.9. The approach is similar to that described in Section 13.6.3.

Given a particular threshold probability $\bar{p}$, we define a classifier $f_{\bar{p}}$ from the binned training data in the following way.

Given a previously unseen observation $\hat{\boldsymbol{x}}$, we first identify the bin S_{ij} in which it lies. Associated with this bin is the Clopper–Pearson confidence interval $I_{ij,\alpha}$, where $1-\alpha$ is the confidence level, as described in Section 13.6. If the upper bound of the confidence interval $\sup(I_{ij,\alpha})$ is greater than the threshold probability $\bar{p}$, then we cannot be confident (at a significance level of $\alpha/2$, since this is effectively a one-tailed rather than two-tailed test) that the probability of extinction within 21 steps is less than or equal to $\bar{p}$. We therefore predict an extinction: $f_{\bar{p}}\ (\hat{\boldsymbol{x}}) = 1$. Otherwise, we can be confident that the probability of extinction within 21 steps is less than or equal to $\bar{p}$, and we do not predict an extinction: $f_{\bar{p}}\ (\hat{\boldsymbol{x}}) = -1$.

In summary:

$$f_{\bar{p}}\ (\hat{\boldsymbol{x}}) = \begin{cases} 1\ , & \text{if } \sup\left(I_{ij,\alpha}\right) > \bar{p}\ , \\ -1\ , & \text{if } \sup(I_{ij,\alpha}) \leq \bar{p}\ . \end{cases} \tag{13.20}$$

where $i, j \in \mathbb{N}$ are chosen such that $S_{ij} \in \hat{\boldsymbol{x}}$.

Using this method, the classifier $f_{\bar{p}}$ is effectively being defined by placing an upper bound of $1-\bar{p}$ on its negative predictive value (see Section 13.2.2). Furthermore, the true negative predictive value of $f_{\bar{p}}$ will generally be considerably higher than $1-\bar{p}$, since the upper bound of the confidence interval $I_{ij,\alpha}$ is an extremely conservative measure of the probability of an extinction occurring within the specified period (21 steps in this case).

By allowing the value of $\bar{p}$ to vary between 0 and 1, we may therefore create a family of classifiers from a set of binned training data, for which the confidence required to make a negative prediction decreases with increasing $\bar{p}$. Naturally, we may also vary the size of the bins in each dimension, to alter the level of resolution of the predictor over the predator–prey population space.

The performances of three such families of predictors on the test data are plotted as ROC curves and precision-recall curves in Figures 13.10 and 13.11. Each of the families is derived from the training data with different sizes of bins. ‘Large’ bins are those visualised in Figure 13.6, while ‘medium’ and ‘small’ bins are a quarter and a sixteenth of the size, respectively, created by successively halving each dimension of the original bins:

Large bins:

$$S_{ij} = \{(x_1, x_2) : x_1 \in \{500i+1, \ldots, 500(i+1)\}, x_2 \in \{100j+1, \ldots, 100(j+1)\}\}$$

Medium bins:

$$S_{ij} = \{(x_1, x_2) : x_1 \in \{250i+1, \ldots, 250(i+1)\}, x_2 \in \{50j+1, \ldots, 50(j+1)\}\}$$

Small bins:

$$S_{ij} = \{(x_1, x_2) : x_1 \in \{125i+1, \ldots, 125(i+1)\}, x_2 \in \{25j+1, \ldots, 25(j+1)\}\}$$

Note that the true positive rate of a particular classifier $f_{\bar{p}}$ decreases as $\bar{p}$ increases, since lower values of $\bar{p}$ imply greater caution of the classifier in terms of the avoidance of false negatives. Bearing this in mind, we may observe from Figure 13.11 that for higher values of $\bar{p}$, using smaller bins results in improved classifier performance, while for lower values of $\bar{p}$, larger bins seem to be associated with improved performance.

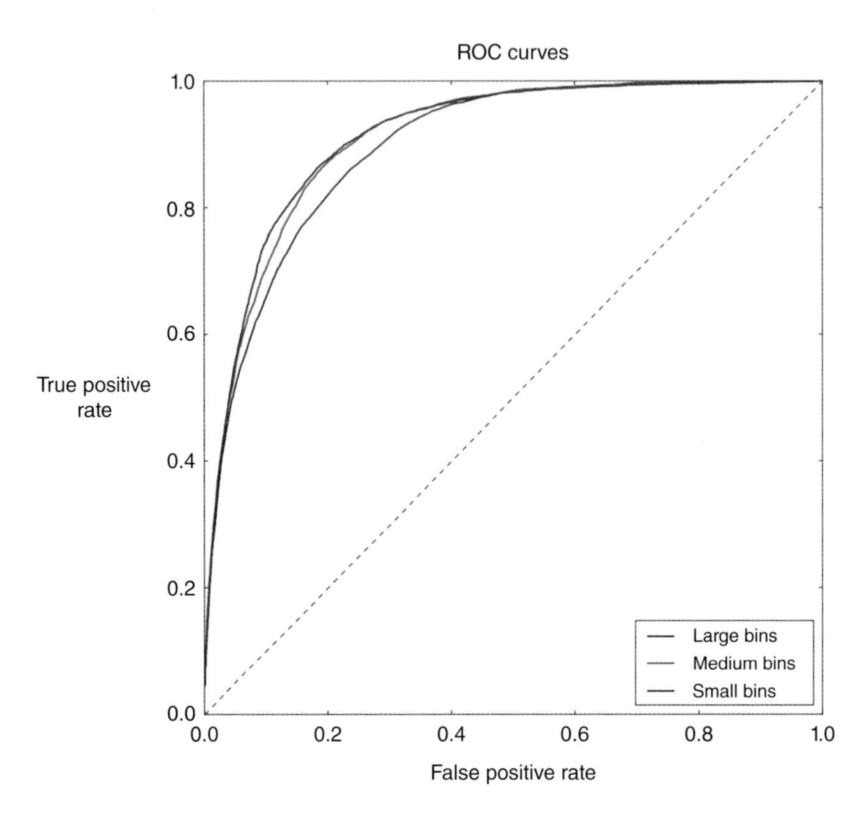

Figure 13.10 ROC curves for three families of classifiers $f_{\bar{p}}$, generated from binned training data, where $\bar{p}$ varies over [0, 1]. Each family of classifiers is generated using bins of different sizes, and performances are measured on the unseen test data. The dotted line represents performances that could be achieved by classifiers making random predictions according to a Bernoulli variable, independently of the data.

This result is to be expected, since small values of $\bar{p}$ imply that a high level of certainty is required to make a negative classification and, consequently, that the decision boundary lies closer to the centre of the region where the density of negatively categorised observations, and indeed of all observations, is highest. The training data would therefore be expected to contain good information on this region for the creation of a fine-scale classifier using small bins.

However, as $\bar{p}$ increases, the decision boundary is pushed further and further out from this high-density region as the classifier becomes increasingly willing to make negative predictions and less concerned with the risk of false negatives. Using small bins in a region of the predator–prey space where observations are sparse means that each bin contains fewer observations, and therefore that the confidence intervals associated with each bin are wider. The information available in this region may not, therefore, be sufficient to produce a good classifier at a fine resolution, and we might thus expect the large-bin predictors to provide better performance.

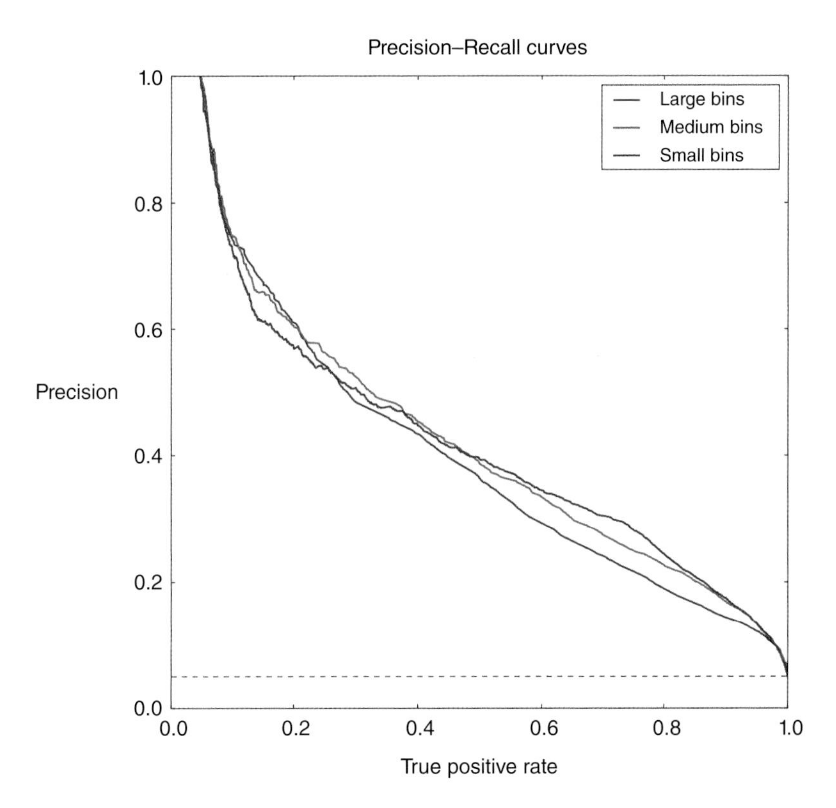

Figure 13.11 Precision-recall curves for three families of classifiers $f_{\bar{p}}$, generated from binned training data, where $\bar{p}$ varies over [0, 1]. Each family of classifiers is generated using bins of different size, and performances are measured on the unseen test data. Note that the term *recall* is synonymous with *true positive rate*. The dotted line at $y = 0.0502$ (the proportion of positively categorised points in the data) represents performances that could be achieved by classifiers making random predictions according to a Bernoulli variable, independently of the data.

13.7 Conclusions

In this chapter, we have described a simple method for the prediction of extinction events in an individual-based model, based on partitioning the phase space into bins and supposing that each bin is associated with a separate independent Bernoulli distributed random variable, which determines whether a microstate of the system that is observed to lie in that bin will evolve to extinction (of the predator species) within a specified time period.

It is clear that the method could be applied to many other types of model to predict a diverse range of events, going far beyond issues of extinction and ecology. The prediction of any outcome that can be unambiguously measured and observed could be attempted in this way. However, the success of such an approach would naturally depend on the nature of the relationship between the descriptors of the system that are tracked and the outcome of interest. In the case considered in this chapter, it was possible to have a reasonable degree of predictive success owing to the clear link between the population densities of the predator and prey species and the occurrence of extinctions of the predator species.

Only two descriptive variables were considered when attempting to predict extinction events: the population densities of the two species that feature in the NANIA model. However, aside from the advantages of visualisation and clarity offered by the restriction to two dimensions, there is no reason why additional descriptive variables could not have been used in an attempt to improve performance, necessitating the use of multidimensional bins. Indeed, there is good reason to believe that the inclusion of variables describing the relative spatial distributions of the two species may well have had a beneficial effect on predictive performance in this case, since global predation efficiency is clearly influenced by the local proximity of prey and predator individuals across the system.

The phase space binning technique that we have considered in this chapter is clearly a fairly simple approach to the problem of event prediction in ABMs. In Chapter 14, a more sophisticated approach will be taken, drawing on mathematical techniques used in the field of image recognition to produce improved classifiers for the prediction of extinctions in the NANIA model.

References

Ackland GJ. (2008). Lotka-Volterra: the classic predator-prey model. Course notes, Mathematical and Spatial Modelling of Ecology, School of Physics, University of Edinburgh. Available at http://www2.ph.ed.ac.uk/~gja/MEID_NANIA.pdf

Ackland GJ (uncredited). (2009). Lotka-Volterra dynamics. NANIA Project. Available at https://web.archive.org/web/20120722125154/http://www2.ph.ed.ac.uk/nania/lv/lv.html

Caron-Lormier G, Humphry RW, Bohan DA, Hawes C and Thorbek P. (2008). Asynchronous and synchronous updating in individual-based models. *Ecological Modelling* 212 (3–4): 522–527.

Clopper CJ, Pearson ES. (1934). The use of confidence or fiducial limits illustrated in the case of the binomial. *Biometrika* 26(4): 404–413.

Cotar C. (2012). Stochastic Processes. Lecture course, London Taught Course Centre (LTCC).

Fawcett T. (2006). An introduction to ROC analysis. *Pattern Recognition Letters* 27(8): 861–874.

Forrest S. (2014). Predator-prey models. Course notes, Introduction to Scientific Modeling, Department of Computer Science, University of New Mexico. Available at https://www.cs.unm.edu/~forrest/classes/cs365/lectures/Lotka-Volterra.pdf

Gretton A. (2013). Adaptive modelling of complex data: kernels. Available at http://www.gatsby.ucl.ac.uk/~gretton/adaptModel

Grimm V, Berger U, Bastiansen F, Eliassen S, Ginot V, Giske J, *et al.* (2006). A standard protocol for describing individual-based and agent-based models. *Ecological Modelling* 198(1–2): 115–126.

Grimm V, Berger U, DeAngelis DL, Polhill G, Giske J, Railsback SF. (2010). The ODD protocol: a review and first update. *Ecological Modelling* 221(23): 2760–2768.

Grimmet G and Stirzaker D. (2001). *Probability and Random Processes*, 3rd ed. Oxford: Oxford University Press.

He S. (2013). Modelling with cellular automata. Course notes, Computational Modelling with MATLAB, School for Computational Science, University of Birmingham. Available at http://www.cs.bham.ac.uk/~szh/teaching/2013backup/matlabmodeling/Lecture12_body_CellularAutomata.pdf

Hogeweg P. (1988). Cellular automata as a paradigm for ecological modeling. *Bioinformatica* 27(1): 81–100.

Idkowiak L and Komosinski M. (2014). Predators and prey: the Lotka-Volterra model. Life Modeling Simulation. Available at http://en.alife.pl/predators-and-prey-the-Lotka-Volterra-model

Internet Archive (1996). Available at https://archive.org

Lotka AJ. (1920). Undamped oscillations derived from the law of mass action. *Journal of the American Chemical Society* 42(8): 1595–1599.

Marion G (principal investigator). (2004). Novel Approaches to Networks of Interacting Autonomes (EPSRC Grant GR/T11777/01): application summary. Available at http://gow.epsrc.ac.uk/NGBOViewGrant.aspx?GrantRef=GR/T11777/01

McKane AJ and Newman TJ. (2004). Stochastic models in population biology and their deterministic analogs. *Physical Review E* 70 (4): 041902.

Murray JD. (1993). *Mathematical Biology*, 2nd ed., vol. 19 of *Biomathematics Texts*. Berlin: Springer.

NANIA. (2009). Novel Approaches to Networks of Interacting Autonomes (EPSRC Grant GR/T11777/01): project website. Available at http://www2.ph.ed.ac.uk/nania

Pineda-Krch M, Blok HJ, Dieckmann U and Doebeli M. (2007). A tale of two cycles – distinguishing quasi-cycles and limit cycles in finite predator-prey populations. *Oikos* 116(1): 53–64.

Ross SM. (1996). *Stochastic Processes*, 2nd ed. Wiley Series in Probability and Mathematical Statistics. New York: John Wiley & Sons.

Sayama H. (2015). Predator-prey ecosystem: a real-time agent-based simulation. Wolfram Demonstrations Project. Available at http://demonstrations.wolfram.com/PredatorPreyEcosystemARealTimeAgentBasedSimulation/

Smith LA. (2001). Nonlinear dynamics and statistics. In *Disentangling Uncertainty and Error: On the Predictability of Nonlinear Systems*. Boston: Birkhäuser, pp. 31–64.

Steinwart I and Christmann, A. (2008). *Support Vector Machines: Information Science and Statistics*. New York: Springer.

Stirzaker D. (2005). *Stochastic Processes and Models*. Oxford: Oxford University Press.

Stroock DW. (2014). *An Introduction to Markov Processes*, 2nd ed. Graduate Texts in Mathematics no. 230. Berlin: Springer.

Tyl TY (uncredited). (2009). Lotka Algorithmic Simulation. JSeed Project. Available at http://jseedsourceforge.net/lotka/

Upton G and Cook I. (2011). *Dictionary of Statistics*, 2nd ed. Oxford Paperback References. Oxford: Oxford University Press.

Volterra V. (1926). Variazioni e fluttuazioni del numero d'individui in specie animali conviventi [in Italian]. *Atti della Reale Accademia Nazionale dei Lincei* 2(3): 31–112. (English translation: Volterra, 1931).

Part Eight
Diffusion Models

14

Urban Agglomeration Through the Diffusion of Investment Impacts

Minette D'Lima, Francesca R. Medda and Alan G. Wilson

14.1 Introduction

Disentangling the reasons behind urban agglomeration is crucial to understanding how a city is the product of competition and collaboration between various urban agencies. Agglomeration can arise from investment in utilities, clustering of firms and proximity of labour and knowledge (Drennan and Kelly, 2011; Rabianski and Gibler, 2007; Rosenthal and Strange, 2004; Storper and Venables, 2004). The implication of this evidence is that the formation of urban agglomeration can be modelled as the creation of peaks of population concentration which are triggered by these different factors. Against this background, we use as the point of departure the model developed by Medda *et al.* (2009): urban concentration patterns are created by the diffusion of rent and transport costs in a continuous urban space. This urban economic model is in turn based on the Turing morphogenetic reaction–diffusion model and incorporates the diffusion of information, for example transport investment, along with its feedbacks. The model demonstrates the emergence of spatial patterns, such as urban agglomeration, as a result of external investment.

In the Medda *et al.* model, the city is assumed to be circular and divided into discrete zones with an additional central business district (CBD). The assumptions on the costs incurred by a household in a given zone are: every month, each household must pay rent P_R and transport cost P_T. Each cost is made up of a fixed term and a variable term. The fixed term in each case is a function of the characteristics of the urban district and the characteristics of the transport system, respectively. The variable terms are a direct function of the number of households in the given zone: the rent term increases with the increase in households, while the transport term decreases with the increase in households. The income of each household is assumed to be constant and equal to Y. Initially, costs and number of households are assumed to be equal in all

Approaches to Geo-mathematical Modelling: New Tools for Complexity Science, First Edition. Edited by Alan G. Wilson.

Companion Website: www.wiley.com/go/wilson/ApproachestoGeo-mathematicalModelling

zones. At time $t = t_1$, in a particular zone x_i, a transport investment reduces the fixed transport cost. This disequilibrium triggers migration. Households from neighbouring zones choose to migrate to where investment has taken place and to benefit from a lower fixed transport cost. This movement has multiple effects, including rent increases where the population grows, and after a certain number of iterations, the model reaches equilibrium. This demonstrates that investment in a particular urban zone can indeed affect urban zones that are distant from the initial investment, and can produce agglomeration effects. However, although the model shows that urban agglomeration can occur, it does not address the conditions necessary for agglomeration.

Our objective therefore is to identify these conditions, and in so doing we develop a discrete model to define the variables and its associated parameters which determine urban agglomeration. The urban assumptions are similar, but by studying the process from a discrete point of view, they allow us to characterise the properties of urban agglomeration. In Section 14.2, we define the model. Thereafter, we present the mathematical interpretation and simulation in Sections 14.3 and 14.4, respectively, and conclude in Section 14.5.

14.2 The Model

The model is discrete and represents urban investment, providing the means to study how this investment can foster agglomeration through diffusion across the urban network. We assume a circular monocentric city with a central business district (CBD). The zones are distributed discretely in radial symmetry. The distance to travel to work (to the CBD) is equal for all zones, and will not influence the decision to migrate to a neighbouring zone. We hold household incomes $\{Y_i\}$ to be constant and equal to Y in all zones. The number of households is constant. There are only two costs incurred (per month) by every household: rent cost and transport cost. Rent cost $\{r_i\}$ and transport cost $\{c_i\}$ are a function of the population in the zone. In fact, the economic dynamics in the city between the nodes follow the same behaviour as in the continuous model: the rent cost increases and the transport cost decreases as population increases. We denote the population in zone i by P_i, and model rent and transport costs using exponential functions:

$$r_i = r_i(P_i) = r\exp(\rho P_i)$$

and

$$c_i = c_i(P_i) = c\exp(-\tau P_i),$$

where r and c depend on the initial values of rent and transport costs, and $\rho > 0$ and $\tau > 0$ are the rates of growth of rent and decay of transport costs, respectively. This ensures that $\frac{\partial r_i}{\partial P_i} > 0$ and $\frac{\partial c_i}{\partial P_i} < 0$. Note that r and c are not the initial values of the rent and transport costs at the start of the experiment, but instead they are the values when the population is zero.

$$\text{Initial rent: } r_0 = r\exp(\rho P_0)$$

$$\text{Initial transport: } c_0 = c\exp(-\tau P_0)$$

For the dynamics, we will consider discrete intervals $t,\ t + \Delta t,\ t + 2\Delta t$, and we denote the values of population, rent cost and transport cost in zone i at time t by $P_i(t)$, $r_i(t)$ and $c_i(t)$, respectively. The household utility in zone i at time t is defined as:

$$U_i(t) = Y_i - r_i(t) - c_i(t) = \mathrm{Y} - r_i(t) - c_i(t).$$

We assume that the changes in rent and transport costs in zone j as perceived in zone I, diffuse in urban space. We incorporate this property into the model by using rent time lag terms $\{\theta_{ij}^R\}$ and transport time lag terms $\{\theta_{ij}^T\}$. These lags increase as the distance between zones increases. We also assume that rent cost diffuses at a faster rate than that of transport cost, or equivalently, the rent time lag is less than the transport time lag, and we model these lags using the following exponential functions:

$$\theta_{ij}^R = \theta_{ij}^R(d_{ij}) = a\exp(\gamma d_{ij})$$

and

$$\theta_{ij}^T = \theta_{ij}^T(d_{ij}) = b\exp(kd_{ij}),$$

where d_{ij} is the distance between i and j; $\gamma > 0,\ k > 0$ are constant; and $\gamma < k$. At time t, the utility of the household that resides in zone j is perceived by a resident in zone i as follows:

$$U_{ji}(t) = Y - r_j(t - \theta_{ij}^R) - c_j(t - \theta_{ij}^T)$$

which introduces utility perception into our model. Given the possible increase in utility in zone j, the number of people who migrate from zone i to zone j in the time interval $(t,\ t\ + \Delta t)$ is denoted by $\{N_{ij}(t,\ t\ + \Delta t)\}$. These household flows are generated by the perceived utility of moving to a neighbouring zone. At time t, we fix a zone i and model the flows out of i to zones $\{j\}$ in the time step $(t, t\ + \Delta t)$. Let $\{m_{ij}\}$ be the transaction cost of moving from zone i to zone j. This cost increases as the distance between i and j increases,

$$m_{ij} = \exp(\mu d_{ij}),$$

where μ is a constant. The perceived benefit of moving from I to j is

$$b_{ij}(t) = \max(U_{ji}(t) - U_i(\mathrm{t}) - m_{ij},\ 0).$$

The flows from i to j are then modelled depending on the current population in i, and the relative attractiveness of the neighbouring zones, in the following manner:

$$N_{ij}(t) = \varepsilon P_i(t) a_i(t) b_{ij}(t),$$

where ε is a constant and represents the propensity of the current population in zone i that chooses to migrate, and
$a_i(t) = \frac{1}{\sum_k b_k(t)}$ represents the competing zones k as a migration choice.

The resulting change in population in zone i is given by the net difference between the inflows and outflows,

$$\Delta P_i(t;\ t\ + \Delta t) = \sum\nolimits_j (N_{ji} - N_{ij})$$

$$P_i(t + \Delta t) = P_i(t) + \Delta P_i(t,\ t + \Delta t).$$

We can then test the dynamic behaviour of households as follows: we introduce a small perturbation in the network by decreasing the transport cost as a result of an investment in an arbitrary zone, thereby increasing the utility of that zone. As this effect diffuses through the city with a time lag, residents choose to migrate based on their perceived utility of all other zones.

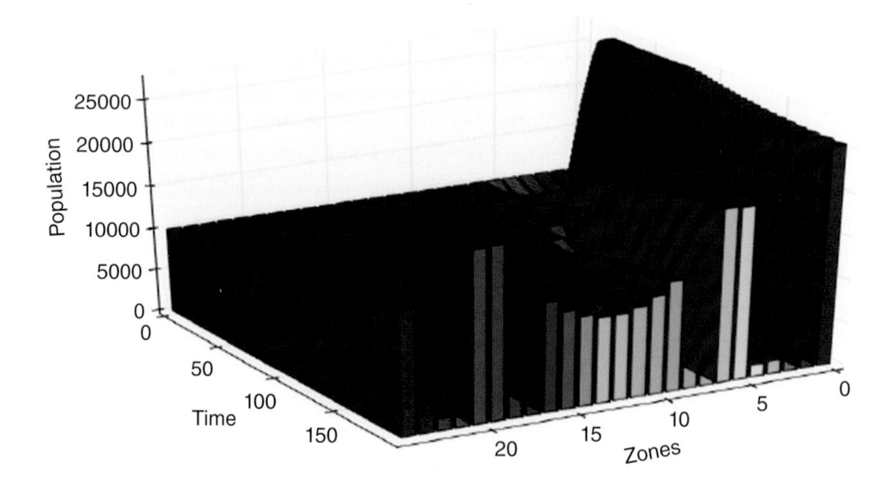

Figure 14.1 Development of population agglomeration zones

The change in population in a zone results in changes in rent and transport costs, and hence the utility of the zone. The initial perturbation leads to multiple ripples, and agglomeration is a consequence of this, corresponding to changes in utility in multiple zones (Figure 14.1).

For the model run which generates Figure 14.1, we assume there are 25 zones in the city. To illustrate the evolution of the urban space structure over time, we have arranged the zones in a line from zone 0 to zone 24. Zones 0 and 24, of course, are adjacent to each other in the circular network city. A transport investment is made in zone 0 which decreases the transport cost, thereby triggering an influx of migrating population as this information of lower transport cost diffuses across the city. In our next step, we analyse the conditions which lead to an urban landscape with multiple agglomeration zones of high population density.

14.3 Mathematical Analysis for Agglomeration Conditions

14.3.1 Introduction

We note that for agglomeration zones of the kind described in Medda *et al.* (2009) to occur, areas with high population density should emerge. We have observed that, indeed, an increase in the household utility is an attractive force that leads to movements of households across the city. The formation of high population density areas is thus evident. Recall that utility of a zone depends on the population in the zone and is defined as follows:

$$U(P) = Income - Rent(P) - Transport(P)$$

$$\mathrm{Rent(P)} = r\exp(\rho \mathrm{P}),$$

$$\mathrm{Transport(P)} = c\exp(-\tau \mathrm{P})$$

where r and c depend on the initial values of rent and transport costs, and $\rho > 0$ and $\tau > 0$ are the rates of growth and decay of rent and transport costs, respectively. Figure 14.2 is a plot of the three curves for selected values of r, ρ, c and τ.

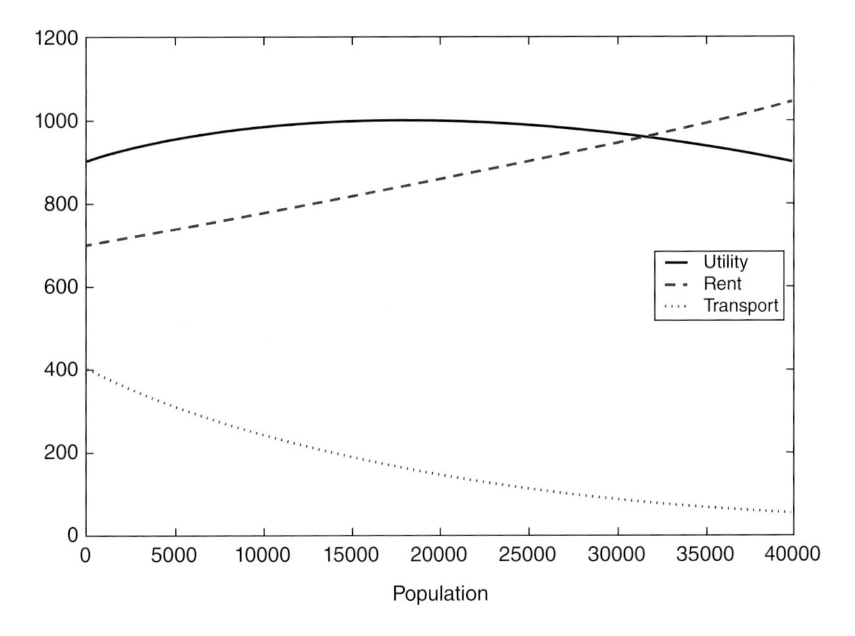

Figure 14.2 Utility, rent and transport curve

From Figure 14.2, we can see that utility is a concave function of population,

$$U(P) = I - r\exp(\rho P) - c\exp(-\tau P),$$

and has a maximum for some value, say, P_{max}:

$$P_{max} = \frac{1}{\rho + \tau}\ln\frac{c\tau}{r\rho}.$$

The implication of this is that, as the population increases beyond P_{max}, the utility in a zone actually decreases. We observe from numerical simulations that the following condition is a prerequisite for agglomeration zones to emerge:

$$P_{max} > P_0$$

where P_0 is the initial population, assumed the same for each zone, at the start of the experiment. We will henceforth refer to this as the *agglomeration condition.*

One could argue that a failure of the urban agglomeration condition would lead to oscillatory behaviour, as high utility attracts larger populations but, conversely, large populations result in low utility if $P > P_{max}$. Therefore, if the agglomeration condition fails at the start of the simulation, then it implies that an increasing population always results in declining utility. The system never gets an opportunity to evolve to a state where some zones have sufficiently high utility to act as an attractive force in comparison to their immediate neighbours. We next examine the agglomeration condition in more detail.

The utility function is determined by two initial conditions r and c, and two parameters ρ and τ. According to the condition for agglomeration given here, we must have

$$P_{max} > P_0$$

$$\frac{1}{\rho + \tau} \ln \frac{c\tau}{r\rho} > P_0$$

At the very least, we must have $\ln \frac{c\tau}{r\rho} > 0$ or, equivalently, $\frac{c\tau}{r\rho} > 1$. We consider the two ratios $\frac{c}{r}$ and $\frac{\tau}{\rho}$. According to the agglomeration condition, the product of the two ratios must be greater than one. Therefore, clearly the combination $\frac{c}{r} \leq 1$ and $\frac{\tau}{\rho} \leq 1$ cannot occur. We use Table 14.1 to examine all the possible cases.

14.3.2 *Case:* r < c

Initial transport cost is greater than initial rent cost in all zones.

Let $r = \alpha c,\ \alpha < 1$. We normalize and fix the value of ρ, $\rho = \frac{1}{kP_0}$ where k is a constant.

1. $$\tau \leq \rho$$

   The rate of decay of transport cost is slower than the rate of growth of rent cost.

   Let $\tau = \beta\rho,\ \beta \leq 1$. The agglomeration condition then translates into

   $$\frac{k}{1+\beta} \ln \frac{\beta}{\alpha} > 1$$

   A suitable choice of $k,\ \alpha,\ \beta$, with $\alpha < \beta$ which satisfies the above condition, can then lead to urban agglomeration. (This result is identified in Table 14.1 with a question mark.)

2. $$\tau > \rho$$

   The rate of decay of transport cost is faster than the rate of growth of rent cost.

   In this case, let $\tau = \beta\rho,\ \beta > 1$. The agglomeration condition then translates into

   $$\frac{k}{1+\beta} \ln \frac{\beta}{\alpha} > 1$$

   A suitable choice of $k,\ \alpha,\ \beta$ which satisfies the above condition can then lead to urban agglomeration.

14.3.3 *Case:* r ≥ c

Initial transport cost is equal to or greater than the initial rent cost in all zones.

Table 14.1 Parameter space classification for urban agglomeration

	$\tau \leq \rho$	$\tau > \rho$
$r < c$	?	?
$r \geq c$	x	?

Here, the initial transport cost is equal to $r = \alpha c,\ \alpha \geq 1$. We normalize and fix the value of $\rho,\ \rho = \frac{1}{kP_0}$:

1. $$\tau \leq \rho$$

   The rate of decay of transport cost is less than or equal to the rate of growth of rent cost. In this case, $\frac{c\tau}{r\rho} \leq 1$, which violates the agglomeration condition.

2. $$\tau > \rho$$

   The rate of decay of transport cost is faster than the rate of growth of rent cost. Let $\tau = \beta\rho,\ \beta > 1$. The agglomeration condition then translates into

$$\frac{k}{1+\beta} \ln \frac{\beta}{\alpha} > 1$$

A suitable choice of k, α, β, with $\alpha < \beta$, which satisfies the above condition, can then lead to urban agglomeration.

Thus, we see that a mathematical analysis of the agglomeration condition leads us to a classification of the parameter space of the utility curve. In Section 14.4, we show that this analysis is borne out in simulations of the model.

14.4 Simulation Results

In order to verify and test our model, we fix the initial transport cost and the exponential rate of growth of rent cost as follows:

$$c = 0.2\,Y,\ \rho = \frac{1}{10P_0}$$

where Y is the income in each zone which remains constant, and P_0 is the initial population in each zone. Bearing in mind the condition for agglomeration, we set $r = \alpha c$ and $\tau = \beta\rho$. We then allow both α and β to vary between 0.2 and 5. Therefore, the initial rent cost is allowed to vary between a fifth and five times the initial transport cost, thus giving us realistic urban range values. Similarly, the decay rate of transport cost is allowed to vary between a fifth and five times the growth rate of rent cost. The analysis in Section 14.2 is validated in the scatterplot in Figure 14.3. To get a uniform plot, we have taken a log scale of both axes.

We make the following observations of the obtained results:

1. If we start in the bottom left quarter, where α and β are both less than one (or equivalently $r < c,\ \tau < \rho$), we see that urban agglomeration can occur when $\ln \beta > \ln \alpha$.
2. Moving clockwise, in the top left, where α is less than one and β is greater than one (or equivalently $r < c,\ \tau > \rho$), we see that agglomeration almost always occurs.
3. In the top right, where α and β are both greater than one (or equivalently $r > c,\ \tau > \rho$), we see that agglomeration can occur when $\ln \beta > \ln \alpha$.
4. Finally, in the bottom right, where α is greater than one and β is less than one (or equivalently $r > c,\ \tau < \rho$), we see that agglomeration never occurs.

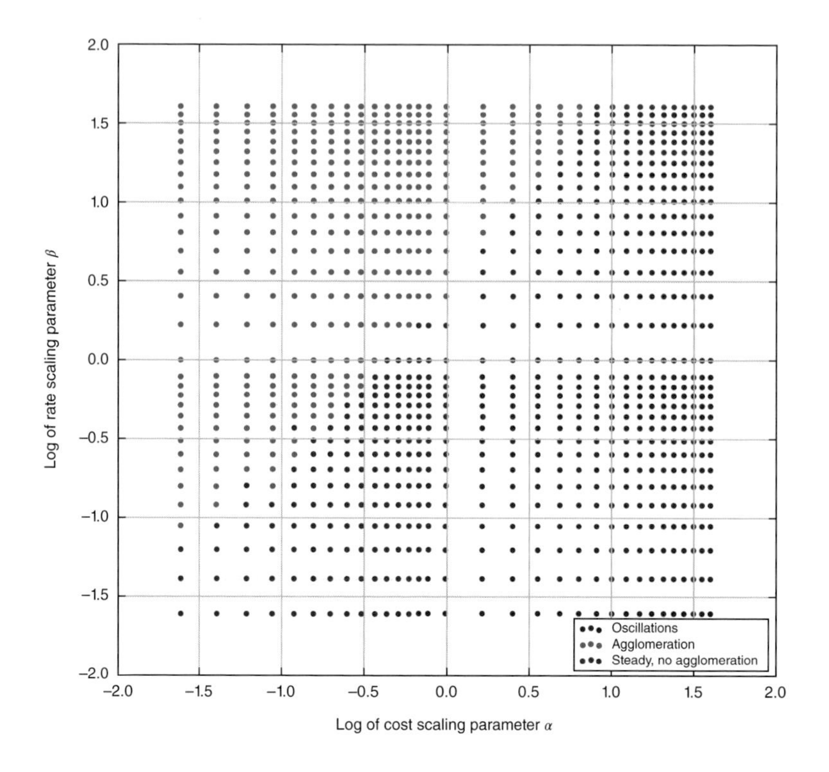

Figure 14.3 Scatterplot of urban agglomeration parameters

Therefore, from our simulations, we have a better understanding of the parameter space where agglomeration occurs, and we can attempt to answer the question marks in Table 14.1.

As the simulation results corroborate our mathematical analysis, we make the following conjecture: *If* $P_{max} > P_0$, *then urban agglomeration will occur.*

The boundary condition $\ln \beta > \ln \alpha$ in observations 1 and 3 (which are indicated by * in Table 14.2) is defined more precisely using condition $\frac{1}{\rho+\tau} \ln \frac{c\tau}{r\rho} > P_0$:

$$\frac{k}{1+\beta} \ln \frac{\beta}{\alpha} > 1$$

$$\frac{k}{1+\beta}(\ln \beta - \ln \alpha) > 1$$

$$\ln \beta - \ln \alpha > \frac{1+\beta}{k}$$

$$\ln \beta - \frac{1+\exp \ln \beta}{k} > \ln \alpha$$

If our conjecture is true, when the above condition $\ln \beta - \frac{1+\exp \ln \beta}{k} > \ln \alpha$ is satisfied, we expect that agglomeration occurs.

We can now interpret the mathematical and simulation results in the urban context of our circular city. Although the condition $r < c$ very likely leads to agglomeration zones, it is very

Table 14.2 Parameter space classification for urban agglomeration

	$\tau \leq \rho$	$\tau > \rho$
$r < c$	*	✓
$r \geq c$	x	*

* The region determined by the condition 2.

unlikely that the initial rent cost will be less than the initial transport cost. These cases are therefore not realistic in the urban setting.

For the purposes of this analysis, we must restrict ourselves to the region $r \geq c$. In this case (Table 14.2), we note that $\tau > \rho$ must be satisfied (i.e. the rate of decay of the transport cost must be faster than the rate of growth of rent cost). In fact, the relationship between τ and ρ is governed by:

$$\frac{1}{\rho + \tau} \ln \frac{c\tau}{r\rho} > P_0$$

This is an implicit function involving τ and ρ, and thus we must have:

$$\ln \frac{c\tau}{r\rho} > 0$$

or, equivalently, $\frac{c\tau}{r\rho} > 1$
or, equivalently, $\tau > \frac{r}{c}\rho$.

In the context of our model, this condition indicates that the rate of decay of transport cost must be faster than $\frac{r}{c}$ times (where $\frac{r}{c} > 1$) the rate of growth of rent cost.

We next examine the stability of our dynamic model. In our simulations, we observe that it can be controlled by varying the parameters of the model. The most important parameter leading to stable systems is the migration factor which is the propensity of the population that chooses to migrate at every time step. (We note that it does not necessarily lead to urban agglomeration.) We have verified that low values of this parameter (less than 0.02) lead to stable final states, whereas higher values lead to oscillations. For example (Figure 14.4), we note that when we fix the migration factor equal to 0.1, we obtain many oscillation points, whereas in the case of a migration factor at 0.01, the model is very stable.

We are also interested in the effect of diffusion rates on the stability of the model. The formulas for diffusion are:

$$\theta_{ij}^{R} = \theta_{ij}^{R}(d_{ij}) = a\exp(\gamma d_{ij})$$

and

$$\theta_{ij}^{T} = \theta_{ij}^{T}(d_{ij}) = b\exp(\kappa d_{ij}).$$

In our initial simulations, we fixed the ratio of the rate at which the rent and transport costs diffused (i.e. we held $\frac{\gamma}{k}$ fixed and varied the speed at which transport costs diffused). However, we found that this did not lead to any significant impact on the urban shape. We then varied the ratio of the two rates, first letting $\frac{\gamma}{k} = 0.25$, and then $\frac{\gamma}{k} = 0.5$. In the latter case, the system was very stable (almost no oscillations) compared to the former.

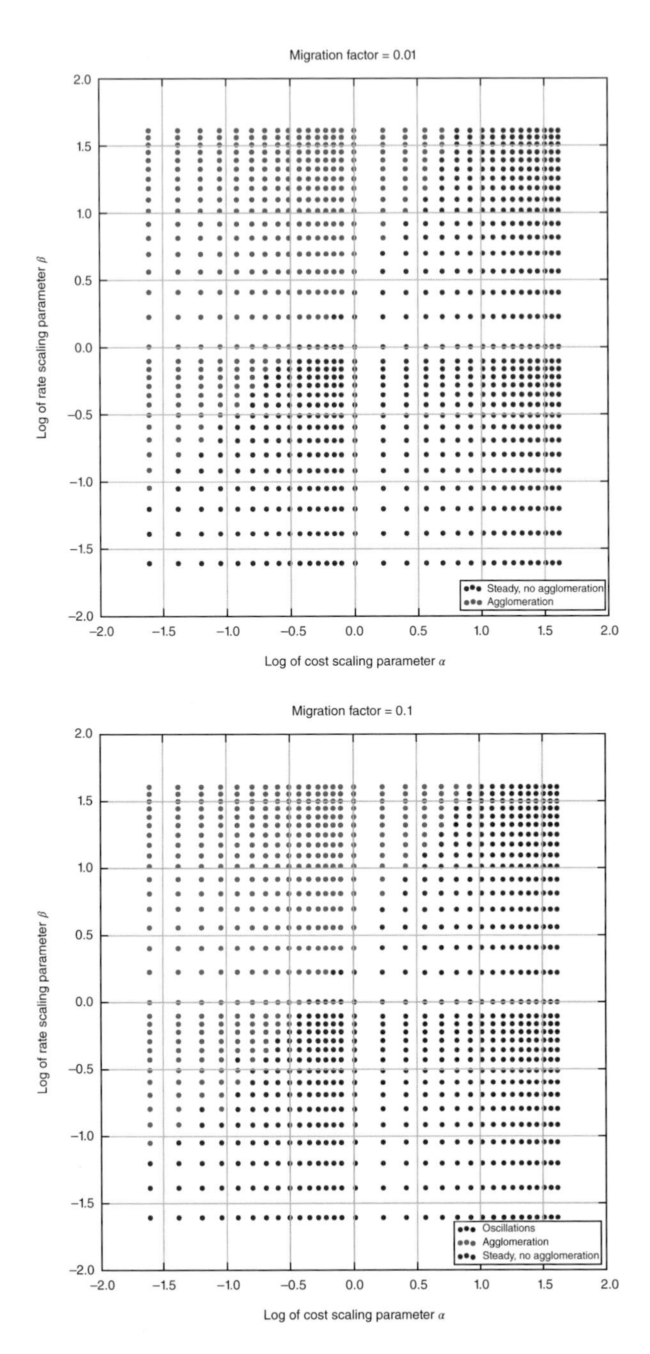

Figure 14.4 Scatterplot of agglomeration with two migration factors (on the left side the migration factor is equal to 0.1, and on the right side the migration factor is equal to 0.01)

In conclusion, given that transport and rent costs diffuse through urban space, which leads to a perception of utility that varies between zones, agglomeration occurs if the pre-existing conditions in the urban space are suitable. We can further link the shape of the utility curve to the emergence of agglomeration zones in the city.

The shape of the utility curve is determined by the values of:

- r – depends on the initial value of rent cost
- ρ – rate of growth of rent cost
- c – depends on the initial value of transport cost
- τ – rate of decay of rent cost.

The analysis and simulation results show that when r, ρ, c and τ obey

$$\frac{1}{\rho+\tau}\ln\frac{c\tau}{r\rho} > P_0,$$

it is likely that urban agglomeration will occur.

14.5 Conclusions

We have defined a model that represents the diffusion of changing transport and rent costs triggered by an urban investment. Households perceive these changes, which affect utility, and they may decide to migrate to other zones within the urban network. Urban agglomeration then occurs. However, agglomeration happens only if the pre-existing conditions in the urban space are suitable. This is an interesting result because it demonstrates how the context of the urban environment where the investment is implemented is fundamental to the success of the investment. Therefore, we have shown in our theoretical model that different cities with the same types of investment may instigate different agglomeration effects. Our future analysis will be to extend the results obtained here by considering a two-dimensional urban network.

References

Drennan MP and Kelly HF. (2011). Measuring urban agglomeration economics with office rents. *Journal of Economic Geography* 11(3): 481–507.

Medda F, Nijkamp P and Rietveld P. (2009). A morphogenetic perspective on spatial complexity: transport costs and urban shapes. In Reggiani A and Nijkamp P. (eds.), *Complexity and Spatial Networks: In Search of Simplicity*. Berlin: Springer.

Rabianski J and Gibler K. (2007). Office market demand analysis and estimation techniques: a literature review, synthesis and commentary. *Journal of Real Estate Literature* 15: 37–56.

Rosenthal S and Strange W. (2004). Evidence on the nature and sources of agglomeration economies. In: Henderson JV and Thisse J-F (eds.), *Handbook of Regional and Urban Economics*, Vol. 4, *Cities and Geography*. Amsterdam: Elsevier.

Storper M and Venables A. (2004). Buzz: face-to-face contact and the urban economy. *Journal of Economic Geography* 4: 351–370.

Part Nine
Game Theory

15

From Colonel Blotto to Field Marshall Blotto

Peter Baudains, Toby P. Davies, Hannah M. Fry and Alan G. Wilson

15.1 Introduction

Colonel Blotto games (attributed to Borel, 1921; translated in Borel, 1953) consider the strategic allocation of resources under competition with an adversary. They require each player to assign a number of their own resources to a series of different *fronts*, which represent battlefields over which a conflict takes place. These fronts can be spatial or non-spatial, depending on the application of the model. Each player specifies an allocation of the entirety of their resources across these fronts, and they do so without knowledge of their opponent's allocation. The more resources each player assigns to a particular front, the higher the probability that they will defeat their opponent at that front. In the original application, all that was required for a player to win a particular front was a greater number of their own resources allocated to that front than their opponent. It is then supposed that the player who wins the most battles over the different fronts wins the game overall.

There are a wide range of possible applications of the Colonel Blotto game, but it is often framed as a military problem where the optimal allocation of troops over a series of conflict zones is the objective of a decision maker (the aforenamed Colonel Blotto). In this case, the opponent may represent either a conventional military force or an insurgency. In the case of an insurgency in particular, it is unlikely that the allocation of insurgents will be known by a counter insurgent force beforehand, and so the use of the Colonel Blotto game appears at first glance to be appropriate. Other examples of possible application of the model include the defence of vulnerable targets from terrorist attacks (e.g. Powell, 2007) and political campaigning, whereby the fronts may correspond to closely fought election constituencies and resources

Approaches to Geo-mathematical Modelling: New Tools for Complexity Science, First Edition. Edited by Alan G. Wilson.

Companion Website: www.wiley.com/go/wilson/ApproachestoGeo-mathematicalModelling

to the time of high-profile campaigners. These examples tend to imply an implicit spatial relationship (conflict zones, for example, often correspond to geographic areas), but there are other applications that do not require a reliance on geography, for example in a business setting when determining the appropriate level of investment in a number of business streams during a time when a competitor might also be considering how to expand their business.

Despite the wide range of possible applications, the Colonel Blotto game fails to capture much of the complexity of real-world competition over different fronts. Golman and Page (2009) argue how fronts are rarely treated in isolation in real-world allocation problems since the allocation of resources to one particular front can influence a battle taking place on a different front. They offer an alternative formulation: the General Blotto game. In this model, the allocation of resources to particular sets of fronts can be just as valuable as the allocation to individual fronts. For example, allocating resources to the hub of an infrastructure network can benefit other fronts since if an infrastructure hub were to be lost to an attacker, then the other fronts would be subsequently lost more easily. Additionally, more central fronts may be more desirable to protect since they enable the transfer of resources to other fronts that might need them.

Dependencies between fronts are likely to be of interest in the case of spatial applications of resource allocation. An example that demonstrates the importance of considering interactions and effects between fronts (rather than simply treating them in isolation) comes from geographically focussed policing initiatives. There is evidence to suggest that the allocation of police resources to certain areas in an effort to reduce crime can also benefit other areas nearby (Bowers et al., 2011). Indeed, proponents of situational crime prevention techniques argue that a diffusion of benefits with respect to crime reduction often occurs, rather than crime displacement, whereby crime simply moves to new areas (Guerette and Bowers, 2009). This example highlights the importance of not only focussing on isolated fronts but also considering the spatial interaction that might occur between them.

Although the General Blotto game proposed by Golman and Page can be applied to incorporate spatial interaction by defining sets of fronts that are considered just as valuable as isolated fronts, their approach does not explicitly model such sets and instead relies on the modeller to specify them. In particular, their formulation is not explicitly spatial. In this chapter, we offer an alternative formulation for extending the Colonel Blotto game to account for spatial relationships between different fronts. To do this, we employ a spatial interaction model of threat, which is introduced in Chapter 9 of *Global Dynamics*. We demonstrate the implications of incorporating a spatial model by using a range of scenarios and highlight the importance of parameter estimation in the model for analysing appropriate strategies. Further investigation of this framework for analysing problems of strategic allocation in spatially explicit environments might offer insight into real-world problems.

In what follows in this chapter, we begin by outlining the Colonel Blotto game and its extensions, using a mathematical formulation. We then review how threat can be modelled with spatial interaction models before proposing a Blotto framework capable of incorporating the spatial configuration of adversaries, where the distance to different fronts plays an important role in determining player strategy. We present the outcome of the game in a number of idealised scenarios, demonstrating both the richness of the model as well as the importance of estimating the parameters of the spatial interaction model. Finally, we conclude by describing future work to ensure the proposed framework can be put to use in determining strategic allocations in the real world.

15.2 The Colonel Blotto Game and its Extensions

The Colonel Blotto game has two players with, respectively, X and Y resources to allocate over N fronts. We denote the set of fronts by $F = \{F_1, F_2, ..., F_N\}$. The original game sets $X = Y$, and the payoff to each player is given by the number of fronts in which they have more resources allocated than their opponents. If $\mathbf{x} = (x_1, x_2, ..., x_N)$ and $\mathbf{y} = (y_1, y_2, ..., y_N)$ are vectors with non-negative entries which give the resources of each player allocated to each front, then the payoff to the player who plays $\mathbf{x}$ against $\mathbf{y}$ is

$$p(\mathbf{x}, \mathbf{y}) = \sum_{i=1}^{N} f(x_i - y_i), \tag{15.1}$$

for a *scoring function* f, where

$$f(z) = \mathrm{sgn}(z) = \begin{cases} -1 & z < 0 \\ 0 & z = 0 \\ 1 & z > 0. \end{cases} \tag{15.2}$$

There are no pure-strategy Nash equilibria to this game when $X = Y$. That is, any allocation of arms from one team can be beaten by the other. Mixed-strategy equilibria, which can occur when each player adopts strategies according to some probability distribution, are possible and have been identified in a number of previous studies both in scenarios where the resources are infinitely divisible (so that $\mathbf{x}, \mathbf{y} \in \mathbb{R}^N_{\geq 0}$) and when they are discrete entities (where $\mathbf{x}, \mathbf{y} \in \mathbb{N}^N$) (e.g. Gross and Wagner, 1950; Hart, 2008; Roberson, 2006).

Golman and Page (2009) promote the Colonel Blotto Game to a class of *General Blotto* games. In particular, they generalise the payoff function so that the margins of victory in each front become important and, as discussed in Section 15.1, allow combinations of fronts to be just as valuable as individual fronts.

To do the former, the scoring function

$$f(z) = \mathrm{sgn}(z)|z|^p, \tag{15.3}$$

is employed for some parameter p. The original Blotto game can be obtained by setting $p = 0$. Golman and Page explore the different characteristics of games for scenarios in which $0 < p < 1$, $p = 1$ and $p > 1$.

To do the latter, the framework of Shubik and Weber (1981) is adopted, who were also concerned with incorporating dependencies between fronts within the Colonel Blotto game. A set S is defined that is composed of all fronts and sets of fronts that are considered to be of value in the game. In the case of the original model, the set S is taken to be the set of individual fronts (so that $S = F$). Golman and Page additionally consider a scenario in which every pair of fronts is of value (as well as the individual fronts) and a scenario in which every possible combination of fronts is of value (in this case, S consists of every possible subset of the set F). For subsets $s \in S$ containing $|s|$ fronts, a *value function* $v_{|s|}$ is defined, which determines how the resources allocated within each zone are aggregated when considering multiple fronts. Then, the payoff to the player who plays $\mathbf{x}$ against $\mathbf{y}$ is

$$p(\mathbf{x}, \mathbf{y}) = \sum_{s \in S} f(v_{|s|}(\{x_i\}_{i \in s}) - v_{|s|}(\{y_i\}_{i \in s})). \tag{15.4}$$

Supposing that an element s of S consists of the fronts given by $s = \{F_{s_1}, F_{s_2}, ..., F_{s_K}\}$, then $|s| = K$ and Golman and Page set

$$v_K(\{x_i\}_{i\in s}) = \prod_{k=1}^{K} x_{s_k}, \tag{15.5}$$

so that, when combinations of fronts contribute as much value to the game as individual fronts, it is the product of resources allocated to each front in that combination that determines the victor for that combination. A number of game specifications are then explored both analytically and numerically. For the numerical case, an example is investigated with five fronts and 10 indivisible resources using replicator dynamics to investigate dominant strategies when both combinations of fronts and the magnitude of victory contribute to the game. A general conclusion is that the addition of higher levels of complexity to the game can, perhaps paradoxically, make it easier to analyse.

The general framework developed in Golman and Page (2009) is not explicitly spatial. Although the model can be applied to spatial problems, the combinations of fronts that contribute to the game must be specified by the modeller beforehand. In this case, consideration of the interactions and dependencies between fronts that add value to the game is performed during a separate phase of the modelling process. It is possible, however, to account for the dependencies between fronts using a model of spatial interaction. This alleviates the need to specify the spatial structure of the game *a priori*. In the next section, we review a spatial interaction model of threat that is introduced in Chapter 9 of *Global Dynamics*. We then go on to consider the implications of incorporating this model within the Colonel Blotto framework.

15.3 Incorporating a Spatial Interaction Model of Threat

In Chapter 9 of *Global Dynamics*, we introduce a general model of threat between two adversaries who are distributed over some spatial domain. The idea is to consider how the spatial distribution of adversaries might influence any resulting conflict. To construct the model in the case of fronts in the Colonel Blotto game, we let d_{ij} be some measure of distance, travel cost or impedance between fronts i and j[1]. Then, as above, assuming that one player plays $\mathbf{x} = (x_1, x_2, ..., x_N)$ and that their opponent plays $\mathbf{y} = (y_1, y_2, ..., y_N)$, the threat that is felt by resources y_j due to the proximity of x_i is given by[2]

$$T_{ij}^{xy} = \frac{x_i y_j^\alpha \exp(-\beta d_{ij})}{\sum_k y_k^\alpha \exp(-\beta d_{ik})}, \tag{15.6}$$

for positive parameters α and β. The model measures the intimidatory effect that the resources x_i have on y_j both by comparing the relative magnitude of y_j against y_k for all other fronts k and by considering the proximity of front i to front j. Fronts with more resources allocated to them invoke a greater threat response from their opponent, since those locations are the ones likely to capture more attention. This feature of the model resonates with strategic allocation

[1] In the present case, we assume this measure is identical for each player, is non-negative and has $d_{ii} = 0$ and $d_{ij} = d_{ji}$ for all i and j.

[2] See Chapter 9 of *Global Dynamics* for a derivation and further explanation.

problems in the real world: for example, if a military chose to place all of their resources at just one front, then they are likely to face considerable threats at that location from opponents located at other fronts.

The total threat on the player with strategy **y** in zone j is then given by

$$\tau_j^y = \sum_i T_{ij}^{xy} \tag{15.7}$$

$$= \sum_i \frac{x_i y_j^\alpha \exp(-\beta d_{ij})}{\sum_k y_k^\alpha \exp(-\beta d_{ik})}, \tag{15.8}$$

and, since j is arbitrary, this holds for fronts $j = 1, 2, 3, ..., N$. The parameter α in equation 15.8 determines the way in which the size of y_j, relative to all other y_k, attracts threat flows from **x**. Larger values of α imply that relatively high levels of resources at a particular front will attract disproportionately larger flows of threat from **x** (subject to how the other components of the model account for the spatial distribution of **x**). The parameter β determines how impedance in the spatial configuration of the model influences the resulting flow of threat. Large values of β imply that impedance plays a stronger role, which implies that the resources at a front cannot exert very much threat over different fronts. Indeed, as $\beta \to \infty$, if d_{ij} is only equal to zero when $i = j$, then the threat will only act within individual fronts with no effect over fronts. Conversely, small values of β would imply that a threat can easily flow across fronts, and, in this case, the game resembles a situation with just one front where all resources are allocated to the same place (it doesn't matter where the resources are placed due to low impedance).

Similarly, the strategy **y** will also exert threat on the strategy **x**, so the total threat on the x_j resources at front j is

$$\tau_j^x = \sum_i T_{ij}^{yx} \tag{15.9}$$

$$= \sum_i \frac{y_i x_j^\gamma \exp(-\delta d_{ij})}{\sum_k x_k^\gamma \exp(-\delta d_{ik})}, \tag{15.10}$$

for similar parameters γ and δ. In the remainder of this chapter, we set $\gamma = \alpha$ and $\beta = \delta$ so that the threat is calibrated similarly for each player. This may be generalised, however, if it is thought that one player has access to lower costs in transporting their resources or if the relatively larger resources at particular fronts are likely to generate different levels of threat from each side.

The spatial interaction model can be incorporated into the Blotto framework by setting the payoff function of **x** against **y** to be

$$p(\mathbf{x}, \mathbf{y}) = \sum_{i=1}^{N} sgn(\tau_i^y - \tau_i^x), \tag{15.11}$$

so that **x** wins front i if it exerts more threat on y_i than **y** exerts on x_i. Using this payoff function within the Colonel Blotto framework thus enables strategic positioning of resources to be accounted for in the game. In particular, the spatial relationship between the fronts becomes important: a player in one zone would need to take into account the resources located in nearby zones of both the other team and its own team. The focus becomes on the ability for the allocated resources to maximise the threat on the opponent, thus potentially extending the range of applications of the model.

15.4 Two-front Battles

We begin our analysis by considering an allocation problem with just two zones. In the original Colonel Blotto game, if $X = Y$, which will be assumed for the remainder of this chapter, a draw will always be the result. It can be shown that this is also true in the game with the payoff function given by equation 15.11. This is because, according to equations 15.8 and 15.9,

$$\sum_j \tau_j^y = X, \quad \sum_j \tau_j^x = Y, \tag{15.12}$$

and, thus, with two zones,

$$\tau_1^y + \tau_2^y = \tau_1^x + \tau_2^x, \tag{15.13}$$

which can be rearranged to

$$\tau_1^y - \tau_1^x = -(\tau_2^y - \tau_2^x); \tag{15.14}$$

thus, if $\tau_1^y > \tau_1^x$, implying victory to $\mathbf{x}$ at front 1, then $\tau_2^x > \tau_2^y$, $\mathbf{y}$ will win zone 2 and the game is a draw.

Nevertheless, it is instructive to investigate the two-front game since determining who wins at each front is not as simple as in the original Colonel Blotto game. Suppose, then, that we have a two-front game and that $X = Y = 4$. For the remainder of this chapter, we consider scenarios in which $\mathbf{x}$ and $\mathbf{y}$ have integer entries and so the resources are not infinitely divisible. We consider the game played with the following strategies:

$$\mathbf{x} = (2, 2) \qquad \mathbf{y} = (1, 3), \tag{15.15}$$

and, since the winner at front 2 is simply the opposite to the winner at front 1, we investigate front 1 only. In the original Colonel Blotto game, since $x_1 > y_1$, strategy $\mathbf{x}$ wins front 1. However, in the case where the payoff function is given in equation 15.11, where the spatial distribution of the resources at other fronts can also play a role depending on the level of impedance (as determined by the parameter β) and the relative attractiveness of fronts with a higher number of resources (as determined by α), we observe a different outcome.

With just two fronts, we suppose the distance measure for $i \neq j$ is equal to one. Then, using equations 15.8 and 15.10, we calculate

$$\tau_1^y = \frac{2}{1 + 3^\alpha e^{-\beta}} + \frac{2e^{-\beta}}{e^{-\beta} + 3^\alpha} \tag{15.16}$$

$$\tau_1^x = \frac{1 + 3e^{-\beta}}{1 + e^{-\beta}}. \tag{15.17}$$

Strategy $\mathbf{x}$ wins front 1 when the threat on y_1 is greater than the threat on x_1, which occurs when

$$\frac{2}{1 + 3^\alpha e^{-\beta}} + \frac{2e^{-\beta}}{e^{-\beta} + 3^\alpha} > \frac{1 + 3e^{-\beta}}{1 + e^{-\beta}}. \tag{15.18}$$

Rearranging, this occurs when

$$3^{\alpha}e^{-3\beta} + (3 \cdot 3^{2\alpha} - 3^{\alpha} - 1)e^{-2\beta} + (3^{2\alpha} + 3^{\alpha} - 3)e^{-\beta} - 3^{\alpha} < 0. \tag{15.19}$$

If $\alpha = 1$, then **x** wins zone 1 when

$$e^{-\beta} \lesssim 0.21, \tag{15.20}$$

or when

$$\beta \gtrsim 1.54. \tag{15.21}$$

Importantly, when β is less than 1.54, in contrast to the original Colonel Blotto game, it is strategy **y** that wins front 1, even though they have a resource deficit at that front. In this case, **x** also wins front 2 with an arms deficit. For these parameter values, the presence of more resources at each front actually increases the level of threat that they experience.

This simple example demonstrates the importance of the parameters α and β. Different values of these parameters can change the winning strategy at a particular front. In the case considered here, the α and β values do not change the overall result of the game. In the next section, we consider an example with five fronts, in which the values of the parameters do change the overall outcome of the game.

15.5 Comparing Even and Uneven Allocations in a Scenario with Five Fronts

In this section, we consider an example with five fronts where each player has 15 resources to distribute between them. This increases the complexity of the game and the influence that the threat model has on its outcome. In particular, we compare two strategies: one in which a player chooses to distribute their resources over the five fronts evenly, and another in which a player chooses to commit higher levels of resources in some fronts, yet still retains resources in all the fronts. Specifically, we consider the strategies

$$\mathbf{x} = (3, 3, 3, 3, 3), \quad \mathbf{y} = (5, 4, 3, 2, 1). \tag{15.22}$$

We explore the game with two distance measures, reflecting two different spatial configurations and relationships between the fronts. We present the distance measure in the form of a matrix where the ij-th entry defines d_{ij}. The first distance matrix supposes that impedances are equal over the different fronts but that there is no impedance within fronts, and it is given by

$$d^{(1)} = \begin{pmatrix} 0 & 1 & 1 & 1 & 1 \\ 1 & 0 & 1 & 1 & 1 \\ 1 & 1 & 0 & 1 & 1 \\ 1 & 1 & 1 & 0 & 1 \\ 1 & 1 & 1 & 1 & 0 \end{pmatrix}. \tag{15.23}$$

In Figure 15.1, we plot the value of the payoff function to **x** for values of α and β between 0 and 10, where the threat measure is calculated using the distance matrix in equation 15.23. Positive values indicate that the strategy **x** wins the overall game, and negative values indicate that the strategy **y** wins across the five fronts.

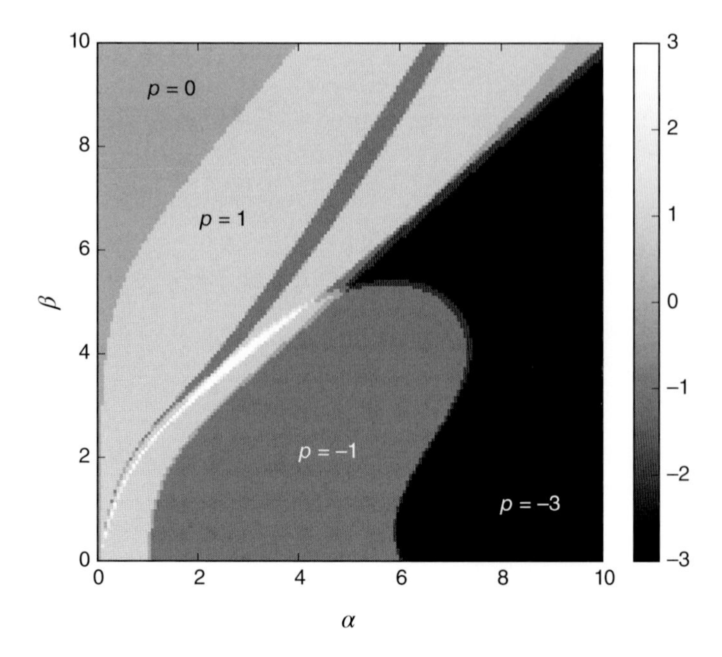

Figure 15.1 The payoff function to strategy $\mathbf{x} = (3, 3, 3, 3, 3)$ against $\mathbf{y} = (5, 4, 3, 2, 1)$. The distance metric is given in equation 15.23

There are some striking patterns that arise from the nonlinearity and complexity in the model. However, there are also some consistencies that can be readily explained. The darker shades on the right-hand side of Figure 15.1 demonstrate how, for large values of α and small values of β, strategies such as $\mathbf{y}$ that cluster resources within a few fronts are more beneficial, owing to the more restricted way in which threat flows can cross boundaries and to the way in which the clustering of resources influences threat flows. In this case, the clustering of resources of strategy $\mathbf{y}$ in fronts 1 and 2 focusses the threat from $\mathbf{x}$ in all fronts onto these zones, leaving them under-resourced to alleviate threats at the fronts at which $\mathbf{y}$ has a deficit of resources. Thus, $\mathbf{y}$ wins overall not because they win fronts 1 and 2, where they have over-committed, but because the presence of more resources at these zones redirects the focus of the resources of $\mathbf{x}$ at the other fronts. Thus, $\mathbf{y}$ actually wins the fronts where they have a deficit because $\mathbf{x}$ is committing too much to the aggregation of resources elsewhere.

For large values of β and small values of α, in the upper left of Figure 15.1, the game is a draw and is an approximation of the original Colonel Blotto game. There is also a large regime in which $\mathbf{x}$ is the winning strategy. This occurs when $\beta > \alpha$ but also when β is not so large that the game becomes an approximation to the Colonel Blotto game. In this scenario, the even distribution of resources of $\mathbf{x}$ is capable of delivering threat in the right proportions to the fronts where it is needed to win the game.

There are also a number of transitions between different outcomes of the game that arise with varying parameter values. These are a likely consequence of the non-linearity of the threat function and warrant further investigation before applying the game to real-world scenarios. This example demonstrates the added complexity to the game and the importance of defining appropriate parameter values in the spatial interaction model of threat.

The second distance matrix we consider is given by

$$d^{(2)} = \begin{pmatrix} 0 & 1 & 2 & 3 & 4 \\ 1 & 0 & 1 & 2 & 3 \\ 2 & 1 & 0 & 1 & 2 \\ 3 & 2 & 1 & 0 & 1 \\ 4 & 3 & 2 & 1 & 0 \end{pmatrix}. \tag{15.24}$$

This matrix provides a scenario in which the distance between any two fronts i and j is given by $|i-j|$. Front 3 is the most central, being only a maximum of 2 units away from all other fronts, while fronts 1 and 5 can be considered to be on the periphery. Figure 15.2 shows the payoff to $\mathbf{x}$ using this distance matrix for values of α and β between 0 and 10.

Figure 15.2 is similar to Figure 15.1. For small values of β and large values of α, it is more beneficial to cluster resources in a few fronts. Similarly, for large values of β and small values of α, the game quickly approximates the original Colonel Blotto game, with the benefits of evenly distributing resources identified somewhere in between these two extremes. However, there are some notable differences between Figure 15.2 and Figure 15.1. There are fewer parameter values which result in a payoff to $\mathbf{x}$ of -3, suggesting that the relative advantage to clustering resources in a few zones is now smaller. This may be due to resources at the periphery of the spatial configuration finding it more difficult to exert threat over the other fronts. Furthermore, even distribution of resources becomes a winning strategy for smaller values of β when compared to Figure 15.1, suggesting that the changed distance metric plays a crucial role in determining the outcome of the game.

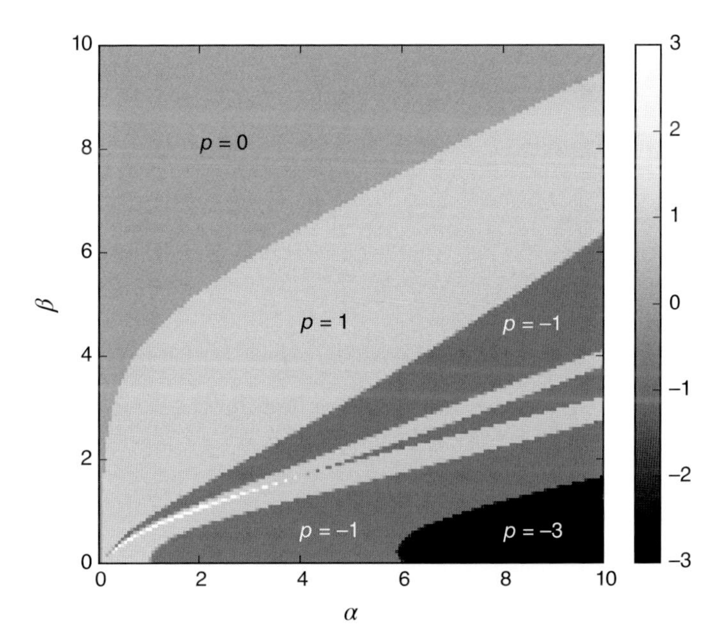

Figure 15.2 The payoff function to strategy $\mathbf{x} = (3,3,3,3,3)$ against $\mathbf{y} = (5,4,3,2,1)$. The distance metric is given in equation 15.24

15.6 Conclusion

In this chapter, we have explored and extended the classic Colonel Blotto game by incorporating a spatial interaction model of threat into the payoff function. This framework has a number of advantages when working with spatial allocation problems. First, as in the General Blotto model of Golman and Page (2009), the model allows for interdependence between fronts: resources allocated to one front may influence the outcome of a battle at a different front. Second, the spatial interaction framework also allows for an advantage to be obtained when more resources are clustered within just a few areas. Finally, in contrast to the framework proposed by Golman and Page – in which sets of fronts that contribute to the game must be specified beforehand – the spatial interactions are modelled explicitly within the game, subject to the specification of just two parameter values: α and β. For this reason, this model is well suited to spatial allocation problems.

The choice of α and β has been shown to dramatically influence the overall outcome of the game. Any practical applications should therefore consider estimation of these parameters as a crucial modelling step, once the spatial Blotto framework has been established. We leave the calibration of α and β in problems of strategic spatial allocation of resources for future work but note that there are a number of similar models with well-established estimation techniques that could be exploited (Batty and Mackie, 1972). Further research might also seek and classify both discrete and continuous pure- and mixed-strategy equilibria of the game. Perhaps the most valuable direction for future research, however, is in the identification of real-world scenarios in which the spatial Blotto framework presented here can be directly applied. Before the model can be applied in real-world contexts, further research is required to fully understand how the non-linearities influence the result. Nevertheless, the framing of the model presented in this chapter provides a starting point for combining Blotto games with spatial interaction models, which may prove to be a useful way of examining problems of strategic spatial allocation.

References

Batty M and Mackie S. (1972). The calibration of gravity, entropy, and related models of spatial interaction. *Environment and Planning* 4: 205–233.

Borel E. (1953). The theory of play and integral equations with skew symmetric kernels. *Econometrica* 21(1): 97–100.

Bowers K, Johnson SD, Guerette RT, Summers L and Poynton S. (2011). Spatial displacement and diffusion of benefits among geographically focused policing initiatives: a meta-analytical review. *Journal of Experimental Criminology* 7: 347–374.

Golman R and Page SE. (2009). General Blotto: games of allocative strategic mismatch. *Public Choice* 138(3–4): 279–299.

Gross O and Wagner R. (1950). A continuous Colonel Blotto game. RAND Corporation RM-408.

Guerette RT and Bowers KJ. (2009). Assessing the extent of crime displacement and diffusion of benefits: a review of situational crime prevention evaluations. *Criminology* 47(4): 1331–1368.

Hart S. (2008). Discrete Colonel Blotto and General Lotto games. *International Journal of Game Theory* 36: 441–460.

Powell R. (2007). Defending against terrorist attacks with limited resources. *American Political Science Review* 101(3): 527–541.

Roberson B. (2006). The Colonel Blotto game. *Economic Theory* 29: 1–24.

Shubik M and Weber RJ. (1981). Systems defense games: Colonel Blotto, command and control. *Naval Research Logistics Quarterly* 28(2): 281–287.

16

Modelling Strategic Interactions in a Global Context

Janina Beiser

16.1 Introduction

Situations that are inherently strategic are often analysed using quantitative data and standard statistical estimators that do not take strategic interaction into account. For example, empirical studies on government repression often test hypotheses using ordinary least squares (OLS) models (see Hill and Jones, 2014, which is also an example). Civil conflict is often analysed using logit models (e.g. Cederman *et al.*, 2010). However, Signorino (1999, 2003) shows in the context of limited discrete outcomes that if the theoretical model that is used to derive empirical expectations is strategic, non-strategic statistical models are going to be misspecified.

Signorino (1999) utilises McKelvey and Palfrey's (1995, 1996, 1998) concept of agent quantal response equilibrium to derive a strategic statistical estimator for discrete outcomes. Initially, this estimator was used for explaining states' interactions in international crises that might lead to full-fledged war (Signorino, 1999)[1]. But strategic models have also been used in other contexts. For example, Carson (2005) uses a strategic estimator to analyse the competition of candidates in US House and Senate elections.

This chapter illustrates strategic statistical estimation as developed in Signorino (1999, 2003) by analysing the interaction between a government and a domestic challenger that might lead to repression and/or rebellion. In this interaction, the government decides on whether to use repressive measures such as torture against the challenger, and the challenger decides on whether to challenge the government. To date, to the best of the author's knowledge, the strategic nature of the situation has not explicitly been taken into account by researchers empirically analysing the connection between civil war and government repression in statistical models.

[1] More specifically, the estimator is applied to a subgame introduced in Bueno de Mesquita and Lalman (1992).

Approaches to Geo-mathematical Modelling: New Tools for Complexity Science, First Edition. Edited by Alan G. Wilson.

Companion Website: www.wiley.com/go/wilson/ApproachestoGeo-mathematicalModelling

However, as will be shown, the complexity of strategic estimation in its current form makes it a tool of limited use in this context.

In order to estimate strategic interactions, a source of stochastic uncertainty needs to be introduced to the theoretical model (Signorino, 1999, 2003). Different types of uncertainty have been discussed in Signorino (2003). I discuss here how these different types of uncertainty can be theoretically justified by factors external to the interaction such as third parties or the international community. In addition, external factors or characteristics of the international system can have effects on players' utilities that current strategic estimators do not take into account. I discuss how international factors can cause problems for current estimators and suggest how they could be taken into account more explicitly in the future.

This chapter proceeds as follows. Section 16.2 discusses the theoretical model of the strategic interaction between the government and the domestic challenger and introduces questions that can be answered by empirically estimating it. The third section discusses strategic estimation as introduced in previous literature. The fourth section derives a strategic estimator in the context of the theoretical model introduced here and discusses how the necessary stochastic uncertainty can be justified using international factors. Section 16.5 discusses illustrative results from the empirical estimation of the model. The final section discusses additional complications that could arise if international influences introduce uncertainty in strategic interactions.

16.2 The Theoretical Model

In the following, I introduce a simple strategic situation. The game is initiated by the government deciding whether or not to use repression against a domestic challenger. The challenger could be a domestic ethnic group or the opposition at large. If this was a purely game-theoretical setting, we might assume that the game ends if the government uses repression as any additional action by the opposition is suppressed. However, in a strategic model, and as will be discussed in more detail below in this chapter, we need to be able to attribute positive probability to every empirically possible outcome (Signorino, 1999, 2003). After the government's move, the challenger decides on whether or not to rebel against the government. This strategic situation is illustrated in Figure 16.1[2].

In a game-theoretical model, utilities are perfectly known by the analyst (Signorino, 2003). As a result, the analyst can solve game-theoretical models by using a solution concept such as subgame perfection. Here, a straightforward subgame perfect equilibrium can be derived if utilities are perfectly known. The challenger fights whenever the utility of fighting at a given information set outweighs the utility of doing nothing. Of course, these utilities need to be specified by the analyst (Signorino, 2003). For example, if the analyst specifies that previous repression has a negative effect on the challenger's utility for fighting, the challenger might not fight after repression but might do so if no repression occurs (also see Beiser, 2014).

Similarly, the government represses whenever the expected utility of the repression scenario is larger than the expected utility of the scenario without repression. This could be the case if the government's utility for fighting is low and repression decreases the challenger's utility for an attack (Beiser, 2014; on the latter point, see e.g. Tilly, 1978, in a more general context as

[2] The game is structurally similar to the game introduced in Beiser (2014) and a subgame presented in Pierskalla (2010).

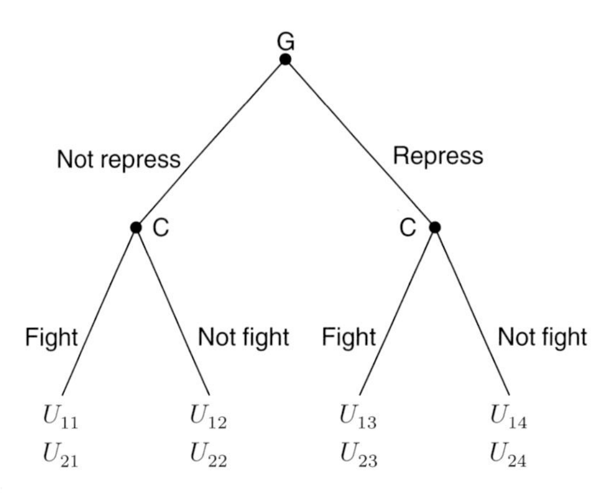

Figure 16.1 The strategic situation.

well as Muller and Weede, 1990). However, repression may also decrease the government's outcome utility via a cost factor (also see Beiser, 2014, following Pierskalla, 2010, and others). In such a case, the government may not have an incentive for using repression.

Here, the interest is less in specifying utilities and solving the game theoretically under these assumptions. Instead, we are interested in using the game as a framework for how challengers and governments interact in the real world. We expect that certain empirically observable variables affect the players' utilities for different outcomes, and we want to test whether these expectations are correct. For this, we must model the strategic interactions explicitly because if we do not take the strategic nature of the interaction into account, we might get the wrong idea about the effect of variables, as has been shown in Signorino (1999, 2003).

The empirical estimation of the game allows answering different questions from what an analytic solution answers. For example, to solve the theoretical model, the analyst needs to specify whether he or she wants repression to serve as a pre-emptive tool by determining whether repression decreases the fighting utility for the challenger. In the empirical model, on the other hand, we can test whether repression can serve as a pre-emptive tool and whether governments use it in such a way as some previous literature has found (e.g. Nordas and Davenport, 2013). For this, we are interested in whether the challenger is less likely to fight after repression than if no repression occurs and whether governments use repression, particularly against challengers with an otherwise high fighting utility (on the latter point, see Beiser, 2014).

16.3 Strategic Estimation

Strategic estimation is a statistical technique that takes the strategic nature of an underlying theoretical model explicitly into account. However, for statistical estimation, it is problematic if outcomes occur with zero probability (Signorino, 1999). For example, we might assume in the theoretical model that a challenger would never fight under repression. As a result, an outcome with repression and war would be expected with zero probability. However, if such a case occurs in the empirical data even though the model predicts it with zero probability, maximum likelihood estimation would be impossible as the likelihood function that multiplies the probability of each observed outcome would be zero irrespective of the values of parameters to be estimated, and thus could not be maximised (Signorino, 1999). As a result, the explicit estimation of strategic models using the maximum likelihood principle requires introducing uncertainty so that all empirically possible outcomes occur with non-zero probability (Signorino, 1999, 2003). Thus, strategic models introduce a stochastic element to players' utilities (Signorino, 1999, 2003).

Drawing on previous work, Signorino (2003) discusses three sources of uncertainty in strategic modelling. In the following three paragraphs, I offer an overview of these three sources of error based on Signorino (2003).

The first type of uncertainty is referred to as a regressor error. In this version, a stochastic element is added to each player's utility for each terminal node. It is assumed that the value of this stochastic element and thus the real utilities are known by players themselves as well as all other players in the game but not by the analyst. The analyst only knows the distribution of each stochastic element, but not its concrete value. Under this assumption, players act on their perfect knowledge, but the analyst can only make probabilistic statements about the course of action as he or she does not fully observe the utilities that players base their strategic calculations on.

The second type of uncertainty is referred to as private information about utilities. Here, the stochastic element is only known to players themselves and is associated with players' utilities for terminal nodes. The analyst and other players, in contrast, do not observe the value of the stochastic element, but they know the distribution it is drawn from. As a result, players must make probabilistic assessments of subsequent players' actions, and the analyst must take these into account as well.

The third type of uncertainty is referred to as agent error. In this version, players and analysts observe all utilities in the game perfectly, but players implement their ideal course of action with error. Thus, the stochastic element is here added to the expected utility of an action as opposed to the utility of terminal nodes. Again, players have to use probabilistic assessments of subsequent players' actions to assess the expected utility of a course of action. The analyst must take these considerations as well as the probability of players' incorrect implementation of strategies into account when assessing the likelihood of outcomes.

As illustrated in Signorino (2003), each of these assumptions produces a different estimator that can result in different estimates. The choice of assumption is thus non-trivial and should be based on theoretical considerations.

In the context of the theoretical model introduced here, all three assumptions about the source of uncertainty can be theoretically justified. Regressor error requires assuming that both the government and the challenger observe each others' utilities perfectly – not unlikely, given they may interact with each other over a long period – but the analyst does not have all data necessary to perfectly measure these utilities empirically (on the latter point, see Signorino, 2003). Agent error can easily be justified by assuming that even though governments and challengers have perfect knowledge and understand what their best response is, they do not perfectly control their repressive apparatus and army, and therefore implementation may differ from the ideal action (also see Esarey *et al.*, 2008). Finally, information about utilities can be assumed to be private. For example, adversaries may not perfectly observe the costliness of fighting for their opponent (Fearon, 1995).

Irrespective of the underlying assumptions about the source of uncertainty, the stochastic elements in the theoretical model are used to derive the probability of each possible action for each player. These probabilities are used to derive the probability of each terminal node in the game being reached. As we observe empirically which terminal node has been reached in each of the observations in the data, we can use the outcome probabilities to estimate parameters for variables that are expected to feed into players' utilities in a maximum likelihood estimation (for details, see Signorino, 1999, 2003). This is discussed in detail in Section 16.4 when the estimator for the theoretical model introduced in this chapter is derived. For this, I introduce theoretical motivations that rely on international factors for two of the types of error introduced in the current section. I illustrate how such expectations about international sources of uncertainty in strategic interactions can be used to derive strategic estimators for the theoretical model on government repression and group rebellion.

16.4 International Sources of Uncertainty in the Context of Repression and Rebellion

Factors external to a strategic interaction can introduce uncertainty. Characteristic for that type of stochastic element is that its actual value is not perfectly known to *any* player in the game at *any* point in time. If external uncertainty is associated with an action, this type of uncertainty is identical to agent error. If this type of uncertainty is associated with a specific outcome, it is akin to private information about utilities. The difference from the usual private utility assumption here is, however, that both players observe neither the other player's *nor* their own stochastic element perfectly. In this section, I illustrate the use of both types of error in the strategic interaction between the government and the domestic challenger introduced in this chapter.

16.4.1 International Sources of Uncertainty Related to Actions

In this subsection, I discuss how international factors can introduce uncertainty about the utilities related to specific actions. Firstly, while a government may observe the cost of fighting perfectly and thus can gauge the utility of ending up in a fight, the cost of repression might be complicated by external factors. A government may observe the opportunity cost of repressing a challenger, but it may not be sure about international consequences of using repression. The international community might implement sanctions, but at the same time such statements may be empty threats. For example, Hafner-Burton (2014) offers an overview of literature assessing in how far aid depends on states' human rights records, and the relationship is far from straightforward. In contrast, investors may expect the government to control the population, and thus there may be a reward related to the use of repression (for an overview of arguments that authoritarianism attracts investment, see Harms and Ursprung, 2002). Thus, it can be assumed that neither the challenger *nor* the government knows how costly the use of repression really is for the government, but they observe the distribution from which this stochastic element is drawn, as suggested in Signorino (2003). If one assumes that this cost and the uncertainty about it are linked to the use of repression as such, *irrespective of what happens afterwards*, then this type of uncertainty is very similar to agent error, as discussed in Signorino (2003)[3]. To illustrate

[3] The only difference here is that I assume the uncertainty to be only related to specific actions, while in Signorino (2003) it is associated with all actions.

this more formally, I add a stochastical element ε_1 to the government's utility in all repression scenarios from Figure 16.1. In the following, $\neg$ stands for an actor not taking the following action.

$$\begin{aligned}
U_{11} &= U_G(\neg r, F)\\
U_{12} &= U_G(\neg r, \neg F)\\
U_{13} &= U_G(r, F) + \varepsilon_1\\
U_{14} &= U_G(r, \neg F) + \varepsilon_1
\end{aligned} \tag{16.1}$$

A similar logic could apply to the challenger's decision. The challenger may not know how the international community reacts to violence, and military interventions are possible. On the other hand, fighting might make the group visible to potential supporters. Again, if this uncertainty is related to the use of violence as such, irrespective of what happens before or afterwards, this is a case of agent error as well. To illustrate the challenger's stochastic element as well, I add ε_2 to the challenger's utility for all outcomes that are achieved by fighting. Again, this element is not known to any player, but its distribution is observed.

$$\begin{aligned}
U_{21} &= U_C(\neg r, F) + \varepsilon_2\\
U_{22} &= U_C(\neg r, \neg F)\\
U_{23} &= U_C(r, F) + \varepsilon_2\\
U_{24} &= U_C(r, \neg F)
\end{aligned} \tag{16.2}$$

This externally based uncertainty is identical to agent error. Thus, the assumption of agent error can be justified if there is an external factor that influences the utility of a player's action.

In the following, I derive the strategic estimator under this type of error, as shown in Signorino (2003). We start by deriving the probability that the challenger fights at each of its information sets. More specifically, we can determine the probability that the utility of challenging outweighs the utility of acquiescing as follows:

$$\begin{aligned}
P_{fight|r} &= P(U_C(r, F) + \varepsilon_2 > U_C(r, \neg F))\\
P_{fight|\neg r} &= P(U_C(\neg r, F) + \varepsilon_2 > U_C(\neg r, \neg F))
\end{aligned} \tag{16.3}$$

Under the assumption that ε_2 follows a normal distribution with mean 0 and standard deviation 1, we can derive these probabilities as

$$\begin{aligned}
P_{fight|r} &= \Phi(U_C(r, F) - U_C(r, \neg F))\\
P_{fight|\neg r} &= \Phi(U_C(\neg r, F) - U_C(\neg r, \neg F))
\end{aligned} \tag{16.4}$$

where Φ denotes the cumulative standard normal distribution. Next, the probability of government repression can be determined. The government uses repression against the challenger if

$$\begin{aligned}
&P_{fight|r}\,(U_G(F, r) + \varepsilon_1) + (1 - P_{fight|r})(U_G(\neg F, r) + \varepsilon_1)\\
&> \quad P_{fight|\neg r} U_G(F, \neg r) + (1 - P_{fight|\neg r}) U_G(\neg F, \neg r)
\end{aligned}$$

$$\Leftrightarrow$$

$$P_{fight|r}\, U_G(F,r) + (1 - P_{fight|r})U_G(\neg F, r) + \varepsilon_1$$
$$> \quad P_{fight|\neg r} U_G(F, \neg r) + (1 - P_{fight|\neg r})U_G(\neg F, \neg r) \tag{16.5}$$

If we assume that ε_1 is distributed identically to ε_2, but is also independent from it, we can derive this probability to be

$$\begin{aligned} P_{rep} &= P(P_{fight|r} U_G(F,r) + (1 - P_{fight|r})(U_G(\neg F, r)) - \\ &\quad (P_{fight|\neg r} U_G(F, \neg r) + (1 - P_{fight|\neg r})U_G(\neg F, \neg r)) > -\varepsilon_1) \\ &= \Phi(P_{fight|r} U_G(F,r) + (1 - P_{fight|r})(U_G(\neg F, r)) - \\ &\quad (P_{fight|\neg r} U_G(F, \neg r) + (1 - P_{fight|\neg r})U_G(\neg F, \neg r))) \end{aligned} \tag{16.6}$$

We can calculate the probability of each terminal node in the game as a function of each of these action probabilities. As we assume that the stochastic components feeding into these probabilities are independent, we can just multiply the action probabilities that lead to any specific outcome.

More specifically, we can specify that

$$\begin{aligned} P(rep, fight) &= P_{rep} * P_{fight|r} \\ P(rep, \neg fight) &= P_{rep} * (1 - P_{fight|r}) \\ P(\neg rep, fight) &= (1 - P_{rep}) * P_{fight|\neg r} \\ P(\neg rep, \neg fight) &= (1 - P_{\neg rep}) * (1 - P_{\neg fight|\neg r}) \end{aligned} \tag{16.7}$$

Now, we can determine a likelihood function that can be used to estimate parameters associated with variables expected to affect actors' utilities for different outcomes as follows:

$$\mathcal{L} = \prod_{i=1}^{n} (P(\neg rep, fight)^{\alpha_1 i} P(\neg rep, \neg fight)^{\alpha_2 i} P(rep, fight)^{\alpha_3 i} P(rep, \neg fight)^{\alpha_4 i}) \tag{16.8}$$

where $\alpha_1 i$ through $\alpha_4 i$ are indicator variables that take a value of 1 if the outcome they are associated with is realised in i, and 0 otherwise.

16.5 International Sources of Uncertainty Related to Outcomes

Moreover, not only the challenger but also the government may be concerned about international intervention. If neither the government nor the challenger can be sure whether external actors will end up supporting or challenging them, both are uncertain about their *own* cost of fighting. In that case, the interesting thing is that both players do not gain knowledge about the actual values of these stochastic terms – including the one affecting their own utility directly – throughout the game when they make decisions. In that sense, this error is different from all types of error discussed in Signorino (2003). It is clearly not agent error, as for the government that moves first it is associated with specific outcomes as opposed to an action. It is not regressor error, as both players do not even know their own error. In addition, it is also not private information, as the values of the stochastic terms are never revealed to any of the players and they act exclusively on their knowledge of their distribution as well. However, as will be seen, in the current setting, the resulting estimator is nevertheless very similar to what would be expected under the private information assumption from Signorino (2003)[4].

I illustrate the type of error discussed here by adding stochastic elements ε_3 and ε_4 to the utility of all fighting outcomes for both actors. In the following, I derive the strategic estimator under this assumption, as shown in Signorino (2003).

$$\begin{aligned}
U_{11} &= U_G(\neg r, F) + \varepsilon_3 \\
U_{12} &= U_G(\neg r, \neg F) \\
U_{13} &= U_G(r, F) + \varepsilon_3 \\
U_{14} &= U_G(r, \neg F) \\
U_{21} &= U_C(\neg r, F) + \varepsilon_4 \\
U_{22} &= U_C(\neg r, \neg F) \\
U_{23} &= U_C(r, F) + \varepsilon_4 \\
U_{24} &= U_C(r, \neg F)
\end{aligned} \tag{16.9}$$

Here, it is assumed that each player knows he or she may receive external support or indeed be challenged by external forces, and they know that the same holds for the other player. However, it is assumed here that these forces act though the cost of fighting as opposed to the actors' probability of winning in a fight. That way, we can assume that ε_3 and ε_4 are independent of each other. If errors affected the probability of winning, this assumption would no longer hold as an increase in one player's probability of winning also results in a decrease in the other player's probability of winning. Considering external support as affecting both players' utilities in adverse ways would be an interesting extension of the model presented here.

Again, we start by deriving the probability that the challenger fights at each of its information sets. As the last actor's decision is not strategic in nature, the assumption about the source of uncertainty does not matter here. As a result, under the assumption that ε_4 follows a normal distribution with mean 0 and variance 1, the probability we derive is identical to the one derived above.

$$\begin{aligned}
P_{fight|r} &= \Phi(U_C(F, r) - U_C(\neg F, r)) \\
P_{fight|\neg r} &= \Phi(U_C(F, \neg r) - U_C(\neg F, \neg r))
\end{aligned} \tag{16.10}$$

Next, we can determine the probability that the government uses repression as the probability that

$$\begin{aligned}
&P_{fight|r}(U_G(F, r) + \varepsilon_3) + (1 - P_{fight|r})U_G(\neg F, r) \\
&> P_{fight|\neg r}(U_G(F, \neg r) + \varepsilon_3) + (1 - P_{fight|\neg r})U_G(\neg F, \neg r)
\end{aligned} \tag{16.11}$$

[4] Again, the difference is that uncertainty is only asssumed to be related to some outcomes.

That is,

$$
\begin{aligned}
P_{rep} &= P(P_{fight|r}U_G(F,r) + (1-P_{fight|r})U_G(\neg F,r) - (1-P_{fight|\neg r})U_G(\neg F,\neg r) \\
&- P_{fight|\neg r}U_G(F,\neg r) \\
&> (P_{fight|\neg r} - P_{fight|r})\varepsilon_3) \\
&= \Phi\left(\frac{P_{fight|r}U_G(F,r) + (1-P_{fight|r})U_G(\neg F,r)}{(P_{fight|\neg r} - P_{fight|r})}\right. \\
&+ \left.\frac{-(1-P_{fight|\neg r})U_G(\neg F,\neg r) - P_{fight|\neg r}U_G(F,\neg r)}{(P_{fight|\neg r} - P_{fight|r})}\right)
\end{aligned}
\tag{16.12}
$$

Here, the difference is that the variance of the distribution differs across observations with the value of $\sqrt{(P_{fight|\neg r} - P_{fight|r})^2} = (P_{fight|\neg r} - P_{fight|r})$. This is because the government's utility for outcomes, which includes the stochastic element ε_3, is moderated by the probabilities with which the government will end up in a fight or not. Subsequently, the likelihood function can be derived by multiplying action probabilities to arrive at outcomes as illustrated above. This is justified by the assumption that the stochastic elements in both players' utilities are independent.

16.6 Empirical Analysis

16.6.1 Data and Operationalisation

In this chapter thus far, two assumptions about error structure that can be justified with international factors were discussed: uncertainty about the utility of specific actions and uncertainty about the utility of specific outcomes. In this section, I illustrate the empirical estimation of the model under the first assumption about uncertainty – the one where uncertainty is related to actions.

However, because of the complexity of the model, the ability to derive results as well as the results themselves depend heavily on the starting values used. For this reason, a range of different starting values at which results could be obtained are considered, and the model with the highest log-likelihood value was chosen for presentation in this chapter. Nevertheless, because of the instability of the model, results presented need to be seen as an illustration as opposed to a basis for inference. In addition, only a limited number of explanatory variables is included here for demonstrative purposes.

I use data on domestic ethnic groups' interactions with the government of a state from the Minorities at Risk Project (2009), as this data includes information about group-specific repression. One case in the data is thus the interaction between a government and a specific domestic ethnic group in a given year. The dependent variable y measures which combination of government repression and group rebellion occurs between the ethnic group and the government in this year. The outcome categories in the data are thus identical to the final nodes in the theoretical model introduced in this chapter.

$$y \in \{r \wedge f, r \wedge \neg f, \neg r \wedge f, \neg r \wedge \neg f\} \tag{16.13}$$

Group rebellion is measured using data from Minorities at Risk (Minorities at Risk Project, 2009) on whether a group is engaged in rebellion in a given year[5]. The variable on fighting measures whether a group started a rebellion in a given year, that is, whether there was no rebellion in at least the previous 2 years[6]. The variable on government repression measures whether the government uses at least one of 16 selected repressive measures in the Minorities at Risk data (Minorities at Risk Project, 2009) against members of an ethnic group that are not involved in collective action[7].

Now, variables that are expected to influence an actor's utility for each of these final outcomes need to be specified.

$U_G(\neg r, F)$: The government's utility for fighting in the absence of repression is modelled as a function of the country's regime type. For this, I use data from Polity 4 (Polity IV Project, 2009). These data are extracted from the Quality of Government database (hereafter, QoG) (Teorell *et al.*, 2013). This data source provides a 21-category indicator where the highest value indicates the highest level of democracy. In addition, I include a constant term here, which is a vector carrying the value of 1 for each observation.

$U_G(\neg r, \neg F)$: This utility is normalised to 0.

$U_G(r, F)$: The government's utility for fighting in the presence of repression is modelled as a function of the country's GDP per capita in hundreds of thousands, using data from Gleditsch (2002) that is also extracted from Teorell *et al.* (2013). A constant term is included here as well.

$U_G(r, \neg F)$: This utility is normalised to 0 as well.

$U_C(\neg r, F)$: This utility is modelled as a function of a group's current status in society. For this, I use a five-category indicator of political discrimination from the Minorities at Risk Project (2009), where the highest category indicates the highest level of discrimination. In addition, I include a constant here as well.

$U_C(\neg r, \neg F)$: I normalise this utility to 0.

$U_C(r, F)$: The group's utility of fighting in the presence of repression is modelled as a function of the government's military strength in relation to the group's population. For this, I include a variable on the government's military personnel per capita (data from World Bank, 2011) as a proportion of the potential challenger's population (data from Minorities at Risk Project, 2009); missing values are linearly interpolated). Again, a constant is added here.

$U_C(r, \neg F)$: I normalise this utility to 0 as well.

---

[5] This variable carries the value of 1 if the group was engaged in the minimum level of rebellion, defined by the data source as "[p]olitical banditry, sporadic terrorism" (Minorities at Risk Project, 90).

[6] The 2 years after an active year of conflict as well as all years of active conflict that are not onsets are excluded here as the probability of an onset is 0 in those cases.

[7] Specifically, the measures of repression that are used here are to determine the presence of repression against civilians from the MAR data are "[f]ew group members arrested", "[m]any group members arrested", "[l]eaders arrested, disappeared, or detained", "[s]how trials of group members or leaders", "[t]orture used to intimidate or interrogate", "[m]embers executed by authorities", "[l]eaders executed by authorities", "[s]ystematic killings by paramilitaries", "[p]roperty confiscated or destroyed", "[r]estrictions on movement", "[f]orced resettlement", "[i]nterdiction of food supplies", "[e]thnic cleansing", "[s]ystematic domestic spying", "[s]tates of emergency" and "[s]aturation of police and military" (Minorities at Risk Project, 91–96).

16.7 Results

In this chapter, two model specifications were introduced: one where the source of error is related to actions and one where it is related to outcomes. In this section, I show results from the first specification, where international uncertainty is assumed to be exclusively linked to actions. Table 16.1 shows the estimated coefficients linked to the variables that are expected to be part of the utilities as introduced here. We can thus derive from the coefficients whether a specific variable has a positive or a negative effect on a player's utility for a specific outcome.

Under this model, if all variables are held at their mean, repression is predicted with a probability of about 9%. Under no repression, group rebellion is predicted with a probability of 2.2%, while under repression it is predicted with a probability of 5.9%.

Discrimination, whose effect on the group's likelihood of rebellion was tested in the no-repression scenario, has a significantly positive effect here. A state's GDP, included as part of the government's utility of the repression and rebellion outcome, has a significantly negative effect. The other variables do not reach statistical significance.

When all variables are held at their mean, governments' utility for fighting in the no-repression scenario is slightly higher than in the repression scenario. This would suggest that repression not helping a government during a fight. Moreover, they are both negative, suggesting that the government generally prefers not to fight in this scenario.

In addition, as a group's probability of fighting is larger under repression, repression seems to have a provoking effect on groups. Moreover, when all variables are held at their mean, groups' utility for fighting is also negative under both scenarios, suggesting they would also prefer not to fight. However, these utilities are much larger than the governments' utilities for fighting, suggesting that the government has an even bigger incentive to prevent a fight.

Table 16.1 The effect of variables on actor utilities. Errors are assumed to be related to actions

	Parameter	p-value
$U_G(\neg r, F)$		
Constant	−34.46	0.03
Regime type	−0.12	0.66
$U_G(r, F)$		
Constant	−30.25	0.00
GDP	−55.78	0.01
$U_C(\neg r, F)$		
Constant	−2.25	0.00
Discrimination level	0.14	0.01
$U_C(r, F)$		
Constant	−1.55	0.00
Military personnel to group ratio	0.00	0.87
N		1310

Thus, in sum, this illustrative model would suggest that repression provokes groups to rebellion and does not offer governments an advantage during a fight.

However, due to issues introduced here, these results can only serve as an illustration of answers that strategic models may be used to obtain. In its current form, strategic estimation is of limited use for analysing the data presented here.

16.8 Additional Considerations Related to International Uncertainty

This chapter illustrates how strategic models can be estimated and how international sources of uncertainty can be used to motivate assumptions about stochastic elements. It has been shown that this tool in its current form is of limited use in the context shown. In addition, if international factors are indeed the source of uncertainty in a strategic model, additional complications may arise that should be taken into account by estimators.

Firstly, if uncertainty stems from a lack of knowledge about the reaction of the international community to the actions of different actors, it is unlikely that different actors' stochastic elements are fully independent. For example, it is more likely that international actors that sanction repression also sanction violence by the challenger and vice versa. Leemann (2014) introduces an estimator that can take correlations between agent errors into account. Similarly, if an actor observes their opponent's action, they may – rightfully or wrongfully – draw a conclusion about the value of the stochastic element that the other player's decision relies on. As a result, they may update their expectation about the value of the uncertain element affecting their own utility. This would need to be taken into account explicitly by models and would be a fruitful avenue for further research. Some have taken actors' ability to learn about their opponent from the opponent's initial move into account (e.g. Esarey *et al.*, 2008). Here, a player's action provides information about their future action and is thus only relevant if this actor acts a second time. In contrast, in the specification introduced here, a player's action might provide information about the *other* player's utility and is thus relevant even if this player does not act repeatedly. This would be particularly the case if external factors affect actors' probability of winning in a fight, quantities that are directly related as they must sum to the value of 1. Moreover, the assumption that costs of fighting are independent of each other that is introduced above is also likely not to hold.

Secondly, the likelihood of international factors introducing uncertainty may be taken into account more explicitly and modelled. For example, the degree to which a state is linked to other states via trade networks may not help predict the value of the stochastic element but may allow inference about the degree of uncertainty. States that are linked to many other states, all of whose reaction to an action is not known, should face more uncertainty than isolated states. Similarly, as international interdependence likely increases with time, internationally based uncertainty might increase with time as well. These changes in the error structure should be analysed and indeed taken into account explicitly if necessary.

16.9 Conclusion

This chapter demonstrates the use of a strategic model in the context of government repression and civil war. It has been found that strategic estimation increases the complexity of the model considerably and is thus only of limited use for the analysis of the data under scrutiny.

The chapter also discusses how uncertainty can be derived from international sources. This derivation provides additional substantive justifications that can help researchers determine what type of uncertainty structure to use. In addition, uncertainty stemming from external sources opens up a number of additional questions that should be explored in future research.

References

Beiser J. (2014). *Targeting the Motivated? The Pre-emptive Use of Government Repression*. Working paper. London: University College London.

Bueno de Mesquita B and Lalman D. (1992). *War and Reason*. New Haven, CT: Yale University Press.

Carson JL. (2005). Strategy, selection, and candidate competition in U.S. House and Senate elections. *Journal of Politics* 67(1): 1–28.

Cederman LE, Wimmer A and Min B. (2010). Why do ethnic groups rebel? New data and analysis. *World Politics* 62(1): 87–119.

Esarey J, Mukherjee B and Moore WH. (2008). Strategic interaction and interstate crises: a Bayesian quantal response estimator for incomplete information games. *Political Analysis* 16(3): 250–273.

Fearon JD. (1995). Rationalist explanations for war. *International Organization* 49(3): 379–414.

Gleditsch KS. (2002). Expanded trade and GDP data. *Journal of Conflict Resolution* 46(5):712–24. Available at http://privatewww.essex.ac.uk/~ksg/exptradegdp.html

Hafner-Burton EM. (2014). A social science of human rights. *Journal of Peace Research* 51(2): 273–286.

Harms P and Ursprung HW. (2002). Do civil and political repression really boost foreign direct investments? *Economic Inquiry* 40(4): 651–663.

Hill DWJ and Jones ZM. (2014). An empirical evaluation of explanations for state repression. *American Political Science Review* 108(3): 661–687.

Leemann L. (2014). Strategy and sample selection: a strategic selection estimator. *Political Analysis* 22(3): 374–397.

McKelvey RD and Palfrey TR. (1995). Quantal response equilibria for normal form games. *Games and Economic Behavior* 10(1): 6–38.

McKelvey RD and Palfrey TR. (1996). A statistical theory of equilibrium in games. *Japanese Economic Review* 47(2): 186–209.

McKelvey RD and Palfrey TR. (1998). Quantal response equilibria for extensive form games. *Experimental Economics* 1(1): 9–41.

Minorities at Risk Project (????). *DATASET USERS MANUAL 030703*. http://www.cidcm.umd.edu/mar/.

Minorities at Risk Project (2009). *Minorities at Risk Dataset*, Center for International Development and Conflict Management, College Park, MD. http://www.cidcm.umd.edu/mar/.

Muller EN and Weede E. (1990). Cross-national variation in political violence: a rational action approach. *Journal of Conflict Resolution* 34(4): 624–651.

Nordås R and Davenport C. (2013). Fight the youth: youth bulges and state repression. *American Journal of Political Science* 57(4): 926–940.

Pierskalla JH. (2010). Protest, deterrence, and escalation: the strategic calculus of government repression. *Journal of Conflict Resolution* 54(1): 117–145.

Polity IV Project (2009) *Polity IV Data Set*. Available at http://www.systemicpeace.org/inscr/inscr.htm

Signorino CS. (1999). Strategic interaction and the statistical analysis of international conflict. *American Political Science Review* 93(2): 279–297.

Signorino CS. (2003). Structure and uncertainty in discrete choice models. *Political Analysis* 11(4): 316–344.

Teorell J, Charron N, Dahlberg S, Holmberg S, Rothstein B, Sundin P and Svensson R. (2013). *The Quality of Government Dataset. Version 15 May*. Gothenburg: The Quality of Government Institute, University of Gothenburg. Available at http://www.qog.pol.gu.se

Tilly C. (1978). *From Mobilization to Revolution*. New York: McGraw-Hill.

World Bank. (2011). World Development Indicators: armed forces personnel, total. Available at http://data.worldbank.org/indicator/MS.MIL.TOTL.P1

17

A General Framework for Static, Spatially Explicit Games of Search and Concealment

Thomas P. Oléron Evans, Steven R. Bishop and Frank T. Smith

17.1 Introduction

In recent years, there has been a growing appreciation of the potential role of game theory as an analytical tool, enabling public and private organisations to make more informed strategic decisions on questions of security and the deployment of resources. Simultaneously, the significant expansion in the field of digital mapping through the development of geographic information systems (GIS) has allowed for an unprecedented degree of spatial sophistication in our approach to a huge variety of real-world problems. The combination of these two areas – the application of game theory to security questions and the development of spatial technologies – offers an opportunity for the development of a new suite of spatially explicit mathematical modelling solutions for security scenarios. However, it also highlights the need for the creation of a broad modelling framework for the field of spatial game theory, which would act as a foundation upon which such new tools could be built.[1]

In this chapter, our goal is take steps towards providing a framework of the sort described here. We will define a general search and concealment game that takes full account of the spatial structure of the space over which it is played; it is suitable for application in a wide range of security scenarios. The game is static in the sense that players do not move, but deploy simultaneously at particular spatial points and receive payoffs based on their relative positions. In this way, the Static Spatial Search Game (SSSG) provides a theoretical foundation for the

[1] The material in Chapter 17 and Appendix D is adapted from the article "Static search games played over graphs and general metric spaces", published in the *European Journal of Operational Research* (Oléron Evans and Bishop, 2013).

Approaches to Geo-mathematical Modelling: New Tools for Complexity Science, First Edition. Edited by Alan G. Wilson.

Published 2016 by John Wiley & Sons, Ltd.
Companion Website: www.wiley.com/go/wilson/ApproachestoGeo-mathematicalModelling

study of the relative strategic value of different positions in a geography. Using the theory of metric spaces, we model situations in which the searching player may simultaneously search multiple locations based on concepts of distance or adjacency relative to the point at which they are deployed.

The SSSG was inspired by the "geometric games" of Ruckle (1983) and particularly by White's "games of strategy on trees" (1994), of which this work may be seen as a significant generalisation. However, while the SSSG certainly builds upon previous work, its simplicity and generality together with its explicit consideration of spatial structure set it apart from much of the literature (see Section 17.3 for a detailed review of related work) and lend it the versatility to describe games over a huge variety of different spaces. The primary contributions of this work are therefore to both propose a highly general model of spatial search and concealment situations, which unites several other games presented in the literature (see Section 17.4.2), and present new propositions and approaches for the strategic analysis of such scenarios.

While the approach presented here is theoretical in nature, the SSSG provides a framework for the analysis of a diverse range of security questions. Aside from explicit search and concealment scenarios, the game may be used to model situations in which some structure or region must be protected against 'attacks' that could arise at any spatial point; for example, the deployment of security personnel to protect cities against terrorist attacks or outbreaks of rioting, security software scanning computer networks to eliminate threats, the defence of shipping lanes against piracy, the protection of a rail network against cable theft or the deployment of stewards at public events to respond to emergency situations.

In this chapter, we provide a brief overview of all necessary game theoretic concepts and a review of the literature on games of search and security, before formally defining the SSSG, examining its relationship to other games in the literature and presenting some initial propositions relating to its strategies. We then restrict our attention to the case of the SSSG played on a graph (the Graph Search Game, or GSG) and identify upper and lower bounds on the value of such games.

Our work on the SSSG is developed further in Appendices C and D.

Appendix C explores several different methods for the analysis and solution of GSGs, starting with an algorithmic approach that employs the iterated elimination of dominated strategies, with a particular focus on games played on trees. Further results include a way to simplify GSGs through consideration of graph automorphisms and an examination of a particular type of strategy for such games, which we describe as an 'equal oddments strategy'. The concept of an equal oddments strategy is then used to find analytic solutions for a particular family of GSGs.

Appendix D extends the SSSG to encompass situations in which the points of the underlying geography can be assigned non-uniform values. This extension is used to determine strategies of optimal random patrol for a player attempting to protect a particular space from an infiltrator.

17.2 Game Theoretic Concepts

The definitions and notation relating to game theory used in this section are adapted from Blackwell and Girshick (1979) and Morris (1994).

When discussing two-player games, we assume the following definition:

Definition 17.1. *A **two-player game in normal form** between Players* A *and* B *consists of:*

- ***strategy sets*** $\Sigma_{\mathrm{A}}, \Sigma_{\mathrm{B}}$
- ***payoff functions*** $p_{\mathrm{A}}, p_{\mathrm{B}}$, *with:*

$$p_{\mathrm{A}} : \Sigma_{\mathrm{A}} \times \Sigma_{\mathrm{B}} \to \mathbb{R}$$
$$p_{\mathrm{B}} : \Sigma_{\mathrm{A}} \times \Sigma_{\mathrm{B}} \to \mathbb{R}$$

If the payoffs are such that for some constant c:

$$p_{\mathrm{A}}(x, y) + p_{\mathrm{B}}(x, y) = c \ , \ \forall x \in \Sigma_{\mathrm{A}}, \forall y \in \Sigma_{\mathrm{B}}$$

then the game is described as a constant-sum game.

The game is played by Players A and B simultaneously choosing **strategies** (described as **pure strategies** in cases where there may be any ambiguity with mixed strategies, described further in this chapter) from their respective strategy sets $x \in \Sigma_{\mathrm{A}}, y \in \Sigma_{\mathrm{B}}$ and receiving payoffs $p_{\mathrm{A}}(x, y), p_{\mathrm{B}}(x, y)$. The objective of each player is to maximise their payoff.

If Σ_{A} and Σ_{B} are finite, with:

$$\begin{aligned} \Sigma_{\mathrm{A}} &= \{x_1, \dots, x_{\kappa_{\mathrm{A}}}\} \\ \Sigma_{\mathrm{B}} &= \{y_1, \dots, y_{\kappa_{\mathrm{B}}}\} \end{aligned} \tag{17.1}$$

for some positive integers κ_{A}, κ_{B}, then it is often convenient to collect the payoffs to each player in **payoff matrices**:

$$\begin{aligned} \mathfrak{P}_{\mathrm{A}} &= (p_{\mathrm{A}}(x_i, y_j))_{i \in \{1, \dots, \kappa_{\mathrm{A}}\}, j \in \{1, \dots, \kappa_{\mathrm{B}}\}} \\ \mathfrak{P}_{\mathrm{B}} &= (p_{\mathrm{B}}(x_j, y_i))_{i \in \{1, \dots, \kappa_{\mathrm{B}}\}, j \in \{1, \dots, \kappa_{\mathrm{A}}\}} \end{aligned} \tag{17.2}$$

Note that, as defined here, the rows of each matrix correspond to the strategies of the player receiving the relevant payoffs, while the columns correspond to the strategies of their opponent.

In certain circumstances, we may allow players to adopt **mixed strategies**, whereby they choose their pure strategy according to a specified probability distribution. If the strategy sets are finite, as given in equation (17.1), the mixed strategies $\boldsymbol{\sigma}_{\mathrm{A}}$, $\boldsymbol{\sigma}_{\mathrm{B}}$ can simultaneously be regarded as vectors:

$$\boldsymbol{\sigma}_{\mathrm{A}} = (\sigma_{\mathrm{A}}[x_1], \dots, \sigma_{\mathrm{A}}[x_{\kappa_{\mathrm{A}}}]) \in [0, 1]^{\kappa_{\mathrm{A}}}$$
$$\boldsymbol{\sigma}_{\mathrm{B}} = (\sigma_{\mathrm{B}}[y_1], \dots, \sigma_{\mathrm{B}}[y_{\kappa_{\mathrm{B}}}]) \in [0, 1]^{\kappa_{\mathrm{B}}}$$

and as functions, which allocate probabilities to pure strategies:

$$\begin{aligned} \sigma_{\mathrm{A}} : \Sigma_{\mathrm{A}} &\to [0, 1] \\ x &\mapsto \sigma_{\mathrm{A}}[x] \\ \sigma_{\mathrm{B}} : \Sigma_{\mathrm{B}} &\to [0, 1] \\ y &\mapsto \sigma_{\mathrm{B}}[y] \end{aligned}$$

$$\sum_{x \in \Sigma_A} \sigma_A[x] = \sum_{y \in \Sigma_B} \sigma_B[y] = 1$$

The following definitions relate to the maximum expected payoff that players can guarantee themselves through careful choice of their mixed strategies:

Definition 17.2. *Given a two-player game, the* **values of the game** u_A, u_B *to Players* A *and* B *respectively are defined as:*

- $u_A = \max_{\tau_A} \min_{\tau_B} \mathrm{E}[p_A(\boldsymbol{\tau}_A, \boldsymbol{\tau}_B)]$
- $u_B = \max_{\tau_B} \min_{\tau_A} \mathrm{E}[p_B(\boldsymbol{\tau}_A, \boldsymbol{\tau}_B)]$

where $\boldsymbol{\tau}_A$ *and* $\boldsymbol{\tau}_B$ *range across all possible mixed strategies for Players* A *and* B *respectively and*

$$\mathrm{E}[p_A(\boldsymbol{\tau}_A, \boldsymbol{\tau}_B)]$$

$$\mathrm{E}[p_B(\boldsymbol{\tau}_A, \boldsymbol{\tau}_B)]$$

represent the expected payoffs to each player, given that they respectively adopt mixed (or pure) strategies $\boldsymbol{\tau}_A$ *and* $\boldsymbol{\tau}_B$.

Definition 17.3. *Given a two-player constant-sum game, where the payoffs sum to* $c \in \mathbb{R}$, *mixed strategies* $\boldsymbol{\sigma}_A$, $\boldsymbol{\sigma}_B$ *for Players* A *and* B *are described as* **optimal mixed strategies** *(OMSs) if and only if:*

- $\min_{\tau_B} \mathrm{E}[p_A(\boldsymbol{\sigma}_A, \boldsymbol{\tau}_B)] = u_A$
- $\min_{\tau_A} \mathrm{E}[p_B(\boldsymbol{\tau}_A, \boldsymbol{\sigma}_B)] = u_B$

where $\boldsymbol{\tau}_A$ *and* $\boldsymbol{\tau}_B$ *range across all possible mixed strategies for Players* A *and* B *respectively.*[2]

For a constant-sum game, where the payoffs sum to $c \in \mathbb{R}$, we have:

$$u_A + u_B = c \tag{17.3}$$

Also, provided that Σ_A and Σ_B are finite, OMSs are guaranteed to exist for both players.

Both of these facts are consequences of the Minimax Theorem (see Morris, 1994: p 102).

Given a constant-sum two-player game with finite strategy sets, a **solution** of the game comprises OMSs $\boldsymbol{\sigma}_A$, $\boldsymbol{\sigma}_B$ and values u_A, u_B for each player.

The following definition allows for a crude comparison of the efficacy of different strategies.

Definition 17.4. *Consider a two-player game with strategy sets* Σ_A, Σ_B *and payoff functions* p_A, p_B. *Given particular pure strategies* $x_1, x_2 \in \Sigma_A$ *for Player* A, *we have:*

[2] Note that, under this definition, it is possible for a mixed strategy to be simultaneously optimal and inadmissible (weakly dominated by some other strategy, in the sense of Definition 17.4). While some authors restrict the definition of an optimal mixed strategy to prohibit such cases, we follow the example of Morris (1994: p 44), who does not.

- x_2 ***very weakly dominates*** x_1 *if and only if:*

$$p_A(x_2, y) \geq p_A(x_1, y), \; \forall y \in \Sigma_B$$

- x_2 ***weakly dominates*** x_1 *if and only if:*

$$p_A(x_2, y) \geq p_A(x_1, y), \; \forall y \in \Sigma_B$$

and $\exists y^* \in \Sigma_B$ *such that:*

$$p_A(x_2, y^*) > p_A(x_1, y^*)$$

- x_2 ***strictly dominates*** x_1 *if and only if:*

$$p_A(x_2, y) > p_A(x_1, y), \; \forall y \in \Sigma_B$$

- x_2 *is* ***equivalent*** *to* x_1 *if and only if:*

$$p_A(x_2, y) = p_A(x_1, y), \; \forall y \in \Sigma_B$$

Since the designation of the players as A and B is arbitrary, obtaining corresponding definitions of strategic dominance and equivalence for Player B is simply a matter of relabelling.

Note that weak dominance, strict dominance and equivalence are all special cases of very weak dominance. Also, strict dominance is a special case of weak dominance.

In this chapter and those that follow, weak dominance is of most relevance. Therefore, for reasons of clarity, the terms "dominance" and "dominated strategies" will be used to refer to weak dominance unless otherwise stated.

Since a player aims to maximise his or her payoff, we would intuitively expect that they should not play any dominated strategies. Indeed, it is known that any strategy that is strictly dominated by some other strategy must be allocated zero probability in an OMS (see e.g. Theorem 2.9 in Morris, 1994: p 49), though this is not necessarily true of weakly or very weakly dominated strategies.

For a general definition of dominance in game theory, see Leyton-Brown and Shoham (2008: pp 20–23), from which the above definition was adapted.

17.3 Games of Search and Security: A Review

17.3.1 Simple Search Games

Games of search and concealment, in which one player attempts to hide themselves or to conceal some substance in a specified space while another player attempts to locate or capture the player or substance, have been widely studied. One of the simplest search games is the well-known high–low number-guessing game in which one player chooses an integer in a given range, while the other player makes a sequence of guesses to identify it, each time being informed whether the guess was too high or too low (Gal, 1974). Continuous versions of the game have also been studied (Alpern, 1985; Baston and Bostock, 1985; Gal, 1978).

Another simple search game involves one player attempting to locate an object that the opposing player has hidden at a location chosen from a finite or countably infinite set with no spatial structure (except a possible ordering). Variants of these games include examples where the searching player has some chance of overlooking the object despite searching the correct location (Neuts, 1963; Subelman, 1981) or where the searcher must simultaneously avoid the location of a second hidden object (Ruckle, 1990).

17.3.2 Search Games with Immobile Targets

A more complicated class of search games is that in which the searching player is mobile and their target is immobile, with payoffs to each player typically (though not universally) being dependent on the amount of time that elapses (or the distance travelled) before the target is located. Such games have been examined over many different types of graph (Alpern, 2008, 2010; Anderson and Aramendia, 1990; Buyang, 1995; Kikuta, 2004; Kikuta and Ruckle, 1994; Pavlović, 1995; Reijnierse and Potters, 1993), though in the most general case the space may be a continuous region (Gal, 1979). While the starting position of the searching player is often fixed, games in which the searching player can choose their position have also been studied (Alpern et al., 2008a; Dagan and Gal, 2008), as have games with multiple searchers (Alpern and Howard, 2000; Alpern and Reyniers, 1994).

17.3.3 Accumulation Games

Accumulation games are an extension of the above concept in which there may be many hidden objects (Kikuta and Ruckle, 1997) or in which hidden objects are replaced with some continuous material that the hiding player can distribute across a set of discrete locations (Kikuta and Ruckle, 2002; Zoroa et al., 2004) or across a continuous space (Ruckle and Kikuta, 2000). The payoffs in these games typically depend on the number of objects or the quantity of material that the searching player is able to locate.

17.3.4 Search Games with Mobile Targets

Adding a further layer of complication, there is the class of search game in which both the searching player and the hiding player are mobile, including so-called ‘princess and monster’ games. Again, the payoffs in such games are typically dependent on the amount of time that elapses before the hiding player is captured and players are typically ‘invisible’ to each other, only becoming aware of the location of their opponent at the moment of capture.

Such games have been considered over continuous one-dimensional spaces such as the circle (Alpern, 1974) and the unit interval (Alpern et al., 2008b), over continuous graphs or networks (Alpern and Asic, 1985; Alpern and Asic, 1986; Anderson and Aramendia, 1992) and over continuous two-dimensional spaces (Chkhartishvili and Shikin, 1995; Foreman, 1977; Garnaev, 1991). In the latter case, it is necessary to introduce the concept of a *detection radius*, with a capture occurring if the distance between the players drops below this value. In some cases, the probability of capture is allowed to vary based on the distance between the players (Garnaev, 1992).

Analyses of search games over discrete spaces in which both searcher and hider are mobile have tended to consider spatial structure in only a very limited way. While this structure may determine the freedom of movement of the players, very little work has been done to introduce an analogous concept to the detection radius to such games. Generally, players move sequentially and may only move to locations that are sufficiently close to their current position (e.g. Eagle and Washburn, 1991), though variants have been considered in which either the searching player (Zoroa et al., 2012) or the hiding player (Thomas and Washburn, 1991) has the freedom to move to any location regardless of adjacency or distance.

Further variations on the search game with mobile searcher and hider include games in which the searching player follows a predetermined path and must decide how thoroughly to search each location visited (Hohzaki and Iida, 2000), games in which the searching player must intercept an opponent attempting to move from a given start point to a given end point (Alpern, 1992) and games with a variegated environment and the possibility that the hiding player will be betrayed by 'citizens' of the space (Owen and McCormick, 2008). Such games have also been used to model predator–prey interactions (Alpern et al., 2011a).

17.3.5 Allocation Games

Allocation games are a related concept, in which the searching player does not move around the space individually, but rather distributes 'search resources' to locate the mobile hiding player. Such games may include false information (Hohzaki, 2007) and may incorporate spatial structure by allowing the influence of resources to spread across space ('reachability'), an area which has seen 'little research' (Hohzaki, 2008).

Variations on this idea include situations in which searching resources are deployed sequentially (Dendris et al., 1997) or both players distribute resources to respectively locate or protect a hidden object (Baston and Garnaev, 2000). Cooperative allocation games, in which multiple players combine their searching resources to locate a moving target, have also been considered (Hohzaki, 2009).

17.3.6 Rendez-vous Games

Rendez-vous games are a parallel concept to games with mobile searching and hiding players, the difference being that these games are cooperative, with both players wishing to locate the other as soon as possible (see Alpern, 2002, for an overview). Typically, in a rendez-vous game, the structure of the space is known to all, with consideration given to the amount of information available to players regarding their relative positions, and their ability to distinguish between symmetries of the space (e.g. whether they have a common understanding of 'North').

Rendez-vous games have been studied over various continuous one-dimensional spaces, such as the line (Alpern and Gal, 1995; Alpern and Beck, 2000; Lim and Alpern, 1996) and the circle (Alpern, 2000); over continuous two-dimensional spaces, such as the plane (Kikuta and Ruckle, 2010) or a general compact metric space (Alpern, 1995); and over discrete spaces, such as lattices (Alpern and Baston, 2005; Ruckle, 2007) and other graphs (Alpern et al., 1999). Costs may also be introduced for movement and examination of particular locations (Kikuta and Ruckle, 2007).

Work has also been done on 'hybrid' games of search and rendez-vous, where, for example, two individuals attempt to meet without being located by a third (Alpern and Lim, 1998) or where the searching player does not know whether the other player is attempting to rendez-vous or to evade capture (Alpern and Gal, 2002).

17.3.7 Security Games

Security games are used to model situations in which some public resource (e.g. airports, transport infrastructure and power facilities) must be protected from attack with limited defensive resources. A good introduction to the topic is provided by Tambe (2012).

Such situations tend to be modelled as Stackelberg games, where it is assumed that the defensive player first commits to some strategy to protect the vulnerable sites and that this strategy is observed by the attacking player, who then chooses an optimal response (Tambe, 2012: pp 4–8). Stackelberg-type security games related to the mobile–searcher–immobile–hider games of Section 17.3.2 have also been proposed to examine optimal patrolling strategies (Alpern et al., 2011b; Basilico et al., 2012).

A related concept is that of the much studied Colonel Blotto game, in which two players must simultaneously distribute a fixed quantity of discrete or continuous resources across a number of sites, each site being 'won' by the player who distributed the greater quantity of resources to it, with payoffs determined by the number of sites that each player wins (see Roberson, 2006). The many extensions of the Colonel Blotto game have included asymmetric versions (Tofias et al., 2007; Hortala-Vallve and Llorente-Saguer, 2012), examples in which resources are allocated to battlefields sequentially rather than simultaneously (Powell, 2009) and examples in which defensive resources are heterogeneous (Cohen, 1966).

Though the deployment sites in such models are often assumed to be wholly separate, with events at one location having no effect on events at other locations, certain security games and Colonel Blotto–type games with strategically interdependent sites have been considered. For example, Shubik and Weber (1981) introduce the concept of a 'characteristic function' for such games, which allocates values to subsets of the sites, thus allowing interdependencies to be captured. Other approaches to modelling such interdependence include an extension of the Colonel Blotto game in which a successful attack on a 'radar site' ensures the success of attacks on other sites (Grometstein and Shoham, 1989), while Hausken (2010, 2011) discusses a classification of the underlying infrastructures of security games based on the interdependence of their sites (series, parallel and complex ...), and Powell (2007) analyses the relative value of defending borders over protecting strategic targets directly.

Though analyses of interdependence in security games and Blotto games may be quite general (e.g. that of Shubik and Weber), interdependence that arises explicitly from the spatial structure of the deployment sites has not been considered in a general setting.

17.3.8 Geometric Games

One of the most general and theoretical analyses of search and concealment type situations is Ruckle's *Geometric games and their applications* (1983). In this book, the author defines a geometric game as a two-player zero-sum game (with players called RED and BLUE) played over a given set S, where the strategy sets for each player Σ_{RED}, Σ_{BLUE} are subsets of the power set $\mathcal{P}(S)$ (the set of subsets of S). Pure strategies for each player are therefore subsets

$R, B \subseteq S$. The payoff to each player is a function of R and B, typically depending directly on the intersection $R \cap B$.

This concept of a geometric game allows Ruckle to model a wide variety of situations of search, ambush and pursuit, as well as a range of abstract games, taking full consideration of the structure of the space S over which the games are played.

Most published work based on Ruckle's ideas (e.g. Alpern et al., 2009, 2010; Baston et al., 1989; Zoroa and Zoroa, 1993; Zoroa et al., 1999a, b, 2001, 2003) has focussed on specific examples of geometric games, rather than on general results.

17.3.9 Motivation for Defining a New Spatial Game

Much of the literature on search games and related concepts has focussed on analysing specific games, rather than attempting to present general frameworks for such situations and identifying more broadly applicable results. While spatial structure may be considered for games in which players are mobile, the geography of the space over which games are played is often given little or limited consideration, particularly in the literature on security games. The concept of *reachability*, as described by Hohzaki (2008), in which a searcher or searching resource deployed at a point has influence over a local neighbourhood, has received very little attention. Games that concentrate purely on the strategic value of a player's chosen position in a space, rather than on strategies for moving through the space for the purposes of search or rendez-vous, have also seen little research, at least since the work of Ruckle (1983).

17.4 The Static Spatial Search Game (SSSG)

The definitions and notation relating to metric spaces used in this section are from Sutherland (1975, pp 19–44).

17.4.1 Definition of the SSSG

The Static Spatial Search Game (SSSG) is a two-player game played over a metric space $M = (\Omega, d)$, where Ω is a set of points x and $d : \Omega \times \Omega \to [0, \infty)$ is the metric or distance, which has the standard properties:

$$
\begin{aligned}
&\text{(M1)}\ d(x,y) \geq 0; && d(x,y) = 0 \Leftrightarrow x = y \\
&\text{(M2)}\ d(x,y) = d(y,x), && \forall x, y \in \Omega \\
&\text{(M3)}\ d(x,y) + d(y,z) \geq d(x,z), && \forall x, y, z \in \Omega
\end{aligned}
$$

The metric d reflects the spatial structure of Ω. In $\mathbb{R}^n$, d may be the Euclidean distance, while in a graph d may be the length of the shortest path connecting two points. However, depending on the interpretation of the game, d could also represent an abstract distance, indicating dissimilarity, difficulty of communication or perceived costs.

In specific cases, it may be sensible to relax some of these conditions. For example, in a graph that is not connected, we could allow infinite distances between vertices that are not joined by a path (yielding an extended metric). Alternatively, to represent a directed graph, we may wish to ignore the symmetry condition (M2) (yielding a quasi-metric; see Steen and Seebach, 1970). However, for the sake of simplicity, we do not consider such cases at this time.

We define a non-negative real number r called the detection radius and use the notation $B_r[x]$ to designate the closed ball centred on x:

$$B_r[x] = \{y \in \Omega : d(x, y) \leq r\}$$

The strategies for Player A (the searching player) and Player B (the concealing player) are specific points of Ω at which they may choose to deploy. In a single play of the game, each player simultaneously picks a point x_A, x_B from their own strategy set, $\Sigma_A, \Sigma_B \subseteq \Omega$.

For the sake of clarity, from this point forward, we use masculine pronouns to refer to Player A and feminine pronouns to refer to Player B.

We define the payoff functions for Player A and Player B respectively as:

$$p_A(x_A, x_B) = \begin{cases} 1, & x_B \in B_r[x_A]; \\ 0, & \text{otherwise.} \end{cases}$$

$$p_B(x_A, x_B) = 1 - p_A(x_A, x_B)$$

This is a constant-sum game and can be analysed accordingly.

In interpreting the game, we imagine that Player B chooses to hide somewhere in Ω, while Player A attempts to locate his opponent. To do this, Player A selects a point of Ω and searches a neighbourhood of this point. If Player B's hiding place falls within the detection radius of Player A's chosen point, the attempt to hide is unsuccessful and Player B is located. Otherwise, Player B remains undetected.

The game is illustrated in Figure 17.1.

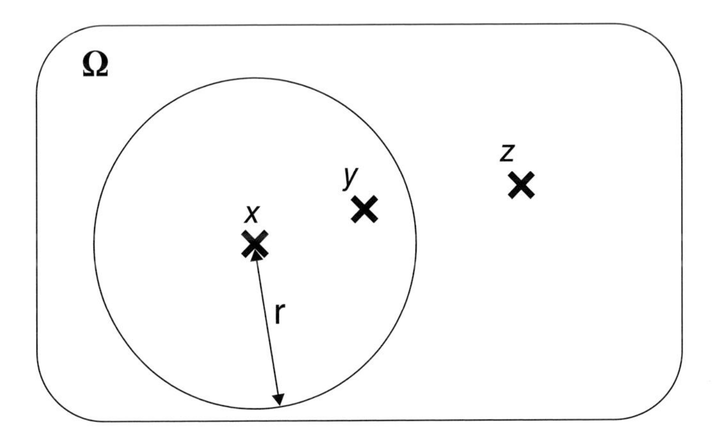

Figure 17.1 The Static Spatial Search Game (SSSG). Players deploy simultaneously at points in some metric space Ω. Suppose Player A deploys at x and can search all points within a radius r. Player B loses the game if deployed at y, but wins if deployed at z. A win results in a payoff of 1; a loss results in a payoff of 0

17.4.2 The SSSG and other Games

As mentioned in Section 17.1, the main sources of inspiration for the SSSG were *Geometric games and applications* (Ruckle, 1983) and, more particularly, *Games of strategy on trees* (White, 1994). Indeed, the SSSG could be seen as occupying a middle ground between the extremely abstract formulation of Ruckle's 'geometric games' and the quite specific family of games studied by White.

It should be noted that the SSSG is not strictly a geometric game by Ruckle's definition (1983: p 2), since it is not zero-sum. However, it could be transformed into a zero-sum game without altering the subsequent analysis, simply by subtracting $\frac{1}{2}$ from all payoffs. The decision that all payoffs should be 0 or 1 has been taken to ensure the clarity of the payoff matrices considered later.

Given this proviso, certain of Ruckle's geometric games can be formulated as particular cases of the SSSG. For example, if transformed to a zero-sum game as described, game AAGV (Ruckle, 1983: p 86; adapted from Arnold, 1962) is an example of the SSSG, with $\Omega = [0, 1] \subset \mathbb{R}$, $\Sigma_A = \Sigma_B = \Omega$ and $d(x, y) = |x - y|$.

Similarly, White's 'games of strategy on trees' (1994) may be seen to be examples of the SSSG, where Ω is the set of vertices of a tree, $\Sigma_A = \Sigma_B = \Omega$, $r = 1$ and d is the length of the shortest path between two vertices.

A game that demonstrates the potential complexity that can arise from apparently simple cases of the SSSG is the 'cookie-cutter' game (or the 'hiding in a disc' game), in which Player A chooses a point in a disc of unit radius and Player B simultaneously places a circular 'cookie-cutter' centred at any point of the disc, winning the game if Player A's point lies within the 'cookie-cutter'. Given appropriate payoffs, this game is an example of the SSSG, where Ω is the closed unit disc, $\Sigma_A = \Sigma_B = \Omega$ and $d(\boldsymbol{x},\boldsymbol{y}) = |\boldsymbol{x} - \boldsymbol{y}|$.

The particular case of this game with $r = 1/2$ was originally proposed by Gale and Glassey (1974), for which OMSs were presented by Evans (1975). The game was extended to all $r > 0$ by Bordelon (1975), who proposed OMSs for all $r > 1/2$, but these results were disputed by Ruckle (1983: p 108). Ruckle's disproof was disputed in turn by Danskin (1990), who showed that Bordelon's results were correct for some values of $r > 1/2$, though false in general. Despite the apparent simplicity of the problem, Danskin was only able to find OMSs for a small range of values of r around $r = 1/2$ and for all $r \geq \sqrt{2}/2$, thus illustrating the hidden complexity of many games of this form.

A particularly simple example of a game that can be represented as an SSSG is 'matching pennies' (see Blackwell and Girshick, 1979: p 13), in which Players A and B simultaneously call 'heads' or 'tails', with Player A receiving a payoff of 1 if the calls are the same and -1 otherwise, and Player B receiving a payoff of -1 if the calls are the same and 1 otherwise. Taking $\Sigma_A = \Sigma_B = \Omega = \{\text{'heads', 'tails'}\}$, $r = 0$, with d as any valid metric, this is an SSSG, again with the proviso that the payoffs must be transformed appropriately.

A more complicated example of a game that can be represented as an SSSG is the graph security game of Mavronicolas et al. (2008) if the number of attackers is restricted to one. This game is played over an undirected graph $G = (V(G), E(G))$ with one defender and (in general) multiple attackers.[3] Simultaneously, the defender chooses an edge, and the attackers each choose a vertex. The defender receives a payoff equal to the number of attackers who

[3] Here and elsewhere in this volume, $V(G)$ and $E(G)$ respectively represent the vertex set and the edge set of G.

choose vertices incident to his chosen edge. Each attacker receives a payoff equal to 0 if their chosen vertex is incident to the defender's edge, and 1 otherwise.

Consider the graph G' obtained by inserting a new vertex at the midpoint of each of the edges of G. Let the set of new vertices created in this way be denoted $V(G')^*$ while the complete vertex set $V(G')$ includes both the new vertices and the original vertices. With the defender as Player A, a single attacker as Player B, $\Omega = V(G')$, $\Sigma_A = V(G')^*$, $\Sigma_B = V(G)$, $r = 1$ and d as the length of the shortest path between two vertices in G', this game is also an example of the SSSG.

The SSSG provides a framework that unites all of these games and allows for a general consideration of the relative strategic values of the different points of a space. It implicitly encompasses the concepts of reachability, interdependence based on spatial structure and the detection radius, as discussed in Section 17.3.

17.4.3 The SSSG with Finite Strategy Sets

Consider an example of the SSSG in which the strategy sets Σ_A, Σ_B are finite. One of the simplest possible mixed strategies available to each player in such a case is the mixed strategy that allocates equal probabilities to all points in a player's strategy set. We denote these mixed strategies by ρ_A and ρ_B for Players A and B respectively.

The following proposition establishes a sufficient condition for ρ_A, ρ_B to be OMSs:

Proposition 17.1. *Consider the SSSG played over a metric space* Ω, *with finite strategy sets* Σ_A, Σ_B *and distance* d. *If there exists a positive integer* χ *such that:*

$$\begin{aligned} |B_r[x_A] \cap \Sigma_B| &= \chi, \quad \forall x_A \in \Sigma_A \\ |B_r[x_B] \cap \Sigma_A| &= \chi, \quad \forall x_B \in \Sigma_B \end{aligned} \tag{17.4}$$

then ρ_A, ρ_B *are OMSs for Players* A *and* B *respectively and* $\chi|\Sigma_A|^{-1} = \chi|\Sigma_B|^{-1} = u$ *is the value of the game to Player* A.

Proof. Suppose that Player A employs the mixed strategy ρ_A which allocates a uniform probability of $|\Sigma_A|^{-1}$ to all points $x_A \in \Sigma_A$. In a particular play of the game, suppose that Player B deploys at point $x_B \in \Sigma_B$.

In this situation, since $|B_r[x_B] \cap \Sigma_A| = \chi$, the expected payoff to Player A is $\chi|\Sigma_A|^{-1}$, the probability that Player A's point x_A lies in $B_r[x_B]$. Therefore, for any mixed strategy σ_B for Player B:

$$\mathrm{E}[p_A(\rho_A, \sigma_B)] = \chi|\Sigma_A|^{-1}$$

and thus:

$$u = \max_{\sigma_A} \min_{\sigma_B} \mathrm{E}[p_A(\sigma_A, \sigma_B)] \geq \chi|\Sigma_A|^{-1} \tag{17.5}$$

Now suppose that Player B employs the mixed strategy ρ_B which allocates a uniform probability of $|\Sigma_B|^{-1}$ to all points $x_B \in \Sigma_B$. In a particular play of the game, suppose that Player A deploys at point $x_A \in \Sigma_A$.

In this situation, since $|B_r[x_A] \cap \Sigma_B| = \chi$, the expected payoff to Player A is $\chi|\Sigma_B|^{-1}$, the probability that Player B's point x_B lies in $B_r[x_A]$. Therefore, for any mixed strategy $\boldsymbol{\sigma}_A$ for Player A:

$$\mathrm{E}[p_A(\sigma_A, \rho_B)] = \chi|\Sigma_B|^{-1}$$

and thus:

$$u = \max_{\sigma_A} \min_{\sigma_B} \mathrm{E}[p_A(\sigma_A, \sigma_B)] \leq \chi|\Sigma_B|^{-1} \tag{17.6}$$

Now, equation (17.4) together with the symmetric property of the distance (M2) imply that $\chi|\Sigma_A| = \chi|\Sigma_B|$ and thus that $|\Sigma_A| = |\Sigma_B|$. By equations (17.5) and (17.6), we therefore have:

$$|\Sigma_A|^{-1} \leq u \leq |\Sigma_B|^{-1} \quad \Rightarrow \quad u = |\Sigma_A|^{-1} = |\Sigma_B|^{-1}$$

and ρ_A, ρ_B are OMSs, by Definition 17.3. ■

17.4.4 Dominance and Equivalence in the SSSG

We can now examine strategic dominance and equivalence (see Definition 17.4) in the context of the SSSG using the notation established in Section 17.4.

Proposition 17.2. *Consider the SSSG played over a metric space* Ω, *with strategy sets* Σ_A, Σ_B *and distance d. For strategies* $x_1, x_2 \in \Sigma_A$, $x_1 \neq x_2$, *for Player* A*:*

- x_2 ***very weakly dominates*** x_1 *if and only if:*

$$(B_r[x_1] \cap \Sigma_B) \subseteq (B_r[x_2] \cap \Sigma_B)$$

- x_2 ***weakly dominates*** x_1 *if and only if:*

$$(B_r[x_1] \cap \Sigma_B) \subset (B_r[x_2] \cap \Sigma_B)$$

- x_2 ***strictly dominates*** x_1 *if and only if:*

$$(B_r[x_1] \cap \Sigma_B) = \emptyset$$
$$(B_r[x_2] \cap \Sigma_B) = \Sigma_B$$

- x_2 *is* ***equivalent*** *to* x_1 *if and only if:*

$$(B_r[x_1] \cap \Sigma_B) = (B_r[x_2] \cap \Sigma_B)$$

This proposition states that for Player A:

- x_2 very weakly dominates x_1 if and only if, when deployed at x_2, Player A can search every potential location of Player B that could be searched from x_1.
- This dominance is weak if there exist potential locations of Player B that can be searched from x_2 but that cannot be searched from x_1 (inclusion is strict).

- Strict dominance only occurs in the trivial case in which no potential locations of Player B can be searched from x_1 while every potential location of Player B can be searched from x_2.
- x_2 and x_1 are equivalent if and only if precisely the same set of potential locations of Player B can be searched from both points.

Proof. We consider each of the four parts of Definition 17.4 and show that, in the context of the SSSG, they are equivalent to the corresponding statements of Proposition 17.2. Recall that p_A takes values in $\{0, 1\}$.

- ***Very weak dominance***

$$
\begin{aligned}
& p_A(x_2, y) \geq p_A(x_1, y), \ \forall y \in \Sigma_B \qquad (*) \\
\Leftrightarrow\ & [p_A(x_1, y) = 1 \Rightarrow p_A(x_2, y) = 1], \ \forall y \in \Sigma_B \\
\Leftrightarrow\ & [y \in B_r[x_1] \Rightarrow y \in B_r[x_2]], \ \forall y \in \Sigma_B \\
\Leftrightarrow\ & (B_r[x_1] \cap \Sigma_B) \subseteq (B_r[x_2] \cap \Sigma_B) \qquad (**)
\end{aligned}
$$

- ***Weak dominance***

  Since $(*) \Leftrightarrow (**)$, it suffices to observe that, if $(**)$ is assumed to be true:

$$
\begin{aligned}
& \exists y^* \in \Sigma_B \ : \ p_A(x_2, y^*) > p_A(x_1, y^*) \\
\Leftrightarrow\ & \exists y^* \in \Sigma_B \ : \ y^* \in B_r[x_2] \text{ and } y^* \notin B_r[x_1] \\
\Leftrightarrow\ & (B_r[x_1] \cap \Sigma_B) \subset (B_r[x_2] \cap \Sigma_B) \qquad [\text{by } (**)]
\end{aligned}
$$

- ***Strict dominance***

$$
\begin{aligned}
& p_A(x_2, y) > p_A(x_1, y), \ \forall y \in \Sigma_B \\
\Leftrightarrow\ & y \in B_r[x_2] \text{ and } y \notin B_r[x_1], \ \forall y \in \Sigma_B \\
\Leftrightarrow\ & \begin{cases} (B_r[x_1] \cap \Sigma_B) = \emptyset, \\ (B_r[x_2] \cap \Sigma_B) = \Sigma_B \end{cases}
\end{aligned}
$$

- ***Equivalence***

$$
\begin{aligned}
& p_A(x_2, y) = p_A(x_1, y), \ \forall y \in \Sigma_B \\
\Leftrightarrow\ & [y \in B_r[x_2] \Leftrightarrow y \in B_r[x_1]], \ \forall y \in \Sigma_B \\
\Leftrightarrow\ & (B_r[x_2] \cap \Sigma_B) = (B_r[x_1] \cap \Sigma_B)
\end{aligned}
$$

■

We now consider dominance and equivalence for Player B:

Proposition 17.3. *Consider the SSSG played over a metric space Ω, with strategy sets Σ_A, Σ_B and distance d. For strategies $x_1, x_2 \in \Sigma_B$, $x_1 \neq x_2$, for Player* B*:*

- x_2 ***very weakly dominates*** x_1 *if and only if:*

$$[x_2 \in B_r[y] \Rightarrow x_1 \in B_r[y]], \ \forall y \in \Sigma_A$$

- x_2 ***weakly dominates*** x_1 *if and only if:*

$$[x_2 \in B_r[y] \Rightarrow x_1 \in B_r[y]], \ \forall y \in \Sigma_A$$

and $\exists y^* \in \Sigma_A$ *such that:*

$$x_1 \in B_r[y^*] \text{ and } x_2 \notin B_r[y^*]$$

- x_2 ***strictly dominates*** x_1 *if and only if:*

$$x_1 \in B_r[y] \text{ and } x_2 \notin B_r[y], \ \forall y \in \Sigma_A$$

- x_2 *is* ***equivalent*** *to* x_1 *if and only if:*

$$[x_2 \in B_r[y] \Leftrightarrow x_1 \in B_r[y]], \ \forall y \in \Sigma_A$$

This proposition states that for Player B:

- x_2 very weakly dominates x_1 if and only if, wherever Player A deploys, if he can search x_2 then he can also search x_1.
- This dominance is weak if there exists a position for Player A from which he can search x_1 but cannot search x_2.
- Strict dominance only occurs in the trivial case in which, wherever Player A deploys, he can search x_1 but cannot search x_2.
- x_2 and x_1 are equivalent if and only if, wherever Player A deploys, he can search x_2 if and only if he can search x_1.

Proof. We consider each of the four parts of Definition 17.4 and show that, in the context of the SSSG, they are equivalent to the corresponding statements of Proposition 17.3. Recall that p_B also takes values in $\{0, 1\}$.

- ***Very weak dominance***

$$\begin{aligned} & p_B(y, x_2) \geq p_B(y, x_1), \ \forall y \in \Sigma_A \qquad (*) \\ \Leftrightarrow\ & [p_B(y, x_1) = 1 \Rightarrow p_B(y, x_2) = 1], \ \forall y \in \Sigma_A \\ \Leftrightarrow\ & [x_1 \notin B_r[y] \Rightarrow x_2 \notin B_r[y]], \ \forall y \in \Sigma_A \\ \Leftrightarrow\ & [x_2 \in B_r[y] \Rightarrow x_1 \in B_r[y]], \ \forall y \in \Sigma_A \qquad (**) \end{aligned}$$

- ***Weak dominance***

  Since $(*) \Leftrightarrow (**)$, it suffices to observe that:

$$\begin{aligned} & \exists y^* \in \Sigma_A \ : \ p_B(y^*, x_2) > p_B(y^*, x_1) \\ \Leftrightarrow\ & \exists y^* \in \Sigma_A \ : \ x_1 \in B_r[y^*] \text{ and } x_2 \notin B_r[y^*] \end{aligned}$$

- ***Strict dominance***

$$p_B(y, x_2) > p_B(y, x_1), \ \forall y \in \Sigma_A$$
$$\Leftrightarrow x_1 \in B_r[y] \text{ and } x_2 \notin B_r[y], \ \forall y \in \Sigma_A$$

- ***Equivalence***

$$p_B(y, x_2) = p_B(y, x_1), \ \forall y \in \Sigma_A$$
$$\Leftrightarrow [x_2 \in B_r[y] \Leftrightarrow x_1 \in B_r[y]], \ \forall y \in \Sigma_A$$

■

The necessary and sufficient conditions for dominance and equivalence for Player B established in Proposition 17.3 can be shown to be equivalent to a simpler set of conditions, clearly analogous to those relating to dominance and equivalence for Player A seen in Proposition 17.2:

Proposition 17.4. *Consider the SSSG played over a metric space Ω, with strategy sets Σ_A, Σ_B and distance d. For strategies $x_1, x_2 \in \Sigma_B$, $x_1 \neq x_2$, for Player* B*:*

- x_2 ***very weakly dominates*** x_1 *if and only if:*

$$(B_r[x_2] \cap \Sigma_A) \subseteq (B_r[x_1] \cap \Sigma_A)$$

- x_2 ***weakly dominates*** x_1 *if and only if:*

$$(B_r[x_2] \cap \Sigma_A) \subset (B_r[x_1] \cap \Sigma_A)$$

- x_2 ***strictly dominates*** x_1 *if and only if:*

$$(B_r[x_1] \cap \Sigma_A) = \Sigma_A$$
$$(B_r[x_2] \cap \Sigma_A) = \emptyset$$

- x_2 *is* ***equivalent*** *to* x_1 *if and only if:*

$$(B_r[x_2] \cap \Sigma_A) = (B_r[x_1] \cap \Sigma_A)$$

Proof. We consider each of the four statements of Proposition 17.3 (which has already been proven) and show that they are equivalent to the corresponding statements of Proposition 17.4.

- ***Very weak dominance***

$$[x_2 \in B_r[y] \Rightarrow x_1 \in B_r[y]], \ \forall y \in \Sigma_A \ (*)$$
$$\Leftrightarrow [y \in B_r[x_2] \Rightarrow y \in B_r[x_1]], \ \forall y \in \Sigma_A \text{ [by (M2)]}$$
$$\Leftrightarrow \quad (B_r[x_2] \cap \Sigma_A) \subseteq (B_r[x_1] \cap \Sigma_A) \qquad (**)$$

- ***Weak dominance***

  Since $(*) \Leftrightarrow (**)$, it suffices to observe that, if $(**)$ is assumed to be true:

$$\begin{aligned} & \exists z \in \Sigma_A \; : \; x_1 \in B_r[z] \text{ and } x_2 \notin B_r[z] \\ \Leftrightarrow & \exists z \in \Sigma_A \; : \; z \in B_r[x_1] \text{ and } z \notin B_r[x_2] \text{ [by (M2)]} \\ \Leftrightarrow & \quad (B_r[x_2] \cap \Sigma_A) \subset (B_r[x_1] \cap \Sigma_A) \qquad \text{[by(**)]} \end{aligned}$$

- ***Strict dominance***

$$\begin{aligned} & x_1 \in B_r[y] \text{ and } x_2 \notin B_r[y], \; \forall y \in \Sigma_A \\ \Leftrightarrow & \, y \in B_r[x_1] \text{ and } y \notin B_r[x_2], \; \forall y \in \Sigma_A \text{ [by (M2)]} \\ \Leftrightarrow & \quad \begin{cases} (B_r[x_1] \cap \Sigma_A) = \Sigma_A, \\ (B_r[x_2] \cap \Sigma_A) = \emptyset \end{cases} \end{aligned}$$

- ***Equivalence***

$$\begin{aligned} & [x_2 \in B_r[y] \Leftrightarrow x_1 \in B_r[y]], \; \forall y \in \Sigma_A \\ \Leftrightarrow & \, [y \in B_r[x_2] \Leftrightarrow y \in B_r[x_1]], \; \forall y \in \Sigma_A \text{ [by (M2)]} \\ \Leftrightarrow & \quad (B_r[x_2] \cap \Sigma_A) = (B_r[x_1] \cap \Sigma_A) \end{aligned}$$

■

While Proposition 17.4 is apparently simpler than Proposition 17.3, note that every part of its proof depends on the symmetric property of the distance (M2). If this condition were to be relaxed, as discussed in Section 17.4.1, Proposition 17.4 would not be valid, and dominance and equivalence for Player B would have to be analysed on the basis of Proposition 17.3.

Definition 17.5. *Consider the SSSG played over a metric space Ω, with strategy sets Σ_A, Σ_B and distance d. For Player* A *or Player* B, *a subset of their strategy set $\hat{\Sigma} \subseteq \Sigma_A$ or $\hat{\Sigma} \subseteq \Sigma_B$ exhibits* ***pairwise equivalence*** *if and only if x is equivalent to y, $\forall x, y \in \hat{\Sigma}$.*

We conclude that a subset $\hat{\Sigma} \subseteq \Sigma_A$ or $\hat{\Sigma} \subseteq \Sigma_B$ exhibiting pairwise equivalence can be reduced to any singleton $\{\hat{x}\} \subseteq \hat{\Sigma}$ without altering the analysis of the game. Since all points in $\hat{\Sigma}$ are equivalent, a player would neither gain nor lose by playing another point in the set over $\hat{x}$.

The following proposition states that if x_2 very weakly dominates x_1 for Player A (and x_1 is adjacent to at least one potential location for Player B), then the distance between the two points must be no greater than $2r$.

Proposition 17.5. *For the SSSG played over a metric space Ω, with strategy sets Σ_A, Σ_B and distance d, if x_2 very weakly dominates x_1 for Player* A *and $B_r[x_1] \cap \Sigma_B \neq \emptyset$, then $x_2 \in B_{2r}[x_1] \cap \Sigma_A$.*

Proof. If x_2 very weakly dominates x_1 for Player A and $B_r[x_1] \cap \Sigma_B \neq \emptyset$, then:

$$\emptyset \neq (B_r[x_1] \cap \Sigma_B) \subseteq (B_r[x_2] \cap \Sigma_B) \text{ [by Proposition 17.4]}$$

$$\begin{aligned}
&\Rightarrow && \exists\, y \in B_r[x_1] \cap B_r[x_2] \\
&\Rightarrow && d(x_1, y) \leq r \text{ and } d(x_2, y) \leq r \\
&\Rightarrow && d(x_1, x_2) \leq 2r \qquad \text{[by (M2), (M3)]} \\
&\Rightarrow && x_2 \in B_{2r}[x_1] \cap \Sigma_A
\end{aligned}$$

■

The condition that $B_r[x_1] \cap \Sigma_B \neq \emptyset$ simply removes trivial strategies that are very weakly dominated by every other strategy.

An analogous result holds for Player B. The proof is similar to that of Proposition 17.6 and is therefore omitted:

Proposition 17.6. *For the SSSG played over a metric space Ω, with strategy sets Σ_A, Σ_B and distance d, if x_2 very weakly dominates x_1 for Player B and $B_r[x_2] \cap \Sigma_A \neq \emptyset$, then $x_2 \in B_{2r}[x_1] \cap \Sigma_B$.*

In this case, the condition that $B_r[x_2] \cap \Sigma_A \neq \emptyset$ removes strategies that very weakly dominate every other strategy.

Note that both of these propositions depend on the symmetric property of the distance (M2) and the triangle inequality (M3) (see Section 17.4.1).

17.4.5 Iterated Elimination of Dominated Strategies

The concepts of dominance and equivalence provide us with a method for reducing the SSSG through an iterative process of removing dominated strategies from Σ_A and Σ_B, reducing pairwise equivalent subsets to singletons and reassessing dominance in the new strategy sets. This is known as the iterated elimination of dominated strategies (IEDS) (see e.g. Berwanger, 2007; Börgers, 1992; Dufwenberg and Stegeman, 2002). Given any game, the aim of IEDS is to identify a simplified game, whose solutions are also solutions of the complete game. These solutions can then be identified using standard techniques (see e.g. Morris, 1994: pp 99–114). The application of this method to games played over graphs is discussed in Appendix C.

It should be noted that because we are considering weak rather than strict dominance, IEDS may not be suitable for identifying all the solutions of a particular game. The results of this form of IEDS are dependent on the order in which dominated strategies are removed (Leyton-Brown and Shoham, 2008: pp 20–23), and some solutions may be lost. It is also necessary to observe that, while IEDS has been shown to be valid for games with finitely many possible strategies, for games with infinitely many possible strategies the process may fail (Berwanger, 2007). Indeed, such infinite games may not have solutions (Ruckle, 1983: p 10).

However, although IEDS is not guaranteed to produce OMSs for the SSSG in such cases, given a pair of mixed strategies $\boldsymbol{\sigma}_A$, $\boldsymbol{\sigma}_B$ obtained by this method, it is straightforward to check whether or not they are optimal by verifying that, for some $u \in [0, 1]$, we have:

$$\begin{aligned}
\inf_{x \in \Sigma_B} \mathrm{E}[p_A(\boldsymbol{\sigma}_A, x)] &= u \\
\inf_{y \in \Sigma_A} \mathrm{E}[p_B(y, \boldsymbol{\sigma}_B)] &= 1 - u
\end{aligned} \tag{17.7}$$

where u is the value of the game to Player A.

This method is described by Blackwell and Girshick (1979: p 65) and used extensively by Ruckle (1983; the method is introduced on pp 8–9) to verify proposed OMSs for geometric games.

17.5 The Graph Search Game (GSG)

The definitions and notation relating to graph theory used in this section are adapted from Bondy and Murty (1976).

17.5.1 Definition of the GSG

Consider a simple graph G, characterised by the symmetric adjacency matrix $\mathfrak{M} = (a_{ij})$, with a set of κ vertices:

$$V(G) = \{v_1, \ldots, v_\kappa\}$$

and a set of edges:

$$E(G) = \{\{v_i, v_j\} : v_i, v_j \in V(G) \text{ and } a_{ij} = 1\}$$

We suppose that G is connected, to ensure that the metric space axioms are fulfilled, but this assumption could be relaxed if we allowed for infinite distances between vertices. We also suppose that all edges of G have unit weight.

The GSG $\mathfrak{G} = (G, \Sigma_\mathrm{A}, \Sigma_\mathrm{B}, r)$ over a finite graph G is defined to be an example of the SSSG, with $\Omega = V(G)$, $\emptyset \neq \Sigma_\mathrm{A}, \Sigma_\mathrm{B} \subseteq V(G)$, r a positive real number, and the distance function $d_G(v, w)$ for $v, w \in V(G)$, $v \neq w$, defined to be the length of the shortest path from v to w in G. We also define $d_G(v, v) = 0$, $\forall v \in V(G)$. The assumption that G is undirected ensures that the symmetry condition $d_G(v, w) = d_G(w, v)$ holds $\forall v, w \in V(G)$.

Setting $r = 1$, we have that $B_r[v]$, the zone that can be searched by Player A when deployed at v, is the closed neighbourhood $N[v]$ of v, the set of all vertices adjacent to v united with $\{v\}$ itself.

The game proceeds by Player A choosing a vertex $v_\mathrm{A} \in \Sigma_\mathrm{A}$ and Player B simultaneously choosing a vertex $v_\mathrm{B} \in \Sigma_\mathrm{B}$. The payoff functions are:

$$p_\mathrm{A}(v_\mathrm{A}, v_\mathrm{B}) = \begin{cases} 1, & \text{if } v_\mathrm{A} \text{ and } v_\mathrm{B} \text{ are equal or adjacent;} \\ 0, & \text{otherwise.} \end{cases}$$

$$p_\mathrm{B}(v_\mathrm{A}, v_\mathrm{B}) = 1 - p_\mathrm{A}(v_\mathrm{A}, v_\mathrm{B})$$

In this game, the pure strategies for each player are particular vertices. In the following analysis, we use the words *strategy* and *vertex* interchangeably, depending on the context.

The GSG is illustrated in Figure 17.2.

At this point, we introduce a particular scenario, which we will return to repeatedly over the course of this chapter and Appendices C and D, as a means of illustrating our concepts and results. Although the setting is necessarily quite artificial, it should provide a comprehensible foundation for the ideas that we will present, while highlighting the potential (though speculative) applicability of this work.

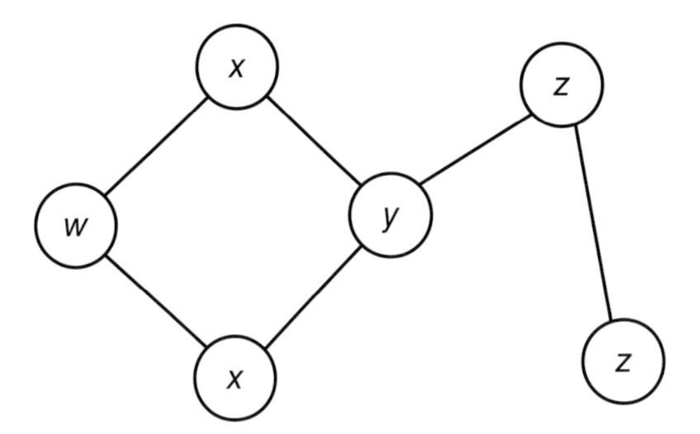

Figure 17.2 The Graph Search Game (GSG). Players deploy simultaneously at vertices of the graph. Suppose that Player A deploys at the vertex marked w. If $r = 1$, Player B loses the game if she deploys at any of the vertices marked w or x, but wins if she deploys at any of the vertices marked y or z. If $r = 2$, Player B loses the game if she deploys at any of the vertices marked w, x or y, but wins if she deploys at either of the vertices marked z. A win results in a payoff of 1; a loss results in a payoff of 0

Example 17.1. *Imagine that the graph depicted in Figure 17.3 represents part of a street network in a city centre location. The area is popular with tourists but is suffering from a growing problem of robbery on the streets. The 15 vertices (which we will label* $v_0, \ldots, v_{14}$*) represent high-risk locations, such as shopping centres, transport hubs and tourist attractions, while the edges between them represent connecting streets.*

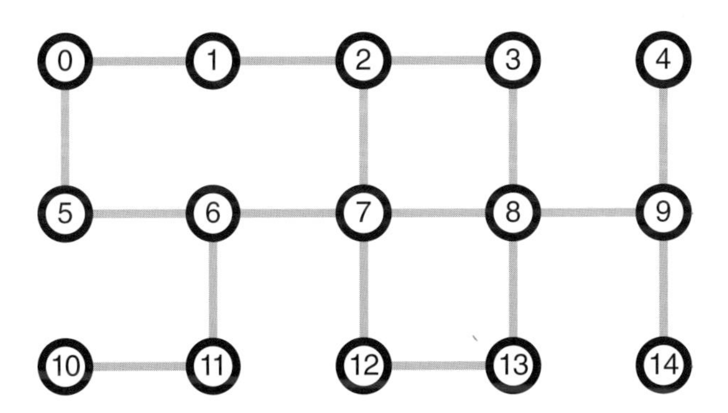

Figure 17.3 A representation of a hypothetical street network, in which high-risk locations (the vertices) must be protected from criminals by a single unmarked police unit

To combat the rising criminality, police have decided to use a single unmarked unit to protect the area, which they will deploy at one of these locations, responding to crimes as they occur. If a robbery occurs at the location at which police are deployed or at any neighbouring location, they will be able to respond quickly enough to catch the offenders. However, if the robbery occurs elsewhere, they will be unable to arrive at the location in time to make an arrest. Each time a robbery occurs, whether successful or not, the police may reconsider their position and move to another deployment location, or they may remain where they are.

The police wish to know where they should deploy to have the best chance of catching offenders. The criminals wish to know where they should commit their crimes to have the best chance of avoiding the police.

If we assume that crimes are never simultaneous and that the police can relocate instantly after a crime has taken place, then this scenario may be modelled as a repeated GSG, $\mathfrak{G} = (G, \Sigma_A, \Sigma_B, r)$*, where G is the graph depicted in Figure 17.3,* Σ_A *and* Σ_B *are equal to* $\{v_0, \ldots, v_{14}\}$ *and* $r = 1$*, with the police as Player* A *and the criminals (treated as a single entity) as Player* B*. For the police, a payoff of 1 represents the successful capture of an offender, while, for the criminals, a payoff of 1 represents the completion of a successful robbery.*

It is reasonable to assume that an OMS would represent an appropriate approach to this game for both players, since any sub-optimal strategy (such as deploying at a particular point permanently or repeating a particular sequence of locations) could be observed and learnt over time, allowing an opponent to employ an exploitative counter-strategy.

17.5.2 The GSG with $r \neq 1$

The restriction to $r = 1$ is not a significant constraint, since any GSG can be reduced to this case by means of a minor alteration.

For situations with $r \neq 1$, we can define:

$$G' = (V(G), E(G'))$$

such that:

$$E(G') = \{\{v_i, v_j\} : v_i, v_j \in V(G) \text{ and } 1 \leq d_G(v_i, v_j) \leq r\}$$

and apply our analysis to G' with $r = 1$.

Equivalently, if $r \in \mathbb{N} \setminus \{1\}$, we can replace the adjacency matrix $\mathfrak{M} = (a_{ij})$ of G with $\mathfrak{M}' = (a'_{ij})$, where:

$$a'_{ij} = \begin{cases} 0, & \text{if } b_{ij} = 0 \text{ or } i = j \\ 1, & \text{otherwise.} \end{cases}$$

with $(b_{ij}) = \sum_{q=1}^{r} \mathfrak{M}^q$, the matrix which shows the number of paths in G of length no greater than r connecting v_i to v_j.

It therefore suffices to exclusively study GSGs with $r = 1$, since these methods for redefining G ensure that such analysis will be applicable to games for any $r \in \mathbb{N}$.

Example 17.2. *Consider the scenario described in Example 17.1. Suppose that, rather than being able to respond to crimes at locations adjacent to their current location in G, police are also able to respond to crimes in locations two steps away. This is equivalent to changing the detection radius from* $r = 1$ *to* $r = 2$*. The method outlined here states that this can be reduced to the* $r = 1$ *case by replacing G with* G'*, where* G' *is the graph created by inserting additional edges between any pair of vertices separated by a path of length 2. The graph* G' *is depicted in Figure 17.4.*

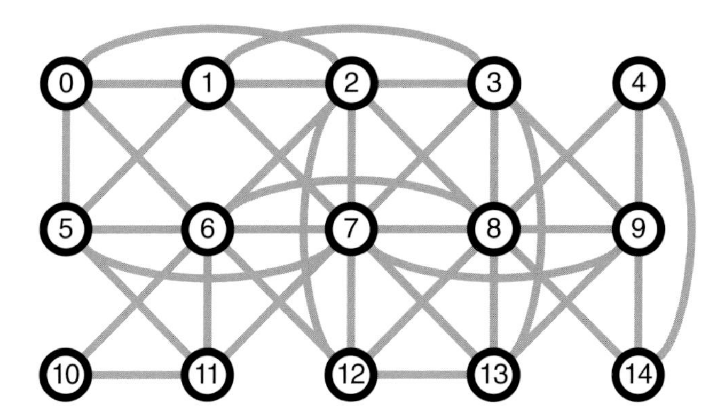

Figure 17.4 The graph G' for a GSG with $r = 2$, played over the graph G that was depicted in Figure 17.3

17.5.3 Preliminary Observations

Analysis of the GSG requires the statement of some preliminary results and definitions. For these results, we use the following notation:

- $N_A[v] = N[v] \cap \Sigma_A$
  The set of all potential positions for Player A in $N[v]$.
- $N_B[v] = N[v] \cap \Sigma_B$
  The set of all potential positions for Player B in $N[v]$.
- $\alpha[v] = |N_A[v]|$
  The number of potential positions for Player A in $N[v]$.
- $\beta[v] = |N_B[v]|$
  The number of potential positions for Player B in $N[v]$.
- $\Delta(G) = \max_{w \in V(G)} [\deg(w)]$
  The maximum degree of the vertices of G.
- $\delta(G) = \min_{w \in V(G)} [\deg(w)]$
  The minimum degree of the vertices of G.

The following proposition states that if Player A can search a globally maximal number of potential positions for Player B from a vertex v, then v is not dominated by any other vertex and is only equivalent to those vertices which have the same closed neighbourhood as v. In a graph in which all vertices have a distinct closed neighbourhood, such as a rectangular grid graph, such a vertex v cannot be very weakly dominated by any other vertex.

Proposition 17.7. *For the GSG $\mathfrak{G} = (G, \Sigma_A, \Sigma_B, r)$, with $r = 1$, consider a vertex $v \in \Sigma_A$ and the subset $\Sigma_A^{(v)} = \{z \in \Sigma_A : N[z] = N[v]\}$. We have that if:*

$$\beta[v] = \Delta(G) + 1 \tag{17.8}$$

then:

1. $\Sigma_A^{(v)}$ *exhibits pairwise equivalence for Player* A*; and*
2. v *is not very weakly dominated for Player* A *by any strategy in* $\Sigma_A \setminus \Sigma_A^{(v)}$.

Proof. To prove (1), observe that for all $w_1, w_2 \in \Sigma_A^{(v)}$, $N[w_1] = N[w_2]$. Hence, $N_B[w_1] = N_B[w_2]$, and therefore w_1 is equivalent to w_2 for Player A, so $\Sigma_A^{(v)}$ exhibits pairwise equivalence for Player A.

To prove (2), suppose for a contradiction that v satisfies equation (17.8) and is very weakly dominated by $\hat{v} \in \Sigma_A \setminus \Sigma_A^{(v)}$.

Observe also that:

$$\begin{aligned} \text{If} \quad & \beta[w] = \Delta(G) + 1, \text{ for some } w \in V(G), \\ \text{then} \quad & N_B[w] = N[w]. \end{aligned} \tag{17.9}$$

and note that $\Delta(G) + 1$ is an upper bound for $\beta[w]$.

Now, we have that:

$$\begin{array}{rrcll} & N_B[v] & \subseteq & N_B[\hat{v}] & \text{[by very weak dominance]} \\ \Rightarrow & \Delta(G)+1 & \leq & \beta[\hat{v}] \quad \leq & |N[\hat{v}]| \quad \text{[by (equation 17.8)]} \\ \Rightarrow & \Delta(G)+1 & \leq & \beta[\hat{v}] \quad \leq & \Delta(G)+1 \\ \Rightarrow & \Delta(G)+1 & = & \beta[v] \quad = & \beta[\hat{v}] \\ \Rightarrow & N_B[v] & = & N_B[\hat{v}] & \text{[since } N_B[v] \subseteq N_B[\hat{v}]] \\ \Rightarrow & N[v] & = & N[\hat{v}] & \text{[by (equation 17.9)]} \\ \Rightarrow & \hat{v} & \in & \Sigma_A^{(v)} & \end{array}$$

This is a contradiction, since we supposed that $\hat{v} \in \Sigma_A \setminus \Sigma_A^{(v)}$. ■

Example 17.3. *Consider the game,* $\mathfrak{G}$*, described in Example 17.1. Observe that* $\Delta(G)$*, the maximum degree of the vertices of* G*, is equal to 4 and that* $\beta[v_7] = \beta[v_8] = 5$ *(note that, in this example, because* $\Sigma_B = V(G)$*,* $\beta[v]$ *is simply the size of the closed neighbourhood of* v*). Since no two vertices of* G *share the same closed neighbourhood, we also have that* $\Sigma_A^{(v_7)} = \{v_7\}$ *and* $\Sigma_A^{(v_8)} = \{v_8\}$*. Therefore, by Proposition 17.7, we deduce that neither* v_7 *nor* v_8 *is very weakly dominated as a deployment location for the police (Player* A*) by any other locations (or by each other).*

The next proposition is a stronger result. It states that, for Player A, given a vertex v that is known not to be very weakly dominated by any vertices outside of a certain subset S, if Player A can search strictly more potential hiding places for Player B when deployed at v than could be searched from any other vertex in S, then v is not very weakly dominated by any other vertex.

Proposition 17.8. *For the GSG* $\mathfrak{G} = (G, \Sigma_A, \Sigma_B, r)$*, with* $r = 1$*, consider a vertex* $v \in \Sigma_A$*. Let* $S \subseteq \Sigma_A$ *be such that* $v \in S$ *and such that* $\Sigma_A \setminus S$ *contains no vertices that very weakly dominate* v *for Player* A*. We have that if:*

$$\beta[v] > \max_{w \in S \setminus \{v\}} (\beta[w]) \tag{17.10}$$

then v *is not very weakly dominated for Player* A *by any other strategy in* Σ_A.

Proof. For a contradiction, suppose that v satisfies equation (17.10) and is very weakly dominated by $\hat{v} \in \Sigma_A$ for Player A. We must have that $\hat{v} \in S$, since vertices outside S cannot very weakly dominate v.

From the very weak dominance, we have:

$$\begin{array}{lllllll} & & & N_B[v] & \subseteq & N_B[\hat{v}] & \\ \Rightarrow & \max_{w \in S \setminus \{v\}} (\beta[w]) & < & \beta[v] & \leq & \beta[\hat{v}] & \text{[by (17.10)]} \end{array}$$

This is a contradiction, since $\hat{v} \in S \setminus \{v\}$. ■

The following proposition is a restatement of Proposition 17.5, reformulated in the context of the GSG. It states that any vertex that very weakly dominates v for Player A can be no more than two steps away from v on the graph. Its proof is identical to that of Proposition 17.5 and is thus omitted.

Proposition 17.9. *For the GSG $\mathfrak{G} = (G, \Sigma_A, \Sigma_B, r)$, with $r = 1$, if w very weakly dominates v for Player* A *and $N_B[v] \neq \emptyset$, then:*

$$w \in B_2[v] \cap \Sigma_A$$

Corollary 17.1. *For the GSG $\mathfrak{G} = (G, \Sigma_A, \Sigma_B, r)$, with $r = 1$, consider a vertex $v \in \Sigma_A$. We have that if:*

$$\beta[v] > \max_{w \in S'}(\beta[w])$$

where:

$$S' = B_2[v] \cap \Sigma_A \setminus \{v\}$$

then v is not very weakly dominated for Player A *by any other strategy.*

Proof. This follows directly from Propositions 17.8 and 17.9, where the set S from Proposition 17.9 is defined as:

$$S = B_2[v] \cap \Sigma_A$$

■

The corollary states that any vertex from which Player A can search strictly more potential hiding places for Player B than could be searched from any other valid vertex lying no more than two steps away cannot be very weakly dominated by any other vertex. These results allow us to considerably narrow down our search for dominated and equivalent vertices.

Analogous results to Propositions 17.8 and 17.9 and Corollary 17.9 hold for very weak dominance for Player B. The proofs of these results are similar to those presented here and are thus omitted.

Proposition 17.10. *For the GSG $\mathfrak{G} = (G, \Sigma_A, \Sigma_B, r)$, with $r = 1$, consider a vertex $v \in \Sigma_B$. Let $S \subseteq \Sigma_B$ be such that $v \in S$ and such that $\Sigma_B \setminus S$ contains no vertices that very weakly dominate v for Player* B. *We have that if:*

$$\alpha[v] < \min_{w \in S \setminus \{v\}} (\alpha[w])$$

then v is not very weakly dominated for Player B *by any other strategy in Σ_B.*

Proposition 17.11. *For the GSG $\mathfrak{G} = (G, \Sigma_A, \Sigma_B, r)$, with $r = 1$, if w very weakly dominates v for Player* B *and $N_A[w] \neq \emptyset$, then:*

$$w \in B_2[v] \cap \Sigma_B$$

Corollary 17.2. *For the GSG $\mathfrak{G} = (G, \Sigma_A, \Sigma_B, r)$, with $r = 1$, consider a vertex $v \in \Sigma_B$, $\alpha[v] \neq 0$. We have that if:*

$$\alpha[v] < \min_{w \in S'}(\alpha[w])$$

where:

$$S' = B_2[v] \cap \Sigma_B \setminus \{v\}$$

then v is not very weakly dominated for Player B *by any other strategy.*

Example 17.4. *Consider the game, $\mathfrak{G}$, described in Example 17.1 and observe that $\alpha[v_{10}] = 2$ (note that, in this example, because $\Sigma_A = V(G)$, $\alpha[v]$ is simply the size of the closed neighbourhood of v). Since the minimum value of $\alpha[v_{10}]$ over all vertices within two steps of v_{10} is equal to 3, Corollary 17.11 tells us that v_{10} is not very weakly dominated as a target for the criminals (Player* B*) by any other location.*

17.5.4 Bounds on the Value of the GSG

A first step in the analysis of a particular GSG is to determine lower and upper bounds on the values of the game (see Definition 17.2).

Recall that for a two-player constant-sum game, it makes sense to restrict discussion of the value to Player A, since this also determines the value to Player B through the condition that the sum of the two values is a fixed constant (see equation (17.3)).

Proposition 17.12. *For the GSG $\mathfrak{G} = (G, \Sigma_A, \Sigma_B, r)$, with $r = 1$, let u represent the value to Player* A*. u is bounded as follows:*

$$\min_{v \in \Sigma_B} \frac{\alpha[v]}{|\Sigma_A|} \leq u \leq \max_{v \in \Sigma_A} \frac{\beta[v]}{|\Sigma_B|}$$

Note that in the case where $\Sigma_A = \Sigma_B = V(G)$, these inequalities become:

$$\frac{\delta(G) + 1}{|V(G)|} \leq u \leq \frac{\Delta(G) + 1}{|V(G)|} \tag{17.11}$$

The proposition derives from a consideration of ρ_A and ρ_B, defined in Section 17.4.3 as the mixed strategies that allocate equal probabilities to all vertices in a player's strategy set. If Player A employs mixed strategy ρ_A, then Player B can do no better than to deploy at the vertex whose closed neighbourhood contains the fewest possible vertices in Σ_A. The value of the game to Player A cannot therefore be less than the sum of the probabilities that ρ_A assigns to these vertices. This reasoning produces the left-hand inequality. The right-hand inequality follows in a similar fashion from an analysis of ρ_{B} as a strategy for Player B. A formal proof follows:

Proof. Suppose that Player A employs the mixed strategy ρ_{A} which allocates a uniform probability of $|\Sigma_{\mathrm{A}}|^{-1}$ to all vertices $w \in \Sigma_{\mathrm{A}}$. In a particular play of the game, suppose that Player B deploys at vertex $v \in \Sigma_{\mathrm{B}}$.

In this situation, the expected payoff to Player A is $\alpha[v]|\Sigma_{\mathrm{A}}|^{-1}$. Therefore, for any mixed strategy σ_{B} for Player B:

$$\mathrm{E}[p_{\mathrm{A}}(\rho_{\mathrm{A}}, \sigma_{\mathrm{B}})] \geq \min_{v \in \Sigma_{\mathrm{B}}} \frac{\alpha[v]}{|\Sigma_{\mathrm{A}}|}$$

and thus:

$$u = \max_{\sigma_{\mathrm{A}}} \min_{\sigma_{\mathrm{B}}} \mathrm{E}[p_{\mathrm{A}}(\sigma_{\mathrm{A}}, \sigma_{\mathrm{B}})] \geq \min_{v \in \Sigma_{\mathrm{B}}} \frac{\alpha[v]}{|\Sigma_{\mathrm{A}}|}$$

The proof of the right-hand inequality is similar. ■

The following corollary is a consequence of equation (17.11):

Corollary 17.3. *Consider the GSG:*

$$\mathfrak{G} = (G, \Sigma_{\mathrm{A}}, \Sigma_{\mathrm{B}}, r)$$

where $G = (V(G), E(G))$ is a regular graph of degree D; $\Sigma_{\mathrm{A}} = \Sigma_{\mathrm{B}} = V(G)$; $r = 1$; and let u be the value of $\mathfrak{G}$ to Player A. *Then:*

- $u = (D+1)/|V(G)|$.
- *The strategy ρ that allocates a uniform probability of $|V(G)|^{-1}$ to all vertices is an OMS for both players.*

u can also be bounded in a different way:

Proposition 17.13. *For the GSG $\mathfrak{G} = (G, \Sigma_{\mathrm{A}}, \Sigma_{\mathrm{B}}, r)$ with $r = 1$, let u represent the value to Player* A *and:*

$$W = \left\{ W' \subseteq \Sigma_{\mathrm{A}} : \bigcup_{w \in W'} N[w] \supseteq \Sigma_{\mathrm{B}} \right\}$$

$$Z = \left\{ Z' \subseteq \Sigma_{\mathrm{B}} : d_G(z_1, z_2) > 2, \forall z_1, z_2 \in Z' \right\}$$

Then u is bounded as follows:

$$\left[\min_{W' \in W} |W'| \right]^{-1} \leq u \leq \left[\max_{Z' \in Z} |Z'| \right]^{-1}$$

The left-hand inequality is derived from consideration of the mixed strategy τ_{A} for Player A that allocates uniform probabilities to a minimal subset of vertices whose closed neighbourhoods cover Σ_{B}.

The right-hand inequality is derived from consideration of the mixed strategy τ_{B} for Player B that allocates uniform probabilities to a maximal subset of vertices with the property that no two vertices are connected by a path of length less than 3.

Proof. First, consider the left-hand inequality.

Consider a subset of vertices $W' \in W$ of minimum cardinality. Suppose that Player A employs the mixed strategy $\boldsymbol{\tau}_\mathrm{A}$ that allocates uniform probability $|W'|^{-1}$ to vertices $w \in W'$ and zero probability to all other vertices.

In a particular play of the game, suppose that Player B deploys at vertex $v \in \Sigma_\mathrm{B}$. Since $W' \in W$, we have:

$$\bigcup_{w \in W'} N[w] \supseteq \Sigma_\mathrm{B}$$

Therefore, $\exists w' \in W'$ such that $v \in N[w']$. Since $\boldsymbol{\tau}_\mathrm{A}$ allocates a probability of $|W'|^{-1}$ to w', the expected payoff to Player A is greater than or equal to $|W'|^{-1}$. Therefore, for any mixed strategy $\boldsymbol{\sigma}_\mathrm{B}$ for Player B:

$$\mathrm{E}[p_\mathrm{A}(\boldsymbol{\tau}_\mathrm{A}, \boldsymbol{\sigma}_\mathrm{B})] \geq \left[\min_{W' \in W} |W'|\right]^{-1}$$

and thus:

$$u = \max_{\sigma_\mathrm{A}} \min_{\sigma_\mathrm{B}} \mathrm{E}\left[p_\mathrm{A}(\boldsymbol{\sigma}_\mathrm{A}, \boldsymbol{\sigma}_\mathrm{B})\right] \geq \left[\min_{W' \in W} |W'|\right]^{-1}$$

Now consider the right-hand inequality.

Consider a subset of vertices $Z' \in Z$ of maximum cardinality. Suppose that Player B employs the mixed strategy $\boldsymbol{\tau}_\mathrm{B}$ that allocates uniform probability $|Z'|^{-1}$ to vertices $v \in Z'$ and zero probability to all other vertices.

In a particular play of the game, suppose that Player A deploys at vertex $w \in \Sigma_\mathrm{A}$. Since $d_G(z_1, z_2) > 2, \forall z_1, z_2 \in Z'$, we clearly have that:

$$|N[w] \cap Z'| \leq 1$$

So, in this situation, the expected payoff to Player A is less than or equal to $|Z'|^{-1}$. Therefore, for any mixed strategy $\boldsymbol{\sigma}_\mathrm{A}$ for Player A:

$$\mathrm{E}[p_\mathrm{A}(\boldsymbol{\sigma}_\mathrm{A}, \boldsymbol{\tau}_\mathrm{B})] \leq \left[\max_{Z' \in Z} |Z'|\right]^{-1}$$

and thus:

$$u = \max_{\sigma_\mathrm{A}} \min_{\sigma_\mathrm{B}} \mathrm{E}\left[p_\mathrm{A}(\boldsymbol{\sigma}_\mathrm{A}, \boldsymbol{\sigma}_\mathrm{B})\right] \leq [\max_{Z' \in Z} |Z'|]^{-1}$$

■

The four bounds on u that have been identified will be labelled as follows:

- $\mathrm{LB}_1 = \min_{v \in \Sigma_\mathrm{B}} \frac{\alpha[v]}{|\Sigma_\mathrm{A}|}$
- $\mathrm{UB}_1 = \max_{v \in \Sigma_\mathrm{A}} \frac{\beta[v]}{|\Sigma_\mathrm{B}|}$
- $\mathrm{LB}_2 = \left[\min_{W' \in W} |W'|\right]^{-1}$
- $\mathrm{UB}_2 = \left[\max_{Z' \in Z} |Z'|\right]^{-1}$

For a particular GSG, each of these bounds may or may not be attained. For example, consider the four graphs shown in Figure 17.5. In each case, consider the GSG $\mathfrak{G} = (G, \Sigma_\mathrm{A}, \Sigma_\mathrm{B}, r)$

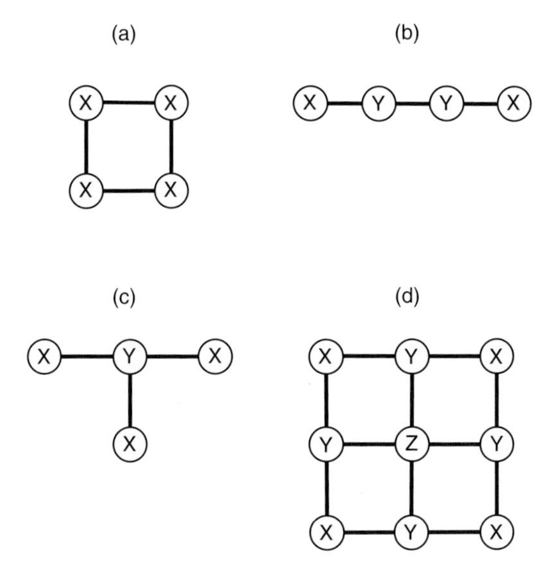

Figure 17.5 Four simple graphs. Solutions of GSGs played over these graphs and the values of the four bounds LB_1, LB_2, UB_1 and UB_2 are summarised in Table 17.1

Table 17.1 Summary of the values of bounds LB_1, LB_2, UB_1 and UB_2; the true values u to Player A; and examples of OMSs for each player for GSGs played over the four graphs shown in Figure 17.5, with $\Sigma_A = \Sigma_B = V(G)$ and $r = 1$. Shaded cells indicate that the relevant bounds are attained. All figures are rounded to two decimal places.

						Optimal strategies					
						Player *A*			Player *B*		
		Bounds				Probability allocated to each vertex marked					
Graph	u	LB_1	LB_2	UB_1	UB_2	X	Y	Z	X	Y	Z
a	0.75	**0.75**	0.50	**0.75**	1.00	0.25	N/A	N/A	0.25	N/A	N/A
b	0.50	**0.50**	**0.50**	0.75	**0.50**	0.00	0.50	N/A	0.50	0.00	N/A
c	1.00	0.50	**1.00**	**1.00**	**1.00**	0.00	1.00	N/A	0.33	0.00	N/A
d	0.40	0.33	0.33	0.56	0.50	0.00	0.20	0.20	0.15	0.10	0.00

with $\Sigma_A = \Sigma_B = V(G)$ and $r = 1$. For such small graphs, OMSs are easy to calculate (e.g. using the method described by Morris (1994, pp. 99–114)). Table 17.1 summarises the OMSs, the true values of u (the value of $\mathfrak{G}$ to Player A) and the values of the bounds for each of the four games.

It should be noted that while LB_1 and UB_1 will generally be easy to calculate, LB_2 and UB_2 may not be, since the minimal and maximal cardinalities of $W' \in W$ and $Z' \in Z$ respectively may be difficult to determine.

Example 17.5. *Consider the game $\mathfrak{G}$, described in Example 17.1. Since $\Sigma_A = \Sigma_B = V(G)$, we can use equation (17.11) to determine the values of* LB_1 *and* UB_1.

We observe that the minimum degree, $\delta(G)$, is equal to 1; the maximum degree, $\Delta(G)$, is equal to 4; and $|V(G)| = 15$. Therefore, by equation (17.11), the value of the game to Player A (the maximum probability of capturing a criminal that the police can ensure through their choice of strategy) lies in the interval $[2/15, 1/3]$.

Example 17.6. *To identify the values of* LB_2 *and* UB_2 *for the game* $\mathfrak{G}$*, described in Example 17.1, we must first identify the sets W and Z and their respective minimal and maximal elements (see Proposition 17.13).*

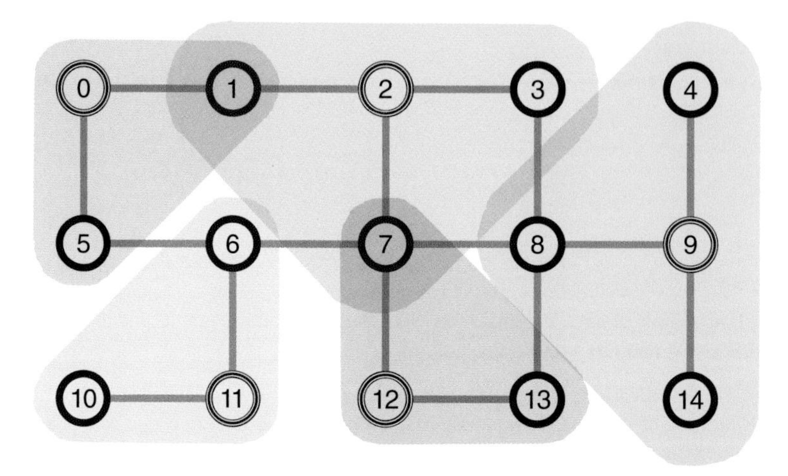

Figure 17.6 The vertex subset $\{v_0, v_2, v_9, v_{11}, v_{12}\}$ (whose members are highlighted in this figure) is an element of the set W (see Proposition 17.13), because the union of the closed neighbourhoods of these vertices (shaded) completely covers the vertices of the graph

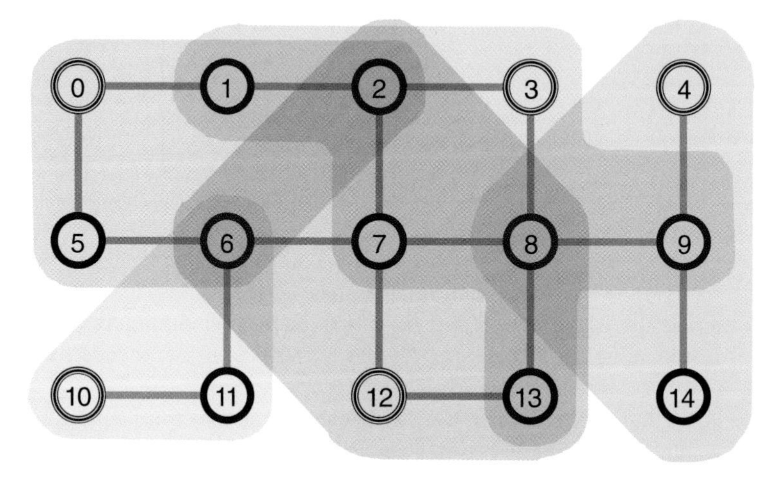

Figure 17.7 The vertex subset $\{v_0, v_3, v_4, v_{10}, v_{12}\}$ (whose members are highlighted in this figure) is an element of the set Z (see Proposition 17.13), because the distance between any pair of these vertices is always at least 3. This is demonstrated in the figure by shading the region $B_2[v]$ around each vertex in $\{v_0, v_3, v_4, v_{10}, v_{12}\}$. Note that no shaded region contains more than one highlighted vertex

W is the set of subsets of vertices whose closed neighbourhoods completely cover $V(G)$ (remembering that $V(G) = \Sigma_A = \Sigma_B$ in this example). One possible minimal element of W is $\{v_0, v_2, v_9, v_{11}, v_{12}\}$, which has cardinality 5 (see Figure 17.6). We can show that this is the minimal cardinality by identifying a subset of vertices that could not be covered by the closed neighbourhoods of any four vertices. The vertex subset $\{v_0, v_3, v_4, v_{10}, v_{12}\}$ satisfies this condition, since there is no vertex in $V(G)$ whose closed neighbourhood contains more than one of the five members of this subset.

Z is the set of subsets of $V(G)$ in which no two vertices are fewer than three steps apart. One possible maximal element of Z is $\{v_0, v_3, v_4, v_{10}, v_{12}\}$, which has cardinality 5 (see Figure 17.7). We can see that this is the maximal cardinality by observing that, given any $Z' \in Z$, the vertices of Z' must have pairwise disjoint closed neighbourhoods (as a result of the minimum distance condition). However, we note that any subset of six vertices satisfying this property could contain no more than two vertices with closed neighbourhoods of cardinality less than 3 (since one of v_4 and v_{14} must be omitted). The sum of the cardinalities of the disjoint closed neighbourhoods of the six vertices would therefore be at least 16, which is a contradiction since there are only 15 vertices in total.

Given all of the above, we observe that $LB_2 = UB_2 = 0.2$. *Hence, the value of the game to the police is* 0.2, *indicating that they can guarantee themselves a probability of at least* 0.2 *of apprehending the criminals if they employ an OMS, and this probability may increase further if the criminals employ a suboptimal strategy.*

17.6 Summary and Conclusions

In this chapter, we have presented a review of the scope and variety of existing search and security games and explained our motivation for defining a new spatial game in this research context. We have defined the SSSG, a general, explicitly spatial game for the modelling of search and concealment scenarios, and we have demonstrated that the SSSG provides a framework to unite other games presented in the literature. We have also derived some initial results on the OMSs of the SSSG and on the values of the game to each player, focusing particularly on games with finite strategy sets and on the special case of the SSSG played over a graph, the Graph Search Game (GSG).

At this stage, the SSSG and GSG remain fairly theoretical and abstract modelling tools. While they may be used to *represent* a particular scenario, it is not yet necessarily clear how these games would aid the *analysis* of such situations. In order to increase the value of the SSSG as a framework for the scrutiny of real-world search and security problems, it will be necessary to investigate and develop methods for the identification of its OMSs. Knowledge of the mathematically "optimal" approach for a given set of strategic circumstances could certainly be of value to policy makers and those charged with the security of certain geographically dispersed resources or networks.

In Appendix C, we will present a number of additional methods to identify the OMSs of a GSG, including an algorithm based on the iterated elimination of dominated strategies, a technique for exploiting symmetries of the graph over which a game is played and a means of identifying exact solutions for games played over a particular family of graphs.

References

Alpern S. (1974). The search game with mobile hider on the circle. In EO Roxin, TP Liu and RL Sternberg (eds), *Differential Games and Control Theory*. New York: Dekker, pp. 181–200.

Alpern S. (1985). Search for point in interval, with high-low feedback. *Mathematical Proceedings of the Cambridge Philosophical Society* 98(3): 569–578.

Alpern S. (1992). Infiltration games on arbitrary graphs. *Journal of Mathematical Analysis and Applications* 163(1): 286–288.

Alpern S. (1995). The rendezvous search problem. *SIAM Journal on Control and Optimization* 33(3): 673–683.

Alpern S. (2000). Asymmetric rendezvous search on the circle. *Dynamics and Control* 10(1): 33–45.

Alpern S. (2002). Rendezvous search: a personal perspective. *Operations Research* 50(5): 772–795.

Alpern S. (2008). Hide-and-seek games on a tree to which Eulerian networks are attached. *Networks* 52(3): 162–166.

Alpern S. (2010). Search games on trees with asymmetric travel times. *SIAM Journal on Control and Optimization* 48(8): 5547–5563.

Alpern S and Asic M. (1985). The search value of a network. *Networks* 15(2): 229–238.

Alpern S and Asic M. (1986). Ambush strategies in search games on graphs. *SIAM Journal on Control and Optimization* 24(1): 66–75.

Alpern S and Baston VJ. (2005). Rendezvous on a planar lattice. *Operations Research* 53(6): 996–1006.

Alpern S, Baston VJ and Essegaier S. (1999). Rendezvous search on a graph. *Journal of Applied Probability* 36(1): 223–231.

Alpern S, Baston V and Gal S. (2008a). Network search games with immobile hider, without a designated searcher starting point. *International Journal of Game Theory* 37(2): 281–302.

Alpern S, Baston VJ and Gal S. (2009). Searching symmetric networks with utilitarian-postman paths. *Networks* 53(4): 392–402.

Alpern S and Beck A. (2000). Pure strategy asymmetric rendezvous on the line with an unknown initial distance. *Operations Research* 48(3): 498–501.

Alpern S, Fokkink R and Kikuta K. (2010). On Ruckle's conjecture on accumulation games. *SIAM Journal on Control and Optimization* 48(8): 5073–5083.

Alpern S, Fokkink R, Lindelauf R and Olsder G-J. (2008b). The "Princess and Monster" game on an interval. *SIAM Journal on Control and Optimization* 47(3): 1178–1190.

Alpern S, Fokkink R, Timmer M and Casas J. (2011a). Ambush frequency should increase over time during optimal predator search for prey. *Journal of the Royal Society Interface* 8(64): 1665–1672.

Alpern S and Gal S. (1995). Rendezvous search on the line with distinguishable players. *SIAM Journal on Control and Optimization* 33(4): 1270–1276.

Alpern S and Gal S. (2002). Searching for an agent who may or may not want to be found. *Operations Research* 50(2): 311–323.

Alpern S and Howard JV. (2000). Alternating search at two locations. *Dynamics and Control* 10(4): 319–339.

Alpern S and Lim WS. (1998). The symmetric rendezvous-evasion game. *SIAM Journal on Control and Optimization* 36(3): 948–959.

Alpern S, Morton A and Papadaki K. (2011b). Patrolling games. *Operations Research* 59(5): 1246–1257.

Alpern S and Reyniers DJ. (1994). The rendezvous and coordinated search problems. In Peshkin M (ed), *Proceedings of the 33rd IEEE Conference on Decision and Control*, vol. 1. Lake Buena Vista, FL: IEEE Control Systems Society, pp. 513–517.

Anderson EJ and Aramendia MA. (1990). The search game on a network with immobile hider. *Networks* 20(7): 817–844.

Anderson EJ and Aramendia MA. (1992). A linear programming approach to the search game on a network with mobile hider. *SIAM Journal on Control and Optimization* 30(3): 675–694.

Arnold RD. (1962). *Avoidance in One Dimension: A Continuous-Matrix Game*. Interim Research Memorandum no. 10 277843. Washington, DC: Operations Evaluation Group.

Basilico N, Gatti N and Amigoni F. (2012). Patrolling security games: definition and algorithms for solving large instances with single patroller and single intruder. *Artificial Intelligence* 184: 78–123.

Baston VJ and Bostock FA. (1985). A high-low search game on the unit interval. *Mathematical Proceedings of the Cambridge Philosophical Society* 97(2): 345–348.

Baston VJ, Bostock FA and Ferguson TS. (1989). The number hides game. *Proceedings of the American Mathematical Society* 107(2): 437–447.

Baston VJ and Garnaev AY. (2000). A search game with a protector. *Naval Research Logistics* 47(2): 85–96.

Berwanger D. (2007). Admissibility in infinite games. In Thomas W and Weil P (eds), *STACS 2007*. Lecture Notes in Computer Science, vol. 4393. Berlin: Springer, pp. 188–199.

Blackwell D and Girshick MA. (1979). *Theory of Games and Statistical Decisions* [reprint of 1954 original]. New York: Dover.

Bondy JA and Murty USR. (1976). *Graph Theory with Applications*. London: Macmillan Press.

Bordelon DJ. (1975). Solutions of elementary problems –E 2469 –editor's comment. *The American Mathematical Monthly* 82(5): 522.

Börgers T. (1992). Iterated elimination of dominated strategies in a Bertrand-Edgeworth model. *Review of Economic Studies* 59(1): 163–176.

Buyang C. (1995). Search-hide games on trees. *European Journal of Operational Research* 80(1): 175–183.

Chkhartishvili AG and Shikin EV. (1995). Simple search games on an infinite circular cylinder. *Mathematical Notes* 58(5): 1216–1222.

Cohen ND. (1966). An attack-defense game with matrix strategies. *Naval Research Logistics Quarterly* 13(4): 391–402.

Dagan A and Gal S. (2008). Network search games, with arbitrary searcher starting point. *Networks* 52(3): 156–161.

Danskin JM. (1990). On the cookie-cutter game –search and evasion on a disk. *Mathematics of Operational Research* 15(4): 573–596.

Dendris ND, Kirousis LM and Thilikos DM. (1997). Fugitive-search games on graphs and related parameters. *Theoretical Computer Science* 172(1): 233–254.

Dufwenberg M and Stegeman M. (2002). Existence and uniqueness of maximal reductions under iterated strict dominance. *Econometrica* 70(5): 2007–2023.

Eagle JN and Washburn AR. (1991). Cumulative search-evasion games. *Naval Research Logistics* 38(4): 495–510.

Evans R. (1975). Solutions of elementary problems –E 2469. *The American Mathematical Monthly* 82(5): 521–522.

Foreman JG. (1977). Differential search games with mobile hider. *SIAM Journal on Control and Optimization* 15(5): 841–856.

Gal S. (1974). A discrete search game. *SIAM Journal on Applied Mathematics* 27(4): 641–648.

Gal S. (1978). A stochastic search game. *SIAM Journal on Applied Mathematics* 34(1): 205–210.

Gal S. (1979). Search games with mobile and immobile hider. *SIAM Journal on Control and Optimization* 17(1): 99–122.

Gale D and Glassey CR. (1974). Problems and solutions –E 2469. *The American Mathematical Monthly* 81(4): 405.

Garnaev AY. (1991). Search game in a rectangle. *Journal of Optimization Theory and Applications* 69(3): 531–542.

Garnaev AY. (1992). A remark on the princess and monster search game. *International Journal of Game Theory* 20(3): 269–276.

Grometstein AA and Shoham D. (1989). Colonel Richard's game. *The Lincoln Laboratory Journal* 2(2): 235–246.

Hausken K. (2010). Defense and attack of complex and dependent systems. *Reliability Engineering & System Safety* 95(1): 29–42.

Hausken K. (2011). Protecting complex infrastructures against multiple strategic attackers. *International Journal of Systems Science* 42(1): 11–29.

Hohzaki R. (2007). Discrete search allocation game with false contacts. *Naval Research Logistics* 54(1): 46–58.

Hohzaki R. (2008). A search game taking account of attributes of searching resources. *Naval Research Logistics* 55(1): 76–90.

Hohzaki R. (2009). A cooperative game in search theory. *Naval Research Logistics* 56(3): 264–278.

Hohzaki R and Iida K. (2000). A search game when a search path is given. *European Journal of Operational Research* 124(1): 114–124.

Hortala-Vallve R and Llorente-Saguer A. (2012). Pure strategy Nash equilibria in non-zero sum Colonel Blotto games. *International Journal of Game Theory* 41(2): 331–343.

Kikuta K. (2004). A search game on a cyclic graph. *Naval Research Logistics* 51(7): 977–993.

Kikuta K and Ruckle WH. (1994). Initial point search on weighted trees. *Naval Research Logistics* 41(6): 821–831.

Kikuta K and Ruckle WH. (1997). Accumulation games, part 1: noisy search. *Journal of Optimization Theory and Applications* 94(2): 395–408.

Kikuta K and Ruckle WH. (2002). Continuous accumulation games on discrete locations. *Naval Research Logistics* 49(1): 60–77.

Kikuta K and Ruckle WH. (2007). Rendezvous search on a star graph with examination costs. *European Journal of Operational Research* 181(1): 298–304.

Kikuta K and Ruckle WH. (2010). Two point one sided rendezvous. *European Journal of Operational Research* 207(1): 78–82.

Leyton-Brown K and Shoham Y. (2008). *Essentials of Game Theory: A Concise, Multidisciplinary Introduction*. Synthesis Lectures on Artificial Intelligence and Machine Learning. San Rafael, CA: Morgan & Claypool.

Lim WS and Alpern S. (1996). Minimax rendezvous on the line. *SIAM Journal on Control and Optimization* 34(5): 1650–1665.

Mavronicolas M, Papadopoulou V, Philippou A and Spirakis P. (2008). A network game with attackers and a defender. *Algorithmica* 51(3): 315–341.

Morris P. (1994). *Introduction to Game Theory*. Universitext. New York: Springer.

Neuts MF. (1963). A multistage search game. *Journal of the Society for Industrial and Applied Mathematics* 2(2): 502–507.

Oléron Evans TP and Bishop SR. (2013). Static search games played over graphs and general metric spaces. *European Journal of Operational Research* 231(3): 667–689.

Owen G and McCormick GH. (2008). Finding a moving fugitive: a game theoretic representation of search. *Computers & Operations Research* 35(6): 1944–1962.

Pavlovic L. (1995). A search game on the union of graphs with immobile hider. *Naval Research Logistics* 42(8): 1177–1189.

Powell R. (2007). Defending against terrorist attacks with limited resources. *American Political Science Review* 101(3): 527–541.

Powell R. (2009). Sequential, nonzero-sum "Blotto": allocating defensive resources prior to attack. *Games and Economic Behavior* 67(2): 611–615.

Reijnierse JH and Potters JAM. (1993). Search games with immobile hider. *International Journal of Game Theory* 21(4): 385–394.

Roberson B. (2006). The Colonel Blotto game. *Economic Theory* 29(1): 1–24.

Ruckle WH. (1983). *Geometric Games and Their Applications*. Research Notes in Mathematics no. 82. Boston: Pitman.

Ruckle WH. (1990). The gold-mine game. *Journal of Optimization Theory and Applications* 64(3): 641–650.

Ruckle WH. (2007). Rendez-vous search on a rectangular lattice. *Naval Research Logistics* 54(5): 492–496.

Ruckle WH and Kikuta K. (2000). Continuous accumulation games in continuous regions. *Journal of Optimization Theory and Applications* 106(3): 581–601.

Shubik M and Weber RJ. (1981). Systems defense games –Colonel-Blotto, command and control. *Naval Research Logistics* 28(2): 281–287.

Steen LA and Seebach JA Jr. (1970). *Counterexamples in Topology*. New York: Springer.

Subelman EJ. (1981). A hide-search game. *Journal of Applied Probability* 18(3): 628–640.

Sutherland WA. (1975). *Introduction to Metric and Topological Spaces*. Oxford: Oxford University Press.

Tambe M. (2012). *Security and Game Theory: Algorithms, Deployed Systems, Lessons Learned*. New York: Cambridge University Press.

Thomas LC and Washburn AR. (1991). Dynamic search games. *Operations Research* 39(3): 415–422.

Tofias M, Merolla J and Munger M. (2007). Of colonels and generals: understanding asymmetry in the Colonel Blotto game. Prepared for MPSA 2007 Panel 34-6 Computational Models. Available at http://www.uwm.edu/∼tofias

White LV. (1994). *Games of Strategy on Trees*. Technical Report Series, TR-94-24. London: Imperial College London.

Zoroa N, Fernández-Sáez MJ and Zoroa P. (1999a). A game related to the number of hides game. *Journal of Optimization Theory and Applications* 103(2): 457–473.

Zoroa N, Fernández-Sáez MJ and Zoroa P. (2004). Search and ambush games with capacities. *Journal of Optimization Theory and Applications* 123(2): 431–450.

Zoroa N, Fernández-Sáez MJ and Zoroa P. (2012). Patrolling a perimeter. *European Journal of Operational Research* 222(3): 571–582.

Zoroa N and Zoroa P. (1993). Some games of search on a lattice. *Naval Research Logistics* 40(4): 525–541.

Zoroa N, Zoroa P and Fernández-Sáez MJ. (1999b). A generalization of Ruckle's results for an ambush game. *European Journal of Operational Research* 119(2): 353–364.

Zoroa N, Zoroa P and Fernández-Sáez MJ. (2001). New results on a Ruckle problem in discrete games of ambush. *Naval Research Logistics* 48(1): 98–106.

Zoroa N, Zoroa P and Fernández-Sáez MJ. (2003). Raid games across a set with cyclic order. *European Journal of Operational Research* 145(3): 684–692.

Part Ten
Networks

18

Network Evolution: A Transport Example

Francesca Pagliara, Alan G. Wilson and Valerio de Martinis

18.1 Introduction

A major challenge for urban modelling has been, and is, the development of dynamic models of the evolution of urban structure. Considerable progress has been made in relation to, for example, retail structure, but modelling the evolution of transport networks is more difficult. There is a higher degree of complexity in this case. In the retail example, the 'structure' is the vector of sizes of retail centres at points in space. In the transport case, the elements are not points but links of a network, and the problem is further complicated by the fact that origin–destination flows are carried on routes which are sequences of links. It is also the case that 'size of link' does not capture the reality of the nature of network evolution: a multilane highway, for example, is grade-separated and of a different nature from lower 'levels' of link.

In the wider literature of the field, many interesting perspectives have been developed, and are still developing, on transport network planning, considering different aspects that vary from a specific modelling evaluation of trip distribution (Tsekeris and Stathopoulos, 2006) and modal split (Ben Akiva *et al.*, 1985), to the analysis of road network structures and geometric classifications for land-use and network development (Levinson and Karamalaputi, 2003; Levinson *et al.*, 2007; Xie *et al.*, 2007b) together with financing aspects (De Palma *et al.*, 2012), and to the characterization of new mobility and urban growth needs and the way to build them into modelling (Bocarejo and Oviedo, 2012; Millot, 2004). Moreover, fundamental reviews have been carried out in recent years in order to better evaluate the specificity of each aspect together with the whole vision of such a complex field (Vitins and Axhausen, 2010; Xie *et al.*, 2007a, 2009). Here, we develop a new approach by extending the simpler dynamic retail structure problem (cf. Harris and Wilson, 1978) by creating levels of a hierarchy and extending these ideas in the context of the equivalent transport network problem. The

Approaches to Geo-mathematical Modelling: New Tools for Complexity Science, First Edition. Edited by Alan G. Wilson.

Companion Website: www.wiley.com/go/wilson/ApproachestoGeo-mathematicalModelling

evolution of a road transport network is highlighted, and issues of complexity are tackled in relation to the activity system and the associated growth in transport demand (Leinbach *et al.*, 2000; Premius *et al.*, 2001; Yerra and Levinson, 2005).

Our approach to transport planning is rooted in a dynamic hierarchical system model of network evolution (Chen *et al.*, 2009; Spence and Linneker, 1994; Wilson, 1983, 2010). We restrict ourselves for illustrative purposes to a road transport network. In this case, the evolution of the network's links is based on performances in terms of travel and saturation rate; that is, the flows–capacity ratio that indicates the congestion level on roads. The dynamics of the system is driven by travel demand that is assumed to increase at each time step. An incremental demand assignment process has been adopted in order to evaluate the effect of each increase on the network and the way it subsequently evolves (Ferland *et al.*, 1975; Janson, 1991); that is, the increased demand at each step is translated into traffic flows that use minimum travel time paths at that time (Train, 2009). The proposed hierarchical algorithm uses as input the level of service reached on each link during the incremental assignment, computed as the saturation rate, and this is coupled with two different operating strategies for network upgrading, with or without budget constraints. In Section 18.2, a hierarchical retail model is outlined, and these concepts are extended for modelling transport network evolution in Section 18.3. In Section 18.4, a new method of transport planning is presented; and in Section 18.5, a case study is presented. In Section 18.6, conclusions are summarised.

18.2 A Hierarchical Retail Structure Model as a Building Block

The original dynamic retail structure model can represent extremes of configuration – large numbers of small centres or small numbers of large centres – and the 'hierarchy' is determined simply by size (Harris and Wilson, 1978). In the case of food shopping, for example, we might then think an appropriate extension would be to identify two 'levels' of centre – 'small' and 'supermarket' – with different cost structures and pulling powers (i.e. to identify two levels of a hierarchy). If we can model the evolution of this kind of system, it will provide insights that will enable the building of a dynamic model of the evolution of a road transport network (in which there are certainly at least three levels of hierarchy).

Suppose in the retail case, we have two hierarchical levels, labelled by h = 1, 2, the second being the 'supermarket'. Let W_j^h be the size of the h-level in zone j. Let K^h be the cost per unit of running an h-level centre, say annually. Let e_i be the unit expenditure by the population of i, and let P_i be the population. Let S_{ij}^h be the set of flows, measured in money units. Then, a suitable model would be

$$S_{ij}^h = A_i e_i P_i (W_j^h)^{\alpha^h} e^{-\beta^h c_{ij}} \quad (18.1)$$

where

$$A_i = 1/\sum_k (W_k^h)^{\alpha^h} e^{-\beta^h c_{ij}} \quad (18.2)$$

to ensure that

$$\sum_{j,h} S_{ij}^h = e_i P_i \quad (18.3)$$

The retail flows from zone i to zone j will be split between h = 1 and h = 2 according to the relative sizes of the Ws and the parameters α^h and β^h. The dynamics, in difference equation form, can be taken as

$$\Delta W_j^h = \varepsilon[D_j^h - K^h W_j^h] W_j^h \tag{18.4}$$

where

$$D_j^h = \sum_i S_{ij}^h \tag{18.5}$$

We will assume that $K^{(2)}$ is less than $K^{(1)}$ because of scale economies. However, these economies will only be achieved if W_j^h is greater than some minimum size. Say:

$$W_j^{(2)} \geq W_j^{(2)min} \tag{18.6}$$

and we expect that $W_j^{(1)}$ is very much smaller than $W_j^{(2)min}$.

The model therefore consists of equations (18.1–18.6) run in sequence. In the first instance, the equilibrium solutions to

$$D_j^h = K^h W_j^h \tag{18.7}$$

should be explored. And then assumptions could be made about the trajectories of exogenous variables so that the dynamics could also be explored.

It would be quite likely that for certain sets of values of exogenous variables, there would be no $W_j^{(2)} > 0$, and as, say, income increased, we could model the emergence of supermarkets. This may be interesting in its own right, but as noted at the outset, the current motivation is to point the way to modelling the evolution of transport networks.

18.3 Extensions to Transport Networks

The argument now follows that in Wilson (1983) but aims first to articulate the simplest transport model that would demonstrate the desired properties and into which we can insert explicit hierarchies. The 1983 paper contains a model that is rich is detail; it ends with a formulation of a simpler model, but one that does not have explicit hierarchies. In the simpler model of that paper, spider networks are deployed to seek to avoid the complexities of detailed real networks (which had been spelled out earlier in that paper). We use the same method here: spider networks are built by connecting nearby zone centroids with notional links.

Let {i} be a set of zones, and these can also stand as names for zone centroids. Let h = 1, 2, 3 label three levels of hierarchy representing different arterial types. We will also use j, u and v as centroid labels. For example, (u, v, h) may be a link on a route from i to j on the spider network at level h. (u, v) ε R_{ij}^{min} is the set of links that make up the best route from i to j. This may involve a mix of links of different levels. Let Γ^h be a measure of the annual running cost per unit of a link of level h – including an annualised capital cost – and let ρ_{uv} be the 'length' of link (u, v). Let c_{ij} be the generalised cost of travel from i to j as perceived by consumers, and let $\{\gamma_{uv}{}^h\}$ be the set of link travel costs (that for simple cases could be considered as travel times) on the level h link (u, v). So:

$$c_{ij} = \sum_{(u,\,v,\,h)\,\in R_{ij}^{min}} \gamma_{u,v}^h \tag{18.8}$$

If O_i and D_j are the total number of origins in zone i and destinations in zone j, and T_{ij} is the flow between i and j, then the standard doubly constrained spatial interaction model is

$$T_{ij} = A_i B_i O_i D_j e^{-\beta c_{ij}} \tag{18.9}$$

with

$$A_i = 1/\sum_k B_k D_k e^{-\beta c_{ik}} \tag{18.10}$$

$$B_j = 1/\sum_k A_k O_k e^{-\beta c_{ki}} \tag{18.11}$$

To model congestion, we need to know the flows on each link. Let q^h_{uv} be the flow on link (u, v, h), and let Q_{uvh} be the set of origin–destination pairs at level h that use the (u, v, h) link. Then

$$q^h_{uv} = \sum_{i,j \in Q_{uvh}} T_{ij} \tag{18.12}$$

γ^h_{uv} is a function of the flow:

$$\gamma^h_{uv} = \gamma^h_{uv}(q^h_{uv}) \tag{18.13}$$

For this simple demonstration model, we first assume that this function is the same for each link in a particular level of the hierarchy. This can be derived from standard speed–flow relationships. Then the equilibrium position of the network can be obtained by following the sequence of equations (18.8)–(18.13) and iterating. This will depend very much on the initial conditions and, in particular, on the links that are in place at higher levels in the hierarchy. These initial conditions represent the 'DNA' of the system at this time (cf. Wilson, 2010). The dynamics, in this case, can be thought of as the addition of higher level links into the network, within a budget, to minimise an objective function – say, consumers' surplus, or for the sake of simplicity, total travel (generalised) costs measured in money units. This would be a mathematical programme of the form:

$$\min C = \sum_{i,j} T_{ij} c_{ij} \tag{18.14}$$

subject to equations (18.8)–(18.13) and

$$\sum_{uv,\, \gamma(t)^h_{uv}} \Gamma^h \rho_{uv} = \Gamma(t+1) \tag{18.15}$$

The summation in equation (18.15) is over possible links that do not exist at time t, so that new links can be added up to an incremental budget spend of $\Gamma(t+1)$.

The dynamic model would then take the form: run equations (18.14), (18.15), insert the new $\gamma_{uv}{}^h$ into equations (18.8)–(18.13) and iterate as usual (as an inner iteration) and then begin the outer iteration again with a repeat of equations (18.14), (18.15). This is easy to formulate conceptually. The mathematical programme of equations (18.14), (18.15) is difficult to handle in practice, and we use a simpler approach in the demonstration model that follows.

This leads to the possibility of a second dynamic model in which each link at each level of the hierarchy is given a capacity, x^h_{uv}, and the link costs then become functions of this capacity as well as the flows:

$$\gamma^h_{uv} = \gamma^h_{uv}(q^h_{uv}, x^h_{uv}) \tag{18.16}$$

We can then think of a difference equation in Δx^h_{uv}. Suppose we take γ^h_{uv} as a measure of congestion – that is, if the unit generalised cost on a link is high, we attribute this to congestion. (There will still be the problem of dealing with non-existent links where this is formally infinite.) We could control the costs in this case by increasing Γ^h in an iterative cycle to ensure that a budget constraint is met. This suggests

$$\Delta x^h_{uv} = \varepsilon^h[\gamma^h_{uv} - \gamma],\ [\gamma^h_{uv} - \gamma] > 0\,;\ \Delta x^h_{uv} = 0\ \textit{otherwise} \tag{18.17}$$

where γ is taken as a threshold and ε is calculated to ensure that the budget constraint holds for this time period:

$$\sum_{uv,\ \gamma(t)^h_{uv}} \Gamma^h \rho_{uv} = \Gamma(t+1) \tag{18.18}$$

$x_{uv}{}^h(t+1)$ would then be fed back into equations (18.8)–(18.13) – with equations (18.16) replacing (18.13).

18.4 An Application in Transport Planning

The modelling of a specific and complex transportation system can be generically broken down into the supply system and demand system (Cascetta, 2009; Ortuzar and Willumsen, 2001). Let $\{i\}$ be a set of zones, and these can also stand as names for zone centroids. Let $h = 1, 2, \ldots$ label different levels of hierarchy representing arterial classes, two-lane highway and multilane highway links. We will also use i, j as centroids labels, and u and v as link nodes labels. For example, (u, v, h) may be a link on a route from i to j on the network at level h, so that $(u, v) \in R_{ij}{}^{min}$ is the set of links that make up the best route from i to j. This may involve a mix of links of different levels. Let Γ^h be a measure of the annual running cost per unit of a link of level h – including an annualised capital cost – and let l_{uv} be the 'length' of link (u, v). Let c_{ij} be the generalised cost of travel from i to j as perceived by consumers, and let $\{\gamma_{uv}{}^h\}$ be the set of link travel costs on the level h link (u, v). We can then use equation (18.8) which is repeated here for convenience:

$$c_{ij} = \sum_{(u,v,h)\in R^{min}_{i,j}} \gamma^h_{u,v} \tag{18.19}$$

We assume that an origin–destination matrix $\{T_{ij}\}$ is loaded onto the network and the array $\{q_{uv}\}$ represents the links flows. By analogy with retail centre dynamics and Lotka–Volterra equations, the dynamic evolution of the network can be described through a set of differential equations with the following form:

$$\partial x^h{}_{uv}/\partial t = \varepsilon\ [\gamma_{uv}(q^h{}_{uv}, x^h{}_{uv}) - \gamma] \tag{18.20}$$

where γ is the average link costs:

$$\gamma = (\Sigma_{uv}\gamma_{uv})/\text{no. of links} \tag{18.21}$$

and ε is a function representing the response of the link capacity to variations from the average. It is convenient, as in the retail case, to transform this continuous relationship into discrete form and to consider the function as constant in the time interval.

Table 18.1 *Highway Capacity Manual* (Transportation Research Board, 2010) characteristics for different facility types, and the facility types used for the case study described in Section 18.5

Facility type		Signals/mile	FFS (miles/H)	SaC (miles/H)
Multilane	Highway	N/A	60	55
Multilane	Highway	N/A	55	51
Multilane	Highway	N/A	50	47
Multilane	Highway	N/A	45	42
Two-lane	Highway	N/A	69	44
Two-lane	Highway	N/A	63	38
Two-lane	Highway	N/A	56	31
Two-lane	Highway	N/A	50	25
Two-lane	Highway	N/A	44	19
Arterial	Class I	0.2	50	33
Arterial	Class I	0.6	50	19
Arterial	Class I	1.6	50	10
Arterial	Class II	0.3	40	25
Arterial	Class II	0.6	40	18
Arterial	Class II	1.3	40	11
Arterial	Class III	1.3	35	11
Arterial	Class III	1.9	35	8
Arterial	Class III	2.5	35	6
Arterial	Class IV	2.5	30	6
Arterial	Class IV	3.1	30	5
Arterial	Class IV	3.8	30	4

N/A, not applicable

$$\Delta x^{h}_{uv} = \varepsilon[\gamma_{uv}(q^{h}_{uv}, x^{h}_{uv}) - \gamma] \tag{18.22}$$

In Section 18.5, we now show how this concept can be applied in a planning context. We consider three different facility types (arterial class I, arterial class II and arterial class III), and we use the characteristics described in the American *Highway Capacity Manual* (Transportation Research Board, 2010), after verifying the similarity with the Italian urban roads. In Table 18.1, we note different recommended values for several facility types from this manual.

For our present purpose, no travel demand evolution models have been applied directly, and only incremental demand has been taken into account. Transportation demand models, like all models used in engineering and econometrics, are schematic and simplified representations of complex real phenomena intended to quantify certain relationships between the variables relevant to the problem under study. Therefore, they should not be expected to give a 'perfect' reproduction of reality, especially when this is largely dependent on individual behaviour, as is the case with transportation demand. Furthermore, different models with different levels of accuracy and complexity can describe the same context. However, more sophisticated models require more resources (data, specification and calibration, computing time etc.) which must be justified by the requirements of the application (Cascetta, 2009).

The models for traffic assignment to transportation networks simulate how demand and supply interact and play central roles in developing a complete model. The results of such models describe a possible state of the system itself, or the 'average' state and its variation if available, and their results, in terms of traffic flows, are the inputs for the design and/or evaluation of projects (CONDUITS, 2011; Levinson and Karamalaputi, 2003; Zhang and Levinson, 2004). For our purpose, we retain the principle represented by equation (18.22) but replace it with a simpler procedure: the facility type is upgraded each time the link Volume Over Capacity (VOC) ratio (also known as saturation rate) reaches the value of 0.85, which we assume to be the start of congested behaviour.

For planning purposes, it is also important to recognise budget constraints. Investment decisions depend on the level of service provided by specific links along with the effective costs evaluated through parametric indicators. Moreover, vehicle tolls can be considered for those roads which have annual revenue from traffic flows (De Palma *et al.*, 2012; Zhang and Levinson, 2004). In this case study, no toll is considered for the arterial type of link (though it can be possible to consider it inside the local taxes amount), while investment for a multilane highway is only reported for private investor analysis. We apply a constraint of €350,000.00 as the budget defined by the local authorities for each iteration that could either allow the whole of the facilities' upgrading needed at that point in time, or, if it infringed, a part of them. To better understand the effects of network upgrading under budget constraints in terms of traffic flows, the model is re-run at each step when one or more links of network reach the limit of 0.85 and the budget limit is reached. We can then see the evolution of the network through assignment of the additional travel demand. In these simulations, only arterial facility types are considered, because the upgrading to two-carriageway roads needs more complex operations on the whole network (e.g. different kind of intersections), and this will be subject to further development. Moreover, Arterial Class IV indicates rural roads that are not considered. In the urban and suburban contexts, Arterial Class III represents those local roads where land access is privileged, Arterial Class I is a facility type where mobility takes precedence over land access and Arterial Class II roads include those roads connecting Class III to Class I (see Figure 18.1).

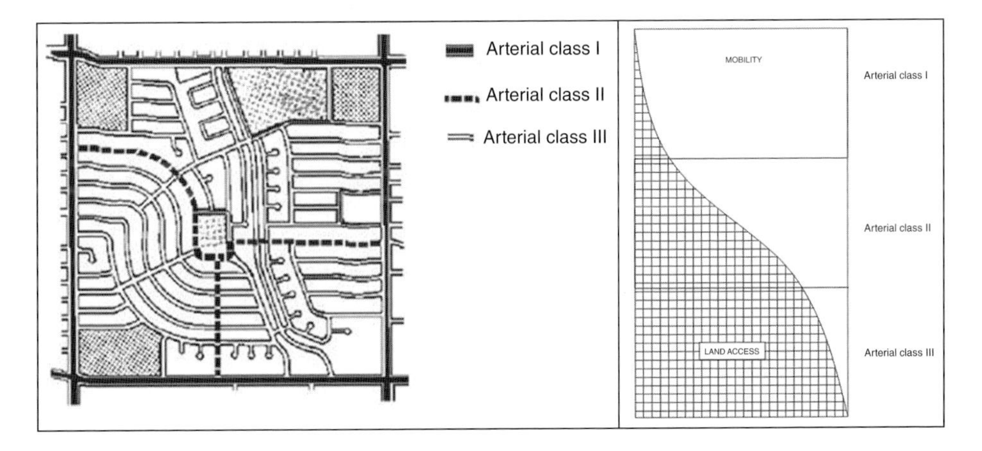

Figure 18.1 Arterial roads classification in urban and suburban areas. Source: Federal Highway Administration, U.S Department of Transportation

18.5 A Case Study: Bagnoli in Naples

This approach has been tested in a real context with a network grid and related traffic and demand data for the area of Bagnoli, at the west end of the city of Naples, where new redevelopment investments are scheduled after the complete disposal of the previous settlement of a steel mill. Moreover, in the near future, the area will have a new harbour, with all the related supply structures, for the America's Cup regattas. Due to these new activities, increasing transport demand is expected.

The modelling of the area is based on GIS files, and the network has been extracted through the TeleATLAS® dataset. In order to clearly represent the results and for better graphics, only the most important roads have been considered (as in Figure 18.2). Moreover, the upgrading of the network has not been considered for those links for which upgrading is not possible due to the particular built-up local context (see Figure 18.3). The transport demand on the network is defined through some centroids – special nodes which generate and/or attract transport demand, located on the most important entrances to the area and here indicated in red with an adjacent red number, while the 'anchor' symbol denotes the planned harbour.

The travel demand in the area of interest has been evaluated from an original travel demand matrix of a generic weekday morning, obtained from previous studies and owned by the Transportation Engineering Department of University of Napoli Federico II. This has been estimated through demand models (Cascetta, 2009) using traffic flows surveys on several significant sections. Internal demand has been ignored due to the position of the selected area in the global urban context of Napoli. It is, in fact, located at the city boundary, and so crossover and inbound-outbound city flows are the most important. All centroids are both travel origins and destinations, except centroids 9 and 30 that represent the two tunnels connecting in one way each selected area to the city centre. The OD matrices built for the incremental demand assignment consist of an additional demand that is indicated at centroid 41 and at the first step is split according to the minimum path travel time in free flow conditions; that is, the less time it takes to go from a given origin to a given destination, the more percentage of demand (500 units) is assigned (see Table 18.2).

The assignment model allows the calculation of performance measures and user flows for each supply element (network link), resulting from origin–destination flows, path choice behaviour and the reciprocal interactions between supply and demand. The adopted assignment algorithm uses an iterative process; in each iteration, network link flows are computed which incorporate link capacity restraint effects and flow-dependent travel times. This assumption can be justified by considering the equilibrium configuration as a state towards which the system evolves (Cascetta, 2009). From this interpretation, it follows that equilibrium analysis is valid for the analysis of the recurrent congestion conditions of the system; in other words, for those conditions that are systematically brought about by a sequence of periods of reference sufficiently large to guarantee that the system will achieve the state of equilibrium and remain in it for a sufficient length of time.

From these starting values which along with the TeleATLAS network set up the initial scenario for the network flows preload, an incremental demand is assumed to be generated or attracted from the new harbour activities, including special events like America's Cup.

Figure 18.2 The test site on Google Earth (top) and the related modelling of the area (bottom) with urban arterials as a continuous black line and the urban highway Tangenziale di Napoli in black dots

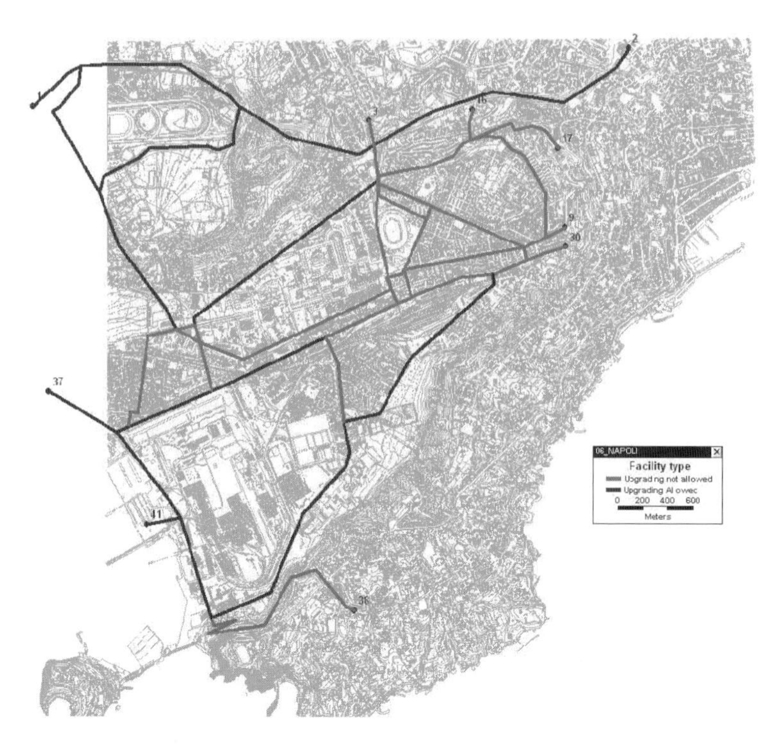

Figure 18.3 Network links with allowed or not allowed upgrading

Table 18.2 OD matrix of actual scenario, with centroid 41 not yet active (in brackets, values of the centroid 41 incremental demand at first step)

	1	2	3	9	16	17	30	37	38	41
1	0	1610	210	0	375	237	362	407	213	0 (180)
2	1818	0	157	0	375	234	470	330	199	0 (148)
3	187	239	0	0	86	74	115	73	41	0 (26)
9	150	168	248	0	233	242	161	102	61	0 (34)
16	101	117	195	0	0	273	73	52	31	0 (18)
17	73	82	121	0	237	0	55	41	26	0 (13)
30	0	0	0	0	0	0	0	0	0	0
37	75	60	71	0	64	58	77	0	64	0 (54)
38	87	85	95	0	88	83	104	142	0	0 (27)
41	0 (180)	0 (148)	0 (26)	0	0 (18)	0 (13)	0 (34)	0 (54)	0 (27)	0

The cost function for a generic link γ_{uv} is assumed to be a BPR (Bureau of Public Roads) cost function (see Cascetta, 2009, for its formulation):

$$\gamma_{u,v} = \frac{l_{u,v}}{v_{Ou,v}} + \alpha_1 \left(\frac{l_{u,v}}{v_{Cu,v}} - \frac{l_{u,v}}{v_{Ou,v}} \right) \left(\frac{q_{u,v}}{x_{u,v}} \right)^{\alpha_2} \tag{18.23}$$

where α_1 and α_2 are assigned values 0.75 and 4, respectively; the links length l_{uv} is given from the TeleATLAS dataset; $q_{u,v}$ is the link flow at generic iteration during the assignment process; and the capacities of the generic link (u, v) x_{uv}, the free flow speed v_{Ouv} and the speed at capacity v_{Cuv} are defined for the specific facility type. We have omitted the hierarchy superscript h for convenience.

For our purpose, we evaluate the transport network evolution for a given increasing of travel demand; and for ease of implementation, the total demand has been increased proportionally, at the end of 12 successive steps, to more than 50% of the starting value. But, of course, any other travel demand evolution model can be applied.

For clarity, only a few steps of the iteration are presented. The initial scenario is shown in Figure 18.4. At the third step of incremental demand assignment (see Figure 18.5), three links have been upgraded because the saturation rate exceeds 0.85. Simulations with and without budget constraints are the same because the budget limit has not been reached. In Figure 18.6, the fifth step of incremental demand assignment, budget constraints, is invoked: there is the first Arterial Class I link, but since the budget limit is reached, only two of three links have been upgraded. In Figure 18.7, the sixth step, the configuration is similar to the *unconstrained* one at the previous step; the third link that needs to be upgraded at the fifth step is now upgraded,

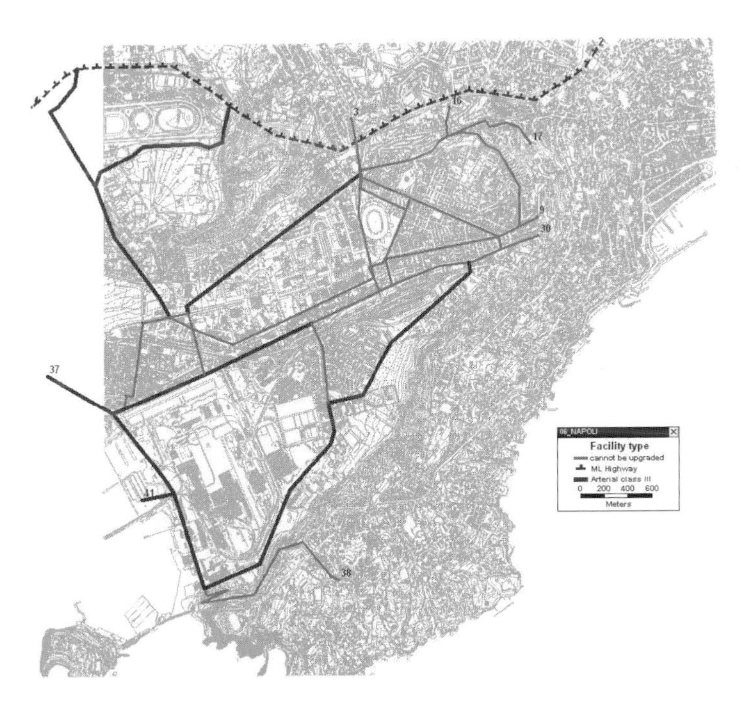

Figure 18.4 Actual scenario facility type

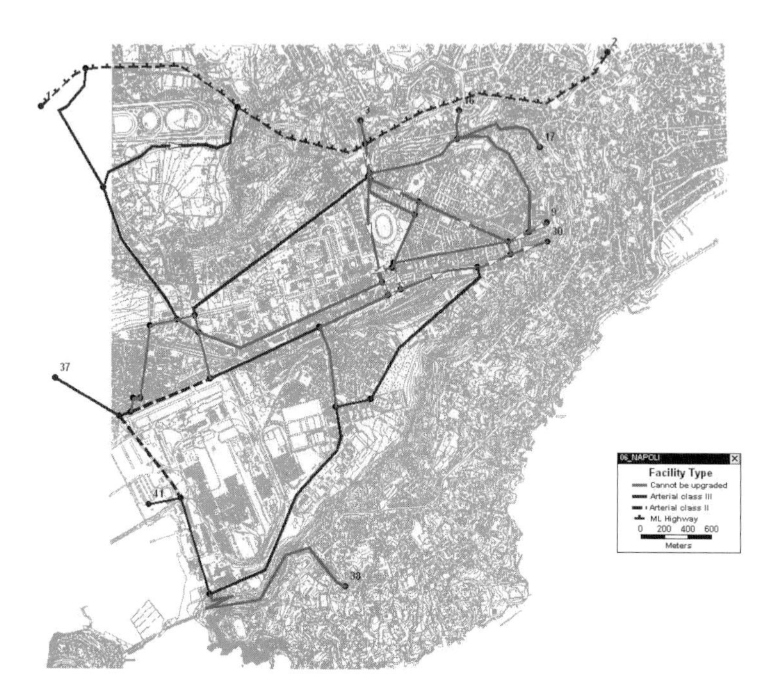

Figure 18.5 Facility types at the third step of the incremental demand scenario

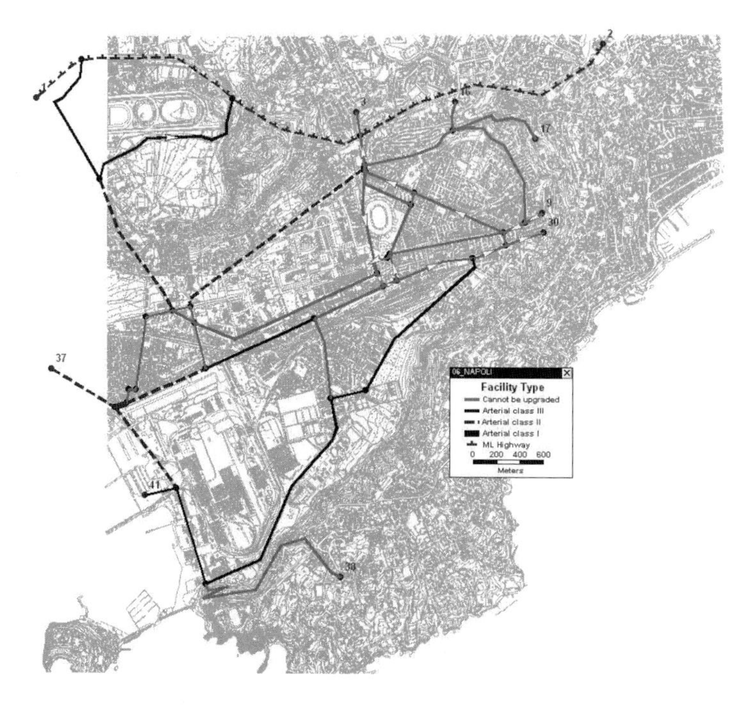

Figure 18.6 Facility types at the fifth step of the incremental demand scenario with budget constraints

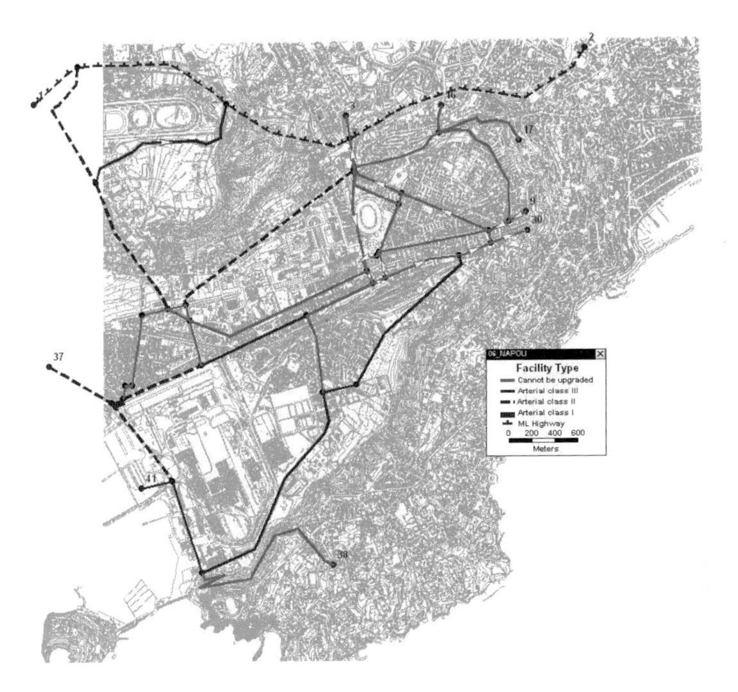

Figure 18.7 Facility types at the sixth step of the incremental demand scenario with budget constraints. (This configuration can be reached without budget constraints at the fifth step.)

but that VOC reached the value of 0.92. At the eighth step, the ML Highway Tangenziale di Napoli is near the 0.85 VOC limit (0.81 exactly). This means that, in the next step, this road will need an addition of one lane for each way. Under budget constraints, a different strategy needs to be implemented in this case because the entire budget is not enough for the upgrading of these links and another decision has to be adopted, for example using private capital and permitting entrepreneurs to participate in the Tangenziale di Napoli company profits and manage the infrastructure. At the eleventh step of incremental demand assignment, the travel demand has been incremented for almost 50% of the initial scenario (see Figure 18.8a and 18.8b). In Figure 18.8a, the budget constraints are not applied; while in Figure 18.8b, a constrained budget is presented. Here, a link has been upgraded before it reaches the VOC limit, in order to ease the future budget-constrained conditions of the 12th step, represented in Figure 18.9, that prefigured the upgrading of three links from Arterial Class II to Arterial Class I.

The final network evolution is represented in Figure 18.10. Continuous lines denote the network upgrading trend, taking into consideration only the maximum saturation rate reached, while dotted lines represent the upgrading under budget constraints. In this last case, the saturation rate could exceed the fixed maximum value. At iteration 9, the multilane highway should be upgraded with one lane for each direction, but this is not done because the available budget of the whole incremental demand process is insufficient for upgrading, and different strategies with other budget sources should be considered. In Figure 18.11, the average values of parametric travel times (i.e. travel time for a generic length) for the specific facility types during the whole iterative process are represented.

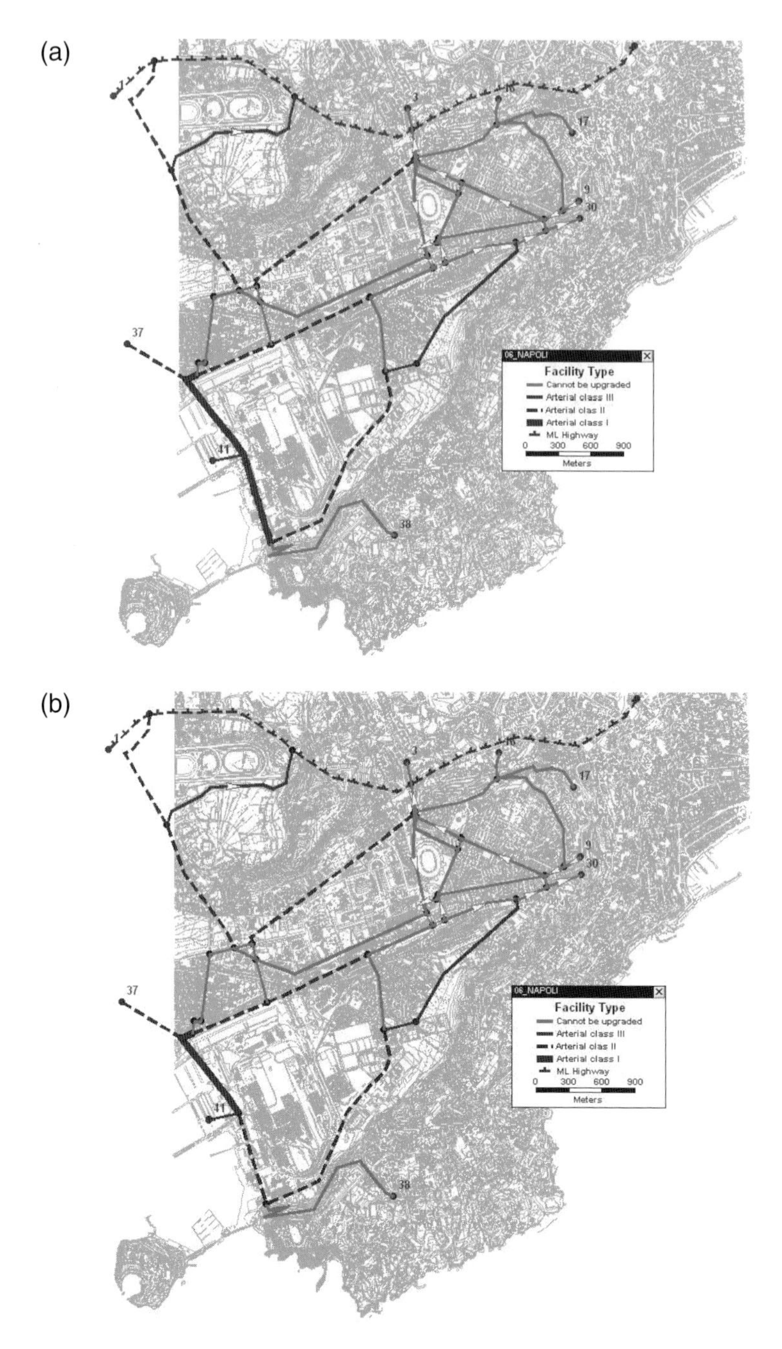

Figure 18.8 (a) Facility types at the 11th step of the incremental demand scenario without budget constraints. (b) Facility types at the 11th step of the incremental demand scenario with budget constraints

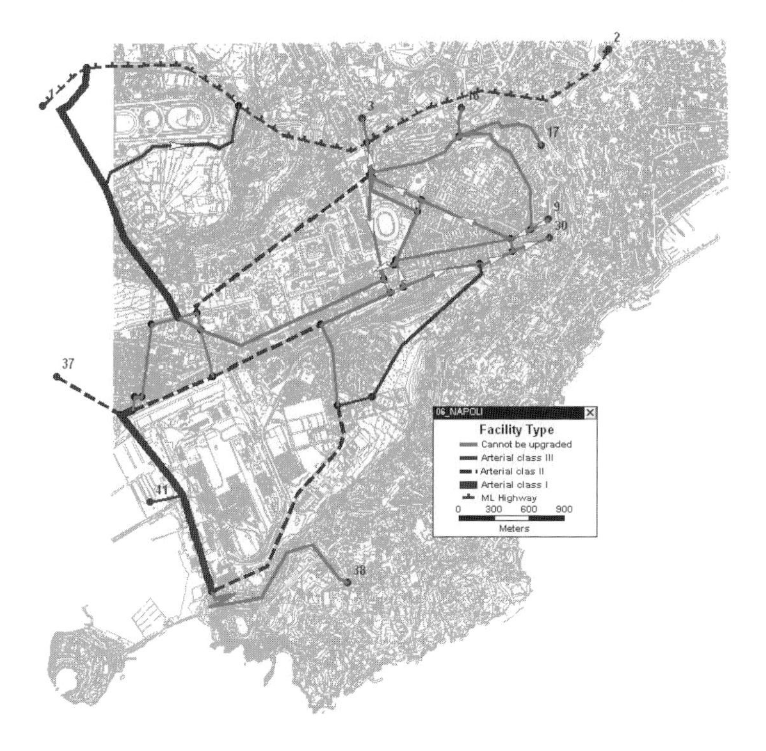

Figure 18.9 Facility types at the 12th step of the incremental demand scenario with budget constraints

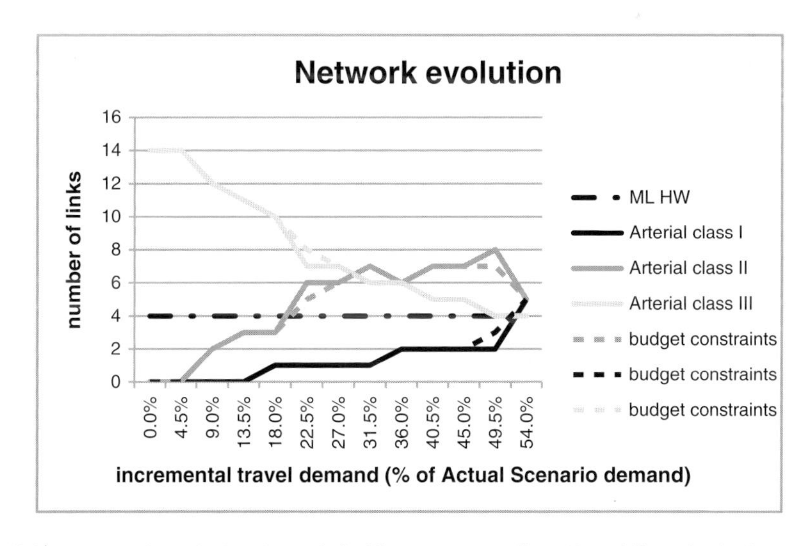

Figure 18.10 Network evolution through facility type upgrading (dotted lines for budget constraints scenario)

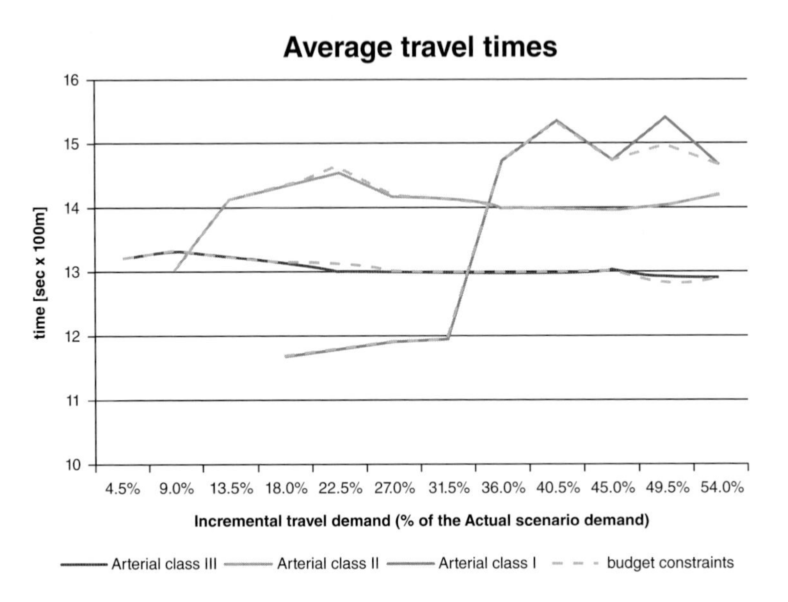

Figure 18.11 Average travel times for different facility types

In Tables 18.3 and 18.4, we report, at a given incremental demand computed as a percentage of the initial scenario demand, the status of the network in terms of number of links per facility type during the incremental demand assignment, both for budget-constrained and unconstrained scenarios; and, as results of the incremental demand assignment, the average travel time from and to centroid 41 and the average saturation rate (Satur column) of the network, intended as a rate between the assigned flows, including preload flows, and the capacity of the network.

Steps 5 and 6 and steps 11 and 12 are highlighted to indicate where budget constraints have been reached. In the first case (steps 5 and 6), it has been decided to postpone the upgrading of one of three links in the following step (step 6); in the second case (steps 11 and 12), it is preferred to realize the upgrading of a link to Arterial Class I a step before (step 11). In Table 18.4, step 9 is underlined; in that step, the multilane highway Tangenziale di Napoli has to be upgraded with one lane for each direction. This particular upgrading operation would give a consistent effect on the network mainly because it is the principal way of connecting different parts of the city, and consequently most of the total traffic flows are on it. As noted in this chapter, the upgrading costs are huge and the budget needed for upgrading is not reachable, so different strategies have to be implemented. In general, with the increasing traffic demand, traffic flows find new paths and, with a consistent preload demand given from the Actual Scenario, the link saturation rate easily reaches high values, while the network upgrading keeps it quite constant with little travel time decreasing. Investment costs have been considered, in order to quantify the operational cost of the network upgrading needed to support the hypothetical increased demand given from new activities.

In Table 18.5, parametric costs are reported. The costs are applied only to the upgrading phases, assuming that the construction costs of the initial scenario are already amortized. In

Table 18.3 Incremental demand assignment results without budget constraints

Iteration number	Incremental demand (%)	Number of used links				Travel time (sec)	Satur (Flows/Q)
		Multilane highway	Arterial Class I	Arterial Class II	Arterial Class III		
0	0.0%	4	0	0	53	303.5	0.51
1	4.5%	4	0	0	53	309.7	0.62
2	9.0%	4	0	2	51	316.9	0.69
3	13.5%	4	0	3	50	321.4	0.73
4	18.0%	4	1	3	49	326.1	0.74
5	**22.5%**	**4**	**1**	**6**	**46**	**323.4**	**0.75**
6	**27.0%**	**4**	**1**	**6**	**46**	**319.1**	**0.74**
7	31.5%	4	1	7	45	324.3	0.76
8	36.0%	4	2	6	45	326	0.75
9	**40.5%**	**4**	**2**	**7**	**44**	**289.3**	**0.68**
10	45.0%	4	2	7	44	291	0.7
11	**49.5%**	**4**	**2**	**8**	**43**	**292.42**	**0.71**
12	**54.0%**	**4**	**5**	**5**	**43**	**294.4**	**0.71**

Table 18.4 Incremental demand assignment results with budget constraints

Iteration number	Incremental demand (%)	Number of used links				Travel time (sec)	Satur (Flows/Q)
		Multilane highway	Arterial Class I	Arterial Class II	Arterial Class III		
0	0.0%	4	0	0	53	303.5	0.51
1	4.5%	4	0	0	53	309.7	0.62
2	9.0%	4	0	2	51	316.9	0.69
3	13.5%	4	0	3	50	321.4	0.73
4	18.0%	4	1	3	49	326.1	0.74
5	**22.5%**	**4**	**1**	**5**	**47**	**323.4**	**0.75**
6	**27.0%**	**4**	**1**	**6**	**46**	**319.1**	**0.74**
7	31.5%	4	1	7	45	324.3	0.76
8	36.0%	4	2	6	45	326.0	0.75
9	**40.5%**	**4**	**2**	**7**	**44**	**323.2**	**0.73**
10	45.0%	4	2	7	44	326.7	0.71
11	**49.5%**	**4**	**3**	**7**	**43**	**320.4**	**0.72**
12	**54.0%**	**4**	**5**	**5**	**43**	**319.9**	**0.71**

Table 18.6, costs for both constrained and unconstrained scenarios are shown. The values in the rows represent the costs in euros needed for upgrading the network to the selected iteration status starting from the previous one. On the right are budget plans with budget constraints in. Highlighted steps in the previous tables are reported here too.

Table 18.5 Network link costs

	Width, m	Cost, €/meter
Arterials Class III	7	(220.11)
Arterials Class II upgrading from III	8	101.70
Arterials Class I upgrading form II	9	141.27
ML Highway (1 lane add)	+3.5	400.0
ML Highway (1 lane add in tunnel)	+3.5	9040.0

ML, multilane. Italian Public Works Authority (2008)

Table 18.6 Network costs at milestones

Iteration	% demand	Upgrading network cost (only Arterial), €	Multilane highway upgrading costs, €	Upgrading network cost with budget constraints, €
0	0.0%	0.00		0.00
1	4.5%	0.00		0.00
2	9.0%	96,048.53		96,048.53
3	13.5%	74,428.13		74,428.13
4	18.0%	78,455.97		78,455.97
5	22.5%	**437,531.71**		**317,426.04**
6	27.0%	0.00		**120,105.67**
7	31.5%	101,259.64		101,259.64
8	36.0%	118,988.90		118,988.90
9	40.5%	83,093.99	13,710,000.00	83,093.99
10	45.0%	0.00		0.00
11	49.5%	**205,504.17**		**320,928.83**
12	54.0%	**458,731.94**		**343,307.29**

18.6 Conclusion

In this chapter, we have shown through the explicit introduction of hierarchies into dynamic models of evolution that the ideas represented in the dynamic retail model can be transferred to the task of modelling the evolution of transport networks. If each iteration is considered to be a time period, the model can then be applied in transport planning. The model can be the basis of a tool that can enable links to be selected for upgrading, and in such a way that budget constraints are satisfied – possibly by upgrading the most congested link first, and so on until the budget for that time period is exhausted. An incremental assignment-based process based on minimum travel time path choice has been used, according to the long-term, macroscopic, simulation scenarios defined by the increasing demand. The proposed strategies based on constrained and unconstrained budgets have been developed with upgrading operation costs used as a reference. It would be useful, and is already taken into consideration for further development, to split the costs for network upgrading and to evaluate the most effective variables through a global sensitivity in order to build different and more appropriate strategies and better exploit the constrained budget. Moreover, it will be useful to compare the reduction of travel generalized costs, externalities such as accidents, pollution and noise costs, with the upgrading network costs in a cost–benefit evaluation.

References

Ben-Akiva M and Lerman S. (1985). *Discrete Choice Analysis: Theory and Application to Travel Demand*. Cambridge, MA: MIT Press.

Bocarejo JP and Oviedo DR. (2012). Transport accessibility and social inequities: a tool for identification of mobility needs and evaluation of transport investments. *Journal of Transport Geography* 24: 142–154.

Cascetta E. (2009). *Transportation System Analysis: Models and Applications*. Berlin: Springer.

Chen S, Peng H, Liu S and Yang Y. (2009). A multimodal hierarchical-based assignment model for integrated transportation networks. *Journal of Transportation Systems Engineering and Information Technology* 9(6): 130–135.

CONDUITS (Coordination of Network Descriptors for Urban Intelligent Transportation System). (2011). Key performance indicators for traffic management and intelligent transport systems. European Research Project No. FP7-TRANSPORT. Available at http://www.transport-research.info/web/projects/project_details.cfm?ID=38146

De Palma A, Proost S and Van Der Loo S. (2012). Network development under a strict self-financing constraint. *Networks and Spatial Economics* 12(1): 109–127.

Federal Highway Administration. (2000). *FHWA functional classification guidelines*. Available at http://www.fhwa .dot.gov/planning/index.htm

Ferland JA, Florian M and Achim C. (1975). On incremental methods for traffic assignment. *Transportation Research* 9(4): 237–239.

Harris B and Wilson AG. (1978). Equilibrium values and dynamics of attractiveness terms in production-constrained spatial-interaction mode. *Environment and Planning A*: 10371–10388.

Italian Public Works Authority. (2008). [Data on network link costs]. Rome: Italian Public Works Authority.

Janson B. (1991). Dynamic traffic assignment for urban road networks. *Transportation Research Part B: Methodological* 25: 143–161.

Leinbach TR. (2000). Mobility in development context: changing perspectives, new interpretations, and the real issues. *Journal of Transport Geography* 8: 1–9.

Levinson D and Karamalaputi R. (2003). Predicting the construction of new highway links. *Journal of Transportation and Statistics* 6: 1–9.

Levinson D, Xie F and Zhu S. (2007). The co-evolution of land use and road networks. In RE Allsop, MGH Bell and B Heydecker (eds), *Transportation and Traffic Theory 2007*. Bingley: Emerald Group, pp. 839–859.

Millot M. (2004). Urban growth, travel practices and evolution of road safety. *Journal of Transport Geography* 12(3): 207–218.

Ortuzar J de D and Willumsen LG. (2001). *Modelling Transport*. Chichester: John Wiley and Sons.

Premius H, Nijkamp P and Banister D. (2001). Mobility and spatial dynamics: an uneasy relationship. *Journal of Transport Geography* 9: 167–171.

Spence N and Linneker B. (1994). Evolution of the motorway network and changing levels of accessibility in Great Britain. *Journal of Transport Geography* 2(4): 247–264.

Train K. (2009). *Discrete Choice Methods with Simulation*. Cambridge: Cambridge University Press.

Transportation Research Board. (2010). *Highway Capacity Manual 2010*. Washington, DC: Transportation Research Board.

Tsekeris T and Stathopoulos A. (2006). Gravity models for dynamic transport planning: development and implementation in urban networks. *Journal of Transport Geography* 14: 152–160.

Vitins BJ and Axhausen KW. (2010). Patterns and grammars for transport network generation. Paper presented at the 10th Swiss Transport Research Conference, Ascona, Switzerland, September.

Wilson AG. (1983). Transport and the evolution of urban spatial structure. In *Atti delle Giornate di Lavoro*. Naples: Guida Editori, pp. 17–27.

Wilson AG. (2010). Urban and regional dynamics from the global to the local: hierarchies, 'DNA' and 'genetic planning'. *Environment and Planning B* 37: 823–837.

Xie F and Levinson D. (2007a). The topological evolution of road networks. Paper presented at the 87th annual meeting of the Transportation Research Board, Washington, DC, January.
Xie F and Levinson D. (2007b). Measuring the structure of road networks. *Geographical Analysis* 39(3): 336–356.
Xie F and Levinson D. (2009). Modeling the growth of transportation networks: a comprehensive review. *Networks and Spatial Economics* 9(3): 291–307.
Yerra BM and Levinson D. (2005). The emergence of hierarchy in transportation networks. *Annals of Regional Science* 39(3): 541–553.
Zhang L and Levinson D. (2004). A model of the rise and fall of roads. Paper presented at the Engineering Systems Symposium, Massachusetts Institute of Technology, Boston, Massachusetts, March.

19

The Structure of Global Transportation Networks

Sean Hanna, Joan Serras and Tasos Varoudis

19.1 Introduction

The scale of analysis in space syntax research has historically been limited to that of the city street network, partly due to the limits of computational resources, due to the availability of maps and data, and for theoretical reasons. This chapter presents the initial stages of a global scale analysis of the structure of international transportation, including roads, but also sea-shipping, train and related networks, which we intend to be the first such study at this scale.

Prior studies of street networks beyond the scale of an individual city either have involved the separate analysis of individual urban graphs, as in the comparative analysis of a set of cities (Figueiredo and Amorim, 2007; Hanna, 2009; Peponis *et al.*, 2007), or have been limited to a particular region within a country, such as the regional mapping of connected cities in the north of England (Turner, 2009). In part, this is a technical and methodological issue of limited computational resources for larger scale analysis. It is also a theoretical issue, as many of the kinds of social phenomena – pedestrian movement, commercial presence, wealth and poverty and so on – that can be observed in the relatively small scale of the city street do not have a clear presence at the much larger scale of an international road network, and more theoretical work would be needed to understand these (Turner, 2009). Nevertheless, these studies have successfully revealed more abstract patterns. Turner (2009), in particular, demonstrated relationships between vehicular movement and measurements of choice in a road network at a regional scale, suggesting that at least some of the local findings of space syntax are replicated at a larger scale, and initiating a methodological framework for such study. This chapter intends to further this line of enquiry, making contributions in methodology that may, in future, lead to extensions of theory.

Approaches to Geo-mathematical Modelling: New Tools for Complexity Science, First Edition. Edited by Alan G. Wilson.

Companion Website: www.wiley.com/go/wilson/ApproachestoGeo-mathematicalModelling

As an extension of methodology, we propose the analysis of global transportation networks as graphs in the manner of street networks. In addition to roads, we include the international railway network, and leave for the future the possibility of including air and sea routes separately or in combination. The data for these latter are currently in preparation. Space syntax measurements, of choice for instance, are highly dependent on the sight and movement constraints of the geometric properties of the network in space, which are expressed in the calculation as, for example, angular weighting. In graph analysis more generally, equivalent measures such as betweenness centrality are strictly topological. As scales increase, and particularly where movement between nodes is independent of the geometry of their intersection such as in air and sea travel, we believe that strictly topological measures will be appropriate, or that other cost factors such as fuel, taxes and route travel speeds may be more relevant. The quantification of these is also left for further research, but we anticipate that they may be accommodated by weighting of the graph.

To position this within a possible future extension of theory, we anticipate that such a methodology may allow the investigation of something analogous to the natural movement of pedestrians within an urban grid that is fundamental to space syntax, but at the larger scale of international trade and economic activity. Although the largest previous studies of road networks (Turner, 2009) have concentrated on the network effect on literal movement because socioeconomic correlations appear to be non-existent, these used as indicators the census data of wealth and poverty, which may have been at too fine a resolution. The work presented here uses much longer range networks and maps the relationship with Gross Domestic Product (GDP) at an international level, and it finds that such correlations become clear. While we are still some way from any theory of *natural economic movement*, our observations suggest that a portion of economic activity at an international scale is determined by the larger transportation network and, more specifically, by the network structure as revealed by the kinds of measures currently used in space syntax analyses at the smaller scale.

19.2 Method

This study uses road and train network data obtained worldwide from the Vector Map Level 0 (VMap0) which was released in 1997 by the National Imagery and Mapping Agency (NIMA) from the United States. This release is the improved version of the map formerly known as the Digital Chart of the World, published in 1992. VMap0 is a 1:1,000,000 scale vector base map of the world. VMap0 has a transportation layer which includes the centre-line representation of major roads and rail lines. The only classification given by the document specification of the data is that the road segments are categorised according to whether they are primary or secondary roads. Looking at the UK road map, it appears that roads represented are mostly highways and some A and B roads. We are considering using more accurate global multimodal transport network representations datasets, such as OpenStreetMap, in future studies.

The analysis of the road and train network data has been performed in DepthmapX 0.24b (Varoudis, 2012). DepthmapX, originally developed as UCL Depthmap by Alasdair Turner, is an open-source multi-platform spatial network analysis software that forked from the original (Turner, 2004). DepthmapX can perform a number of spatial network analyses used in the research domain known as *space syntax* and is used by a wide community of academics and practitioners. Space syntax is a set of theories and techniques which apply graph network analysis to study the configuration of spatial networks in urban design and transport planning, and it originated from research by Bill Hillier and Julienne Hanson (Hillier and Hanson, 1984). In recent (Turner, 2007) space syntax network models, each street segment, or portion of straight street between intersections, is drawn to represent a node in the graph network. The road and rail centre-line models extracted from VMap0 have been used to generate the graph structure for DepthmapX.

Space syntax theorises the relationship between space and movement where a proportion of pedestrian movement is determined by the spatial configuration itself (Hillier *et al.*, 1993). Empirically, studies have found a strong correlation between the space syntax measure of 'Choice' with pedestrian flows and vehicular flows (Penn *et al.*, 1998). Recent advances suggest that the use of angular segment choice explains aggregate movement better than topological or metric choice (Hillier and Iida, 2005). The choice measure in space syntax was first presented in Hillier *et al.* (1987). Choice measures the quantity of movement that passes through each segment on the shortest trips between all pairs of origin–destination segments in a system, and so is analogous to the mathematical graph measure of betweenness centrality that is widely used in network sciences. In detail, angular segment choice is calculated by summing the number of 'angular weighted' shortest paths that pass from a segment. More formally, choice (or betweenness centrality) measures how many times shortest paths overlap between all pairs of origins and destinations, where $g_{jk}(p_i)$ is the number of geodesics between node p_j and p_k which contain node p_i and where g_{jk} is the number of all geodesics between p_j and p_k (equation 19.1).

$$C_B(P_i) = \sum_j \sum_k g_{jk}\left(P_i\right) / g_{jk}(j < k)^1 \tag{19.1}$$

Angular weighting of graph links implicitly associates a travel cost with turning direction; metric weighting represents the cost of distance travelled, and unweighted graphs capture network topology only. It is likely, as scales increase to the point where route choice decisions are assisted by maps, marked routes and GPS instead of visibility, that distance plays an increasingly more important role compared to turn angle; however, factors such as terrain and its effect on travel speed are also relevant, and are not presently available in the data. Due to similar lengths of segments in the map, angular weighting was used as a rough proxy for terrain and travel speed differences, and the space syntax measure of 'angular segment choice' was used throughout this study. This also allows for potential continuity with smaller scale studies of urban movement. Initial tests of different measures on samples of the network did not reveal significant differences between weighting methods; however, we feel this warrants more detailed study.

High values of choice indicate that a segment lies on the shortest paths for a high number of nodes in the system, and it is significant to the function of the network as it can be said to be a 'through segment'. During the analyses, four cut-off radii were also set: 50 km, 100 km, 500 km and 1000 km. The cut-off is essentially excluding the shortest path search for paths longer than the cut-off value in order to expose network structures that are more dominant at these scales. The number of segments (graph vertices) reachable from a source within a certain cut-off radii represents the 'node count' measure of that source segment.

[1] Equation 19.1 on betweenness centrality is from Hillier and Iida (2005), referring to Freeman (1977).

An issue to be considered while dealing with networks at a global scale is the role of map projections to minimise the error in the representation of networks. VMap0 uses the World Geodetic System from 1984 (WGS 84) which uses a spheroidal reference surface to represent the network; however, DepthmapX requires coordinates to be on a plane. For our analyses, we therefore projected the sample area onto a single plane. This is a relevant issue as it introduces changes in the angles from the original projection to the planar projection; this is visually apparent looking at the shape of England on the network figures from this article. Future studies will examine more accurate methods for translation of angles to better deal with three-dimensional representations for very large sample areas.

We have extracted GDP in US dollars (USD) and GDP per capita in USD from the World Bank's World Development Indicators data. It is also important to note that we have decided to use the values relating to the year 1992 so that the comparisons between the network and the economic indices are as fair as possible. Even though the network data were released in 1997, we believe that the year for which one can account for good representation is still 1992. In any case, we also believe that the slow speed of change of the transport network itself will not lead to very different results when using GDP values in a 5-year range.

19.3 Analysis of the European Map

The full study is working towards a complete global map, with terrestrial transportation within continents linked together by sea and air routes. A number of issues are still to be resolved with respect to the treatment of both sea and air transport, which currently exist only as relatively long-range origin–destination pairs and so are very different from segments on land routes; for this reason, we focus in the current analysis entirely on the combined road and rail networks, restricting ourselves to a distinct geographical region in which these routes predominate. The portion of the global map containing Europe and the immediate surroundings (including a portion of North Africa) was selected for analysis. This has the advantage of continuity with other studies and also the availability of related data, used in Section 19.4. The sample area for the transport network selected spans through 50 countries, but in this analysis, we have only considered those countries whose network coverage was complete: in other words, some countries located at the boundaries of the sample area were only partially represented by our map, and so they were removed. We have removed nine countries from the analysis, leaving a total of 41 countries.

Where the two maps are used together, rail and road segments are treated as a single network with any two segments linked at their endpoints regardless of their type. This assumes the free movement from rail to road (or the reverse) at any such link, which, in the absence of more detailed information, approximates the most likely conditions at stations on the rail routes.

DepthmapX analyses of angular choice over the European region are displayed in Figure 19.1 and Figure 19.2. At first glance, the network appears to behave similar to that of an urban region, albeit at a much larger scale. For relatively small radii of 50 km, equivalent to the local neighbourhood of an urban centre, the map (Figure 19.1) resolves itself into clearly defined 'hotspots' representing major cities. London, Paris and Frankfurt are clearly marked, as are large conurbations such as the English West Midlands and around Cologne, Dusseldorf and Essen in Germany. At much larger scales of radius 1000 km, equivalent to distances of international travel between countries, the network (Figure 19.2) is resolved into a clear

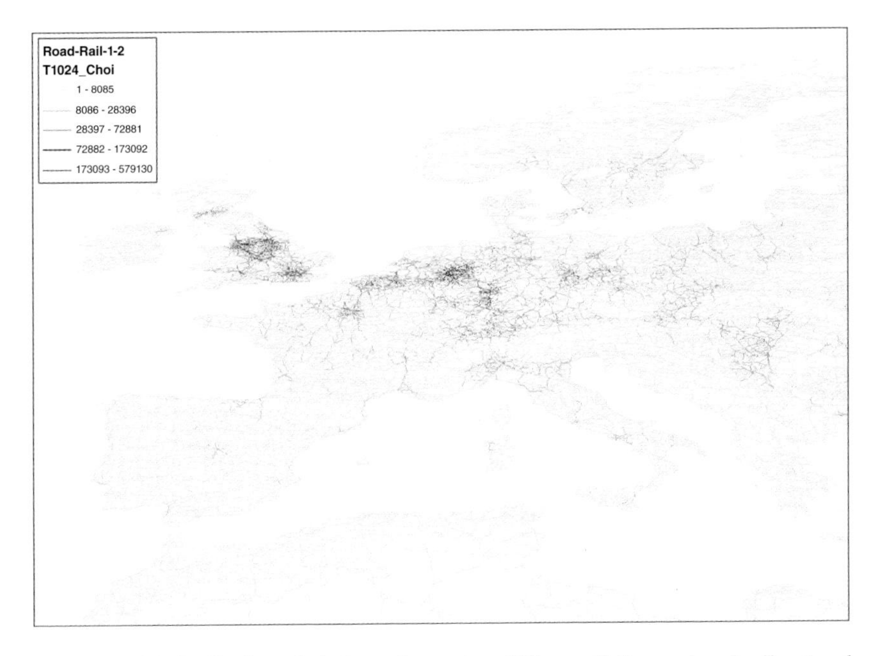

Figure 19.1 Spatial distribution of choice values using 50 km radii for road and rail networks. Five link classes have been calculated using the natural breaks (Jenks) technique, and zero values have been excluded from the visualisation

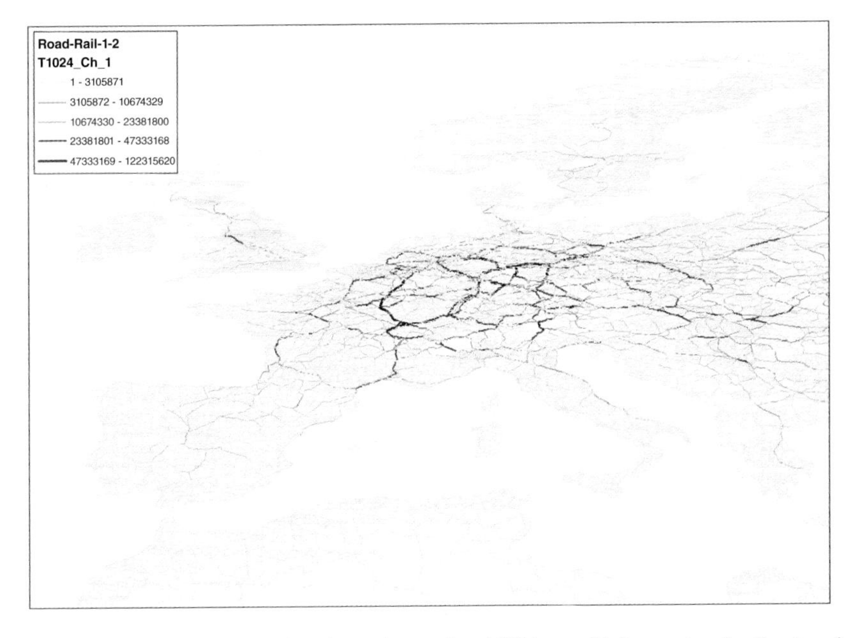

Figure 19.2 Spatial distribution of choice values using 1000 km radii for road and rail networks. Five link classes have been calculated using the natural breaks (Jenks) technique, and zero values have been excluded from the visualisation

foreground of major routes between these centres, and a generally homogeneous background. This is, at least in appearance, similar to the foreground and background networks seen in cities (Hillier *et al.*, 2012).

This suggests that some principles of the grid may be universally consistent across scales. Where local urban through-movement within a city network, as represented by choice at radii that are the scale of the neighbourhood (e.g. 1000 m), highlights the centres of local retail and commercial activity, the equivalent 'local' scale on the global map (e.g. 50 km) highlights urban centres that are likely to be the financial and industrial equivalents. Where through-movement across urban networks, as represented by choice at radii approaching the full scale of the city, displays major linear transportation routes, the same is true in the European map. The identification of major financial centres such as London and Frankfurt along these highlighted routes suggests that larger economic activity may correlate in a similar way, to be confirmed by economic data.

19.4 Towards a Global Spatial Economic Map: Economic Analysis by Country

If a relationship between the transportation network and economic activity should prove to exist across scales, we would expect to be able to model and predict a variety of spatially located economic outputs, from city, to regional, to national scales. In the analyses that follow, we test the assumption that measurements on the sum of segments within a particular region provide an indication of the economic output of that region. At present, our most reliable economic data are limited to the resolution of countries, and we test the plausibility of such a model with national GDP statistics.

Highlighted regions of high choice in maps in Figure 19.3 would appear to correspond to zones of economic activity, indicating as a plausible initial hypothesis that the sum total of segment choice values within a given region correlates to its overall economic output as expressed by GDP. This was tested for all countries in the dataset, correcting for the skewed distribution in the scale of countries by plotting each on a log–log scale. In doing so, two outliers are immediately obvious. Malta, with choice values far below any in the dataset, appears to have a disproportionately high GDP. This is as should be expected given that, as a small, isolated island, it will not be appropriately connected by our transportation networks until sea and possibly air routes are represented. Czechoslovakia, with a disproportionately low GDP, was in the process of dissolution during the year of the data (1992). Pending further analysis, these might be considered exceptions that lend credence to the rule. For the remaining countries, it can be seen that there is indeed a clear relationship between segment choice and GDP, with correlations to r^2 of 0.609, 0.584, 0.425 and 0.353 at 50 km, 100 km, 500 km and 1000 km scales, respectively (see Figure 19.3).

The distribution of choice values at these scales, however, results in very few segments of high value, with a vastly larger number of low choice 'background' segments, the latter of which actually dominate the overall sum. It is the number of total segments, rather than their values, that characterise the measurement of national total choice. This is evidenced by taking the correlations as above between a simple count of segments and GDP, in which a similar result holds with r^2 of 0.638, 0.611, 0.486 and 0.416 at 50 km, 100 km, 500 km and 1000 km scales. Two problems therefore exist with this basic measure: (a) it is dominated largely

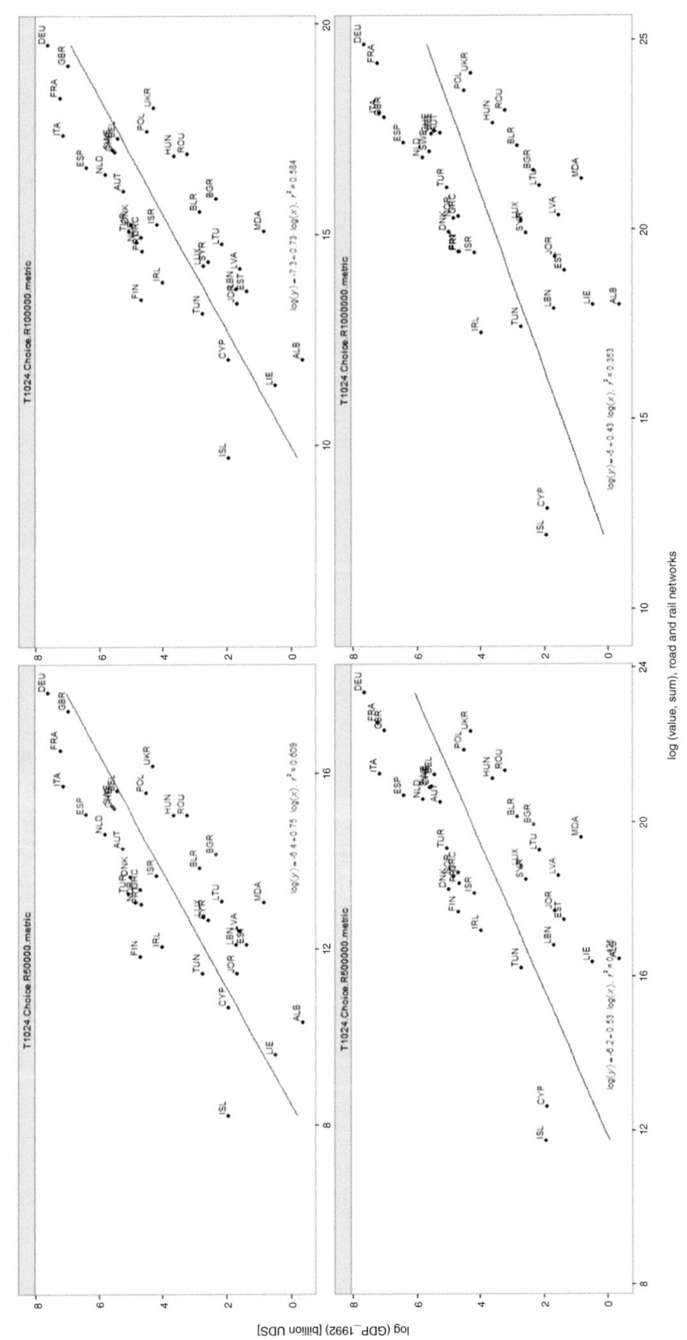

Figure 19.3 Comparison between the (log) sum of choice measures for road and rail networks against (log) total GDP for different radii

by the simple sizes of countries, and skewed by relatively few large ones; and (b) it indicates only a simple relationship between the size of a country's transportation infrastructure and its economic wealth, while saying nothing about how the network is connected.

The first issue would ideally be addressed by a finer aggregation of economic data, but in the absence of this we divide both the GDP and total choice values by the geographical area of each country. Figure 19.4 indicates the correlation between countries' GDP per km^2 and total choice per km^2, resulting in clear correlations of $r^2 = 0.517$, 0.485, 0.314 and 0.231 at 50 km, 100 km, 500 km and 1000 km scales. This normalisation has the secondary benefit of mediating some of the skew in data due to the uneven distribution of nation sizes, such that, if plotted on a non-logarithmic scale, the same linear correlations rise to r^2 values of up to 0.719 (for a 50 km radius).

To say something meaningful about the shape of the network – where infrastructure is built and how it is connected – we examine the portions of the network that have values of choice over a given threshold. We exclude the bulk of the 'background' network that contributed to the general node densities above by counting only the top 10% of segments by choice value, and excluding the 90% below this threshold value. This results in a much more selective map that represents only the apparent urban 'hot spots' of the 50 km radius map, or the prominent international routes of the 500 km map. The count of these segments within a region therefore indicates not simply network density but also how well that region is connected within the network.

Plots of total GDP against total choice of the filtered networks are given in Figure 19.5. In this case, the anticipated uneven distribution of the filtered portion of the network does not appear to justify the normalisation by unit of area (as in Figure 19.4), but to mediate the uneven distribution of data we have again plotted correlations using log–log scales. These indicate that countries with more segments within this upper 10% choice band do indeed have greater GDP. Correlations are only slightly lower than those between total GDP and overall node density, with $r^2 = 0.587$ and 0.581 for choice radii of 100 and 500 km, respectively. If the plots (Figure 19.5) are examined in succession, this correlation is strongest at 100 km and 500 km distances, which suggests that it is the major routes at distances roughly between international centres that may play the more significant role towards each countries' economic ranking.

Continuing under the assumption that economic performance is strongly tied to network choice, the distance above and below the line of regression also indicates countries that might be considered to be over- or underperforming in their economic activity. The UK, France and especially Germany appear to be over-performers in this sense, while Poland and Ukraine appear to underperform economically.

Both overall node density and locations of highest choice provide very good indications of a country's economic standing in terms of GDP, but the two have different geographical distributions and so provide two different stories. No specific causal direction can be definitively inferred in either case. Certainly, it is plausible that overall economic capacity (GDP) allows more roads or rail lines to be built within a country, thereby increasing node density. It is also plausible that such density is a generator of economic activity. In the case of the relationship between high choice and economic activity, the suggestion that it is the much larger scale radii (corresponding to international routes between countries) lends at least some weight to the possibility that the network plays a causal role, as these routes span several countries and therefore cannot be created as a result of economic productivity of any single country. The attraction of

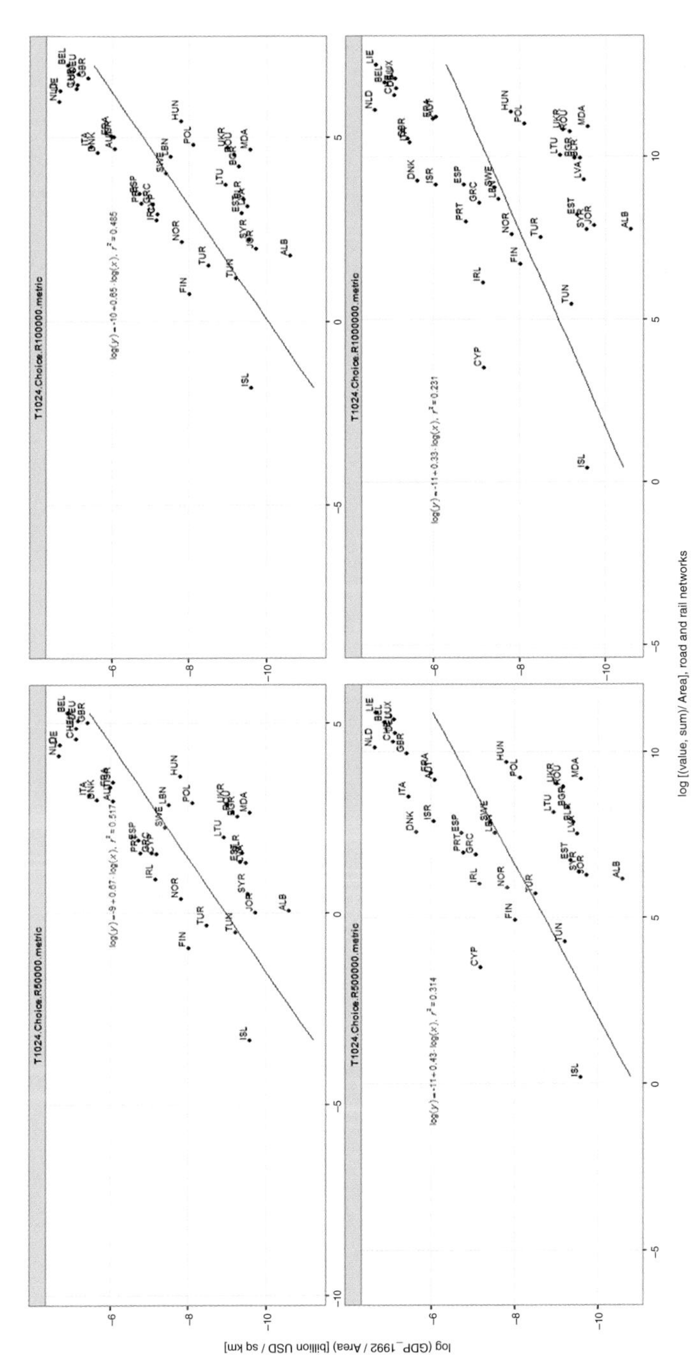

Figure 19.4 Comparison between the (log) total choice measures for road and rail networks against (log) total GDP, both normalised by unit of area, for different radii

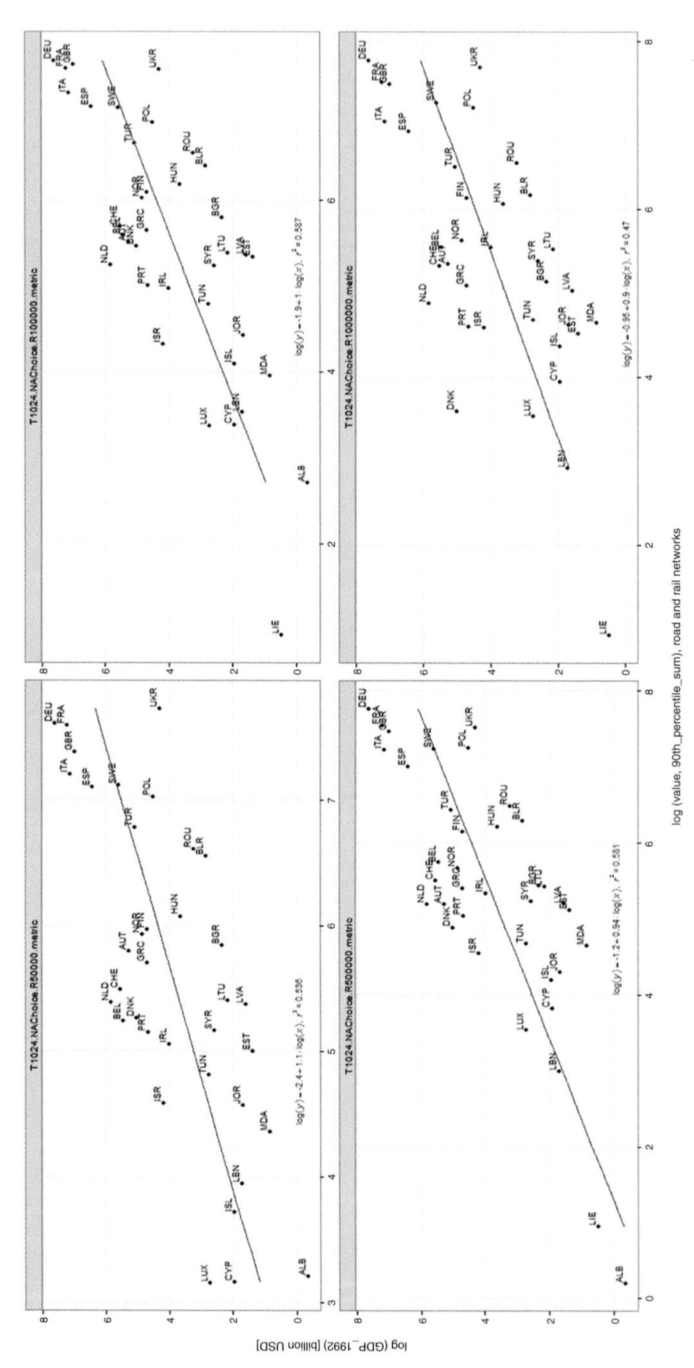

Figure 19.5 Comparison between the sum of the top 10% (log) choice measures for road and rail networks against (log) total GDP for different radii

neighbouring regions to a potential market, however, must certainly be acknowledged. In the absence of further evidence, it seems plausible that the transportation network and the international economy are both contributors to a cycle of feedback, with road and train links both creating economic opportunity and being shaped by the outcome of this activity.

19.5 An East-west Divide and Natural Economic Behaviour

Many factors contribute to a nation's total GDP, including agriculture and resources that might be highly dependent on land, so it is reasonable to expect larger GDP values for larger countries, and this is likely to be part of the explanation for the correlations between overall choice and total GDP as expressed (Figure 19.3). A nation's population is also a contributing factor to overall wealth. Of possibly greater relevance than the total economic outputs examined in Section 19.4 is the effect of the network on a country's potential for economic activity relative to this population – how much opportunity does the network afford to each person, with respect to their position within it?

To answer this, we look at the GDP per capita of a country against the mean value of choice within that country. This follows methodologically from space syntax studies of urban movement assessing the relative count of pedestrians in each spatial unit of the city (e.g. Hillier *et al.*, 1993), in that we look at the distribution of economic wealth of each unit of the international population. The underlying hypothesis is that the greater choice levels within a given region indicates greater economic potential, and will therefore correlate with greater average wealth of each person living in that region.

Using the same set of countries as a sample, we initially find correlations with values of r^2 ranging between 0.31 and 0.36 for choice radii from 50 km to 1000 km. This is not completely insignificant, but it is much lower overall than the simple tests of total GDP in this chapter. There are, of course, a number of factors other than transportation networks that impact economic output, not the least of which is the difference in economic history of each country, so this should not be entirely unexpected. The road and train networks are persistent over a relatively long period of time, change slowly, and resulting choice levels may remain more stable than the fluctuations in any local or national economy. A potentially greater such factor, in the 20th century at least, is the east-west division of the former communist countries of Eastern Europe and free-market economies of Western Europe, resulting in two very different and largely incompatible economic systems.

To control for this, we split our whole sample into two distinct sets for analysis (Figures 19.6 and 19.7). It is clear immediately that these form two distinct clusters, linearly separable on a plot of GDP per capita against mean choice. It is perhaps not surprising that the cluster representing Western Europe uniformly outperforms that of Eastern Europe. More relevant, however, is the difference in how the network measures of these two zones correlate with wealth. In Eastern Europe, no significant correlation can be found with economic output at any radius. Western Europe, by contrast, displays high correlations of $r^2 = 0.60$ and 0.61, peaking at radii 500 km and 1000 km. These are in the same range as those for overall GDP and node density, but indicate the potential economic productivity of each person in the nation.

It appears relevant that the radius distances at which the correlations peak are similar to the distances seen to be most relevant when looking at total choice values (Figure 19.3), and that these are roughly the distance between international centres. This implies that the range of

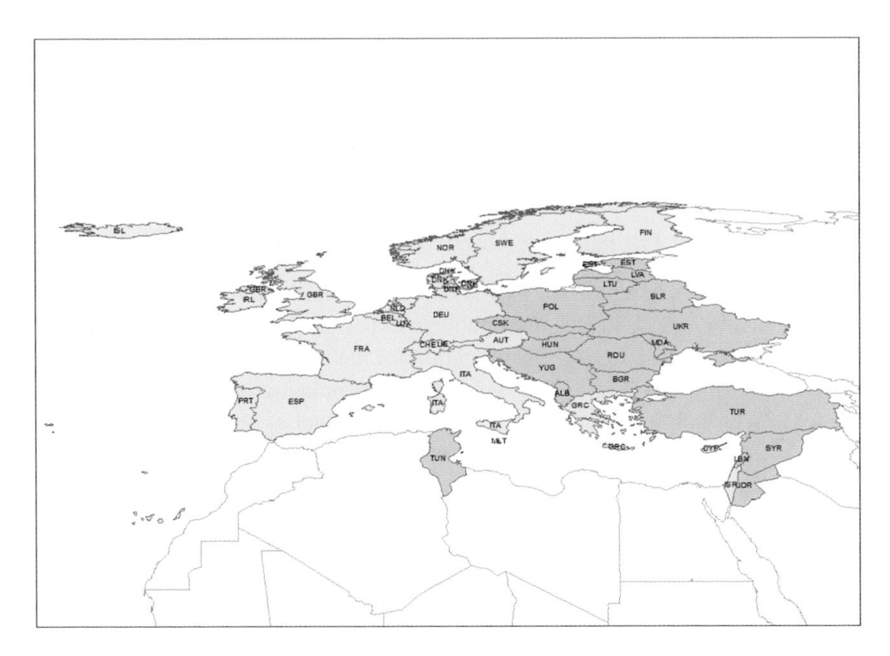

Figure 19.6 Map layout showing how we have defined countries belonging to Eastern (purple) and Western (green) Europe

impact of the configuration of the network is not limited to one particular country, but is an international effect involving organisation across regions.

To the extent this relationship between network and GDP is causal, it suggests that market economies in Western Europe have taken advantage of these transportation networks to engage in economic activity. Countries in Eastern Europe, by contrast, have not. It seems equally possible that in more controlled economies, there is not such opportunity for the network to guide economic activity by allowing markets to freely self-organise, resulting in an absence of correlation with GDP. Where natural movement of pedestrians is the movement determined by the configuration of the street grid in the absence of other limiting factors (Hillier *et al.*, 1993), we would propose here that there is a form of larger scale economic behaviour influenced by the much larger scale network we have examined. Although the details and mechanism cannot yet be discerned, the observation suggests that there is at least an economic activity that is 'natural' inasmuch as it is free to exploit the network capacity, and therefore determined by this only when external political control is absent.

The details of international-scale economic activity may be quite different from those of pedestrian movement. Movement is unlikely to be based on sight and the cost of angular turns, but on the topology or different weightings of travel within the network. Transactions will be based on longer term planning than immediate colocation. Also, phenomena have so far been observed only against economic, not other social factors. Each of these can be tested empirically. Regardless of these differences, it would appear that the topology of transportation networks has a clear capacity to order behaviour, to the extent that agents are otherwise free and self-organising, at multiple scales, and further research is warranted at the level of the global economy, as has already been valuable in these other systems.

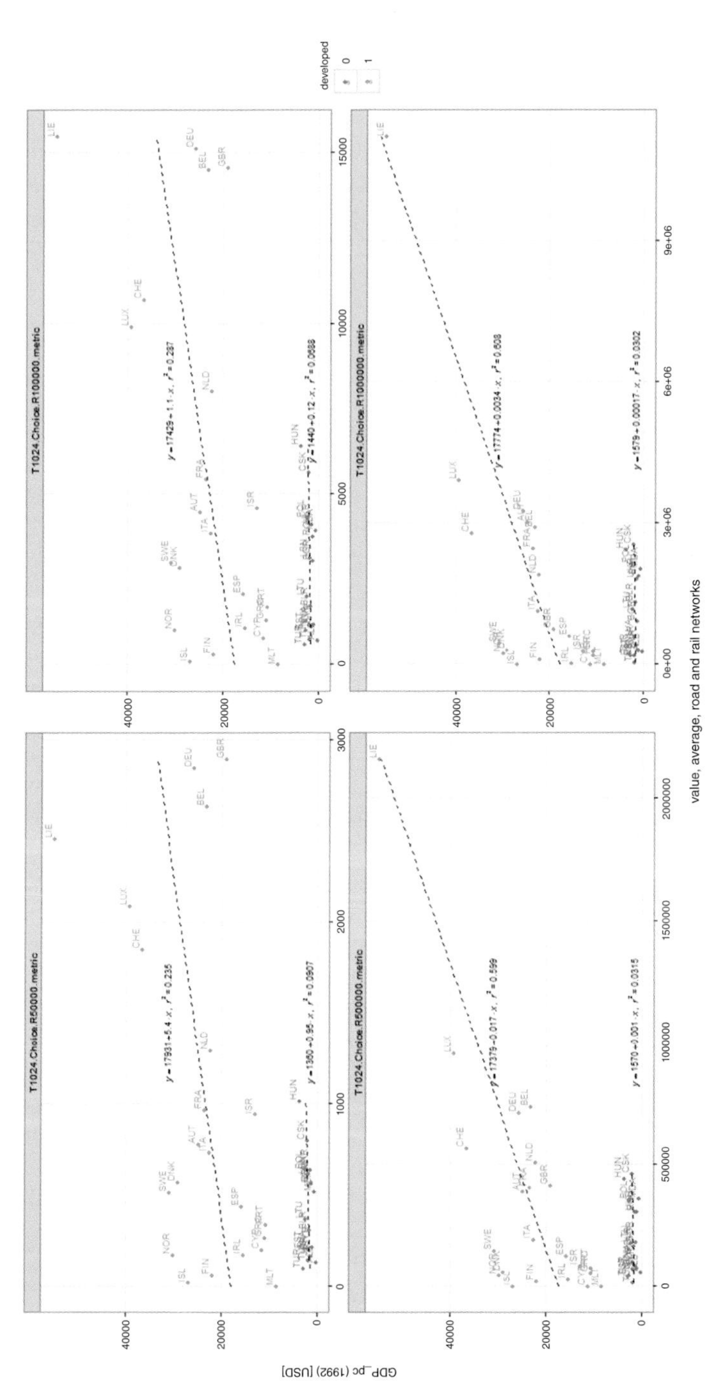

Figure 19.7 Comparison between the average choice measures for road and rail networks against GDP per capita for Eastern (red) and Western (blue) Europe countries and for different radii

19.6 Conclusion

The analyses here are presented as an initial study of the feasibility of such a global-scale analysis, but they indicate that some of the phenomena noted in smaller space syntax work do apply at this level. GDP has been used as the primary economic indicator, aggregated at the scale of individual countries but considered both as a total and per capita. GDP may be considered an indictor of the general state of a national industry. This was found to correlate well with the total number of segments within a country, and also to a threshold number of high choice segments at radii between 50 km and 500 km. While the former may appear to be caused by economic activity, the scales at which choice routes appear suggests at least some feedback at a regional level.

Per capita GDP may be considered the ability to maximise each individual person's output, so a relationship to the network would appear more likely to be an effect of the network on social economic activity. Strong correlations were observed with national mean choice, this time peaking around 500 km and 1000 km, but only for the countries in Western Europe, whereas eastern countries displayed no correlation. We suggest that this is a result of the influence of the network on free economic behaviour, a freedom which in the western countries has historically been greater, in markets constrained only by the physical network. We anticipate that a similar study of historical economic data will help to refine this, and we plan this in the near future.

The context of the work in this chapter is part of a much larger study of global dynamics, including trade, migration, security and development aid. The ENFOLDing project is an interdisciplinary collaboration that seeks to model these and their effects on one another, as the complexity of such interactions is little understood but crucial to international relations. In providing a basis for understanding at least the economic effect of physical infrastructure, it is our hope to allow further understanding of global social phenomena such as trade and development. The measures used here have counterparts in the graph methods used in economic analysis (Blochl *et al.*, 2011) such as the input–output models of different sectors of an economy, also represented as a graph. In these, measures of centrality have been used to understand the structure and behaviour of economic movement completely decoupled from space, but it appears to be methodologically possible to begin to analyse a combined network of spatial and economic transactions at several different levels, and we are currently working to do so. The next phase of analyses will include networks we are assembling for air and sea transportation, which should allow the separate industries represented by these economic models to be unpacked when compared against each relevant mode of transportation, and thus better understood in a spatial context.

We anticipate that the coupling of the spatial system with political and economic networks will show promise for the extension of these methods. An understanding of this network, and its most likely routes of concentrated activity as revealed by measurement of high choice or similar, would indicate points most susceptible to shock or intervention. Scenarios for disruption of the network can be tested against known disasters or development. As such, we would hope to provide understanding of phenomena crucial to international relations and policy decisions.

References

Blochl F, Theis FJ, Vega-Redondo F and Fisher EON. (2011). Vertex centralities in input-output networks reveal the structure of modern economies. *Physical Review E* 83.

Figueiredo L and Amorim L. (2007). Decoding the urban grid: or why cities are neither trees nor perfect grids. In *Proceedings of the 6th International Space Syntax Symposium, Istanbul 2007*.

Hanna S. (2009). Spectral comparison of large urban graphs. In Koch D, Marcus L and Steen J (eds.) *SSS7: Proceedings of the 7th International Space Syntax Symposium*. Royal Institute of Technology (KTH), Stockholm, Sweden.

Hillier B, Burdett R, Peponis J and Penn A. (1987). Creating life: or, does architecture determine anything? *Architecture et Comportement/Architecture and Behaviour* 3 (3): 233–250.

Hillier B and Hanson J. (1984). *The Social Logic of Space*. Cambridge: Cambridge University Press.

Hillier B and Iida S. (2005). Network and psychological effects in urban movement. In Cohn AG and Mark DM (eds), *COSIT 2005, LNCS 3693*. New York: Springer, pp. 475–490.

Hillier B, Penn A, Hanson J, Grajewski T and Xu J. (1993). Natural movement: or configuration and attraction in urban pedestrian movement. *Environment and Planning B: Planning and Design* 20: 29–66.

Hillier B, Yang T and Turner A. (2012). Normalising least angle choice in Depthmap – and how it opens up new perspectives on the global and local analysis of city space. *Journal of Space Syntax* 2012: 155–193.

Penn A, Hillier B, Banister D and Xu J. (1998). Configurational modelling of urban movement networks. *Environment and Planning B: Planning and Design* 24: 59–84.

Peponis J, Allen D, Haynie D, Scoppa M and Zhang Z. (2007). Measuring the configuration of street networks: the spatial profiles of 118 urban areas in the 12 most populated metropolitan regions in the US. In *Proceedings of the 6th International Space Syntax Symposium, Istanbul 2007*.

Turner A. (2004). Depthmap 4. In: *A Researcher's Handbook*. http://www.vr.ucl.ac.uk/depthmap/handbook/depthmap4.pdf

Turner A. (2007). From axial to road-centre lines: a new representation for space syntax and a new model of route choice for transport network analysis. *Environment and Planning B: Planning and Design* 34(3): 539–555.

Turner A. (2009). Stitching together the fabric of space and society: an investigation into the linkage of the local to regional continuum. In Koch D, Marcus L and Steen J (eds), *Proceedings of the 7th International Space Syntax Symposium*. Stockholm: Royal Institute of Technology (KTH), p. 116.

Varoudis T. (2012). DepthmapX – multi-platform spatial network analyses software. https://github.com/varoudis/depthmapX

20

Trade Networks and Optimal Consumption

Robert J. Downes and Robert G. Levy

20.1 Introduction

This chapter is concerned with the notion of rebalancing the economy of a nation, that is, altering the consumption pattern of an economy for the purposes of economic growth.

From an economic perspective, sectoral rebalancing at a national level is a challenging task: the structure of any economy is extremely complicated, and dynamical changes are hard to predict with certainty (Ball, 2012). Different economic sectors are inextricably linked, and any change will have a broad impact (Miller and Blair, 2009). Beyond the myriad political difficulties (Frieden, 2009, 2010), the specifics of such a sectoral rebalancing are unclear (Yeo and Zedillo, 2010). Furthermore, the economy of any one country is integrated with those of other countries: import and export partners, regional political groupings and ethno-social relationships all contribute to this mutual dependence. Taking account of global integration is essential in any discussion of designed change or rebalance via policy intervention (Levy *et al.*, 2016). Discussion of policy intervention to achieve rebalancing has become common in British politics, despite the latitude of meaning inherent in the term (Froud *et al.*, 2011). In an editorial in early 2013, the *Financial Times* 2013 addressed the British Government's proposed economic and industrial strategy in light of the recent 'Great Recession':

> Ministers think the state's involvement can assist in rebalancing the economy away from finance and towards manufacturing … boost[ing] Britain's recent growth performance, which at best has been disappointing.

Tapping into the heated debate taking place in the wake of the recession, the editorial expresses the view that rebalancing is both necessary and desirable for future economic growth. As emphasised in Gardiner *et al.* (2013), this debate is nothing new:

Approaches to Geo-mathematical Modelling: New Tools for Complexity Science, First Edition. Edited by Alan G. Wilson.

Companion Website: www.wiley.com/go/wilson/ApproachestoGeo-mathematicalModelling

> ... there has long been a debate over the role of sectoral structure in shaping economic development. In development economics ... [there] are those who argue that national economic development is best promoted by 'balanced growth'.

The British Government itself favours a policy of 'balanced growth' as outlined by Gardiner *et al.* (2013), intending to create an economy

> that is not so dependent on a narrow range of sectors ... that [is] more evenly balanced across the country and between industries.

In this chapter, we follow the definition of rebalancing provided by Gardiner *et al.* (2013): 'the simultaneous and coordinated expansion of several sectors, i.e. sectoral diversification'. We contribute to the ongoing debate by bringing to bear a formal approach using a global economic model based on newly available data.

Our data set consists of 40 nations representing more than 80% of global economic output; see Section 20.2 for a discussion of data sources used in this chapter. The consumption patterns of each country, derived from this data set, determine the sectoral breakdown of each nation's economy. We view the set of all such sectoral breakdowns as a *feasible set*. From this set, we select a specific *base* country. We then perturb the base country's consumption pattern towards the consumption pattern of each other *target* country in the feasible set. Using a global economic model introduced by Levy *et al.* (2016), we evaluate the impact of each perturbation upon the base country's GDP. We repeat for each combination of base and target country in our data, ranking nations by the extent to which each target country's consumption pattern increases the GDP of each base country.

Throughout, we consider:

1. Whether rebalancing is a worthwhile objective as indicated by the *Financial Times* (2013), Her Majesty's Government (2013) and the Department for Business and Innovation & Skills (2010);
2. What a successful rebalancing would entail in broad macro-economic terms.

Furthermore, we introduce a new mathematical means of perturbing a base country's consumption pattern and use several concepts from graph theory to elucidate our findings. While the model framework utilised by our global economic model may be familiar to many readers, this particular implementation, added to the perturbation technique and our use of graph theory to analyse the global economic system, is the source of novelty in this piece.

The remainder of this chapter is structured as follows. In Section 20.2, we discuss the underlying model framework and data sources. Section 20.3 outlines the perturbation process. We present our findings in Section 20.4, along with a discussion of the metrics underpinning our conclusions and recommendations, which complete the chapter in Section 20.5.

20.2 The Global Economic Model

20.2.1 Introduction

The global economic model underlying our approach is introduced in Levy *et al.* (2016). In this model, the economies of individual countries are represented as input–output models, each of which is linked through trade flows representing imports and exports. A comprehensive overview of input–output modelling can be found in Miller and Blair (2009); while most input–output studies have had a regional focus (Levy *et al.*, 2016) examples of multiregional or global models include Lenzen *et al.* (2013), Tukker *et al.* (2013) and Ianchovichina and Walmsley (2012).

20.2.2 Data Sources

Relying on recently published data from the World Input-Output Database (WIOD) (Timmer *et al.*, 2012) and the UN COMTRADE database (United Nations, 2010), Levy *et al.* (2016) model linkages between 40 countries at a 35-sector resolution over 17 years, from 1995 to 2011. The internal economic structure of each country is based on WIOD data, while trade linkages are determined by UN COMTRADE data for each year. According to Levy *et al.* (2016), this approach is one of the first to leverage WIOD's unparallelled resolution of the global economy based on real economic information. This resolution combined with the relatively parsimonious parameter requirement motivate our choice of this model compared with alternatives. Additionally, the central role of distinct and quantifiable economic sectors in Levy *et al.* (2016) links well with our topic of investigation. In this chapter, we use only data from 2011: this restriction allows us to focus on developing appropriate methods of analysis for the perturbation process.

20.2.3 Model Overview

The structure of each national economy is characterised by 35×35 *technical coefficients*, representing the input demanded per-unit output of each national sector; these are determined from WIOD data.

These are supplemented by 35 *import ratios*, determining the proportion of input demanded per-unit output in each of the 35 sectors imported rather than produced internally; these are determined from WIOD data. Finally, the proportion of goods imported by each sector from each of the 40 countries is determined by 35×40 *import propensities*; these are determined from UN COMTRADE data. Together, the technical coefficients, import ratios and import propensities determine the global economic structure. The model is then driven by the specification of the *final demand* for each country, a 35-dimensional vector representing the goods demanded by final markets: in effect, national consumption (i.e. total production minus goods exported or used as inputs to different sectors at a national level). Final demand is determined from WIOD data.

20.3 Perturbing Final Demand Vectors

20.3.1 Introduction

Holding import ratios, import propensities and technical coefficients constant is equivalent to holding the structure of each sector constant and fixing the proportion of goods imported from abroad. Changes to the final demand vector under these assumptions alter the sectoral make-up of the national economy under consideration *without* modifying the internal structure of each sector (e.g. in terms of technological efficiency); this is the approach we take here, and, hence, the final demand vector is our basic object of consideration. The set of all 40 countries' final demand vectors forms the *feasible set* of consumption patterns.

Initially, we specify a *base* country whose final demand vector we will perturb. Then, every remaining country in the feasible set becomes a *target* country: we perturb the base country's consumption pattern (i.e. final demand vector) towards that of each target country, ensuring that total consumption remains fixed for each base country. Finally, we evaluate the impact of the perturbation on the base country's GDP by re-running the global model using the perturbed consumption pattern. For every base-target country pair, we then record the magnitude of the change in base-country GDP produced by perturbing the base country's consumption pattern towards that of the target country; this may be positive or negative.

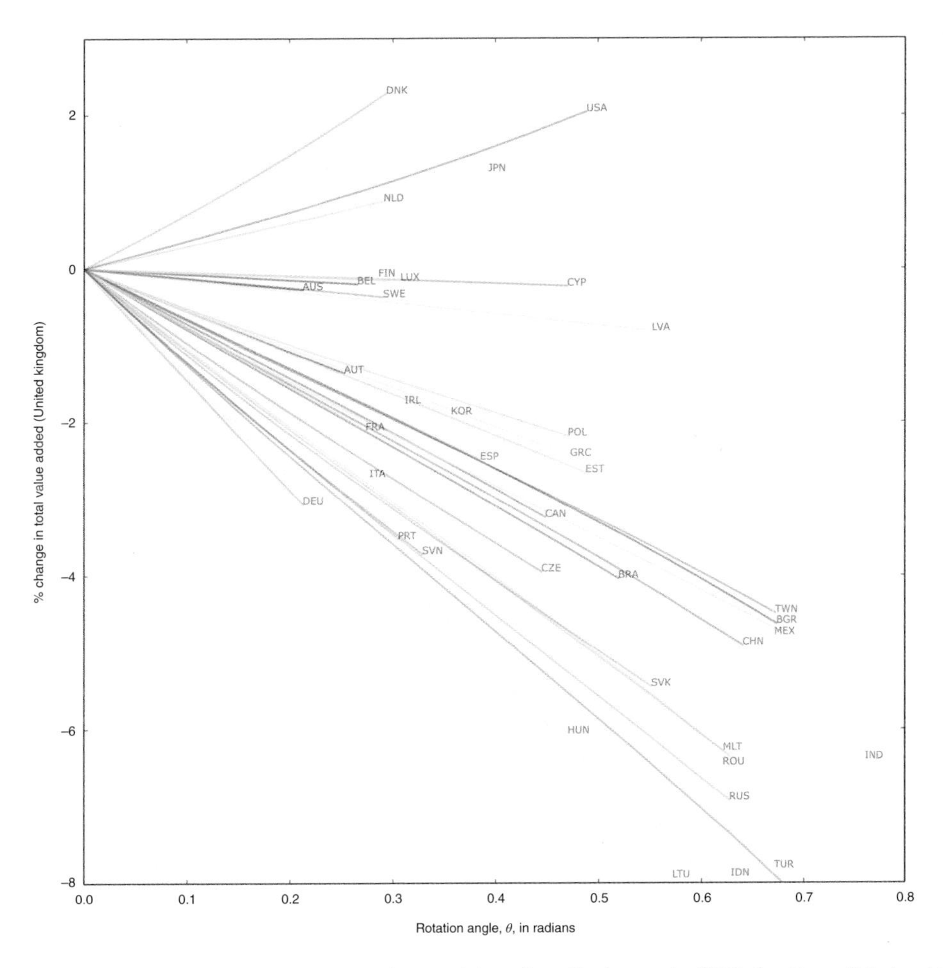

Figure 20.1 Perturbation distance (rotation angle) against % change in GDP for Great Britain. All trajectories originate at Great Britain's consumption pattern ($\theta = 0$). Destinations are the consumption patterns of each target country in the feasible set

For clarity, a sample model run is given in Figure 20.1. The base country is Great Britain, and the % change in GDP resulting from each perturbation is recorded for each target country in the feasible set. The horizontal axis may be thought of as the *distance* between the consumption patterns of base and target countries; units are explained in Section 20.3.2.

Perturbations towards Great Britain's nearest target country, Germany (DEU), lead to a decrease in GDP; if Great Britain consumes exactly like Germany (the destination of the trajectory; see Figure 20.1), the magnitude of this decrease is approximately 3%. Conversely, consuming exactly like Denmark (DNK) increases Great Britain's GDP by 2%.

When running the model in the sequel, we standardise the perturbation distance to 1 deg throughout our analysis[1]; this provides a measure of comparison across all countries.

For this fixed distance, we rank target countries by the number of base countries they improve (i.e. the number of perturbations towards each target country resulting in a positive change to the base country's GDP). We also rank base countries by the number of target countries by which they are improved.

20.3.2 *Perturbation Process*

Here, we present the detail behind the perturbation process outlined in Section 20.3.1. Notation relating to final demand vectors is taken from Levy *et al.* (2016) for ease of comparison. While this section is essential for a full understanding of the novel method we employ in this chapter, readers without a mathematical background will be forgiven for skipping forwards.

The purpose of the perturbation process is to rotate the final demand vector (consumption pattern) of the base country, $\mathbf{f}^{(i)}$, in the direction of that of a given target country, $\mathbf{f}^{(j)}$, by a specified (small) angle θ.

Mathematically, the setting for this process is 35-dimensional Euclidean space $\mathbb{R}^{35}$: this reflects the structure of the final demand vectors and, ultimately, the fact that the underlying data set represents economies with 35 sectors. Final demand vectors are expressed in terms of the standard basis $E = \{\mathbf{e}_i\}_{i=1}^{35}$ (Roth, 2014).

The process of vector rotation is a standard mathematical technique consisting of five steps:

1. Use the two final demand vectors to generate a new basis U for $\mathbb{R}^{35}$ via the Gram–Schmidt process[2]. The input is the linearly independent set

$$\{\mathbf{f}^{(i)}, \mathbf{f}^{(j)}, \mathbf{e}_3, \mathbf{e}_4, \dots, \mathbf{e}_{35}\}.$$

   The first two vectors of the basis U generated by this process, $\mathbf{u}_1$ and $\mathbf{u}_2$, will span the plane spanned by $\mathbf{f}^{(i)}$ and $\mathbf{f}^{(j)}$.
2. Form the change of basis[3] formula B to map the new basis U into the standard basis E:

$$B = \begin{pmatrix} \uparrow & \uparrow & \cdots & \uparrow \\ \mathbf{u}_1 & \mathbf{u}_2 & \cdots & \mathbf{u}_n \\ \downarrow & \downarrow & \cdots & \downarrow \end{pmatrix} \tag{20.1}$$

[1] In radians, the units of Figure 20.1, 1 deg $\equiv \pi/180$ rad.

[2] The Gram–Schmidt process may be thought of as a tool for generating orthonormal bases. Taking an independent set of vectors as an input, the process outputs an orthonormal basis for the space in question (Weisstein, 2014).

[3] Any two bases of $\mathbb{R}^n$ can be related by a *change of basis*; this allows us to express one basis in terms of another (Rowland, 2014).

The action of the matrix B maps a vector expressed in the new basis U into a vector expressed in the standard basis E. Conversely, the action of the inverse of this matrix, B^{-1}, maps a vector expressed in the standard basis E into a vector expressed in the new basis U.

3. Form the Givens rotation[4] matrix $G(\theta)$:

$$G(\theta) = \begin{pmatrix} \cos(\theta) & -\sin(\theta) & 0 & 0 & \cdots & 0 & 0 \\ \sin(\theta) & \cos(\theta) & 0 & 0 & \cdots & 0 & 0 \\ 0 & 0 & 1 & 0 & & \vdots & \vdots \\ 0 & 0 & 0 & 1 & & \vdots & \vdots \\ \vdots & \vdots & & & \ddots & & \\ 0 & 0 & \cdots & \cdots & & 1 & 0 \\ 0 & 0 & \cdots & \cdots & & 0 & 1 \end{pmatrix}$$

This will rotate vectors expressed in the new basis U by an angle θ in the plane spanned by the first two vectors of the basis U, $\mathbf{u}_1$ and $\mathbf{u}_2$.

4. Form the *final rotation matrix* $R(\theta)$, defined as

$$R(\theta) := B\ G(\theta)\ B^{-1}.$$

This matrix will map a vector expressed in the standard basis E into a vector expressed in the new basis U, rotate the vector in the plane spanned by $\mathbf{u}_1$ and $\mathbf{u}_2$ by an angle θ, then map the resulting vector back into the standard basis E.

5. Apply the final rotation matrix to the final demand vector of the base country, $\mathbf{f}^{(i)}$.

The perturbation process will then produce a new vector

$$\tilde{\mathbf{f}}^{(i)} = R(\theta)\mathbf{f}^{(i)},$$

the base country's final demand vector $\mathbf{f}^{(i)}$ rotated by an angle θ towards the target country's final demand vector $\mathbf{f}^{(j)}$.

Finally, this perturbed vector must be rescaled so that the sum of the components is the same as the original, unperturbed vector:

$$\hat{\mathbf{f}}^{(i)} = \frac{\tilde{\mathbf{f}}^{(i)}}{|\tilde{\mathbf{f}}^{(i)}|}|\mathbf{f}^{(i)}|$$

where

$$|\mathbf{v}| := \sum_{i=1}^{35} |v_i|,$$

is the sum of the absolute values of the components of the vector $\mathbf{v}$.

Economically, this final rescaling holds total consumption constant throughout the perturbation process: the rescaled perturbed vector $\hat{\mathbf{f}}^{(i)}$ and the unperturbed final demand vector $\mathbf{f}^{(i)}$ represent the same total consumption.

[4] A *Givens rotation* is a rotation in a plane spanned by two coordinate axes. As a rotation is a linear transformation, this can be represented in the form of a matrix acting on the vector we wish to rotate (Venkateshan and Swaminathan, 2013).

This means that the total consumption of the base country's economy is held fixed, while the sectoral breakdown varies throughout the perturbation process.

We note that the perturbation process produces monotonic changes to the share of final demand enjoyed by each of the 35 economic sectors, for small angle rotations. This means that our selection of 1 deg for the fixed perturbation distance is not privileged: alternative choices would lead to changes of different magnitudes, but ultimately of the same sign (positive or negative).

20.4 Analysis

20.4.1 Introduction

For each base-target country pair, the perturbation process detailed in Section 20.3 produces a perturbed consumption pattern. The result of evaluating this perturbed consumption pattern using the global model given in Section 20.2 is a modified GDP figure, the outcome of the base country perturbing its consumption pattern towards the target country's consumption pattern. For brevity, we refer to this process as the base country consuming *like* the target country.

20.4.2 A Directed Network Representation

If the GDP of country i increases when it consumes like country j, we say that 'i is improved by j'. Then, to each country i is associated with a set of countries by which it is improved and a set of countries by which it is *not* improved. We can arrange these relationships in a matrix G, with i and j indexing rows and columns, respectively:

$$G = \begin{cases} 1 & \text{if } j \text{ improves } i \\ 0 & \text{Otherwise} \end{cases} \tag{20.2}$$

This matrix G can then be thought of as the asymmetric adjacency matrix[5] of an induced *network of improvements*: countries are the nodes of this network, and an edge exists between i and j if and only if j improves i.

As we work with a directed network, there are two basic quantities associated with each node, the *in-degree* and the *out-degree*, which have the following interpretation: the in-degree, k_i^{in}, is the total number of countries that are improved by consuming like country i; and the out-degree k_j^{out} is the total number of countries that improve country j.

Table 20.1 shows countries ranked by in- and out-degree, respectively.

We see from these results that countries which are good improvers (high in-degree) tend to be improved by few (low out-degree), and *vice versa*, but that the correlation is not perfect. Figure 20.2 shows this relationship explicitly: there is a clear negative correlation ($p \approx 10^{-11}$) between in- and out-degree metrics.

In addition, we observe that countries with high in-degree also tend to be richer than those with low in-degree. To determine whether in-degree merely reflects a country's wealth,

[5] The key concepts from network theory utilised in the sequel are given in Newman (2010); we direct the interested reader to this text for an accessible introduction to the subject.

Table 20.1 Results of consumption pattern perturbation for all base-target country pairs. In Table 20.1a, target countries are ranked by the number of base countries they improve (in-degree). Conversely, Table 20.1b shows base countries ranked by the number of improving target countries they have (out-degree).

(a) Target countries ranked by in-degree

Country	In-degree
USA	36
Denmark	36
Japan	35
Netherlands	30
Great Britain	27
⋮	⋮
Hungary	3
Russia	3
Lithuania	2
Indonesia	1
Turkey	1

(b) Base countries ranked by out-degree

Country	Out-degree
Turkey	39
Lithuania	38
Russia	37
Hungary	32
Malta	31
⋮	⋮
Luxembourg	2
China	1
Indonesia	1
Japan	1
USA	1

we perform a statistical analysis after defining a means of measuring the diversity of an economy.

20.4.2.1 Sectoral Diversity

The central thesis of economic rebalancing is that a more diverse economy leads to stronger economic growth (Murphy *et al.*, 1989). Under this assumption, we would expect a measure of sectoral diversity to be a good predictor of whether a given country is an improver.

For clarity, a measure of sectoral diversity will determine how *balanced* an economy is with respect to its consumption pattern (i.e. its final demand vector).

We use the standardised M1 index of qualitative variation (Gibbs and Poston, 1975) which, in this implementation, expresses the tendency of an economy to be dominated by a small number of large sectors:

$$\mathrm{M1}(i) := 1 - \frac{\sum_{k=1}^{35} p_k^2 - 1/35}{1 - 1/35} \tag{20.3}$$

where p_k is the share of final demand held by sector k of country i when the economy is partitioned into 35 sectors; for each country, p_k is determined directly from the corresponding final demand vector.

Note that $\mathrm{M1} = 1$ when the share of the national economy held by each sector is uniform; qualitatively, this is a perfectly balanced economy. Conversely, $\mathrm{M1} = 0$ when the economy is dominated entirely by a single sector; qualitatively, this is a completely unbalanced economy.

The M1 index is related to the Simpson diversity index (Simpson, 1949) and the Herfindalh Hirshman index (Hirschman, 1964), commonly used to identify company market share within a given sector.

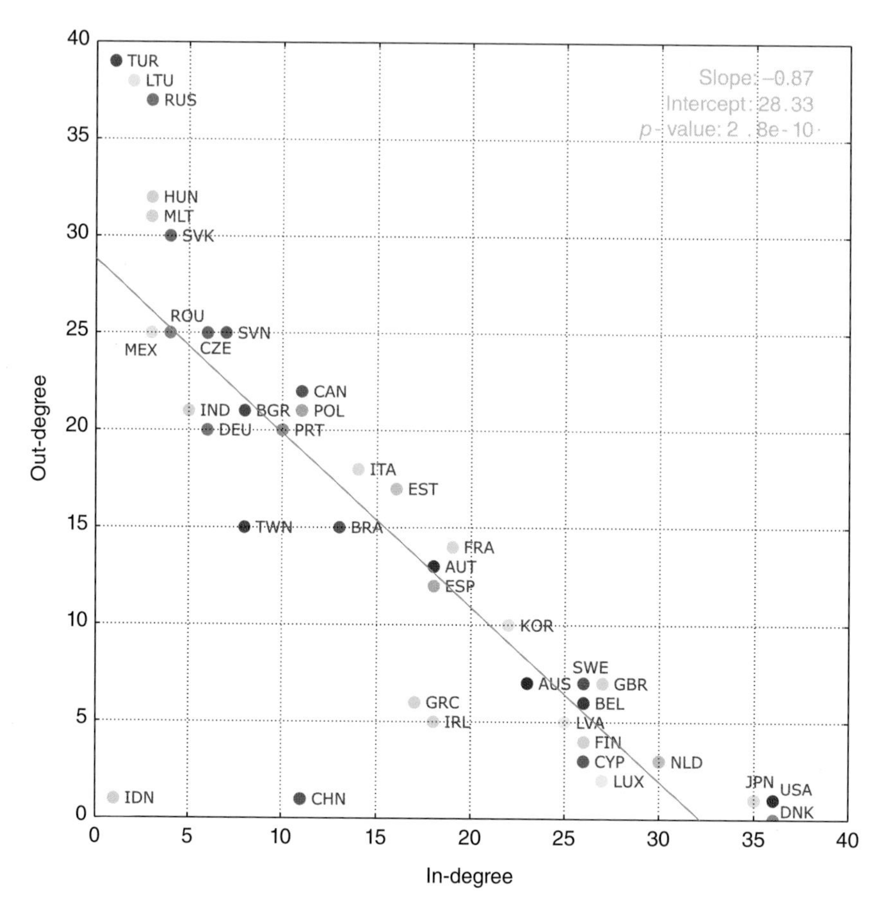

Figure 20.2 In-degree, a count of the number of countries who would benefit from consuming like the country in question, plotted against out-degree, a count of the number of countries the country in question would benefit from consuming like. An ordinary least squares (OLS) regression suggests $y = -0.75x + 26$ with a p-value of the order 10^{-11}

20.4.2.2 Determinants of a Balanced Economy

The relationship between sectoral diversity and in- and out-degree, respectively, is investigated in Figure 20.3. As anticipated, we can identify a significant ($p \approx 10^{-4}$) positive relationship between M1 and in-degree.

By inspection, countries with larger values of M1 (USA, Denmark (DNK) and Japan (JPN)) tend also to be rich. The relationship between M1 and in-degree may therefore reflect an underlying dependence on other macroeconomic variables such as GDP per capita.

To exclude this possibility, we control for country-specific relationships by running a simple linear regression of the form:

$$k_i^{in} = \alpha + \beta \mathrm{M1}_i + \boldsymbol{\gamma}_i \boldsymbol{x}_i + \epsilon_i \tag{20.4}$$

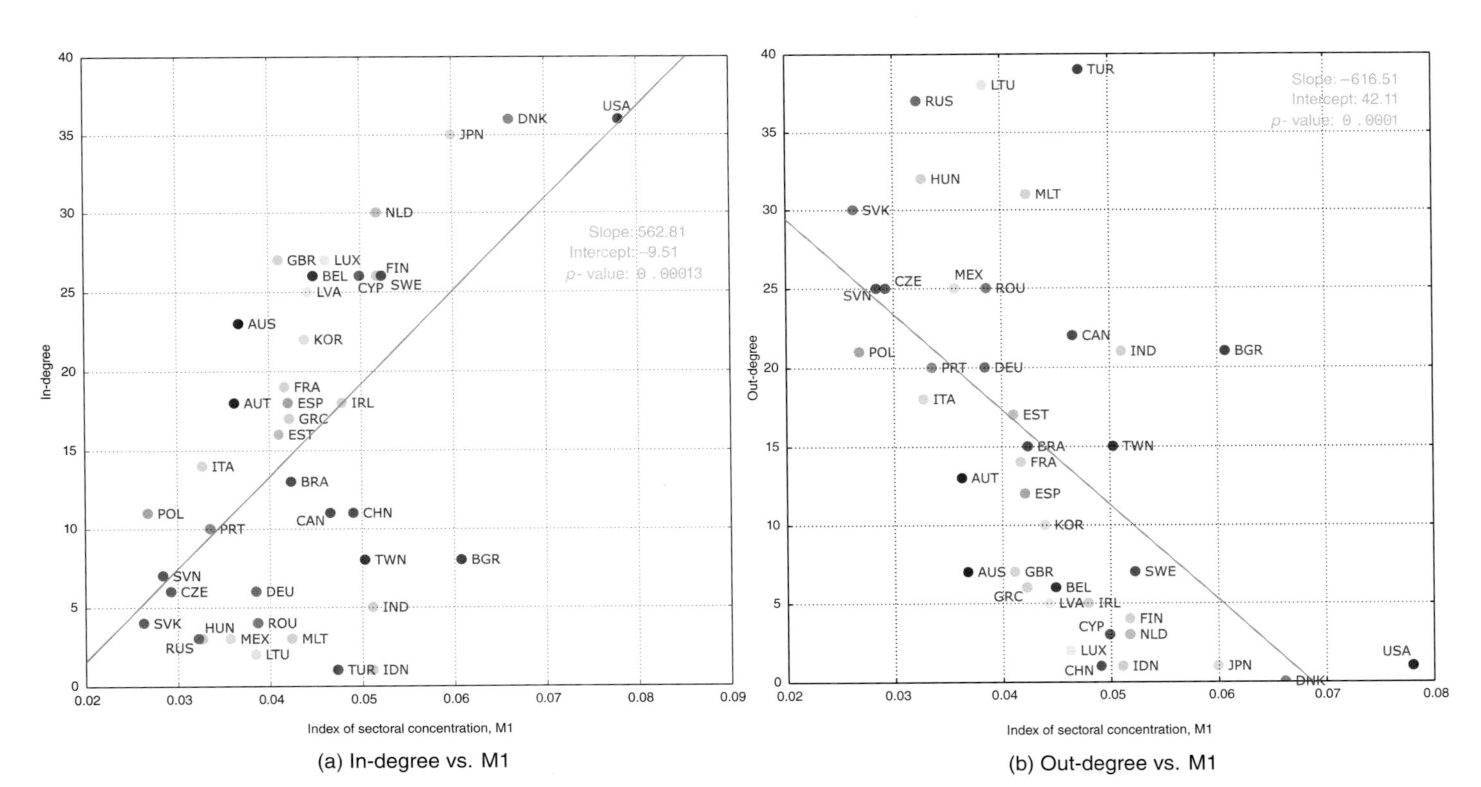

Figure 20.3 Figure 20.3a shows in-degree plotted against the M1 metric of sectoral diversity; Figure 20.3b shows out-degree plotted against the M1 metric of sectoral diversity. An OLS regression line has been added to each plot. The *p*-value in both cases is of the order 10^{-4}

where M1 is the measure of sectoral diversity (equation 20.3), and $\boldsymbol{x}_i$ is a vector of country-specific variables.

The thesis that a balanced economy vector makes a country more likely to be an improver implies $\beta > 0$, which should be robust to the addition of arbitrary elements of $\boldsymbol{x}_i$.

We give three different regression model specifications:

1. includes M1 only as the independent variable;
2. includes, in addition to M1, GDP per capita and population[6];
3. includes, in addition to the three regressors of specification 2, two indices describing the bias of country i's final demand vector in favour of services and manufacturing sectors. These indices are calculated as the ratio of, respectively, services and manufacturing consumption to total consumption in country i. See Section 20.5 for our categorisation of the 35 economic sectors as services or manufacturing (or neither).

The summary of this analysis, presented in Table 20.2, identifies the M1 sectoral diversity measure as highly significant in all three specifications ($p \approx 10^{-4}$). This supports the thesis that balanced consumption is an important criterion for an economy to be an improver.

GDP per capita is highly significant in specification 2, while population is not. Therefore, size alone is not sufficient to make a country a strong improver: that country must be rich in terms of GDP per capita, as indicated in Section 20.4.2.

In specification 3, both economic indicators are highly significant, but with opposing correlation coefficients: being services-biased increases improver status, while being manufacturing-biased decreases it. The addition of these indices greatly reduces the point estimate against GDP per capita, suggesting that what may appear to be a wealth effect is in

Table 20.2 OLS regressions. Dependent variable is IN-DEGREE$_i$ in all specifications. Standard errors are shown in brackets below each point estimate. $^{*}p < 0.1$; $^{**}p < 0.05$; $^{***}p < 0.01$.

	(1)	(2)	(3)
Sectoral diversity, M1 (%)	5.6^{***}	4.7^{***}	2.6^{***}
	(1.3)	(1.1)	(0.59)
GDP per capita (10,000s of US$)		2.4^{***}	0.63^{*}
		(0.56)	(0.36)
Population (billions)		−3.5	0.84
		(4.3)	(2.2)
Services Index (%)			0.76^{***}
			(0.13)
Manufacturing Index (%)			-0.98^{***}
			(0.28)
R^2	0.32	0.62	0.91
Observations	39	39	39

[6] The order of magnitude multiplier against each independent variable is chosen simply to result in point estimates of convenient magnitude.

reality a bias-towards-services effect. Despite this, GDP per capita remains significant at the 5% level.

20.4.3 A Weighted Directed Network Representation

We can extend the directed network representation of Section 20.4.2 by considering not only whether one country improves another, but also the extent to which it does so.

Formally, denote by g_{ij} the percentage change in base country i's GDP resulting from a perturbation towards target country j. As before, we arrange the g_{ij} as a matrix G', setting negative entries to zero:

$$G' = \begin{cases} g_{ij} & \text{if } j \text{ improves } i \\ 0 & \text{Otherwise} \end{cases} \tag{20.5}$$

G' may then be interpreted as the adjacency matrix of a corresponding weighted directed graph, the network of percentage GDP increases (if any) resulting from the perturbation process. The weighted directed network of improvements is given in Figure 20.4; edge weight is proportional to g_{ij}.

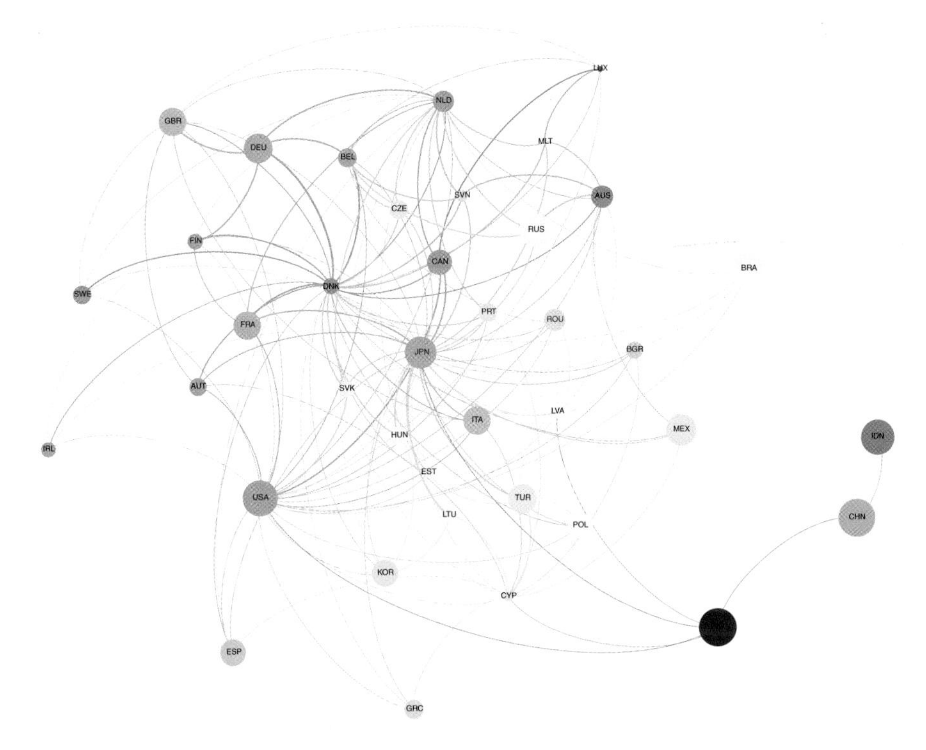

Figure 20.4 A network of improvements with countries as nodes and G' defining edges; see equation (20.5). Outgoing links lie counter clockwise of the direct line between two nodes and are limited in this diagram to the five largest. Node size is proportional to population; colour is defined by GDP per capita set along a scale where red and blue indicate lowest and highest GDP per capita, respectively

We note that connectivity in this graph appears to be dependent upon GDP per capita; countries with high GDP per capita appear tightly clustered on the left of Figure 20.4, while lower GDP per capita countries appear on the right.

20.4.3.1 Centrality and In-degree

Beyond visual inspection, we adopt network theoretic tools to investigate the structure of this network in a quantitative manner. To begin, we consider the *eigenvector centrality* of the nodes in our network.

Generally speaking, centrality measures the relative influence of a node in a network in terms of connections to other nodes (Newman, 2010). Eigenvector centrality is a refined version of this concept: the measure determines whether a node is strongly connected to the parts of the underlying network which are themselves strongly interconnected.

In our network, a country with high eigenvector centrality improves countries which themselves improve many others. Low eigenvector centrality implies not only that a country improves few other countries, but also that these other countries themselves improve few countries. Explicit centralities are presented in Table 20.3.

Comparing Table 20.3 with Table 20.1a, we see that the former offers a higher fidelity perspective on improvers. For example, we are able to distinguish between the relative influence of the USA and Denmark which, in the unweighted case (Table 20.1a), were scored identically. In contrast, in Table 20.3, Denmark achieves a significantly higher centrality score than the USA.

20.4.4 Communities in the Network of Improvements

While centrality measures provide information related to the relative influence of nodes within a network, they do not provide detailed structural information about the network as a whole.

Table 20.3 Eigenvector centrality by country for the network defined in Section 20.4.3. Only the 10 largest are shown

Country	Eigenvector centrality
Denmark	0.960
USA	0.273
Japan	0.049
Netherlands	0.040
Luxembourg	0.011
Belgium	0.007
Cyprus	0.005
UK	0.005
Finland	0.004
Latvia	0.004

In order to investigate network structure further, we utilise cluster analysis, which seeks to identify groups of nodes which are strongly mutually connected while simultaneously being weakly connected to the rest of the network.

Cluster analysis will allow us to locate each country within a specific group, the nodes of which share stronger relationships than with the rest of the network. We then investigate the structure of each cluster in detail to determine whether the constituent countries share common economic attributes. In particular, the measure of sectoral diversity and indices given in Section 20.4.2.1 form the basis for this analysis.

20.4.4.1 Method Overview

We follow the method outlined in Reichardt and Bornholdt (2006) whereby the minimisation of a certain *quality function* produces a nodal clustering which minimises the overall energy of the system. This is shown to be equivalent to the more usual modularity method presented in Newman and Girvan (2004); we select Reichardt and Bornholdt (2006) for computational purposes.

According to Reichardt and Bornholdt (2006), the quality function should 'follow the simple principle: group together what is linked, keep apart what is not'. Assessing the quality of a given clustering of nodes is carried out relative to an underlying null model, determined from the in- and out-degree of each node in the network (Reichardt and Bornholdt, 2006).

The relationship between the actual network and the null model is mediated by a single parameter, γ. We select an appropriate γ value using a parameter sweep that is designed to produce two clusters of countries for ease of analysis.

Utilising Hahsler *et al.*, (2014), we produced a co-occurrence matrix to assess the fidelity of the clustering produced by the aforementioned process across multiple runs. See Buchta *et al.* (2008) for further details. This allows for the visual inspection of the optimal clustering of our network.

20.4.4.2 Identifying Clusters

The result of the cluster analysis is shown in the co-occurrence matrix in Figure 20.5. Figure 20.5 shows the ordered co-occurrence matrix: we see two clusters, one each in the upper-left and lower-right portions of the matrix. Visual inspection suggests a GDP per capita disparity between the two clusters, the uppermost having high GDP per capita relative to the lower cluster. A logit regression confirms that GDP per capita is a strong predictor of cluster membership, robust to specifications (2) and (3) of Table 20.2

In addition, we are able to identify several *transition* countries without clear cluster membership: Korea (KOR), Indonesia (IDN) and Greece (GRC) are the most clear-cut examples. Further analysis suggests there are eight countries without a 'perfect' affinity for either cluster[7]. We restrict our consideration here to the three aforementioned transition countries[8].

[7] These countries are Ireland (IRL), Korea (KOR), Indonesia (IDN), Czech Republic (CZE), Cyprus (CYP), Slovenia (SVN), Portugal (PRT) and Malta (MLT). We note that this is likely an artefact of our insistence on only two clusters.

[8] This can be achieved by selecting a minimum value of dissimilarity between a country and either cluster. Selecting a value of 0.1 with respect to the underlying co-occurrence matrix, we find the three countries in question are distinct from either cluster.

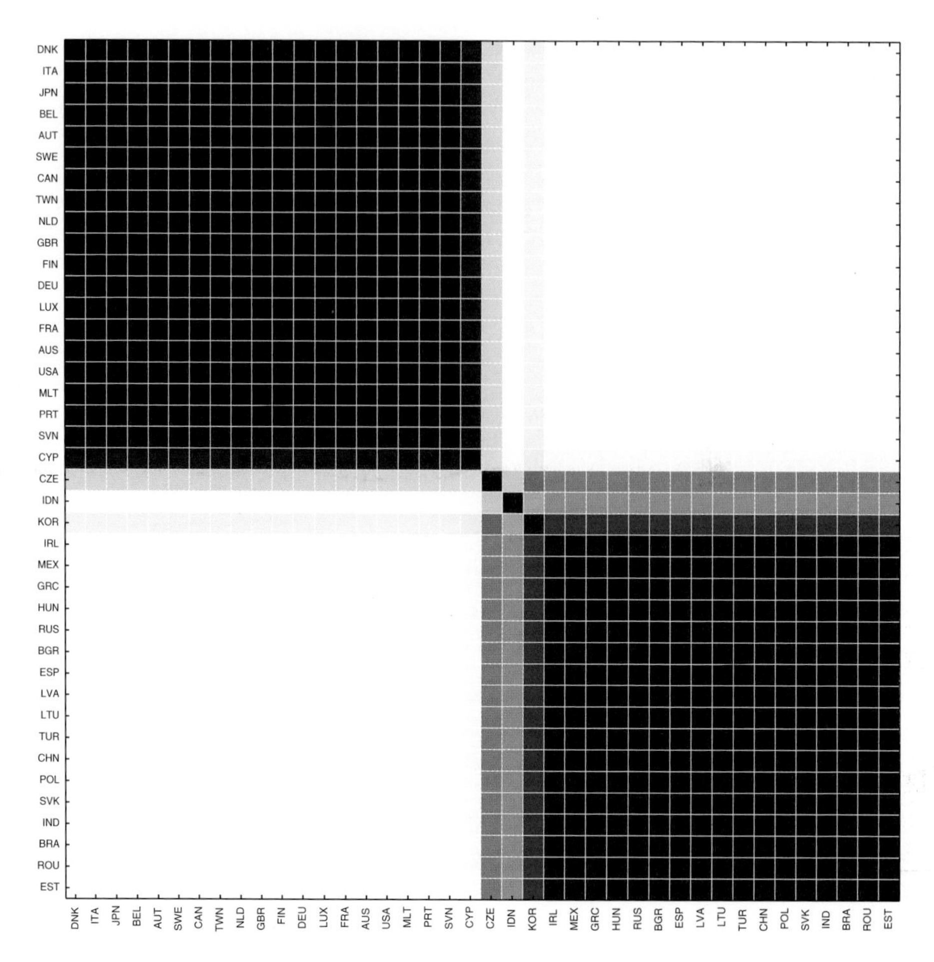

Figure 20.5 Co-occurrence matrix of the optimal clustering ordered similarity. White areas are very dissimilar; dark areas are highly similar. We see two well-defined clusters, along with a small number of countries without clear cluster membership

The role of the transition countries can be discerned by interrogating a diagram of the network; see Figure 20.6. Indonesia (IDN) is a transition country in the sense that it is only loosely tied to the low-GDP-per-capita cluster.

Conversely, the Czech Republic (CZE) and Korea (KOR) are tied to both the low- and high-GDP-per-capita clusters. Consideration of the edges connecting these nodes to the rest of the network suggests they may play different roles. Korea (KOR) appears to have mainly incoming egdes: this suggests it acts as an improver for countries belonging to both the low- and high-GDP-per-capita clusters.

By comparison, the Czech Republic (CZE) improves countries in the low-GDP cluster but is improved by countries in the high-GDP cluster. In this sense, it acts as a bridge between the two clusters.

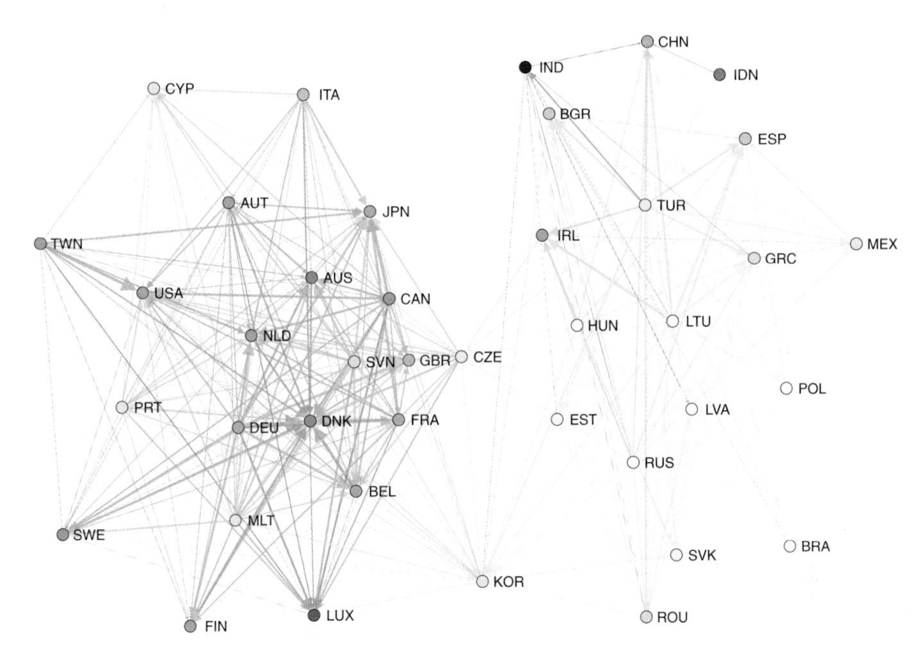

Figure 20.6 A network of improvements showing the two clusters of Figure 20.5. Edges between the clusters have been removed, with the exception of the three transition countries Korea (KOR), Indonesia (IDN) and Greece (GRC). Only improvements producing changes greater than 0.1% of GDP are shown. Node colouring reflects underlying GDP per capita as before; edge colouring reflects the averaged GDP per capita of source and destination country

20.5 Conclusions

Within the set of feasible consumption patterns demonstrated by our data set, we have shown that a country is likely to increase GDP as a result of diversifying its economy. In this sense, economic rebalancing is a desirable objective from a policy-making perspective.

As a corollary, we emphasise that our model indicates that even in the case when total demand is fixed, sectoral rebalancing alone can lead to economic growth.

Statistical consideration of the role of economic diversity suggests that, while always significant, other macro-economic quantities contribute to the success of a country as an improver for other nations. In particular, GDP per capita is a significant explanatory variable, although point estimates are significantly reduced by the inclusion of services and manufacturing indices. In particular, what may initially be observed as a wealth effect is likely a bias-towards-services effect.

Introducing higher fidelity methods based on network analysis, we show that the eigenvector centrality of a nation in the induced network of improvements is closely linked to the likelihood of being an improver.

Having applied community detection analysis to our network, we are able to identify two clear clusters, with a small number of transition countries. Membership of one cluster is shown to be positively correlated with GDP per capita and the proportion of the economy held by the services and manufacturing sectors.

Crucially, cluster membership is not significantly correlated with economic diversity. While countries rich in GDP per capita terms may dominate ranked in-degree and centrality tables, a subtler dynamic exists: countries with similar GDP per capita are tightly clustered by improver–improved-by relationships; hence, for almost every country, there exists a multiplicity of improvers who are close in GDP-per-capita terms to the country under consideration.

From a policy perspective, changes in economic diversity based on countries with similar GDP per capita may be more feasible than changes based on countries with dissimilar GDP per capita, as this may, for example, reflect analogous levels of economic development.

Furthermore, we have shown that the manufacturing index is significantly and negatively correlated with improver status. This suggests that modifying consumption patterns such that manufacturing comprises a greater share of total consumption is unlikely to increase GDP, assuming total consumption is held fixed.

This finding suggests further analysis is needed to demonstrate that expansion of the manufacturing base is a desirable aim for countries rich in GDP per capita, in particular where such a policy is a regularly stated aim of successive administrations.

Acknowledgements

The authors would like to thank Hannah Fry, Peter Baudains, Alan Wilson, Thomas Oléron Evans and David Zentler-Munro for their constructive comments during the preparation of this chapter, and acknowledge the financial support of the Engineering and Physical Sciences Research Council under the grant ENFOLDing[9].

References

Ball P. (2012). *Why Society Is a Complex Matter*. Berlin: Springer.

Buchta C, Hornik K and Hahsler M. (2008). Getting things in order: an introduction to the r package seriation. *Journal of Statistical Software* 25(3): 1–34.

Department for Business and Innovation & Skills. (2010). *Local Growth: Realising Every Place's Potential*, vol. 7961. London: The Stationery Office.

Financial Times. (2013). Editorial: lifting industry. Available at http://www.ft.com/cms/s/0/5a89f04a-9099-11e2-a456-00144feabdc0.html

Frieden JA. (2009). *Global Imbalances, National Rebalancing, and the Political Economy of Recovery*. Council on Foreign Relations, Center for Geoeconomic Studies and International Institutions and Global Governance Program Working Paper. Washington, DC: Council on Foreign Relations.

Frieden JA. (2010). The political economy of rebalancing. In Claessens S, Evenett S and Hoekman B (eds), *Rebalancing the Global Economy: A Primer for Policymaking*. London: CEPR.

Froud J, Johal S, Law J, Leaver A and Williams K. (2011). *Rebalancing the Economy (or Buyer's Remorse)*. Manchester: CRESC, University of Manchester.

Gardiner B, Martin R, Sunley P and Tyler P. (2013). Spatially unbalanced growth in the British economy. *Journal of Economic Geography* 13(6): 889–928.

Gibbs JP and Poston DL. (1975). The division of labor: conceptualization and related measures. *Social Forces* 53(3): 468–476.

Hahsler M, Buchta C and Hornik K. (2014). Seriation: Infrastructure for Seriation. R package version 1.0-13. Available at http://www.cran.r-project.org

[9] Explaining, Modelling, and Forecasting Global Dynamics, reference no. EP/H02185X/1.

Her Majesty's Government. (2013). Reducing the deficit and rebalancing the economy. Policy statement. Available at https://www.gov.uk/government/policies/reducing-the-deficit-and-rebalancing-the-economy
Hirschman AO. (1964). The paternity of an index. *The American Economic Review*: 761–762.
Ianchovichina E and Walmsley TL. (2012). *Dynamic Modeling and Applications for Global Economic Analysis*. Cambridge: Cambridge University Press.
Lenzen M, Moran D, Kanemoto K and Geschke A. (2013). Building eora: a global multi-region input–output database at high country and sector resolution. *Economic Systems Research* 25(1): 20–49.
Levy RG, Evans TPO and Wilson AG. (2016). A global inter-country economic model based on linked input-output models. Submitted.
Miller RE and Blair PD. (2009). *Input–Output Analysis: Foundations and Extensions*. Cambridge: Cambridge University Press.
Murphy KM, Shleifer A and Vishny RW. (1989). Industrialization and the big push. *Journal of Political Economy* 97(5): 1003–1026.
Newman M. (2010). *Networks: An Introduction*. Oxford: Oxford University Press.
Newman ME and Girvan M. (2004). Finding and evaluating community structure in networks. *Physical Review E* 69(2): 026–113.
Reichardt J and Bornholdt S. (2006). Statistical mechanics of community detection. *Physical Review E* 74(1): 016–110.
Roth A. (2014). Standard basis. *MathWorld – A Wolfram Web Resource, created by Eric W. Weisstein*. Available at http://mathworld.wolfram.com/StandardBasis.html
Rowland T. (2014). Orthonormal basis. *MathWorld - A Wolfram Web Resource, created by Eric W. Weisstein*. Available at http://mathworld.wolfram.com/OrthonormalBasis.html
Simpson EH. (1949). Measurement of diversity. *Nature* 163: 688.
Timmer M, Erumban A, Gouma R, Los B, Temurshoev U, de Vries G and Arto I. (2012). The world input-output database (WIOD): contents, sources and methods. WIOD background document. Available at http://www.wiod.org
Tukker A, de Koning A, Wood R, Hawkins T, Lutter S, Acosta J, *et al.* (2013). Exiopol–development and illustrative analyses of a detailed global mr ee sut/iot. *Economic Systems Research* 25(1): 50–70.
United Nations. (2010). United Nations commodity trade statistics database. Available at http://comtrade.un.org
Venkateshan S and Swaminathan P. (2013). *Computational Methods in Engineering*. London: Academic Press.
Weisstein EW. (2014). Gram-Schmidt orthonormalization. *MathWorld –A Wolfram Web Resource, created by Eric W. Weisstein*. Available at http://mathworld.wolfram.com/Gram-SchmidtOrthonormalization.html
Yeo S and Zedillo E. (2010). Foreword. In Claessens S, Evenett S and Hoekman B (eds), *Rebalancing the Global Economy: A Primer for Policymaking*. London: CEPR.

Appendix

Table A.1 A list of all sectors in the model. A ✓ indicates the sector is regarded as either services or manufacturing in this paper.

	Services	Manufacturing
Agriculture		
Air Transport	✓	
Business Services	✓	
Chemicals		✓
Communications	✓	
Construction		
Education	✓	
Electricals		
Financial Services	✓	
Food		
Fuel		
Health	✓	
Hospitality	✓	
Inland Transport	✓	
Leather		
Machinery		✓
Manufacturing		✓
Metals		✓
Minerals		
Mining		
Other Services	✓	
Paper		
Plastics		✓
Private Households	✓	
Public Services	✓	
Real Estate	✓	
Retail Trade	✓	
Textiles		✓
Transport Services	✓	
Utilities	✓	
Vehicle Trade	✓	
Vehicles		✓
Water Transport	✓	
Wholesale Trade	✓	
Wood		✓

Part Eleven

Integration

21

Research Priorities

Alan G. Wilson

In Chapter 1, we introduced the idea of a toolkit of methods for geo-mathematical modelling under the following headings:

- Estimating missing data: bi-proportional fitting and principal components analysis (Part 2)
- Dynamics in account-based models (Part 3)
- Space–time statistical analysis (Part 4)
- Real-time response models (Part 5)
- The mathematics of war (Part 6)
- Agent-based models (Part 7)
- Diffusion models (Part 8)
- Game theory (Part 9)
- Networks (Part 10).

Because the methods have been illustrated by examples of use, in almost all cases, combinations of tools are involved and there are more primitive underlying elements. For example, nearly every illustration involves some kind of spatial interaction modelling – a feature of the 'geo' focus of the book. The elements of the various systems of interest are usually 'counted', and we have shown how the formal construction of accounts is helpful and guarantees consistency. The use of the statistics toolkit is evidenced throughout. When we consider time – the dynamics – we formulate differential or difference equations in various ways. We show applications at different scales, and the micro scales in particular take us to the elements of the toolkit – notably, agent-based models and game theory. We have already noted the ubiquity of spatial interaction – the flows – and these flows are carried on networks, and hence the examples of Part 10. A characteristic of the field, shared by many other fields, is for researchers to become experts in particular areas and therefore for the expertise to be structured in this way – even to the extent of specialist journals. A principal objective of this book has been to demonstrate that it is worth knowing something about the full contents of the toolkit so that a

Approaches to Geo-mathematical Modelling: New Tools for Complexity Science, First Edition. Edited by Alan G. Wilson.

Companion Website: www.wiley.com/go/wilson/ApproachestoGeo-mathematicalModelling

comprehensive approach can be brought to bear on any particular problem. It is an argument for moving beyond silos. The remarks in this concluding chapter, therefore, focus on expanding the use of the toolkit in integrated ways and in reviewing the associated ongoing research challenges.

The first step is to review systematically the potential areas of application and hence to explore the full range of geo-mathematical modelling. There are many possible systems of interest that can be defined in terms of their elements and the scales at which they are represented – these latter in terms of groups or sectors, spatial units and time. The implication of 'groups and sectors' is that it will not usually be possible to model at the fine scale of the individual element – an exception being some instances of agent-based modelling. The elements might (for example) be people, firms or government organisations; they may be energy and materials' flows; or they may be plants and animals in an ecosystem. In the case of space, there is a spectrum from treating space continuously and locating elements by coordinates to having some kind of zone system and locating by zone. A 'zone' might be a country, a city, or a neighbourhood within a city. Time can also be treated as continuous – now relevant to 'real-time' data as part of the 'big data' agenda – but it is more likely to be discrete: one year, five-year or 10-year units, for example.

To fix ideas, think of the countries of the world as 'zones' in a global system; or think of the cities of a country as a 'system of cities'; or a city in its region, divided into zones that are neighbourhoods. In any of these cases, the key elements might be the people, the economic units of production and the infrastructure. The questions which we ask in our examples mainly relate to global migration, trade, development aid and security. But these repeat themselves at national or subnational scales. Migration becomes 'internal migration' rather than 'international'; trade becomes trade between cities – which exposes a research question to be discussed further here. Development aid becomes 'welfare payments' or 'economic development grants'. The security questions are largely the same, but a change of scale shifts them from the military to the police and security services.

In relation to this wide array of systems of interest, there are many policy questions – demonstrating the possibilities of model-based policy development and planning contributions. For example, "How and where will people live?" is a question that applies across the scales. In addition, what policies effectively support future economic development? What will the future of public services be? How are the challenges of climate change and resource depletion to be met (i.e. the sustainability agenda)? What is the best use of land in different situations and circumstances in the future? What energy policies will support various ways of life and economic development? What should future transport systems look like? How will the future be governed?

It becomes clear very quickly that these questions are interdependent. How and where people will live will depend on their incomes, and their incomes will depend on their employment (in general). Parts of these incomes will be used to pay for their consumption of basic utilities, and if there is a shortfall in income, this will create problems in this territory. Economic development will only be sustained if there is an effective education system, and the economy must be sufficiently successful for the Government's tax take to fund this.

It is possible to build models for the analysis that support policy development to address these kinds of questions. They can be used directly for short-run policy development and planning. For the long run – and consideration of the long run is vital for the sustainability agenda – forecasting is not possible. There are too many uncertainties and too many nonlinearities, the latter creating path dependence and possible phase changes. What is possible in this case is the development of future scenarios on a variety of assumptions – in effect, the exogenous variables of the models – which can then be explored through model-based analysis.

In this book, and its sister book, Global dynamics, we have provided examples of modelling migration, trade flows, security resources (e.g. in relation to piracy, insurgency and riots) and development aid. We have used input–output modelling at the global scale to underpin analyses of economic development and the impact on this of trade, migration and aid. This begins to expose some challenging research questions relating to the use of this toolkit in the future. For example, it would be very interesting and important to have an interregional input–output model for the cities within a country. This has not been done since cities, unlike countries, do not have any record of imports and exports and trade flows. Estimating this 'missing data' becomes the new research problem – and the kinds of methods used in Part 2 of this book can be brought to bear here.

This kind of analysis of future (and fruitful) research challenges can be extended systematically – reviewing all the possible systems of interest and the associated policy challenges – together with the range of scales at which the systems can be defined. The 'sector' scale will move us between, say, agent-based modelling and spatial interaction modelling in the toolkit; we have already noted the range of spatial scales, from the global to the neighbourhood. What we haven't noted, but should, is that an exploration of the time spectrum will take us into modelling historical periods (and associated systems of interest) – in effect, a new branch of history which is now being developed.

There is a conjecture which provides a suitable conclusion to this book: if a system of interest (with its spatial dimensions), and its associated policy challenges, can be well defined, then there is a geo-mathematical model-building toolkit which will provide the basis for analysis and exploration!

Index

Approaches to Geo-mathematical Modelling: New Tools for Complexity Science, First Edition. Edited by Alan G. Wilson.

Companion Website: www.wiley.com/go/wilson/ApproachestoGeo-mathematicalModelling